PARTICLE PRODUCTION NEAR THRESHOLD

CONFERENCE PROCEEDINGS NO. **221**

PARTICLES AND FIELDS SERIES 41

PARTICLE PRODUCTION NEAR THRESHOLD

NASHVILLE, IN 1990

EDITORS:

HERMANN NANN
EDWARD J. STEPHENSON
INDIANA UNIVERSITY

American Institute of Physics **New York**

L.C. Catalog Card No. 91-55134
ISBN 0-88318-829-5
DOE CONF-9009332

Printed in the United States of America.

CONTENTS

SESSION D: MESON–MESON AND MESON–BARYON INTERACTIONS

SESSION E: FUTURE EXPERIMENTAL DEVELOPMENTS

SESSION F: THE PRODUCTION OF STRANGE PARTICLES

SESSION G: FUTURE MACHINE DEVELOPMENTS

CLOSING

Preface

The Topical Conference on "Particle Production Near Threshold," sponsored jointly by the Indiana University Cyclotron Facility, Bloomington, Indiana, U.S.A., and the Forschungszentrum Jülich, Jülich, West Germany, was held on 30 September through 3 October 1990 at the Brown County Inn, Nashville, Indiana, U.S.A. The scope of this conference covered a growing field of nuclear physics research at intermediate energies stimulated by the construction of a new class of particle acceleration and storage rings utilizing the technology of electron and stochastic cooling. Particle production close to threshold provides a special kinematic regime in which to test predictions of the standard model of strong and electroweak interactions since some degrees of freedom are frozen out, thus limiting the ways in which the kinetic energy of the entrance channel can be converted into hadronic matter. The reaction mechanism for this conversion is closely connected to the coupling of nuclear and subnuclear degrees of freedom within hadrons. Understanding the basic mechanism whereby mesons are produced and interact with nucleons and nuclei may not only provide the framework within which more exotic phenomena can be identified, but also guide experimentalists in choosing the most sensitive observables and configurations for testing the limits of the standard model. We hope that the information gathered by this conference will be helpful toward this goal.

Beam cooling techniques have been used for many years at LEAR for the production of antiproton beams. The first such ring utilizing light ion beams, the Cooler at the Indiana University Cyclotron Facility, has recently been commissioned and has completed its first experiments. More such accelerators are under construction and will begin operation soon, including the CELSIUS ring at Uppsala, Sweden, and the COSY ring at Jülich. These storage rings offer features previously unavailable for nuclear physics research at intermediate energies, including the use of ultrathin, windowless targets and exceptional energy resolution. The high energy resolution is expected to have significant impact on measuring particle production rates near threshold, and a significant fraction of the proposed research at these storage rings is devoted to this topic.

The conference evolved to include many diverse specialties: rare and forbidden decays, electroproduction and photon-induced reactions, meson production with hadronic beams, meson–meson and meson–baryon interactions, production of strange particles, and future experimental and machine developments. We hoped this conference would bring together people working on experiments in particle production near threshold as well as those with new ideas for physics questions to be addressed with these new experimental facilities and their possible expansion to higher beam energies. The conference was attended by 112 physicists from laboratories and universities around the world. The conference included plenary talks as well as short contributions describing specific experimental efforts.

These proceedings contain the written versions of the talks that were received by December 1990. We hope that they capture the enthusiasm of the conference.

Many people deserve acknowledgment for their work in the conference. We would like to extend our appreciation for the success of the conference to the speakers who presented their carefully prepared papers, to the session chairmen who ran their sessions most effectively, and to the participants for their questions and expert comments which livened up the discussions not only in the lecture hall but also in the hallways during the breaks. We would like to express our deep appreciation to the members of the program committee and to the following individuals, J. Arvieux, P. Barnes, A. Bernstein, L. Bland, J. Cameron, R. Eisenstein, G. Emery, R. Holt, N. Isgur, A. Johansson, T. Londergan, T. Mayer-Kuckuk, R. Mensday, B. Nefkins, J. C. Peng, R. Redwine, J. Speth, and C. Wilkin, for helping to identify and select speakers. We especially thank the members of the conference arrangements committee. Their expertise and dedication were indispensable. Finally, we thank the many IUCF graduate students for their help during the conference.

Hermann Nann
Edward J. Stephenson
Bloomington, Indiana
January 1991

on behalf of

Program Committee:

Hermann Nann, chair, Indiana University
Kurt Kilian, Forschungszentrum Jülich
Hans-Otto Meyer, Indiana University
Peter Schwandt, Indiana University
Edward J. Stephenson, IUCF
Steven E. Vigdor, Indiana University

Conference Arrangements Committee:

Edward J. Stephenson, co-chair, IUCF
Peter Turek, co-chair, Forschungszentrum Jülich
Charles Foster, IUCF
Diana McGovern, IUCF
Hermann Nann, Indiana University
Peter Schwandt, Indiana University
Philip Thompson, IUCF
Becky Westerfield, IUCF
Robert Woodley, IUCF

SESSION A

Rare Decays

RARE AND FORBIDDEN DECAYS OF LIGHT MESONS[1]

John N. Ng
Theory Group, TRIUMF, 4004 Wesbrook Mall, Vancouver, B.C., Canada V6T 2A3

ABSTRACT

An up-to-date review of the use of light mesons, in particular the η-meson, as a probe of physics beyond the standard model is given. Emphasis is given to lepton number violation and CP violation (CPV) processes.

INTRODUCTION

Recently there is a rising interest in the experimental community to measure rare and forbidden decays of light mesons such as the π^0, η and η'. This is due to the possibility of producing large fluxes of these particles. For instance recent advances at Saclay raise the hope that η-flux several orders of magnitude higher than in any previous experiment achieved.[1]

On the theoretical side, it is now very clear that the standard model of SU(2)×U(1) gauge theory of electroweak interactions is an accurate description of physics below the 100 GeV scale. It is equally clear that it cannot be the final theory since it contains an embarrassingly large number of free parameters. Furthermore, the mechanism of gauge symmetry breaking and mass generation remains ill understood. The standard model has one Higgs doublet field to break the symmetry and provide masses to both the gauge bosons *and* fermions. This results in a very successful prediction of the ratio of gauge boson masses; however, it leaves the fermion masses totally arbitrary. The strength of the coupling of quarks and leptons to the weak charged current is parametrized in three mixing angles and one CP-violating phase known as the Cabibbo-Kobayashi-Maskawa (CKM) mechanism or paradigm. None of these parameters can be predicted within the model. Most physicists now feel that some new physics beyond the standard model must exist that enables us to answer these questions. Unfortunately, after years of trying and numerous attempts we do not have a totally satisfactory model of physics beyond the standard model.

Clearly the ball is now in the experimentalists' court. We have to test the standard model in every conceivable way. In the low energy domain where light mesons play an important role this means stringent tests of the CKM paradigm. The kaon is particularly valuable for this. The η meson is useful in searching for CP violation[2] beyond the standard model as well as new flavor conserving (i.e. $\Delta S = 0$) neutral current which has scalar or pseudoscalar Lorentz structure.[3]

---

[1]Invited talk given at the Conference on "Particle Production near Threshold", Nashville, Indiana, 1990

Another arena where new physics can be looked for is in the lepton number violating decays. A prime example of this is the decay $\eta \rightarrow \mu e$. This decay is forbidden in the standard model and has occupied a central role in kaon physics. The significance of these studies is illustrated by the decay $K_{0L} \rightarrow \mu e$. It has placed a very important constraint in technicolor model building where the gauge symmetry is dynamically broken. The importance of these searches cannot be overemphasized.

Before we enter into specific discussions of what kind of new physics can be probed by η decays, it is instructive to define the boundaries of this study. It is clear that exotic physics that induce lepton number violation decays processed via the exchange of exotic particles of mass M which is larger than the Fermi scale. It is commonly assumed that this new scale is 0 (TeV). Hence, their rate will behave like $1/M^4$ and is naturally small. One must be concerned with the limit set by rare K decays which is usually more stringent. Thus, the usefulness of η meson will lie in $\Delta S = 0$ new interactions. A second possibility is in probing relatively light exotic particles that can induce rare or forbidden decays of η. Examples are η to scalar, (non-standard Higgs) pseudoscalar (invisible axion) or "new-photon". Here we exclude the standard model Higgs since the current limit from LEP for Higgs mass is greater than 40 GeV. Its effect on η meson decays is negligible.

In this talk I will begin with a discussion of tests of discrete symmetry. The most interesting being CP violating decays. The second part will concentrate on flavor violation decays. My discussion will focus on leptonic, or semileptonic decays since they involve less uncertainty than hadronic decays. This will allow us to reduce the already considerable amount of theoretical uncertainties that exist when we try to probe physics beyond the standard model using hadrons.

PSEUDOSCALAR MESON TO LEPTON-ANTI-LEPTON PAIRS

The decay of a pseudoscalar meson into a lepton pair, in particular $\mu\bar{\mu}$ pair, is a benchmark rare decay mode. It has been studied extensively as a test of unitarity and a place where new sources of CP violation can be searched for.

Our discussion will concentrate on

$$\eta \rightarrow \mu\bar{\mu} \ , \tag{R1}$$

decay. One can easily replace η by π^0, η' or other light mesons. The amplitude, M, for the decay has the following general form,

$$M = a\bar{u}i\gamma_5 u + b\bar{u}v \ , \tag{1}$$

where a and b are form factors and are in general complex. The partial width is easily calculated to be

$$\Gamma_{\mu\bar{\mu}} = \frac{M_\eta}{8\pi} \ \beta(|a|^2 + \beta^2|b|^2) \ , \tag{2}$$

where $\beta \equiv \left(1 - \frac{4m_\mu^2}{M_\eta^2}\right)^{1/2}$, and M_η is the mass of the pseudoscalar meson. For (R1), the η decay, the latest value from the Particle Data Group[4] puts it at

$$\Gamma(\eta \to \mu\bar{\mu}) = 1.19 \times (6.5 \pm 2.1) \times 10^{-6} \text{ keV} . \tag{3}$$

It is well-known that a lower bound on $\Gamma_{\mu\mu}$ is given by the two-photon intermediate state, i.e. it proceeds via the chain $\eta \to \gamma^*\gamma^* \to \mu\bar{\mu}$. (See Fig. 1a.) This process only contributes to a and the imaginary part, $Im\, a_{\gamma\gamma}$, can be calculated by unitarity; i.e. with two photons placed on-shell. The branching ratio normalized to $\eta \to \gamma\gamma$ can be written as

$$\frac{\Gamma(\eta \to \mu\bar{\mu})|_{\gamma\gamma}}{\Gamma(\eta \to \gamma\gamma)} = \frac{\alpha^2}{2\beta} \frac{m_\mu^2}{M_\eta^2} \left[\ln \frac{1+\beta}{1-\beta}\right]^2 . \tag{4}$$

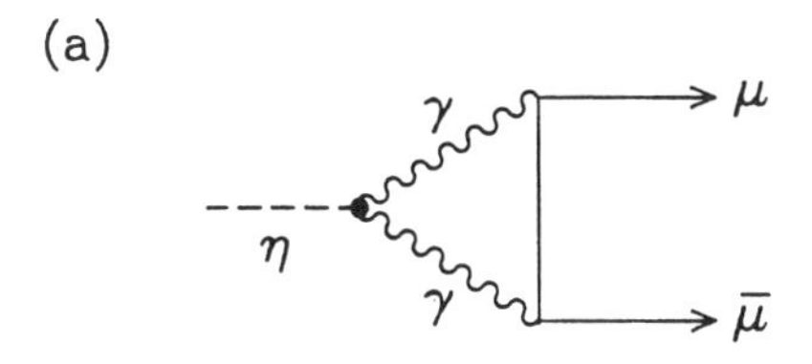

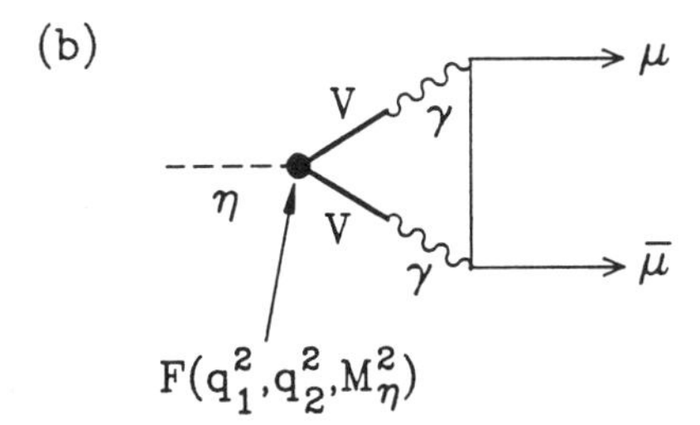

Fig. 1. Feynmann diagrams for $\eta \to \mu\bar{\mu}$ decays. (a) Two photon intermediate states, (b) vector dominance model (VDM) for the decay.

Using the experimental value for the branching ratio[4] of $\text{BR}(\eta \to \gamma\gamma) = 0.389 \pm 0.005$ we get $\text{BR}(\eta \to \mu\bar{\mu})|_{2\gamma} = (4.30 \pm 0.06) \times 10^{-6}$ and this is to be compared with the experimental value of $\text{BR}(\eta \to \mu\bar{\mu})_{\text{exp}} = (16.5 \pm 2.1) \times 10^{-6}$. Clearly a factor of ten improvement in the branching ratio (R1) would be very welcome. Note that unitarity bound is calculated reliably with no theoretical uncertainty. Table I gives an indication how well the unitarity bound is saturated in π^0, η and K-decay.

Table I

Reaction	Unitarity bound	Experiment
$\mathrm{BR}(\pi^0 \to e^+e^-)$	$4.632 \pm 0.002 \times 10^{-8}$	$< 1.3 \times 10^{-7}$
$\mathrm{BR}(\eta \to \mu^+\mu^-)$	$4.304 \pm 0.055 \times 10^{-6}$	$6.5 \pm 2.1 \times 10^{-6}$
$\mathrm{BR}(K_{0L} \to \mu\bar{\mu})$	$(6.83 \pm 0.28) \times 10^{-9}$	$7.28^{+0.59}_{-0.57} \times 10^{-9}$

The experimental values for π^0, η decays are taken for PDG,[4] whereas the K_L^0 decay includes new Brookhaven results.[5] The theoretical implications of this decay for the standard model is reviewed in Ref. 6.

The error bars in the unitarity bound are due solely to experimental uncertainties in the 2γ branching ratios. Evidently, the imaginary part of the 2γ amplitude dominate, we can write

$$\mathrm{Br}(\eta \to \mu\bar{\mu}) = \mathrm{Br}(\eta \to \mu\bar{\mu})|_{\gamma\gamma} \left\{ 1 + \left| \frac{Re\ a_{\gamma\gamma}}{Im\ a_{\gamma\gamma}} \right|^2 \right\} , \tag{5}$$

where we have neglected the b term. We shall return to b later. From Table I we obtain

$$\left| \frac{Re\ a_{\gamma\gamma}}{Im\ a_{\gamma\gamma}} \right|^2 = 0.512^{+0.26}_{-0.48} . \tag{6}$$

The real part of the 2γ amplitude can only be calculated with the aid of phenomenological hadronic models. A number of such models and their prediction for $\left| \frac{Re\ a_{\gamma\gamma}}{Im\ a_{\gamma\gamma}} \right|^2$ is given in Table II.

It appears that these models all give $\left| \frac{Re\ a_{\gamma\gamma}}{Im\ a_{\gamma\gamma}} \right|^2$ of the order of ten percent or less. Improvement on the measurement of R1 will give us a quantitative test of how well these vector dominance plus QCD models work. (See Fig. 1b.) This is very important and interesting for hadron physics. Using this knowledge we can predict $\pi^0 \to e\bar{e}$ and $K_L^0 \to \mu\bar{\mu}$ decays with more certainty. It is also important to note how model dependent $Rea_{\gamma\gamma}$ is. The theoretical uncertainty is about a factor of ten. This uncertainty limits the use of (R1) for a test of new physics beyond the standard model.

Table II. Theoretical models for $\dfrac{|Re\, a_{\gamma\gamma}|^2}{|Im\, a_{\gamma\gamma}|^2}$ for $\eta \to \mu\bar{\mu}$.

Model	Form factor for $P_{\gamma\gamma}$ vertex $F(q_1^2,\, q_2^2\, M_p^2)$	$\left\lvert\dfrac{Re\, a_{\gamma\gamma}}{Im\, a_{\gamma\gamma}}\right\rvert^2$
2 virtual γ coupled to vector mesons[7] (VDM)	$F = \left(1 - \dfrac{q_1^2}{M_{v^2}}\right)^{-1} \left(1 - \dfrac{q_2^2}{M_{v^2}}\right)^{-1}$	0.06 $V = \rho$ 0.09 $V = \omega$ 0.21 $V = \phi$
VDM and SU(3)[7]	$F = \dfrac{\sum_{\rho,\omega,\phi} \frac{g_{\rho v\gamma}}{2g_{v\gamma}} \left(1 - \frac{q_1^2}{M_v^2}\right)^{-1} \left(1 - \frac{q_2^2}{M_v^2}\right)^{-1}}{\sum_{\rho,\omega\psi} \frac{g_{\rho v\gamma}}{2g_{v\gamma}}}$	~ 0.01
VDM and QCD[8]	$F = \dfrac{1}{1 - \frac{q^2}{m_{eff}^2}}$ $m_{eff} \simeq 830$ MeV High q^2 $F \to \dfrac{m_{eff}^2}{q^2}$ Low q^2 $F \to$ VDM	0.12

The decay (R1) can also be a probe of CP violation beyond the CKM model. CP violation manifests itself in the longitudinal polarization asymmetry, P_L of the outgoing muon.[9] Explicitly, P_L is defined by

$$P_L \equiv \frac{N_R - N_L}{N_R + N_L}\ , \tag{7}$$

where $N_{R(L)}$ is the number of right- (left)-handed outgoing muons. A non-vanishing P_L implies a CP violating component decay of (R1). This is seen by examining the quantum numbers of the dimuon system which is given by $CP = (-1)^{S+1}$ and $P = (-1)^{l+1}$. If the $\mu^+\mu^-$ system is in a $^1S_0$ state then it is CP odd and P odd. On the other hand, if it is in $^3P_0$ state it is CP even and P

even. Hence, if the decay conserved CP then parity must be even and $P_L = 0$. However, if the decay has a CP violating component it will show up as a parity violating measurement and $P_L \neq 0$.

P_L can be calculated from Eq. (1) in terms of a, b and it is given by

$$P_L = \frac{M_\eta^2\, \beta^2\, Im(ba^*)}{4\pi\Gamma_{\mu\mu}} \,. \tag{8}$$

Putting in the measured numbers leads to the following expression for P_L:

$$P_L \simeq 5.31 \times 10^9 \quad b\, Im\, a_{\gamma\gamma} \,, \tag{9}$$

where we have used the fact that the form factor a is dominated by the two photon mechanism. The only unknown quantity is b. The real part $Re a_{\gamma\gamma}$ is not important here. Different CPV models will predict different values of b and hence P_L.

In the standard model, b is zero to at least the one-loop level. Hence P_L is unmeasurable if there are no CPV sources other than the KM phase.

In a recent paper[9] we have investigated four different classes of gauge models of CPV where new sources of CPV can occur. This is in addition to the usual CP violating KM phase whose origin is in the complex quark mass matrices or Yukawa couplings. It seems to us reasonable to assume that the quark masses are complex in the weak eigenbasis and the observed CPV in the kaon system is mainly due to this phase. However, this does not necessarily mean that no other CP violation source exists. In particular, if there are explicit CPV terms in the Higgs potential or there are more than one Higgs doublet around that suffer spontaneous symmetry breaking then their vacuum expectation values can in general be relatively complex. This latter case is known as spontaneous CPV (SCPV) and is first pointed out by Lee[10] and Weinberg.[11] However, the two-Higgs-doublet model of Lee suffers from the existence of flavor changing neutral currents (FCNC). The branching ratio of $K_L^0 \to \mu\bar{\mu}$ forces the mediator of this FCNC to be heavier than 5 TeV and it is doubtful that such a heavy spin-0 meson is a theoretically meaningful object. A well-known way to avoid this is to couple one Higgs doublet, ϕ_1, to u_R and the other ϕ_2 to d_R. Now one requires at least 3 Higgs doublets to achieve SCPV.[11] Recently, we have shown that one can achieve this and with fewer Higgs fields by employing one or more Higgs singlet fields.[12] The minimum number of Higgs fields required is two doublets plus one singlet. This simplest model accommodates only one additional source of CPV, namely mixing between scalars and pseudoscalars, other than the KM phase. Incidentally, this Higgs singlet field can also be used to generate large Majorana neutrino masses if their vacuum expectation values are much larger than the weak scale. Hence, they are interesting objects in their own rights in addition to being a possible new source of CP violation.

In these models CPV is due to the phenomenon of scalar-pseudoscalar mixing which we shall refer to as $R - I$ mixing. The later nomenclature reflects the fact that the Higgs potential after spontaneous symmetry breaking will contain terms that mix the real and the imaginary parts of different Higgs fields. This, of course, does not happen in the standard model where only one Higgs doublet field exists. When $R - I$ mixing takes place the mass eigenstates of these spin-0 fields can couple to a given fermion, q, with both $\bar{q}q$ and $\bar{q}i\gamma_5 q$ terms. Thus, this leads to CP violation involving fermions. Estimates of P_L in various models are given in Table III.

Table III. Estimates of $P_L(\eta \to \mu\bar{\mu})$ in various models. (Details see Ref. 9.)

Model	Constraint	Estimates $P_L(\eta \to \mu\bar{\mu})$	Estimates $P_L(K_L^0 \to \mu\bar{\mu})$
Extended Higgs (Weinberg,[11] Geng & Ng[2,9,12])	EDM of neutron and electron $\frac{X}{M_\phi^2 v^2} < 1.2 \times 10^{-4}$	$< 1 \times 10^{-3}$	$< 4 \times 10^{-3}$
Leptoquark (Barr & Masiero[13])	BR of $\mu \to e\gamma$ $M_\phi > 300$ GeV	$< 10^{-8}$	$< 10^{-8}$
SUSY Models E_6	BR of $\mu \to e\gamma$	$< 10^{-8}$	< 0.07
Left-right symmetric models (Frère & Liu[14])	BR of $K_L \to \mu\bar{\mu}$	$< 10^{-8}$	< 0.02

We conclude this section by noting that $P_L(\eta \to \mu\bar{\mu})$ is a good probe of CPV in the extended Higgs model. Other sources of CPV all give very small P_L. Independent tests of these mechanisms can be obtained by examining EDM of muon, electron and the neutron. It goes without saying that it is of paramount importance to investigate these phenomena even if we accept the CKM paradigm as the source of CPV in kaons which is not completely established yet.

LEPTONIC FLAVOR VIOLATING DECAYS

It is easy to see from Eq. (9) and the subsequent discussions that aside form CPV $\eta \to \mu\bar{\mu}$ decays would not be a fertile ground to search for new physics. The main reason being the uncertainty involved in calculating $Re a_{\gamma\gamma}$. However, this is not the case for lepton number violating decays such as

$$\eta \to \bar{\mu}e,\ \bar{e}\mu\ , \tag{R2}$$

$$\eta \to \pi^0\bar{\mu}e,\ \pi^0\bar{e}\mu\ , \tag{10}$$

which are forbidden in the standard model.

These decays necessarily probe new interactions mediated by exotic particles. We will focus on leptoquarks and scalar or pseudoscalars such as horizontal Higgs bosons that couple to μe. Examples of theories featuring leptoquarks are grand unified theories, technicolor models and composite models of quarks and leptons. These type of leptoquarks we consider to be spin 1 or spin 0 objects carrying both lepton and quark numbers. Leptoquarks that couple to pairs of baryons must be very heavy in order for them not to mediate proton decay. They are not relevant for our discussions. Of relevance to us are leptoquarks that do not couple to baryons and do not change quark flavors.

Horizontal interactions that connect between families of quarks and leptons have long been speculated upon. Recently there are attempts to embed them in superstring models. The particles mediate interfamily interactions and they can be spin 1 gauge bosons or spin 0 Higgs-like objects. Although it is generally believed that Higgs particles couple with strength proportional to the fermion masses, in our case muon and electron masses, it need not be true for horizontal Higgs bosons. The reason being that these guys may not have anything to do with fermion mass generation. In the interest of generality we will keep their couplings as arbitrary constants.

In the limit of heavy masses, these interactions can be represented by point-like four-fermi contact terms. See Fig. 2. The general effective coupling between quarks and μe pairs can be written in the form

$$L = \frac{g_{uu}g_{\mu e}}{M_u^2}\,\bar{u}u\bar{\mu}e + \frac{g_{dd}g_{\mu e}}{M_d^2}\,\bar{d}d\bar{\mu}e + \frac{g_{ss}g_{\mu e}}{M_s^2}\,\bar{s}s\bar{\mu}e\ , \tag{11}$$

arising from the large mass limit of Fig. 2. If the heavy particle is spin 1, then we will have $\bar{q}\gamma^\mu\bar{\mu}\gamma_\mu e$ structure. For the purpose of this discussion the difference is not important.

The first two terms of (Eq. 11) will be stringently constrained by $\mu^- N \to e^- N$ experiment which has a current limit of $<4.6 \times 10^{-12}$. On the other hand $\frac{g_{ss}g_{\mu e}}{M_s^2}$ does not receive any such constraint. Notice that $K_L^0 \to \bar{\mu}e$ will be

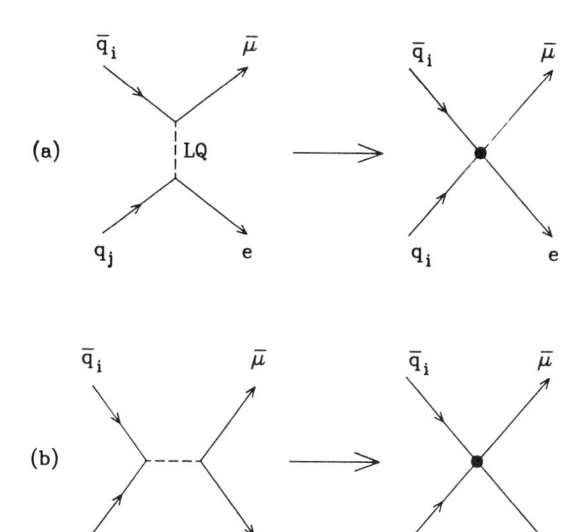

Fig. 2. Effective four Fermi interactions $\bar{q}_i q_j \bar{\mu} e$ where i, j are flavor indices which can arise from the exchange of (a) leptoquarks, (b) horizontal Higgs bosons.

sensitive to $\frac{g_{sd}\, g_{\mu e}}{M^2}$. A branching ratio measurement of $< 10^{-10}$ will translate into a limit

$$\frac{M^2}{g_{ss} g_{\mu e}} > 4 \text{ TeV}^2 \,.$$

This is very competitive with $K_L^0 \to \bar{\mu} e$ decays.

There is another η decay which is interesting from the standpoint of new physics; namely, the β-decay of η:

$$\eta \to \pi^+ e^- \nu_\mu \quad or \quad \pi^+ e \nu_\tau \,. \tag{R3}$$

Notice that (R3) are lepton numbers violating decay. Experimentally it is indistinguishable from the decay into $\pi^+ e^- \nu_e$ which involves a second class vector current which cannot be accommodated in the usual gauge theories. The decay (R3) can proceed through first class scalar or pseudoscalar currents. The decay width is given by[13]

$$\Gamma\left(\eta \to \pi^+ e^- \nu\right) = \frac{g_{ud}^2 g_{e\nu^2}}{M^4} \frac{m_\eta^5}{96\pi^3} f\left(\frac{m_\pi^2}{m_\eta^2}\right) \,,$$

where $f(z) = 1/2 - 4z + 4z^2 - 1/2z^4 - 6z^2 \ln z$. Simple arithmetic translates this into the following limit

$$M > 457\,(g_{ud} g_{e\nu})^{1/2} \text{ GeV} \,,$$

for a 10^{-10} branching ratio measurement.

If such accurate measurements can be achieved it will give strong bounds on the mass coupling constant ratios for lepto-quarks or lepton number violating Higgs bosons. Optimistically these decays will be discovered and a new frontier of particle physics will then be opened. We eagerly look forward to these measurements being done.

REFERENCES

1. B. Mayer, Saclay preprint DPh-N/Saclay 2565 (1980). See also Proceedings of International Workshop on Rare Decays of Light Mesons, Gif-sur-Yvette, France, 1090; (Edition Frontières).

2. C.Q. Geng and J.N. Ng, Phys. Rev. Lett. **62**, 2645 (1989). X.G. He and B.H. J. Mckellar, U of Melbourne preprint 90/04.

3. W. Buchmüller and D. Wyler, Phys. Lett. **117B**, 377 (1986). P. Herczeg, Ref. 1.

4. Particle Data Group, Phys. Lett. **239B**, 1 (1990).

5. W. Morse, Proceedings of the 25th Int. Conf. on High Energy Physics, Singapore, 1990, to be published.

6. J.N. Ng, *ibid*, 1990.

7. C. Quigg and J.D. Jackson, UCRL-18487, (1968).

8. K.S. Babu and E. Ma, Phys. Lett. **119B**, 449 (1982).

9. For an up-to-date discussion of this and other CPV reactions see C.Q. Geng and J.N. Ng, Phys. Rev. **D42**, 1509 (1990) and references therein.

10. T.D. Lee, Phys. Rev. **D8**, 1226 (1973).

11. S. Weinberg, Phys. Rev. Lett. **37**, 657 (1976).

12. C.Q. Geng and J.N. Ng, Phys. Lett. **211B**, 111 (1988).

13. S.M. Barr and A. Masiero, Phys. Rev. Lett. **58**, 187 (1987).

14. J-M. Frère and J. Lien, Nucl. Phys. **324B**, 333 (1989).

15. P. Singer, Phys. Rev. **B139**, 483 (1965).

RARE DECAYS OF THE NEUTRAL PION

by

H Calen, A Johansson, S Kullander, H Rubinstein, B Trostell
University of Uppsala, Uppsala, Pa 75105
L Bergström
University of Stockholm, Stockholm, Pa 11346
Z Pawlowski
Warsaw Technical University, Warsaw, Pa 00665
J Stepaniak, I P Zielinski
Institute for Nuclear Studies, Warsaw, Pa 00681
J Zlomanczuk, Z Wilhelmi
University of Warsaw, Warsaw, Pa 00681

ABSTRACT

An experimental program aiming at a study of very rare processes is being prepared in Uppsala. The production and decays of the pseudoscalar mesons, π and η, will be studied using the cooler storage ring, CELSIUS and a wide angle 4π detector named WASA. Here we discuss some issues connected with the rare decays of the neutral pion and we give the status of the WASA project.

INTRODUCTION

The rare decays of the hadrons are a precious source of information on elementary particles and their interactions. The reasons for rarity may be of many different kinds. Included in the notion of rare decays are also some decays that are forbidden by conservation laws implied by present day theory, the main example of this class being proton decay which has been extensively searched for, so far without positive result. The theoretical reason for looking for such forbidden decays is that the present "standard model" of elementary particles and their interactions, although extremely successful in accounting for a vast amount of experimental data, is quite unsatisfactory as a fundamental theory. It contains a large number of free parameters, it provides no real unification of the known gauge interactions, and there is no room for gravity in it. Therefore, the main goal of particle physics today, both theoretical and experimental, is to find ways to reach beyond the standard model. In theory, the presently preferred road to follow, involves concepts such as supersymmetry and superstrings. These theories predict new particles which would have direct experimental consequences if the particles are not too heavy.

Our cooler-storage ring, CELSIUS (maximum proton kinetic energy 1.3 GeV) is well suited for the study of rare decays of the neutral pion since the experimental background should be very small. The ultra-thin internal hydrogen or deuterium targets greatly reduce the secondary interactions of the decay particles in the target and hence improve the experimental conditions compared to that with external beams and thicker targets. The decay of a pion into an electron-positron pair is treated in detail partly because it exhibits the virtues of a cooler storage ring with thin targets, partly because this decay mode is interesting from a theoretical viewpoint and badly known experimentally. Any deviation from quantum electrodynamics may be a sign of a new interaction perhaps mediated by pseudoscalar particles in the s-channel. It may also be a sign of a low-energy effective interaction induced by new physics at a much higher mass scale. A high-statistics study of this decay allows us to probe new physics - indeed at the TeV mass scale.

Besides the electron-positron decay mode one may also search for charge conjugation violating decays into three photons, lepton number violating decays into muon and electron, etc. The fact that one may tag the pions by the outgoing protons, in the production reaction, means that one may even search for normally invisible decay modes of the pion into non-standard neutrinos or photinos (the supersymmetric partners of the photon).

Also decays of the eta meson to an accuracy several orders of magnitude better than any previous experiment will be studied. An experimental study of rare eta decays may be rewarding, since the higher mass and the contents of strange quarks could enhance certain decay modes involving exotic particles. In fact, a combined study of rare pion and eta decays would push the limits of the masses and couplings of many hypothetical new interactions by orders of magnitude.

Apart from the study of rare decays and searches for prompt photons and electron pairs, there is also a possibility to study in detail the pion production processes, especially near the threshold. More information is needed about such important quantities in particle physics as isospin violation, particle mixing, quark mass and sigma term in quantum chromodynamics (QCD). There is also great interest in the astrophysics community, for accurate measurements of pion and photon production cross sections at low energy, since they enter into the calculations of star evolution and composition and distribution of cosmic rays.

For studying these and other topics we presently build a general purpose instrument, named "WASA" (Wide Angle Shower Apparatus). It consists of a pellet target system providing high luminosity, a vertex detector having an axial magnetic field and a forward detector. WASA will identify charged particles and photons and measure their momenta with high precision over a solid angle close to 4π. The knowledge of the complete kinematics will enable a precise determination of the reaction channel by using the four constraints in momentum-energy. It is hoped that our approach will create pathways to some relevant problems in subatomic physics which require the knowledge of identified processes on the picobarn level.

ELECTRON-POSITRON DECAYS OF THE NEUTRAL PION

The rare decays of the neutral pion are interesting but also difficult to study. The neutral pion is short-lived (about 10^{-16} sec) and the decay particles, the photons and the electrons, produce showers in their interaction with matter and they are hard to measure with precision.

The various decays of the neutral pion, that have been measured or searched for so far, are listed below together with their present experimental branching ratios or upper limits [1].

Table I . Rare decay modes of the neutral pion (1990)

Mode		Fraction (Γ_i/Γ)	
2γ		(98.798± 0.032) %	
$e^+e^-\gamma$		(1.198±0.032) %	
$e^+e^+e^-e^-$		(3.14±0.30)	$\times 10^{-5}$
e^+e^-		< 1.3	$\times 10^{-7}$
4γ		< 2.0	$\times 10^{-8}$
$\nu\bar{\nu}$		< 6.5	$\times 10^{-6}$
$\nu_e\bar{\nu}_e$		< 1.7	$\times 10^{-6}$
$\nu_\mu\bar{\nu}_\mu$		< 3.1	$\times 10^{-6}$
$\nu_\tau\bar{\nu}_\tau$		< 2.1	$\times 10^{-6}$
Charge conjugation (C) or Lepton Family number (LF) violating modes			
3γ	C	< 3.1	$\times 10^{-8}$
μ^+e^-	LF	< 1.6	$\times 10^{-8}$

The $\pi^o \rightarrow e^+e^-$ decay mode, this much debated, still elusive, decay mode, requires the highest degree of sophistication of the apparatus. The WASA detector is designed to meet these requirements and could give important information also on the other decays listed.

The present upper limit for the $\pi^o \rightarrow e^+e^-$ decay given in Table I is factor of two larger than the value calculated from the quantum electro-dynamics contribution through a two-photon loop. The calculated rate of (0.63 ± 0.15) $\times 10^{-7}$ is weakly model dependent and determined mainly by the imaginary part of the amplitude which can be calculated using unitarity [2,3].

Attempts to measure the branching ratio for the decay $\pi^o \rightarrow e^+e^-$ were so far made in four experiments. The first was performed at CERN by Fischer et al [4,5], using π^os tagged from $K^+ \rightarrow \pi^+\pi^o$ decays. Only seven e^+e^- events were observed with an invariant mass greater than 0.96 m_π. The branching ratio was B = (2.23 +2.4/-1.1) $\times 10^{-7}$. In the experiment done at Los Alamos by Mischke et al [6] the reaction $\pi^-p \rightarrow \pi^o n$ was used to generate the π^o. The incident momentum was chosen to 300 MeV/c to minimize the background and to increase the acceptance of the apparatus. Due to the dimensions of the liquid hydrogen target employed (5 cm diameter and 25 cm long), the multiple scattering of electrons and positrons and the external conversion of photons caused severe problems. After background subtraction,

59±29 e^+e^- events remained in the π^o mass region, corresponding to a branching ratio, (1.7±0.6±0.3)x10^{-7}. Lately, the $\pi^o \rightarrow e^+e^-$ decay has been studied at two laboratories. From one of these, performed at CERN by the Omicron Collaboration, with essentially the same technique as in the Los Alamos experiment, only an upper limit of 5.3 × 10^{-7} was derived [7]. At SIN the pion was produced from the charge exchange reaction $\pi^-p \rightarrow \pi^o n$ at rest. The neutron was measured by time-of-flight techniques and the lepton pair in an axial field spectrometer with multiwire proportional chambers and plastic scintillators for the tracking. The hydrogen target had a thickness of 1cm which is about 1000 times thicker than that anticipated in WASA. The experiment resulted in an upper limit, 1.3 × 10^{-7} for the branching ratio [8].

The aim of our experiment using CELSIUS and WASA is to have 3000 $\pi^o \rightarrow e^+e^-$ per week of running. Any deviation from QED could be due to an exotic contribution exchange of a pseudoscalar heavy particle in the s-channel, or a leptoquark interaction in the t-channel. The suppression of the electromagnetic contribution due to the high order in the coupling and the helicity suppression would permit even a small exotic component to contribute visibly. It should be noted that all models which contain supersymmetry predict an enlarged Higgs sector containing at least one neutral pseudoscalar Higgs particle. This is in particular true for most known compactifications of the currently fashionable superstring theories. In fact, the $\pi^o \rightarrow e^+e^-$ decay with the present experimental value interpreted as an upper bound, has recently been used to constrain the parameters of a particular string-inspired model [9], and also in another context to bound leptoquark couplings [10].

The electromagnetic contribution to the $\pi^o \rightarrow e^+e^-$ decay involves vector couplings and is therefore helicity suppressed. The best candidate for a new interaction visible in this channel is then a helicity-flipping pseudoscalar particle of large mass in the s-channel. It should be possible to explore masses up to 600 GeV. It must be noted, though, that for Higgs-like particles, the Yukawa couplings, after spontaneous symmetry breakdown, become proportional to the fermion masses and are therefore still helicity suppressed. This need not be the case, however, in technicolour schemes and in other composite models.

Also for Higgs-like particles, the $\pi^o \rightarrow e^+e^-$ decay may still be sensitive, namely for light particles, e g with couplings such as those of an axion. As an example, the present experimental value, interpreted as an upper bound for the decay, has already been used to exclude some non-standard axion models recently proposed to explain the electron-positron peaks observed in heavy ion collisions at GSI in Darmstadt [11].

To summarize the physics situation for the $\pi^o \rightarrow e^+e^-$ decay: it is badly known experimentally and several attempts to measure it have either failed or given an uncertain value based on low statistics. The theoretical prediction in the standard model is firm and based on QED and unitarity. If the present disagreement between theory and data persist, there are several possible new interactions, well founded in theory, that may account for the difference. An accurate measurement of the branching ratio is therefore urgent and the WASA-CELSIUS combination seems ideally suited to settle this question in a few months of running time at nominal luminosity.

DALITZ DECAYS OF THE NEUTRAL PION

Another interesting decay is the single Dalitz decay $\pi^o \to e^+e^-\gamma$. Since this decay probes the vector form factor of the pion, it may hint the off-shell continuation of the anomalous triangle diagram. The lepton mass distribution is usually calculated for a point-like pion from ordinary QED. The effect of the pion structure is introduced by a form factor which modifies the pointlike distribution.

It is customary to introduce the dimensionless variable $x=s/m_\pi{}^2$ where s is the invariant mass squared of the lepton pair and parametrize the form factor as a linear function of x:

$$F(x) = 1 + ax$$

The slope parameter, a has been measured in six investigations [11-16] with poor agreement inbetween the various experiments (see Table II) even with regard to the sign of the slope, which should be positive if standard physics is involved [2,3]. The QED radiative corrections to the process seem to be undestood now and they are important when the slope parameter is extracted from the data.

Table II Slope parameter, of the form factor, F. Data in brackets are corrected for radiative processes.

author laboratory	experimental method	number of events	slope parameter	π^o production method
Kobrak (EFI)	Hydrogen bubble chamber	not given	-0.15±0.1	$\pi^- p \to \pi^o n$
Samios (COLU,BNL)	Hydrogen bubble chamber	3071	-0.24±0.16	$\pi^- p \to \pi^o n$
Devons (COLU,ROMA)	Spark chambers	2200	0.01±0.11	$\pi^- p \to \pi^o n$
Fisher (CERN,SLAC)	Magnetic spectr. MWPC, Cerenkov	30000	0.05±0.03 (0.10±0.03)	$\pi^- p \to \pi^o n$
Gumplinger (OREGON)	NaI,wire chambers	-	0.01+0.08/-0.06	$K^+ \to \pi^- \pi^+$
Fonvieille (PASC,CBER, SACLAY)	Magnetic spectr. drift chambers	32000	-0.11±0.03	$\pi^- p \to \pi^o n$

The behaviour of the form factor was derived through different theoretical models such as vector meson dominance model, baryon loop model, QCD bound state calculation dispersion relations and quark triangle mechanism.The contribution from two-photon exchange has also been recently evaluated.The more precise data are however still needed to distinguish between different predictions.

With WASA, we expect that a great improvement over previous experiments can be made due to a close to 4π acceptance and the detection of both the leptons and the photon. The expected statistics collected as a by-product of the search for the $\pi^0 \to e^+e^-$ decays will be at least one order of magnitude larger than in the most precise experiment so far [5]. It is then conceivable that even finer details of the form factor can be tested. In some models, there is a deviation from linearity that depends on details of the pion wave function [17, 18].

The double Dalitz decay, $\pi^0 \to e^+e^-e^+e^-$, was historically important to determine the parity of the neutral pion. Since then there has been no new measurement improving on the 206 events collected at that time. With WASA, statistics would improve by several orders of magnitude, making possible, e.g., a search for a small parity violating component in the electromagnetic interaction which would show up in the angular correlation between the planes defined by the two e^+e^- pairs. This correlation is well-known in shape for the main parity conserving part, since an exact calculation has been performed using QED [19].

OTHER RARE PROCESSES

Before its decay, a π^0 or η may interact strongly with a second proton of the target. Such an interaction will result in three protons and two photons in the final state. Since the target length is very short, given by the meson decay length, the probability for this process is negligibly small in work with external beams and thick targets. In a cooler storage ring, the circulating intensity is very high and the process is detectable. In this way, information on the scattering properties of these mesons may be obtained.

Also violation of charge conjugation invariance may be searched for in the π^0 decays, namely by looking for the C-violating decay $\pi^0 \to 3\gamma$. The experimental upper limit is presently 3.1×10^{-8} . WASA at CELSIUS is a natural detector to detect $\pi^0 \to 3\gamma$ and the allowed, but in QED highly suppressed by angular momentum barrier, decay $\pi^0 \to 4\gamma$ which could receive contributions for an axion-like particle decaying into two photons.

Muon number violating rare decays are interesting research topics because they evidently lie outside of the standard model. The experimental effort has been concentrated on muon and kaon decays [20]. For the $\pi^0\mu e$ decay, a limit is of the order of 10^{-8} has been measured [21], and this limit could certainly be improved.

The alrge acceptance of WASA and the fact that the produced pions may be tagged by the two forward-going protons means that one may also perform a sensitive search for events containing missing energy, as would be expected from π^0 decays into neutrinos or photinos. A neutrino with a mass in the 10 eV range would give a truly negligible contribution. The maximum value for a standard model Dirac neutrino is obtained for a mass around 50 MeV and corresponds to a branching ratio around 3×10^{-9}. It is, however, difficult to reconcile such a neutrino mass with constraints from cosmology. In principle, a heavy Majorana neutrino or a photino could also show up. As an illustration, for the latter process, it has been estimated that the branching ratio may be of the order of 10^{-5} for a photino mass of 50 MeV and a scalar quark (squark) mass around 20 GeV [22] Also in this case, however, there are contradicting arguments related to cosmology for mass parameters in that range.

A process which has not been investigated at all experimentally, is the decay $\pi^o \rightarrow \gamma$ + nothing. This is a channel where it is difficult to find particles in presently discussed extensions of the standard model which could contribute. The standard model process $\pi^o \rightarrow \nu\nu\gamma$ can be estimated to give only a branching ratio of 10^{-18}, so the observation of $\pi^o \rightarrow \gamma$ + nothing would truly imply non-standard physics. With the properties of WASA, as specified in the detector proposal, we believe that a useful limit of at least 10^{-4} can be put for such processes. Again we remark that at present there is no limit at all in the Particle Data Tables.

As we have shown in this section, the rare decays of the neutral pion may be a window to new interactions outside the standard model and in this sense represent a useful complement to the dedicated rare muon, kaon and charged pion decay experiments performed elsewhere in the world. It may at this point be appropriate to remind of the crucial role, the neutral pion has already played in the development of many concepts in particle physics, such as the SU(3) colour gauge theory and the anomalies with their deep connection to topology. In fact, even to this day one of the most convincing arguments for the number of colours being three comes from the successful prediction for the $\gamma\gamma$ decay rate as calculated from the anomalous triangle diagram.

CELSIUS

The CELSIUS ring is an accelerator ring for storage and cooling of ions from the Gustaf Werner synchrocyclotron in Uppsala. Its circumference is 82 metres, corresponding to a revolution frequency of 3.4 MHz for 1.8 GeV protons. The ring is intended for nuclear and particle physics using thin internal targets interacting with the stored beam.

An electron cooler with a maximum electron energy of 300 keV corresponding to an energy of the stored beam of 550 MeV per nucleon, is now installed and has cooled low-energy protons successfully. For higher proton energies the experiments will be made without the cooling.

So far 4×10^{10} protons have been stored in the ring using stripping injection. A beam of H_2^+-ions converts to protons in a 20 $\mu g/cm^2$ thick carbon foil which is seen by the injected particles at their entry inside the bending magnet of the ring.

The maximum energy of 1.36 GeV for protons will be increased to 1.8 GeV after an upgrade of magnet power supplies. The importance of this increase can be judged from Table III showing the threshold energies for producing some of the long-lived mesons.

Table III. Threshold energies in MeV of protons for production of mesons in reactions of the type $p+A \rightarrow p+A+X$.

X	A = p	A = d	A = α
π^0	279.6	207.3	171.4
η	1258.1	903.6	727.3
ω	1891.6	1337.4	1061.8
η'	2403.8	1681.0	1321.6
$2K^{\pm}$	2494.0	1741.0	1366.6

The cluster-jet target at CELSIUS [23] is to a large extent based on the design of the UA6 target at CERN. It is aligned vertically with the differentially pumped beam source at the top, the scattering chamber where the circulating beam interacts with the target beam at the center and the target-beam dump at the bottom.

The cross section of the target beam at the intersection with the circulating CELSIUS beam is situated 250 mm from the nozzle aperture and defined by the oval skimmer with a length of 8 mm along the circulating beam and 5 mm across. The target-beam profiles are shown to be well bounded with a rapid increase to maximum values. Target thicknesses of 3×10^{14}, 5×10^{13} and 2.4×10^{13} atoms/cm^2 have been obtained for hydrogen, nitrogen and argon, respectively, at a pressure of 10^{-7} mbar in the scattering chamber. The target-related partial half-lives of the circulating proton beam have been shown to be about 20 s at 50 MeV and 20 min at 600 MeV.

PELLET TARGETS

The research on the refueling of fusion tokamak reactors by injection of hydrogen isotopes has lead to the development of several types of solid hydrogen pellet generators. One of these, developed at the University of Illinois at Urbana-Champaign [24], has a promising potential of being adaptable as a generator of hydrogen targets for cooler storage rings. A test set-up of a hydrogen pellet generator has now been built in Uppsala and work is in progress to demonstrate the feasibility of continuous injection into vacuum by using differential pumping along the path of the pellets. The generator is a modified version of the University of Illinois design whereas the vacuum injector is unique. The principle of the pellet target system as shown in fig. 1

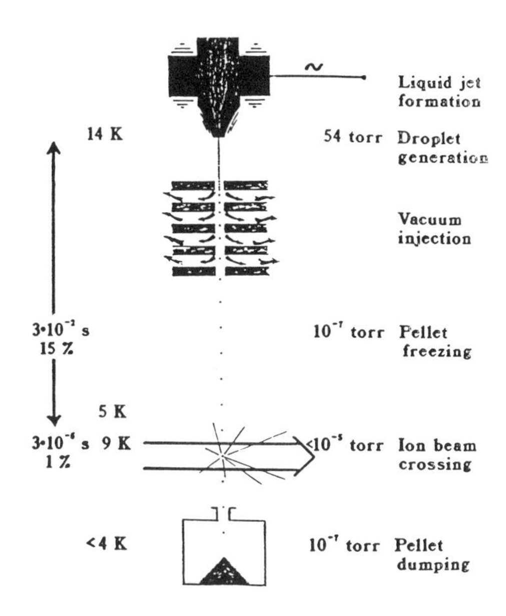

Fig 1. Principle of the pellet target system.

is that a pure liquid hydrogen jet, emerging through a glass nozzle with an inner diameter of 15 μm at near triple point conditions is broken up into uniformly sized and spaced droplets by means of acoustical excitation of the nozzle. The velocity of the jet is 10 m/s and the production rate about 200 kHz for the generation of droplets having diameters of 20 μm. By keeping the pressure in the droplet formation region slightly below the triple point pressure a frozen shell is allowed to develop on the droplets prior to their entry into vacuum through a two-stage differentially pumped section where the velocity is increased to about 100 m/s due to gas drag acceleration. Purely liquid droplets would be shattered in the interaction with the gas.

By charging the droplets, electric fields can be used to deflect a fraction of them into a dump volume, thereby reducing the frequency of targets crossing the beam to about 35 kHz i.e. so that just one target at the time will be present in a 3 mm wide beam. With 10^{10} stored protons in the ring, the experimental luminosity is 10^{32} $cm^{-2}s^{-1}$.

When entering the proton beam, some ten milliseconds after their creation, the targets have cooled down to temperatures of 6 K and lost about 20 % of their original mass due to evaporation [25]. The pellet surface temperature evolves with time according to figure 2.

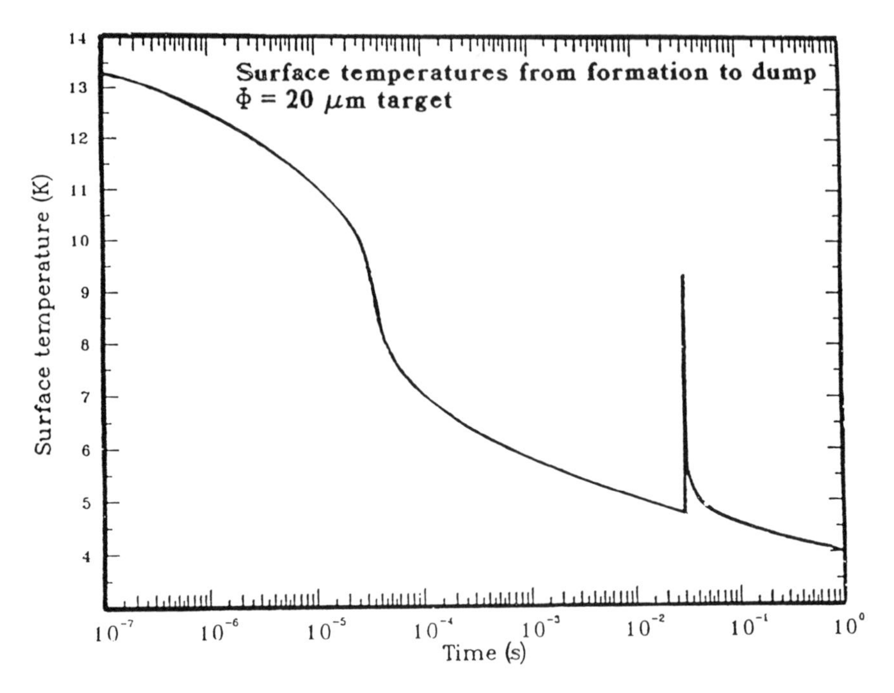

Fig 2. Pellet surface temperature versus time.

When passing through the beam, they are tracked by means of optical techniques and since their frequency of appearance has been matched to the beam size, the reaction vertex can be assigned to the only pellet in the beam when the appropriate triggers are alerted.

The pellets are heated to about 9 K when passing through the proton beam giving rise to a gas load rate of about 4×10^{-4} Pa m^3 s^{-1} in the interaction region [25]. The pellets are finally dumped in a two-stage differentially pumped collector.

WASA

To exploit the exciting possibilities to study rare reactions at CELSIUS, an apparatus is being prepared by which inelastic reactions between light beam particles and light targets can be measured accurately over a large solid angle. The apparatus, shown in figure 3, is named WASA [26] (acronyme for Wide Angle Shower Apparatus) and was originally proposed by scientists from Uppsala, Warsaw, Studsvik, Stockholm and Osaka. During the last year groups from KEK, the Japaneese National Lab, Novosibirsk, Dubna and ITEP, Moscow have joined the collaboration. The apparatus should be capable of:

- handling luminosities around 10^{32} cm^{-2} s^{-1} using spatially well defined targets of hydrogen and deuterium, in a coasting beam of 10^{10} protons

- measuring and identifying charged particles and photons over a solid angle close to 4π with high accuracy in energy, charge and track coordinates

- keeping photon conversion in the target, in the beam tube and in the vertex detector as small as possible

- providing a fast trigger for the selection of rare events.

WASA consists of a pellet-target system, a vertex detector for particles scattered isotropically and a forward detector for particles scattered at angles below 25 degrees. The vertex detector consists of four major parts:

- mini drift chamber (MDC) for the measurement of the coordinates of particle tracks for polar angles between 25° and 155°

- plastic scintillator barrel (PSB) consisting of 48 scintillators to provide fast logic signals for the trigger and analogue signals for the energy loss

- superconducting solenoid to provide an axial magnetic field in the volume of a mini drift chamber

- scintillator electromagnetic calorimater (SEC) with a high granularity to measure the energy and angle of photons and charged particles

Beam particles interact with the pellet targets in the centre of the vertex detector. The recoiling target nuclei and beam particles are emitted preferentially in the forward direction whereas the mesons and associated decay particles are emitted more isotropically. The reaction products enter the MDC through a thin-walled beam tube (0.5 mm beryllium). Momentum measurements of charged particles are accomplished in the MDC by determining the track curvatures in an axial magnetic field of 1.3 T generated by the superconducting solenoid. It consists of NbTi-wire embedded in aluminium matrix and is built in collaboration with the KEK experts that have conceived this concept. The cesium iodide scintillators fro the electromagnetic calorimeter are provided by the Novosibirsk group. They are doped with sodium that gives better resistance against radiation and shorter time for the light decay compared with thallium. In a June test run 24 scintillators of this type were used successfully for the measurement of protons and photons emitted in proton-hydrogen interactions at 1100 MeV.

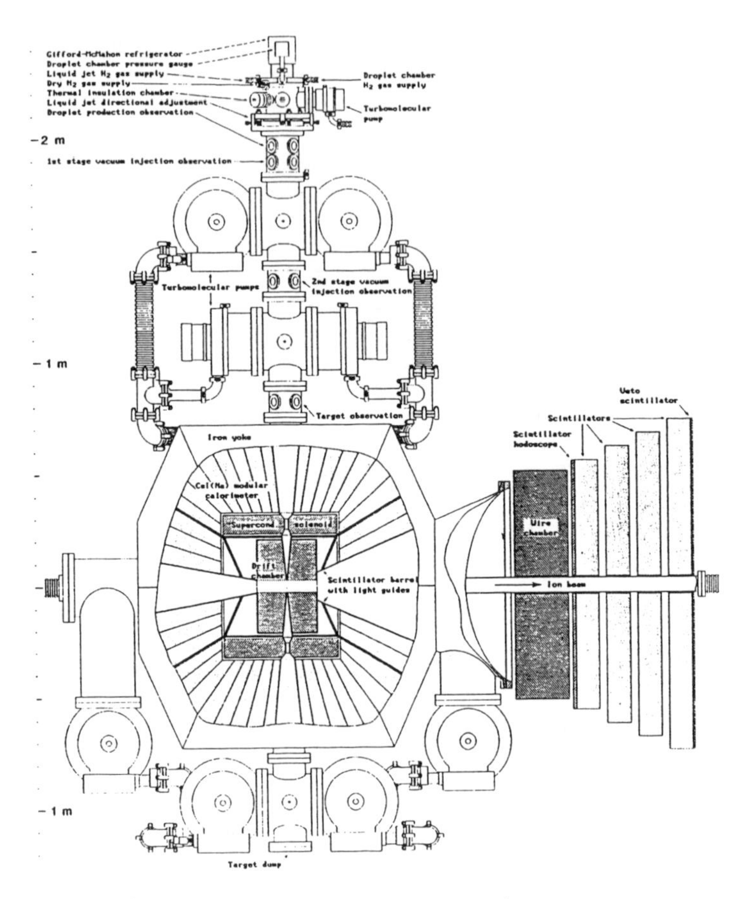

Fig 3. View of the proposed WASA experimental set-up.

Measurements of momenta, energies and energy losses are combined to give full identification of particles. The acquisition trigger for rare events is based on the multiplicity of charged particles and on the measured energy losses. In addition, some more sophisticated selection criteria, based on the kinematics, may be used to limit the amount of stored data. Work is presently in progress on the construction of forward spectrometer, scintillator electromagnetic calorimeter and pellet target system, on tests of components for the drift chamber and on a study of the most suitable solenoid to be used.

REFERENCES

1. Review of Particle Properties, Particle Data Group, Phys. Lett. 239B, (1990)
2. L. Bergström, Invited talk at "Rare decays of Light Mesons" Workshop Saclay, France, March 29-30, 1990 and references therein
3. L.G. Landsberg, Phys.Rep. 128, 301 (1985) and references therein
4. J. Fisher et al., Phys.Letters 73B, 359 (1978)
5. J. Fisher et al., Phys.Letters 73B, 364 (1978)
6. R.E. Mischke et al., Phys.Rev.Lett. 48, (1982) 1153
7. A.G. Zephat et al., J.Phys. G13, 1375 (1987)
8. C. Niebuhr et al., Phys Rev. D40, 2796 (1989)
9. M.E. Zeller, private communication
10. B.A. Cambell et al., Int.J.Mod. Phys. A2, 831 (1987)
11. E. Ma, Phys.Rev. D34, 293 (1986)
12. S. Devons et al., Phys. Rev. 184, 1356 (1969)
13. J. Fisher et al., Phys.Lett. 73B, 359 (1978)
14. P. Gumplinger , Doct. Thesis, Oregon State University (1987)
15. H. Fonvieille, Phys Lett. B233, 65 (1989)
16. H. Kobrak et al, Nuovo Cimento 20, 1115 (1961)
17. L. Bergström and H. Snellman, Z.Phys. C8, 363 (1981)
18. G.A. Kozlov, Z. Phys. C21, 63 (1983)
19. T. Miyazaki and E. Takasugi, Phys.Rev. D8, 205 (1973)
20. M.D. Cooper, Proc. HEP85, Bari, Italy, eds L. Nitti and G. Preparata
21. J. McDonough, Phys. Rev. D38, 2121 (1988)
22. M.K. Gaillard et al., Phys. Lett. 123B, 241 (1983)
23. C. Ekström, Topical Conf. on Electronuclear Physics with Internal Targets, Stanford, U.S.A., 1989. To be published by World Scientific Publishing Company
24. C.A. Foster, K. Kim, R.J. Turnbull and C.D. Hendricks, Rev. Sci. Instrum. 48, (1977) 625.
25. B. Trostell, Proc. 1st European Particle Accelerator Conf., Rome, Italy, June (1988).
26. Osaka-Stockholm-Studsvik-Warsaw- Uppsala collaboration, Proposal CA 4, The Svedberg Laboratory, December 1987.
27. G. Tupper, Phys. Rev. D35 (1987) 1726

η DECAYS AT SACLAY

B.Mayer

SEPN, CEN Saclay, 91191 Gif-sur-Yvette, France

Saclay (DPhN and LNS) - UCLA - PSI - Washington - Beersheva - Dubna collaboration.

ABSTRACT

A facility dedicated to the production of η mesons has been installed at the Saturne synchrotron with the purpose of investigating rare decays of this meson. The η are produced by the pd $\to$ ^{3}Heη reaction near threshold and tagged by the detection of ^{3}He in a magnetic spectrometer (SPES2). A rate of 10^5/s tagged η can be achieved. In the first experiment, $\eta \to \mu^+\mu^-$, the μ will be detected in range telescopes. Magnetic spectrometers for lepton detection are considered for future experiments.

1. INTRODUCTION

The search for rare (and forbidden) decay modes of light mesons provides stringent tests of symmetries and conservation laws. While μ, π and K mesons are abundantly produced in various laboratories and their decay modes studied at the level of 10^{-12} and even lower, the data on η decays are scarce and very unprecise because until now η mesons were produced with low flux, essentially through the π^-p $\to$ ηn reaction. Recent experiments at Saturne have shown a remarkably high production rate of η mesons in the pd $\to$ ^{3}Heη reaction, with a low level of background [1]. This discovery openned the possibility to study rare decay modes of the η meson with a flux several orders of magnitude higher than in any previous experiment. Another advantage is that the η mesons are tagged. Now a facility dedicated to the production of tagged η and including a magnetic spectrometer (SPES2) has been installed at Saturne. It is described in the next paragraph. Then we discuss the Physics involved in the study of various decay modes of the η meson into charged particles. The first experiment $\eta \rightarrow \mu^+ \mu^-$, planned for the fall of 1990, is discussed next. The last paragraph presents a detection system based on magnetic spectrometers to detect leptons for future experiments.

2. THE SATURNE η FACTORY

The η mesons are produced by the pd → $^3He\eta$ reaction near threshold. Both ^3He and η are emitted in a forward cone, with a narrow dispersion in angle and momentum. The ^3He are detected with the magnetic spectrometer SPES2 (Fig. 1) which has various advantages: high acceptance in angle (±3° horizontally, ±6° vertically) and in momentum (±18%). SPES2 would detect ^3He without loss of acceptance for incident energies up to 8 MeV above threshold. SPES2 is made of two dipoles through which the incident proton beam passes, which is a favorable condition to minimize the background.

The η production cross section was measured with a liquid deuterium target, as a function of the incident energy, with 1 MeV steps. The detection consisted in two chambers, each containing two planes of MWPC; three planes of scintillators provided dE/dx and time-of-flight identification and a trigger. The ^3He could be easily selected on line and particle identification was done essentially with dE/dx from only one hodoscope plane.

The measured cross section, shown in Fig. 2, is between 0.4 and 0.5 μb within a range of at least 5-6 MeV of the incident energy. This allows the use of a liquid deuterium target as thick as 10 cm without losing in production rate, taking into account the proton energy loss which is 0.37 MeV/cm. A production rate of at least 10^{10} η/day is therefore expected with the full beam intensity of 10^{12} protons/burst.

The fact that the η are produced near threshold has a few advantages; in particular when the η decays into two particles, there is a strong correlation between the aperture angle of the two particles and the η (or ^3He) energy.

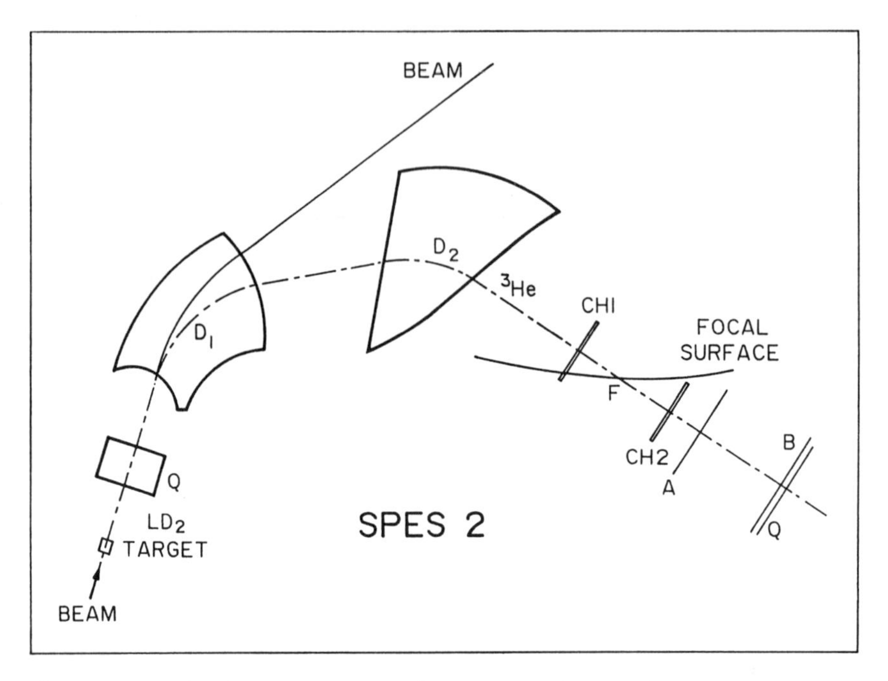

Fig. 1: The SPES2 spectrometer is used to detect ^{3}He from pd → ^{3}Heη.

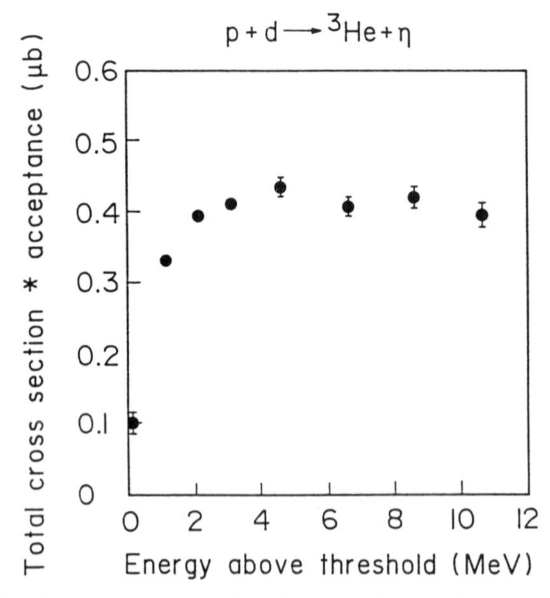

Fig. 2: Cross section x SPES2 acceptance for pd → ^{3}Heη, as a function of the incident energy above threshold.

Furthermore, since the η is produced in a narrow cone, the two particles emitted from the decay are nearly coplanar with the beam direction. Those kinematical features are very useful in helping identify the decay process against background.

A background of 4-25 % was observed under the η peak in the focal plane, after empty target subtraction. This background is of course dependant on the resolutions momentum and angular measurements. It is minimal at 1 MeV above threshold. This background may consist, at least partly, in ρ contamination to the η production. That would make very difficult the identification of certain decay modes of the η. In order to further investigate the ρ production, we have detected 2 π in coincidence with the η (the ρ decays essentially into π^+ π^-). The π^+ π^- pairs were identified by their kinematic correlations with the ^3He momentum. The correlated 2 π background was found to be ~ 2% of the η production, and this constitutes an upper limit to the ρ contamination.

3. PHYSICS MOTIVATIONS FOR INVESTIGATING η (AND π^0) RARE DECAYS

Since the η meson is isoscalar and non strange, its decay modes provide information complementary to those of the π^0, which is isovector and of the K mesons which are strange. We are discussing here only decay modes that we are considering measuring by detecting charged perticles. Detailed reviews of the Physics of η decays may be found in proceedings of workshops [2].

1) $\eta \to e^+e^-\gamma$ and $\eta \to \mu^+\mu^-\gamma$ give the transition form factor, an information on the meson structure. The branching ratio (BR) for $\eta \to \mu^+\mu^-\gamma$ has been already measured [3] but the data need improvement to be selective among the various theoretical predictions [4].

2) $\eta \to \mu^+\mu^-$, $\pi^0 \to e^+e^-$ and $\eta \to e^+e^-$ are primarily second order electromagnetic processes, suppressed also by helicity conservation. The imaginary part of the amplitude is calculable from unitarity [5]. A substantial deviation of the branching ratio from the theoretically expected ones, would imply exotic contributions [6]. A measurement of the BR allows to set limits on couplings with neutral currents [7].

The BR for $\pi^0 \to e^+e^-$ has been measured in various experiments [8]. The most accurate value from LAMPF is three times larger than expected from

QED, a result which has triggered speculations about exotic contributions. The last result of PSI sets a more conservative upper limit. A more precise measurement is clearly needed in order to draw a reliable conclusion.

The electromagnetic contribution for $\eta \to e^+e^-$ is more than one order of magnitude lower than for $\pi^0 \to e^+e^-$; this makes the η even more interesting if one hopes to unravel exotic contributions. The current experimental value ($BR(\eta \to e^+e^-) < 3 \times 10^{-4}$) is far from the unitary limit and could be largely improved. However the ρ contamination to η production makes this experiment very difficult since the ρ has a BR to e^+e^- that is expected to be four orders of magnitude higher than $BR(\eta \to e^+e^-)$.

3) $\eta \to \mu e$ is not allowed if lepton number is conserved. Till now no violation of this conservation law has been observed and various experiments allow to set limits of hundreds of TeV on mass scales for lepton number violating interactions [9]. However most extensions of the Standard Model predict such violation at levels depending on the specific processes, but generally quite low. Experimentally, no upper limit has been set on $BR(\eta \to \mu e)$, but an upper limit of 10^{-10} has been derived phenomenologically from experimental limits of other processes, mainly muon conversion into electron in nuclei [10]. Such a level of sensitivity is reachable with the production rate achieved at Saturne.

4) $\eta \to \mu^+\mu^-\pi^0$ and $\eta \to e^+e^-\pi^0$ will test C conservation in electromagnetic processes down to the level of $10^{-8} - 10^{-9}$ where one can expect a contribution from two-photon exchange. The current experimental limits are two or three orders of magnitude higher.

5) Measuring the muon longitudinal polarisation in $\eta \to \mu^+\mu^-$ would be of great interest since this polarisation can result only from CP violation. In the Standard Model, with CP violation arising from a phase in the Cabibbo-Kobayashi-Maskawa matrix, the muon polarisation in $\eta \to \mu^+\mu^-$ is 0. However other sources of CP violation exist in models beyond the Standard Model. Limits on muon polarisation in those models have been evaluated ($P_L < 0.1-4$ %) [10, 11]. A measurement with an accuracy of 2 % seems to be achievable at Saturne.

4. FIRST EXPERIMENT: $\eta \to \mu^+\mu^-$

Various models suggest that the real part of the amplitude increases by about 30% the decay width given by the unitary limit [4]). The current experimental value [12]), consistent with most theoretical models, is of low accuracy (27 events). We intend to measure the BR for $\eta \to \mu^+\mu^-$ with an accuracy of 6%.

The muon detection system (Fig. 3) will consist in a double-arm apparatus covering the angular region 50°-75°, on the left and right sides of a 7 mm thick liquid deuterium target. A first wedge-shaped iron absorber will stop most of the protons coming from the target and slow down the muons in such a way that the muon energy will be independant of angle. Then a hodoscope of narrow scintillators provides a measurement of angle with 20 mrad resolution, while still tolerating a high counting rate. It is followed by another iron absorber that will stop most of the pions, and a hodoscope plane for triggering. Finally six scintillators planes, 5 cm thick, measure the energy losses of a particle until it is stopped, while a veto counter at the end will signal particles that traverse the whole system. The acceptance of the system for $\eta \to \mu^+\mu^-$ is 3.8%, so that about 130 events/day are expected for a 4×10^{11} protons/s beam.

The background comes from correlated $\rho \to \pi^+\pi^-$ events as well as accidentals arising from the large hadronic production rate in the deuterium target. The iron absorbers eliminate all the correlated pion background as well as most of the accidentals. The remaining background consists essentially of $\rho \to \pi^+\pi^-$ events followed by the decay of both pions, and accidentals, mainly from pions produced by quasi free nucleon-nucleon reactions. The reduction of this background is achieved by applying kinematical constraints:

a) Total energy conservation: the range telescopes provide measurement of the two muons energies with a resolution of 12 MeV. The sum of the muons and ^{3}He energies will be compared with the total incident energy.

b) Muon pair invariant mass: should be equal to the η mass.

c) coplanarity of η, μ^+ and μ^- trajectories.

d) Muon angular correlation: the openning angle between the two muons depends only on the ^3He energy.

e) Further μ/π identification is provided by the successive energy losses in the six range counters.

Monte Carlo simulation shows that the contribution from $\rho\to\pi^+\pi^-\to\mu^+\mu^-$ (+ 2 neutrinos) is 0.5%, after applying the kinematical cuts. The accidental/signal ratio is about 1%.

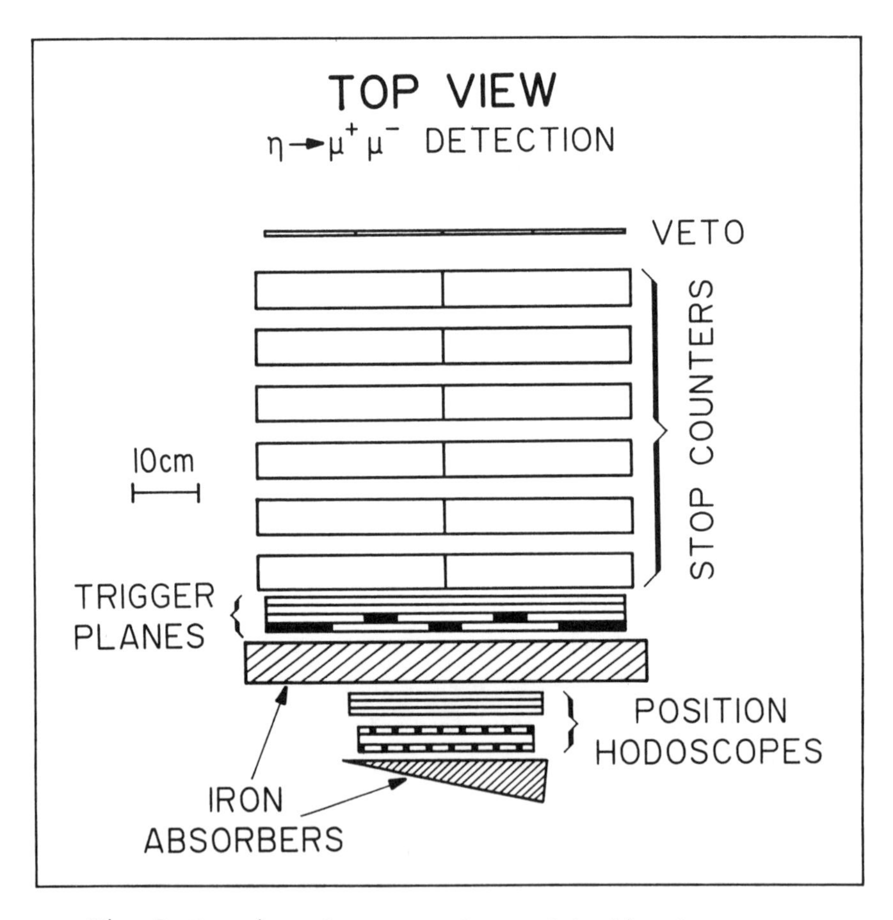

Fig. 3: Top view of one arm of muon detection for $\eta\to\mu^+\mu^-$

5. AN EXPERIMENTAL SET-UP FOR FUTURE EXPERIMENTS

To detect the decay products of rarer processes, magnetic spectrometers are necessary. Their first function is to filter the hadronic background, with the consequence that the angular deviation must be large enough so that the detection is not in direct view of the target. Their second function is to allow a determination of particle momenta that permits the identification of the process through kinematical constraints. We propose to build two spectrometers that will detect particles symetrically with respect to the beam (at 62° for muons). The magnetic field is in the same plane as the beam (Fig. 4) so that the dispersive plane (Fig. 5) is decoupled from the longitudinal dimension of the target. One can therefore use a long target and reconstruct the vertex with a resolution of 5 mm, by measuring the trajectory in the direction perpendicular to the dispersive plane. With a gap of 20 cm and a vertical aperture of ± 20°, the acceptance is 1%, assuming that both charges are detected in each magnet.

The detection consists in two small gap MWPC. Those chambers should be able to work at a rate of 3×10^7 particles/s. MWPC with 1.5 mm gap, anode wires 20 μm thick, using a mixture of 80% CF_4 and 20% isobutane, are being developped at Dubna. A counting rate of $10^7/cm^2/s$ has been achieved. In order to reduce accidentals, an iron absorber will be placed in the focal plane. It will eliminate most of the pions since pions have a shorter range than muons, for a given momentum. One has to take in account, however, the decay of pions into muons; a detailed Monte Carlo simulation is in progress.

This experimental set-up could be used to perform the Physics program discussed in paragraph 3. But alternative set-ups for studying neutral decay modes may be considered also.

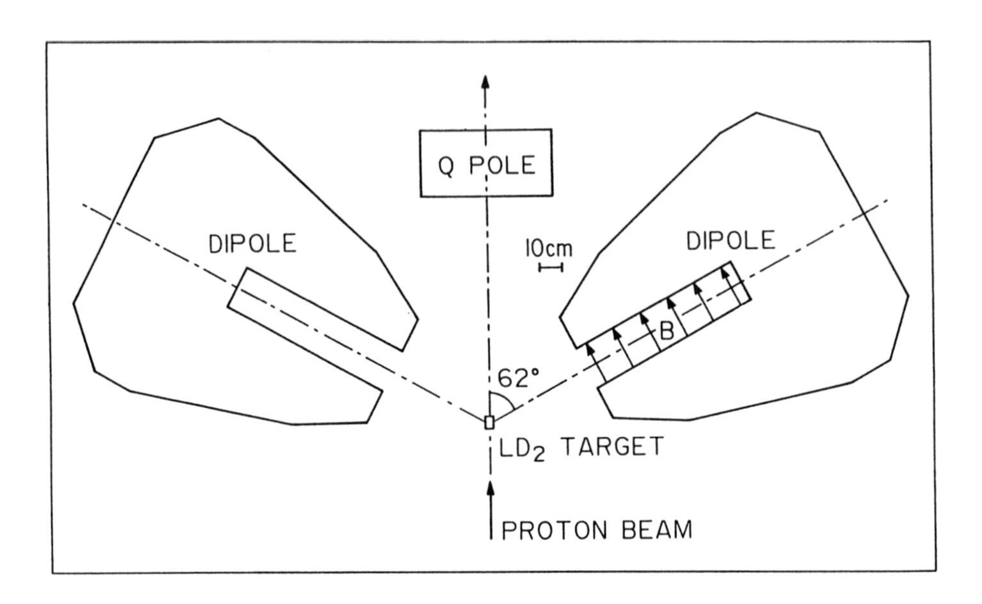

Fig. 4: Two magnets for the detection of decay products of the η.

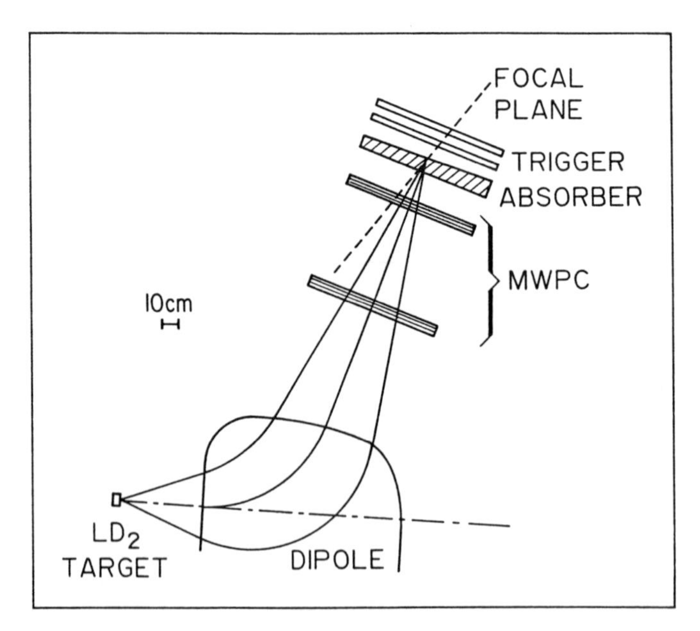

Fig. 5: The dispersive plane of each magnet.

REFERENCES

1. J.Berger et al., Phys. Rev. Lett. 61(1988)919
2. Physics with Light Mesons & the Second International Workshop on Pi-N Phys., ed. W.R.Gibbs & B.M.K.Nefkens, LA-11184-C, 1987; Workshop on Production and Decay of Light Mesons, march 1988, Paris, ed. P.Fleury, World Scientific; Workshop on Rare Decays of Light Mesons, 29-30 march 1990, Gif-sur-Yvette, ed. B.Mayer, Editions Frontières
3. R.I.Dzhelyadin et al., Phys. Lett. 94B (1980) 548
4. L.G. Landsberg, Phys. Rep., Vol 128, No. 6 (1985) 301
5. S.M.Berman and D.A.Geffen, Nuovo Cimento 18 (1960) 1192; D.A.Geffen and B.L.Young, Phys. Rev. Lett. 15 (1965) 316
6. L.Bergstrom, Z.Phys.C Particles and Fields 14 (1982) 129
7. P.Herczeg, Workshop on Production and Decay of Light Mesons, march 1988, Paris, ed. P.Fleury, World Scientific;
8. J.S.Frank et al., Phys. Rev. D28(1983)423; Collaboration Omicron, J.Phys.G. Nucl. Phys. 13(1987)1375; C.Niebuhr et al., Phys. Rev. D 40 (1989) 2796
9. P.L.Mélèse, Comments Nucl. Part. Phys. 19 (1989) 117
10. P.Herczeg, Workshop on Rare Decays of Light Mesons, 29-30 march 1990, Gif-sur-Yvette, ed. B.Mayer, Editions Frontières
11. J.N.Ng, this conference; C.Q.Geng and J.N.Ng, Phys. Rev. Lett. 62(1989)2645; X.G.He and B.H.J.McKellar, Workshop on Rare Decays of Light Mesons, 29-30 march 1990, Gif-sur-Yvette, ed. B.Mayer, Editions Frontières
12. R.I.Dzhelyadin et al., Phys. Lett. 97B (1980) 471

SESSION B

Electro-Production and Photon-Induced Reactions

J/ψ and Y(1s) Production*

Walter H. Toki
Stanford Linear Accelerator Center
Stanford University, Stanford, California 94309

Abstract

Physics topics from decays of the J/ψ, ψ ', and Y(1s) mesons are discussed. Comparisons between these particles and old and new puzzles are reviewed.

INTRODUCTION

The J/ψ and the Y(1s) are unique particles being produced below flavor threshold. They form bound states of $c\bar{c}$ and $b\bar{b}$ quarks respectively that exist below open charm and bottom threshold and thereby have unusual decay properties. In this paper we review current problems or puzzles in various decays of these mesons. The topics that will be discussed are;

- Radiative decays of the J/ψ
- J/ψ and ψ ' comparisons
- J/ψ and Y(1s) comparisons
- Evidence for anomalous J/ψ decays from the Y(1s)
- Rare decays of the J/ψ

The last topic will discuss possible results from a very high luminosity machine that is being proposed. Before starting the main discussion on the above topics we begin with a short section on e^+e^- colliders and a brief description of our current understanding of heavy quarkonium decays. The material is introductory and descriptive with an emphasis towards the nuclear physics audience specializing in much lower energy reactions near threshold which are being discussed in this conference on Particle production near threshold.

Invited talk presented at the topical conference on Particle Production near Threshold, Brown County Inn, Nashville, Indiana, Sept. 30-Oct. 3, 1990, sponsored by the Indiana University Cyclotron Facility and the Forschungszentrum Julich GmbH

*Work supported in part by the National Science Foundation and by the Department of Energy contracts DE-AC03-76SF00515

Table 1. Resonance Production in e+e- collisions

Particles	CMS(GeV/c^2)
ω/ρ	.770
ϕ	1.020
J/ψ	3.097
$e^+e^- \to \tau^+\tau^-$	3.56
$\psi' \to \gamma\chi_c(1^3P_{0,1,2})$, $\gamma\eta_c(1^3S_0)$	3.680
$\psi'' \to D^+D^-, D^\circ\overline{D}^\circ$	3.770
$e^+e^- \to DD^*, Ds\overline{Ds}, D^*s\overline{Ds}$	4.0-4.4
$e^+e^- \to \Lambda_c\overline{\Lambda}_c, \Sigma_c\overline{\Sigma}_c$	5-6
$\Upsilon(1s)$	9.46
$\Upsilon(2s) \to \gamma\chi_b(1^3P_{0,1,2})$	10.355
$\Upsilon(3s) \to \gamma\chi'_b(2^3P_{0,1,2})$	10.355
$\Upsilon(4s) \to B^+B^-, B^\circ\overline{B}^\circ$	10.575

Table 2. Summary of Low energy e+e- storage rings

Machine	Detector	Data
SPEAR	MARKII Crystal Ball	1.3-2M J/ψ ~17K $D\overline{D}$ pairs 1M ψ'
ORSAY	DM2	8M J/ψ
SPEAR	MARKIII	5.8M J/ψ 50K $D\overline{D}$ pairs ~6.3K DsDs pairs .3M ψ'
Beijing	BES	3-4 timesSPEAR Luminosity
Spain? 1996-7?	TauCharm	10^9 J/ψ /month 5×10^9 ψ' /month 4×10^7 $\tau\tau$/year 2×10^8 $D\overline{D}$/year 1×10^7 $Ds\overline{Ds}$/year

Low Energy e+e- storage rings

The decays of the J/ψ and the $\Upsilon(1s)$ are studied largely in e+e- colliders where they can be produced at high rates. The e^+e^- colliders have the advantageous property of directly producing the vector particles such as the ω, ρ, ϕ, J/ψ and $\Upsilon(1s)$. In addition they produce pairs of tau leptons, charm mesons and baryons. The various particles and pairs of particles produced in e^+e^- production are listed in Table 1. The machines in the 3-6 GeV center of mass energy region for the production of charm particles and charmonium particles are listed in Table 2. The Crystal Ball, Mark II, DM2 and the Mark III detectors are no longer running whereas the Beijing machine has just began to take data. The Beijing Electron Positron Collider (BEPC) has approximately 5 times the luminosity of SPEAR. The Beijing Spectrometer detector[1] (BES) has logged 3M J/ψ events. The proposed Tau-Charm machine or Tau-charm factory is being discussed at a site in Spain. The slightly higher energy machines in the 9-12 GeV region for the production of bottomonium and B mesons and baryons are are CESR and DORIS machines with the CLEO and ARGUS detectors. New high luminosity machines for B physics or B factories are being proposed at SLAC, Cornell and KEK for the study of CP violation. These machines or factories will provide the opportunity to study extremely large samples of events. In the case of the tau-charm factory the number of J/ψ's produced in a few months of running is expected to be 10^9.

PHYSICS

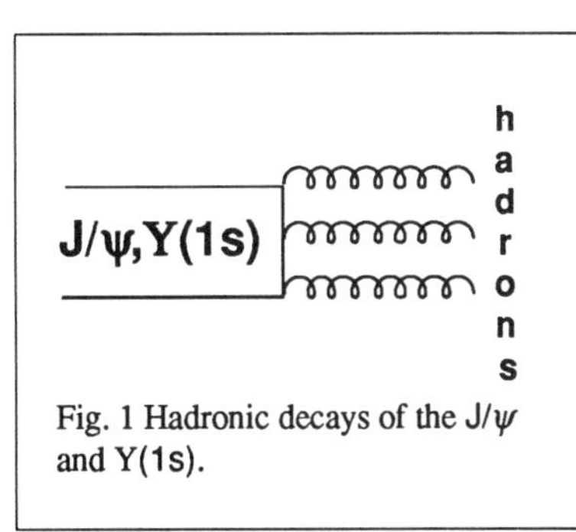

Fig. 1 Hadronic decays of the J/ψ and Y(1s).

Since both the J/ψ and the Y(1s) are below open threshold and cannot decay into mesons with open charm or bottom flavors, they are expected to decay via 3 gluons as shown in the diagram in Fig. 1. This prediction successfully determines the total width with a reasonable estimate of α_s. The electromagnetic decays of the $c\bar{c}$ (charmonium) and the $b\bar{b}$ (bottomonium) systems are very well understood in the non-relativistic quark model. The models use simple potentials and predict E1 and M1 transitions. The radiative decays are calculated with a quark spin flip as drawn in Fig. 2 and accurately predict the partial decay rates, angular distributions and the mass splittings between the 1^1S_0, 1^3S_1, 2^3S_1, 1^3P_0, 1^3P_1, and the 1^3P_2 states. The evidence for the 1^1P_1 and 2^1S_0 states is limited to one experiment each and not yet confirmed. The energy level splitting for the lowest lying charmonium system is drawn in Fig. 3. Charmonium and bottomonium spectroscopy of the radiative transitions of the χ states has provided compelling proof of the simple quark model.

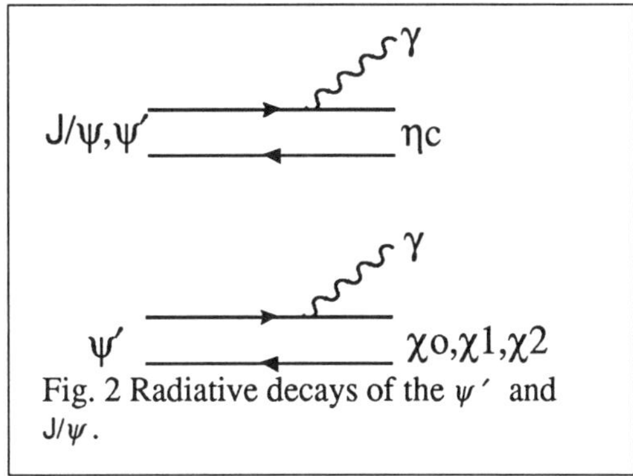

Fig. 2 Radiative decays of the ψ' and J/ψ.

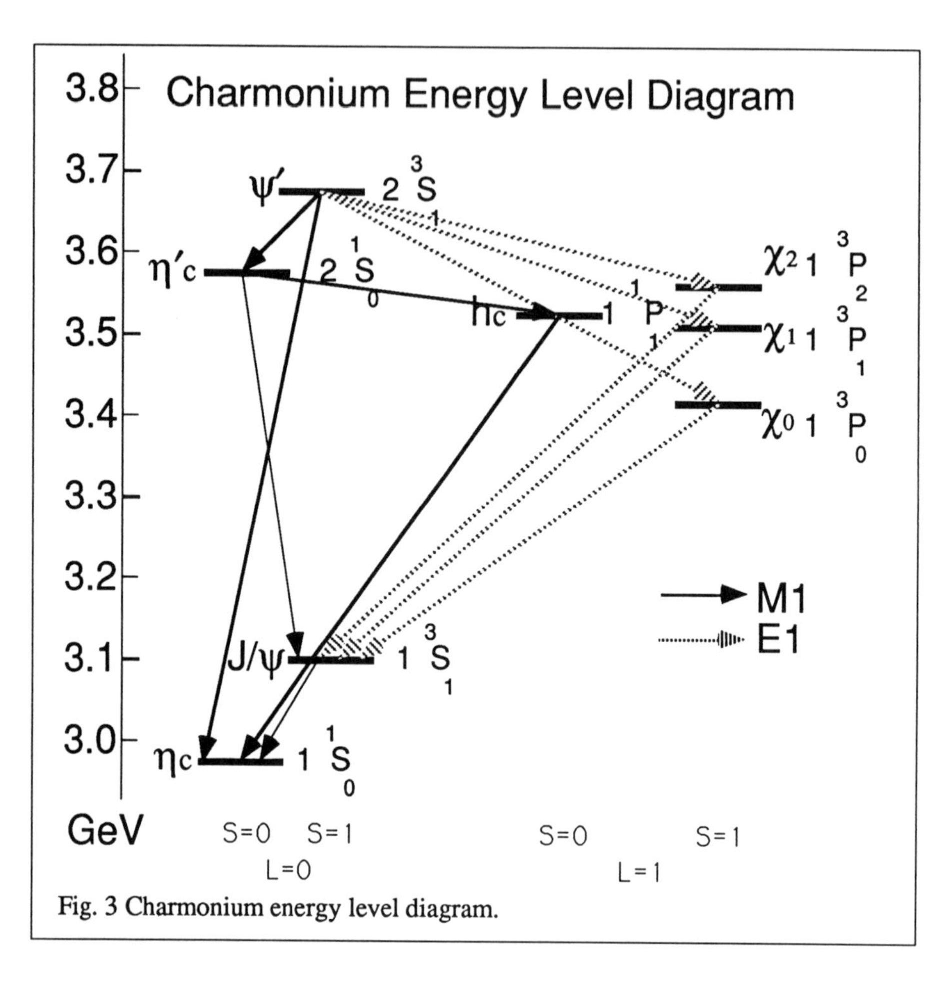

Fig. 3 Charmonium energy level diagram.

Radiative and hadronic J/ψ decays

The radiative decays of the J/ψ have been an area of intense study because it is expected to be a source of 2 gluons as shown in Fig. 4. The abelian nature of QCD predicts gluon self couplings and consequently a pair of gluons could resonate or “stick” together to form gluonium or glueballs. Hence any unusual resonances observed in radiative decays could be a glueball candidate. The central prediction of QCD lattice gauge theories is the existence of the lowest lying scalar glueball. Predictions[2] have previously centered around a mass of 1 GeV and recently moved up into the 1.5-2 GeV mass region. Also the mass of the tensor glueball is estimated in these theories to be a factor ≅1.5 larger than the scalar glueball mass. The characteristics of these glueball candidates are;

- They form isoscalar and SU(3) flavor singlet decays
- They have no radiative nor 2 photon decays
- Their lowest lying modes should have $J^{PC} = 0^{++}$ and 2^{++} for L=0 and 0^{-+} for L=1
- They should appear in hard gluon channels such as radiative J/ψ decays

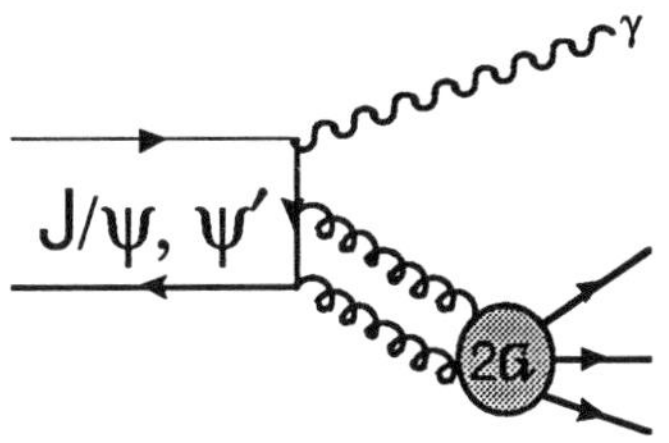

Among the different candidates seen in J/ψ decays are the iota and the theta resonances, now called the $\eta(1440)$ and the $f_2(1720)$ by the Particle data group. In this paper they will be denoted as the $\iota/\eta(1440)$ and the $\theta/f_2(1720)$.

Fig. 4 J/ψ radiative decays.

The properties of $\iota/\eta(1440)$ are listed below;

- Observed in $J/\psi \to \gamma\ \iota/\eta(1440)$, $\iota/\eta(1440) \to K\bar{K}\pi$
- Mass $\approx$ 1430 MeV/c2
- width $\approx$ 60 MeV/c2
- Spin $\approx 0^{-+}$
- BR $\approx$ .5%, very large

In a recent Mark III publication[3] the $\iota/\eta(1440)$ has been partial wave analyzed into quasi-two body decay modes. The main modes resolved are 0^{-+} and 1^{++} components which could be a radial excitation of the $\eta'(958)$ and the $f_1(1420)$ previously known as the E(1420). An alternate analysis by the DM2 group[4] however has different results. These Mark III results indicate that the $\iota/\eta(1440)$ could be largely a mixture of other conventional mesons.

The properties of the $\theta/f_2(1720)$ are listed below;

- Observed in $J/\psi \to \gamma\theta/f_2(1720)$, $\theta/f_2(1720) \to K\bar{K}$
- Mass $\approx$ 1720 MeV/c2
- width $\approx$ 138 MeV/c2
- Spin $\approx 2^{++}$
- BR $\approx$ .1%, large

In another recent Mark III result[5], the $\theta/f_2(1720)$ has been partial wave analyzed using a complete moments analysis. The results indicate that the $\theta/f_2(1720)$ is composed largely of spin 0, although an underlying spin 2 component cannot be excluded. This result alters a previously published Mark III analysis[6] which was based on a smaller sample and which tested the spin parity by treating the entire $\theta/f_2(1720)$ region as totally spin 0 or spin 2 and not a combination of both overlapping spins.

This result leads to a very interesting *conjecture* about the nature of the $\theta/f_2(1720)$. One could possibly identify the $\theta/f_2(1720)$ as the scalar glueball as predicted by Lattice Gauge Theories. This would set the mass scale for a constituent gluon at roughly that of the $\rho°$ meson. This would then place the mass of the vector glueball (~3 times the $\rho°$ mass) around 2250 MeV/c2 and the tensor glueball around the same region.

Table 3. Comparison of ψ' and J/ψ decays

Mode	$B(J/\psi \to mode)$	Ratio($\psi' \div J/\psi$ in %)
$p\bar{p}$	0.22 %	8.6 ± 2.4 %
$p\bar{p}\pi^+\pi^-$	0.60	15.1 ± 4.1
$K^+K^-\pi^+\pi^-$	0.72	22.2 ± 9.0
$p\bar{p}\pi°$	0.11	14.0 ± 6.3
5π	3.4	9.5 ± 2.7
7π	2.9	13.0 ± 7.0
$\gamma f(1270)$	0.14	9.0 ± 3.0
$\rho\pi$	1.28	<0.63
K^*K	0.75	<2.07
$\gamma\eta$	0.86	<1.8
$\gamma\eta'$	0.42	<2.6

J/ψ and ψ′ comparisons

The ψ' is the radial excitation of the J/ψ and is expected to hadronically decay like the J/ψ via the 3 gluon mode shown in Fig. 1. If we compare the 3 gluon widths, they can be related to the leptonic and total widths through the following relations;

$$\frac{B(\psi' \to \text{hadrons})}{B(J/\psi \to \text{hadrons})} = \frac{B(\psi' \to ggg)}{B(J/\psi \to ggg)} = \frac{\Gamma(\psi' \to e^+e^-)\,\Gamma(J/\psi)}{\Gamma(J/\psi \to e^+e^-)\,\Gamma(\psi')} = (12.2 \pm 2.4)\%$$

If we list the the actual measurements of the J/ψ and ψ' and form the ratios of the modes, we obtain the list shown in Table 3. We observe that with in the errors, the baryon modes and the multipion modes roughly follow the 12% prediction, however there is a large suppression of the modes with vector-pseudoscalar combinations. This result[7] has led to the development of a number of theoretical models.

• *Nambu* & *Freund* Model (1975)[8]

In this early model the puzzle is not the suppression of the $\rho\pi$ rate in the ψ' decays, but the existence of the $\rho\pi$ rate from the J/ψ. The rate from the J/ψ should in fact be suppressed, but it is being produced by a constructive interference from a nearby resonance. This resonance is a Pomeron daughter called the "ϑ" meson which is an SU(4) singlet vector meson that can mix with the ω, ϕ and J/ψ. It is expected to decay into $\rho\pi$ and KK^* and very little into e^+e^- or $K\overline{K}$. The model also predicts the "ϑ" mass around 1.4-1.8 GeV with a width of 50-100 MeV.

• *Hou* & *Soni* Model (1982)[9]

As in the previous model, a vector gluonia, "ϑ" interferes with the J/ψ enabling vector-pseudoscalar decays. The mass of this resonance is predicted to lie around 2.4 GeV if the ratio of the $\rho\pi$ branching ratios is 1.25% instead of the 12%. The resonance mass will also increase higher if the suppression is larger. The resonance is expected to appear in the reactions J/ψ, $\psi' \to (\eta,\eta',\pi\pi) + \vartheta$, $\vartheta \to \rho\pi, K^*K$.

In a recent paper[10] by Donnachie and Clegg a new possible connection may be made to vector gluonia in a puzzle in diffractive photoproduction. In this puzzle the relative rates of photoproduction and e^+e^- production of isovector vector mesons is well understood, but the isoscalar rate in final state modes such as $\rho\pi$ in diffractive photoproduction is anomalously high. An explanation may lie in the existence of 3 gluon isoscalar vector states in the 2 GeV mass region being photoproduced.

• *Brodsky, Lepage* & *Tuan* Model (1987)[11]

In this model, a vector gluonium resonance is proposed to lie within 100 MeV of the J/ψ and the interference produces the vector pseudoscalar decays of the J/ψ. This explains the puzzle as to why QCD hadron helicity conservation[12] fails to suppress the large $\rho\pi$ decay of the J/ψ. In addition this may explain why $J/\psi \to \phi S^*$ and not $\delta\pi$ is observed, since the ϑ mixes with the ϕ and enhances a mode that would otherwise be suppressed.

• *Slaugher* & *Oneda* Model (1988) [13]

In this model the vector glueball that mixes with the J/ψ is proposed also to explain why the decay $J/\psi \to \gamma\eta_c$ is so small (1.3%). In the charmonium model the M1 transition of J/ψ is expected to be a factor of three larger than observed.

• *Tornquist* & *Chaichian* Model (1988)[14]

In this model the puzzle is explained by introducing a hadronic form factor that exponentially decreases the two meson decays of the ψ' relative to the J/ψ by $\exp[-(m(\psi')^2 - m(J/\psi)^2/4K^2]$ where K is a parameter fitted to the data. This prediction explains the decay rate for $\phi \to \rho\pi$ and predicts a large suppression for many two meson modes. In a recent Mark III result, evidence[15] for the decay, $\psi' \to b_1(1235)\pi$, is found and consequently this could contradict this model.

• *Pinsky* Model (1989)[16]

In this paper it is pointed out that the radiative decay of the η from the ψ' is a hindered M1 transition. The radiative transitions to the η are predicted to scale as the η_c rates corrected for the phase space factors as;

$$\frac{B(\psi' \to \gamma\eta)}{B(J/\psi \to \gamma\eta)} = \left(\frac{p_{\eta}^{\psi'}}{p_{\eta}^{J/\psi}}\right)^3 \left(\frac{p_{\eta_c}^{J/\psi}}{p_{\eta_c}^{\psi'}}\right)^3 \frac{B(\psi' \to \gamma\eta_c)}{B(J/\psi \to \gamma\eta_c)} = 0.2\%$$

This predicts a small rate for the radiative η decay from the ψ'. Using a phenomenological model and estimates for the OZI amplitudes, the predicted rate for $\psi' \to \rho\pi$ is slightly less than 10^{-5}.

J/ψ and $Y(1s)$ comparisons

Estimates can be made to compare the radiative and hadronic decays of the J/ψ and the $Y(1s)$. The radiative rates will scale as,[17]

$$\frac{\Gamma(Y \to \gamma gg)}{\Gamma(J/\psi \to \gamma gg)} \cong \left(\frac{q_b}{q_c}\right)^2 \left(\frac{M_\psi}{M_Y}\right)^2 = \frac{1}{40}$$

and the hadronic rates will scale as,[18]

$$\frac{BR(Y \to AB)}{BR(J/\psi \to AB)} \cong \frac{BR(Y \to ggg)}{BR(J/\psi \to ggg)} \left(\frac{M_\psi}{M_Y}\right)^{4-6} = \frac{1}{100-1000}$$

In Table 4 are listed measurements from LENA[19], ARGUS[20], CLEO[21] and the Crystal Ball[18][22] groups. In general the rates are not yet sensitive enough to test the predictions.

Table 4. exclusive Υ(1s) decays

Mode	Group	BR	Prediction	BR(J/ψ→mode)
$\Upsilon\to\rho\pi$	LENA	$<2.1\times10^{-3}$	1.3×10^{6}	127000×10^{-6}
	CLEO	$<1.3\times10^{-4}$	1.3×10^{6}	127000×10^{-6}
	CrystalBall	$<2.2\times10^{-3}$	1.3×10^{6}	127000×10^{-6}
$\Upsilon\to a_2\pi$	CrystalBall	$<.5\times10^{-3}$	1.3×10^{-4}	84×10^{-5}
$\Upsilon\to\gamma\eta$	CrystalBall	$<3.5\times10^{-4}$	2×10^{-5}	86×10^{-5}
$\Upsilon\to\gamma\eta'$	CrystalBall	$<13\times10^{-4}$	11×10^{-5}	420×10^{-5}
$\Upsilon\to\gamma f_2(1270)$	CrystalBall	$<8.1\times10^{-4}$	5×10^{-5}	160×10^{-5}
	CLEO	$<4.8\times10^{-5}$	5×10^{-5}	160×10^{-5}
	ARGUS	$<13\times10^{-5}$	5×10^{-5}	160×10^{-5}

Anomalous Υ(1s) decays

The Υ(4s) is a radial excitation of a $b\bar{b}$ vector meson above threshold and is expected to decay 100% into B mesons. Recent results from the CLEO group[23] and the ARGUS group[24] however have shown an anomalous excess of J/ψ decays from the Υ(4s). These decays are found to have a momentum high enough such that they are not possible to be secondaries from the B mesons that are produced in pairs from the decay of the Υ(4s). For X>.378, the observed branching ratio is $BR(\Upsilon(4s)\to J/\psi+X)=(.22\pm.06\pm.04)\%$. There are observed J/ψ decays from the Υ(1s) and they have a partial width of $\Gamma(\Upsilon(1s)\to J/\psi+X)\approx50$ eV. The partial width of the Υ(4s) decay is $\Gamma(\Upsilon(4s)\to J/\psi+X)>50$ KeV or roughly a thousand times larger than expected. This is perhaps the most serious violation of the OZI rule. Other searches by the Mark III[25] have been performed in the decay $J/\psi\to\phi+X$ and an excess is found but was not conclusive due to background uncertainties. There has been some evidence for anomalous exclusive decays of the ψ″ reported in an unpublished Mark III thesis.[26] The models developed to explain this puzzle are ;

• *Khodjamirian, Rudaz* & *Voloshin* Model (1990) [27]

In this model the Y(4s) is predicted to be an admixture of $b\bar{b}$ plus perhaps a gluon. The would force the $b\bar{b}$ system to be in a color octet. This would lead to a large partial width which could produce J/ψ decays. In addition this model predicts a large radiative decay of the Y(4s). Tests of this model should be forth coming from Cornell.

• *Atwood, Soni* & *Wyler* Model (1990)

Conventional quarkonium spectroscopy is suggested to explain the excess J/ψ events through decays such as $\Upsilon(4s) \to h_b(1P) + \eta$ and $\eta_b + h_1(1170)$. If this is correct then other modes with $D\bar{D}$ content should also be seen and this could conceivably affect semileptonic estimates of V(ub).

• *Lipkin* Model (1985)[28]

This model was developed to explain a possible OZI violation in ψ″ decays which was later determined to be experimentally incorrect.[29] The idea is that the decays occur into a pair of D mesons which subsequently annihilate into non-charm states. This would possibly explain the large partial width of $\phi \to \rho\pi$.

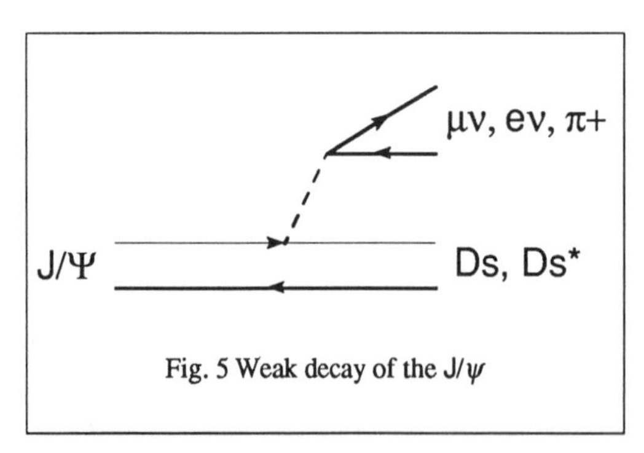

Fig. 5 Weak decay of the J/ψ

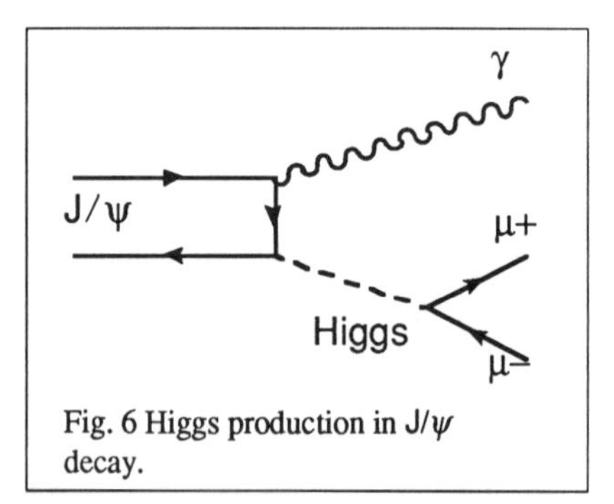

Fig. 6 Higgs production in J/ψ decay.

Rare J/ψ decays

In this last section we study models which may be possible to probe with very high luminosity tau charm factories. Although there exists no evidence for decays from these models, it is nevertheless interesting to consider them because a sample of 10^9 J/ψ's will eventually be realizable in a tau charm factory.

• *Verma-Kamal-Czarnecki* Model (1990)[30]

The possibility of probing 10^9 J/ψ events allow the study of weak decays of vector mesons that predominately decay strongly. The rate can be roughly estimated by comparing the total width to the decay rate of charm quark as estimated from D meson lifetimes.

$$BR(J/\psi \rightarrow weak\ decays) = \frac{\hbar/(4 \times 10^{-13})}{2 \times \Gamma(J/\psi)} \cong 10^{-7}$$

Hence we obtain a total BR of 10^{-7}. In a model with BSW factorization estimates for exclusive decays such as $J/\psi \rightarrow Ds\pi$ and $D^{\circ}K^{\circ}$ yield branching ratios of a few times 10^{-10}. Of course any evidence larger than these rates would indicate a breakdown of the Standard Model.

• *Wilczek* Model (1979)[31]

The 8 billion dollar question in U.S. (and maybe European) high energy physics is where is the Higgs scalar. Although we might expect a heavy Higg mass, there are models for a light Higgs. A model to produce them in radiative decays of vector mesons was developed by Wilczek. The model predicts the absolute rate, once the mass is known.

$$B(V \rightarrow \gamma H^{o}) = \frac{G_F M_V^2}{4\sqrt{2}\pi\alpha}\left(1 - \frac{M_{H^o}^2}{M_V^2}\right) B(V \rightarrow \mu^- \mu^+)$$

Upper limits have been set by the Argus[20], CUSB[32] and Cleo[21] groups which rule out this minimal standard model in their accessible mass range. However extensions or modifications could easily alter the branching ratio predictions, hence it is still important to achieve more sensitive measurements.

SUMMARY

In conclusion, we have reviewed several topics in J/ψ and Y(1s) decays. Our understanding of the detailed aspects of the hadronic decays of the J/ψ and ψ′ are still not understood. The comparisons of the J/ψ and Y(1s) to test simple models are not yet achieved. The anomalous J/ψ decays of the Y(1s) are a big puzzle and could lead to some surprises. In the future tau-charm and B factories could provide new and exciting results in this area of physics.

REFERENCES

[1] Y. Minghan, invited talk, Lepton Photon Symposium, Stanford University, August 7-12, 1989.Z. Zheng, talk at the Singapore Conference, August 1990.

[2] For recent reviews see papers by S. Sharpe and G. Schierholz, *Proceeding of the BNL Workshop on Glueballs, Hybrids and Exotic Hadrons*, August 29-September 1, 1988, BNL, Associated Universities, Inc., Upton, New York 11973 and calculations by C. Michael and M. Teper, Nucl. Phys. B314, 349 (1989) and G. Schierholz, preprint, DESY 88/172, 1989.

[3] Z. Bai *et al.*, SLAC-PUB-5275, June 1990, accepted for publication by Phys. Rev. Lett.

[4] L. Stanco, talk at the Hadron90 conference at St. Goar, Germany, Sept. 1990.

[5] Liang Ping Chen, SLAC-PUB-5378, November 1990.

[6] J. Becker, *et al.*, Phys. Rev. **D35**, 2077, (1987).

[7] Parts of this review appeared in W. Toki, SLAC-PUB-5093, September 1989.

[8] P. Freund and Y. Nambu, Phys. Rev. Lett. **34**, 1645 (1975). see also J. Bolzan, W. Palmer, S. Pinsky, Phys. Rev. **14D**, 3202 (1976).

[9] W. Hou and A. Soni, Phys. Rev. Lett. **50**, 569 (1983).

[10] A. Donnachie and A. Clegg, preprint M/C-Th-90/8, May 1990.

[11] S. Brodsky, P. Lepage and S.F. Tuan, Phys. Rev. Lett. **59**, 621 (1987).

[12] S. Brodsky and G. Lepage, Phys. Rev. **24D**, 2848 (1981). Note that this rule is violated by the electromagnetic decay $J/\psi \rightarrow \omega\pi$.

[13] S. Oneda and M. Slaughter, preprint LA-UR-88-617, February 1988.

[14] M. Chaichian and N. Tornqvist, preprint HU-TFT-88-11, March 1988.

[15] L. Parrish, talk at the Rencontre de Moriond Symposium, March 1990.

[16] S. Pinsky,Phys. Lett., **B236**, 479(1990).

[17] H. Tye, Proc. 1982 DPF Summer Study on Elementary Particle Physics and Future Facilities, Snowmass, Colorado, Donaldson, Gustafson Paige (eds). J.G. Korner et al., Nucl.Phys. B229,115(1983).

[18] B. Renger, Phd. thesis, Carnegie Mellon University, Sept. 1987, unpublished.

[19] B. Niczyporuk et al., Z. Phys. **C17**, 197 (1983).

[20] H. Albrecht et al., Z.Phys.,**C42**,349(1989).

[21] A. Bean et al., Phys. Rev.,**34D**, 905 (1986).
[22] P. Schmitt *et al.*, Z. Phys. **C40**, 199, (1988).
[23] J. Alexander *et. al.*, Phys. Rev. Lett., **64** (1990) 2227.
[24] M. Danilov, talk at the Rencontre de Moriond Symposium, March 1990.
[25] G. Gladding, preprint UIUC-HEPG-90-59, 1990.
[26] Y. Zhu, PhD thesis, California Institute of Technology, 1988, unpublished.
[27] A. Khodjamirian, S. Rudaz, M. Voloshin, preprint TPI-MINN-90/14-T, March 1990.
[28] H. Lipkin, Phys. Lett. **B179**, 278 (1986).
[29] J. Adler, *et al.*, Phys. Rev. Lett. **60**, 89(1988).
[30] R. Verma, A. Kamal, A. Czarnecki, preprint ALBERTA-THY-13-90, June 1990.
[31] F. Wilczek, Phys.Rev.Lett.,**39**,1304 (1979).
[32] P. Franzini et al., Phys. Rev., **35D**, 2883 (1987).

NUCLEAR BREMSSTRAHLUNG NEAR THE ONE- AND TWO-PION THRESHOLDS

S. Brand, H. Freiesleben, J. Krug, E. Kuhlmann,
P. Ringe, M. Steinke, R. Werding
Institut f. Experimentalphysik I, Ruhr-Universität
Bochum, 4630 Bochum, F.R. Germany

P. Michel, K. Möller, L. Naumann
ZfK Rossendorf, 8051 Dresden, F.R. Germany

P. Cloth, V. Drüke, D. Filges, K. Kilian, H. Machner,
H.P. Morsch, N. Paul, E. Roderburg, M. Rogge, T.
Sefzick, P. Turek
Institut für Kernphysik, Forschungszentrum Jülich GmbH,
5170 Jülich, F.R. Germany

ABSTRACT

The experiment on pp-bremsstrahlung presented here aims at a detailed study of off-shell effects in the strong interaction. A detector set-up will be developed which allows measurements at maximum kinematically allowed photon energies, where the strongest deviations from the soft-photon approximation (which incorporates only on-shell information) are to be expected. The detector mainly consists of a large time-of-flight spectrometer covering the full solid angle.

INTRODUCTION

Several terms contribute to the NN-bremsstrahlung such as convection and magnetization currents as well as two-body (mainly π-exchange) currents. Depending on the type of the nucleons involved (as e.g. p, n or $\bar{p}$) one or the other of these currents dominate. A rather pure case is found in the pp-system where most of the cross section is governed[1)] by just the magnetization current alone (fig. 1). Exchange terms are almost negligible and, due to the absence of a dipole moment, convection currents enter only through higher, yet weaker multipoles. Furthermore the symmetry of the system only allows for states having either spin-singlet even-L or spin-triplet odd-L values. The strongest impact of off-shell effects is observed near the maximum allowed photon energy (fig. 1) where the two protons are emitted with zero relative momentum and opposite (in the CM-system) to the photon. Interestingly in all previous experiments on pp-bremsstrahlung[2,3] this high energy regime has not been investigated; on the contrary the two outgoing protons were always detected on different sides of the beam line with relative angle $\delta\theta \geq 25°$.

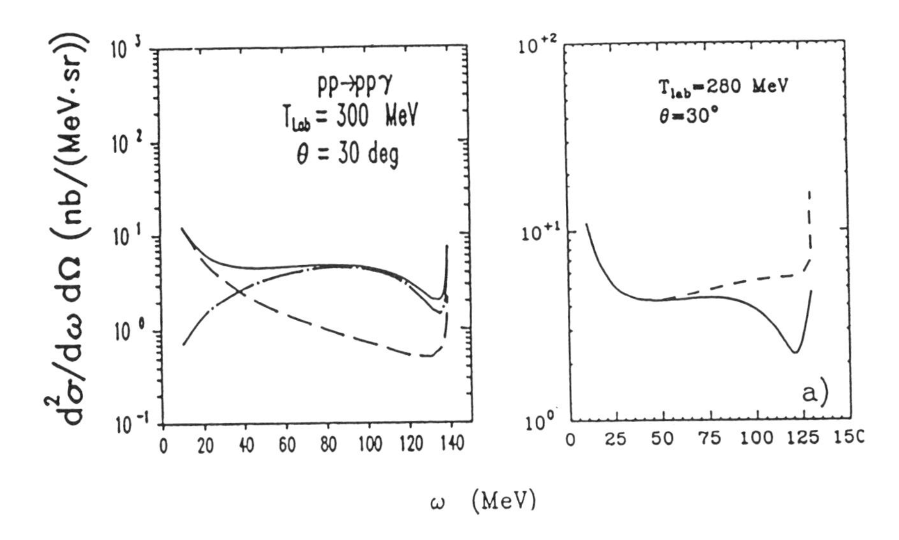

Fig. 1 left: Predictions of the inclusive cross section of the pp-bremsstrahlung with convection (dashed) and magnetization (dot-dashed) current contributions;
right: Inclusive cross section with (solid) or without (dashed) off-shell contributions.
The fig. is adapted from ref. 1.

EXPERIMENTAL APPARATUS

Exclusive cross sections of the reaction pp->ppγ are to be measured at COSY by determining the 4-momenta of both protons with a time-of-flight (TOF) spectrometer covering 4π. A thin scintillator arrangenemt in immediate vicinity of the liquid H_2-target serves as a start-detector. Stop signals will be generated by protons traversing three layers of 5mm thick scintillator material mounted to the interior of a large cylindrical vacuum tank (3m Ø x 7.5m); dE/dx measurements for particle discrimination will also be performed. To keep the maximum length of any scintillator element at about 4m, three identical barrels are foreseen, one of them being shown in fig. 2 together with the endcap which covers the forward direction. Of utmost importance to the performance of the detector is the way the individual scintillator elements are cut and arranged to form a grid-like pattern of triangular pixels. Coincidences yield fast and reliable θ, Φ-information of the impinging protons (fig. 2). Obvious advantages of this concept are: i) the rotational symmetry of the detector design which yields a minimal number of θ-and Φ-values for a given number of pixels and ii) the fact that all elements within each subdetector cover the same fraction of solid angle and hence are exposed to the same flux of emerging particles. Further details can be found elsewhere[4,5].

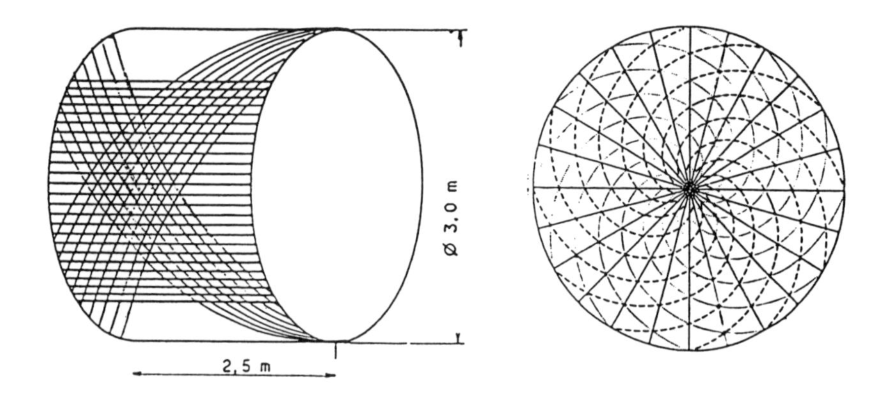

Fig. 2: Sketch of the TOF spectrometer showing the grid-like pattern in one of the three barrels (left) and the endcap (right).

STATUS

Extensive Monte Carlo calculations as well as experimental tests on individual scintillator elements have shown that the proposed TOF spectrometer is well suited for the study of pp-bremsstrahlung in the region of maximum allowed photon energy. The high spatial resolution of the detector has been demonstrated at LEAR, where a small version of the endcap has been installed: first signs of the reaction $p\bar{p} \to 16\pi$ (threshold momentum p_{th}=1.44 GeV/c) were seen at p_{lab}= 1.66 GeV/c by observing a multipicity of 16.[6] It should be noted that at this momentum 50% of all pions are expected within a cone of halfangle 13° and the maximum kinematically allowed angle is only around 40°, hence a high sensitivity close to the beamline is necessary.

The large dimensions of the present detector greatly affect timing and dE/dx signals. Due to the attenuation of the photon-shower while travelling along the individual scintillator strips (up to 4m long) towards the photomultiplier tubes a strong position dependence of either signal is observed. It can be corrected for in the off-line analysis without major problems since from the layout of the scintillator elements the place of origin is known for any signal. Excellent timing-resolution is indispensable and is even more critical, since all scintillator elements will be read out at one end only. By compromising between experimental demands and financial constraints the presently achieved and near to optimum values are in the range $\delta T(FWHM) = 0.6-1.5$ ns depending on the place of particle impact.

The $pp \rightarrow pp\pi^{\circ}$ reaction with its consecutive π°-decay into two hard photons is the most problematic background reaction. Monte Carlo simulated missing mass reconstructions of the unobserved third particle have shown that with the planned spatial and timing resolution (about 1000 pixels in the endcap and ca. 3000 pixels in each of the barrels) the pp-bremsstrahlung can unambiguously be identified up to about the two-pion threshold at 1.2 GeV/c. For an extension of these studies to yet higher beam momenta an additional large volume shower detector will be necessary.

This work has been funded by the BMFT under Contract No. 06 BO 171.

REFERENCES

1) V. Herrmann and K. Nakayama, preprint and priv. communication
2) B.M.K. Nefkens et al., Phys. Rev. C19, 877 (1979)
3) K. Michaelian et al., Phys. Rev. D41, 2689 (1990)
4) T. Sefzick, Diplomarbeit Jülich, Jül-Spez-480 (1988)
5) A. Empl, Diplomarbeit, Jülich, Jül-Spez-510 (1989)
6) W. Oelert, private communication

STUDY OF $\gamma p \rightarrow n\pi^+$ VERY CLOSE TO THRESHOLD

E. Korkmaz, D.A. Hutcheon, N.R. Kolb, D. Mack, W.J. McDonald, W.C. Olsen, and N.L. Rodning
TRIUMF and Nuclear Research Centre, University of Alberta,
Edmonton, Alberta T6G 2N5, Canada

J.C. Bergstrom, H.S. Caplan, and D.M. Skopik
University of Saskatchewan and Saskatchewan Accelerator Laboratory,
Saskatoon, Saskatchewan S7N 0W0, Canada

ABSTRACT

We are planning to use the Tagger facility at the Saskatchewan Accelerator Laboratory (SAL) to measure the total cross section for the reaction $\gamma p \rightarrow n\pi^+$ very close to threshold. The purpose is to extract the electric dipole amplitude $E_{0^+}(\pi^+)$ to high accuracy. The experimental techniques involve the use of a liquid hydrogen target and the detection of the reaction neutrons in coincidence with tagged photons.

Pion photoproduction at threshold is a fundamental process in low energy pion-nucleon physics. Using the conservation of the electromagnetic current and the partially conserved axial current (PCAC) hypothesis, the low energy theorems (LET) allow the expression of the s-wave electric dipole amplitudes for pion photoproduction at threshold in terms of the strong πNN coupling constant and the static properties of the nucleon and the pion. Accurate knowledge of these amplitudes is a direct test of our understanding of chiral symmetry breaking effects at the hadronic level. (For a thorough treatment see Ref. 1.)

Until quite recently, measurements of $\gamma N \rightarrow \pi N$ near threshold appeared to be in reasonable agreement with the LET predictions. The LET prediction for the electric dipole amplitude of $\gamma p \rightarrow n\pi^+$ is[1] $E_{0^+}(\pi^+) = 27.4 \times 10^{-3}/m_\pi$, while the accepted experimental value is[2,3] $(28.3 \pm 0.5) \times 10^{-3}/m_\pi$. In the chiral symmetry limit the amplitude $E_{0^+}(\pi^o)$ of $\gamma p \rightarrow p\pi^o$ is zero; the LET prediction is $E_{0^+}(\pi^o) = -2.5 \times 10^{-3}/m_\pi$, based on a chiral-symmetry-breaking factor of relative size $(1/\sqrt{2})m_\pi/M_p$, while the accepted experimental value was[4] $(-1.8 \pm 0.3) \times 10^{-3}/m_\pi$.

This situation was badly shaken by recent measurements[5,6] at tagged-photon facilities of $E_{0^+}(\pi^o)$ which yielded a best-value of $E_{0^+}(\pi^o) = (-0.4 \pm 0.1) \times 10^{-3}/m_\pi$. At the moment it is unclear whether or not the LET's can be 'rescued' by more careful theoretical treatment of the neutral pion amplitude. What is clear is that the $\gamma p \rightarrow p\pi^o$ cross section is highly suppressed near threshold.

Re-examination of the experimental value of $E_{0^+}(\pi^+)$ shows it to be based on quarter-century-old measurements with Bremsstrahlung beams. Those closest to threshold (151.4 MeV) were taken under adverse conditions (that of Ref. 3 at 154 MeV reported with 30% uncertainty). The others required extrapolation from considerably higher energies.

The Low Energy Theorems relate $E_{0^+}(\pi^+)$ to pion-nucleon scattering lengths a_1 and a_3 deduced from π-N elastic scattering and charge-exchange reactions. Many analyses find that the accepted value of $E_{0^+}(\pi^+)$ is consistent with LET expectations, based on fitted values for a_1 and a_3. However, a recent analysis[7] which seems to reconcile some long-standing discrepancies amongst low-energy π-N scattering results, predicts a value of $E_{0^+}(\pi^+)$ at threshold which is 8% above the accepted value.

There is other evidence suggesting that the accepted charged pion amplitude may, in fact, be too large. It is possible to link the reactions pp $\rightarrow$ dπ^+ and np $\rightarrow$ dπ^o to $\gamma p \rightarrow n\pi^+$ through a chain of relationships.[8] Some time ago, a value of $E_{0^+}(\pi^+) = (27.0 \pm 2.1) \times 10^{-3}/m_\pi$ was thus deduced, based on $\pi^+ d \rightarrow pp$ measurements near threshold.[8] Despite its apparent agreement with the LET prediction, this result was challenged by recent very accurate measurements of the np $\rightarrow$ dπ^o cross section very close to threshold.[9] The results of these measurements fall below those of Ref. 8 and, assuming charge independence, predict $E_{0^+}(\pi^+) = (23.7 \pm 0.3) \times 10^{-3}/m_\pi$. This value falls so far below the accepted one that one must seriously question other links in the chain before drawing a definite conclusion (see Refs. 8-10 for discussion).

Furthermore, the amplitudes $E_{0^+}(\pi^+)$ and $E_{0^+}(\pi^o)$ are intimately related to the amplitude for $\gamma p \rightarrow \gamma p$, Compton scattering from the proton. Some years ago, a simultaneous partial wave analysis[11] of existing $\gamma p \rightarrow \gamma p$, $\gamma p \rightarrow n\pi^+$, and $\gamma p \rightarrow p\pi^o$ data, suggested a possible strong disagreement between theoretical and experimental $\gamma p \rightarrow \gamma p$ amplitudes in the region E_γ=150-200 MeV.

In view of this situation, we feel that it is time to make a direct measurement of the $\gamma p \rightarrow n\pi^+$ cross section close to threshold, using modern facilities.

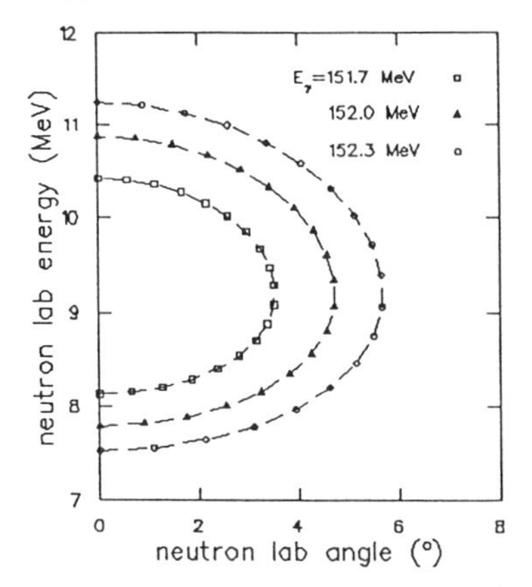

Fig. 1: Kinematics of $\gamma p \rightarrow n\pi^+$ very close to threshold.

In our experiment,[10] we aim for an accurate measurement of the $\gamma p \rightarrow n\pi^+$ cross section within 1 MeV of threshold, using the tagged-photon (Tagger) facility at SAL. While the Tagger permits accurate knowledge of beam flux and energy distribution, we will get very close to threshold by detecting the reaction neutrons thus sidestepping the many outstanding problems of pion detection near threshold. As shown in Fig. 1, neutrons within 1 MeV of threshold are confined to small forward angles and maintain energies well above zero — readily measured by time of flight with a 'start' from the Tagger. Simultaneous measurement of T_n and θ_n is sufficient to define E_γ. Independent measurement of E_γ in the Tagger serves then to suppress background.

A schematic picture of the experimental apparatus is shown in Fig. 2. Photons of 150 to 155 MeV are produced in the Tagger. A collimator ensures a halo-free beam and is followed by a sweeping magnet to deflect charged particles. The beam will then strike a 5-cm thick liquid hydrogen target. Three meters beyond the target the reaction neutrons will be detected in a well-shielded, position-sensitive, liquid-scintillator counter, consisting of a stack of 12 rectangular bars, $6 \times 6 \times 80$ cm^3 (a gap being left in the middle to allow free passage of the beam). The detector will cover most of the angular range of interest (0.5°-7.5°) with 1° resolution, and provide an average energy resolution of 0.3 MeV. Thin plastic scintillators in front will be used to veto charged particles, and a combination of pulse-shape-discrimination and time-of-flight techniques will serve to eliminate photon events.

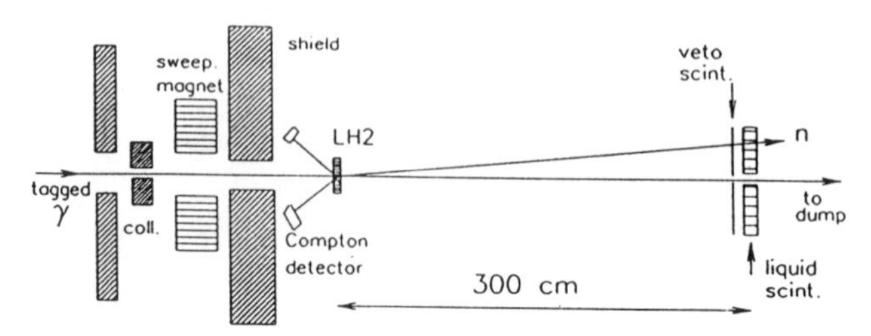

Fig. 2: Schematic of the experimental apparatus.

The detection efficiency (including detector acceptance and neutron attenuation) as a function of neutron energy and angle will be measured on site and under the same conditions as for the actual measurement, using a calibrated ^{252}Cf fission source. Complementary efficiency measurements will utilize the rectaion $\gamma d \rightarrow pn$ at 90° c.m. with the proton serving as a neutron tagger. We will aim for efficiency determination with an accuracy of better than 2%.

In addition to frequent measurements of the tagging efficiency during the run, we will monitor the (beam flux)×(target thickness) normalization by detecting, simultaniously with the measurement, backward-scattered Compton photons from the target. Over a week of running time, this normalization should be determined to within 1% accuracy.

Background subtraction will be done via empty-target measurements. Yet, achievement of good cross section accuracy clearly depends on an adequate signal to background ratio. We have carried out test measurements aimed at estimating the overall neutron background rate to which the measurement will be subjected. The results[10] clearly indicated that this rate is relatively low and should be of no concern.

We will need about two weeks of production time for 2% statistical accuracy. This, coupled with the efficiency and normalization uncertainties cited above, should yield 4% accuracy on the cross section, or 2% on $E_{0^+}(\pi^+)$. We expect to be ready to run in the first half of 1991.

REFERENCES

1. T. Ericson and W. Weise, Pions and Nuclei, Clarendon Press, Oxford, 1988.
2. M.I. Adamovich et al., Sov. J. Nucl. Phys. 9, 496 (1969).
3. D.A. McPherson et al., Phys. Rev. 136, B1465 (1964).
4. B. B. Govorkov et al., Sov. J. Nucl. Phys. 4, 265 (1967).
5. E. Mazzucato et al., Phys. Rev. Lett. 57, 3144 (1986).
6. R. Beck et al., these Proceedings, 1990.
7. P.B. Siegel and W.R. Gibbs, Phys. Rev. C33, 1407 (1986).
8. C.M. Rose, Phys. Rev. 154, 1305 (1967).
9. D.A. Hutcheon et al., Phys. Rev. Lett. 64, 176 (1990).
10. SAL Proposal No 27, J.C. Bergstrom and E. Korkmaz, spokespersons, 1989.
11. W. Pfeil, H. Rollnik and S. Stankowski, Nucl. Phys. B73, 166 (1974).

ETA PHOTOPRODUCTION NEAR THRESHOLD*

M. Benmerrouche and Nimai C. Mukhopadhyay
Department of Physics, Rensselaer Polytechnic Institute
Troy, New York 12180-3590

ABSTRACT

The eta photoproduction near threshold has a number of similarities and differences in comparison to the pi-zero photoproduction near threshold. Both processes probe Born amplitudes involving the s- and u- channel nucleon exchanges. While eta production is also sensitive to the vector meson contribution in the t-channel and excitation of several baryon resonances the pi-zero process is not. The former has a large s-channel $S_{11}(1535)$ excitation amplitude, but the latter has a relatively small contribution from the $\Delta(1232)$ excitation. We discuss the implications of our analysis on the recent Bates data presented at this conference, and report the the value of the $A_{1/2}$ amplitude for the $S_{11} \rightarrow N\gamma$ transition that we can extract from the existing eta photoproduction experiments. The latter is of interest to the test of hadron models.

INTRODUCTION

The eta photoproduction reaction

$$\gamma + p \rightarrow \eta + p, \tag{1}$$

is of considerable theoretical interest[1] near threshold, just as the reaction

$$\gamma + p \rightarrow \pi^0 + p \tag{2}$$

is. These processes have threshold at the photon lab energy $E_\gamma = 709.3$ and $144.7 MeV$ respectively. There has been considerable excitement from the recent experiments at Saclay[2] and Mainz[3] on the second process as the measured cross-section near threshold is found to be considerably less than what is theoretically predicted[4]. In particular, the E_{0+} multipole, dominant near threshold, is found

*** Presented by N.C. Mukhopadhyay. Invited paper at the Conf. on Particle Production near Threshold, Nashville, IN (1990).**

to be considerably smaller than what the simple theory predicts. It has been believed that a careful treatment of the final state interaction would explain the difference, but we have heard from Blankleider[5] in this conference that this is not so, at least in the calculation he just reported. There are speculations[6] that corrections due to chiral symmetry breaking resulting from the non-zero current quark masses might explain the discrepancy. All of these add to the interest in studying the process (1). The η meson exhibits a subtle SU(3) flavor mixing[7]:

$$\eta = \eta_8 cos\theta - \eta_1 sin\theta, \tag{3}$$

where the octet η_8 and the singlet η_1 are given by

$$\begin{aligned} \eta_1 &= (u\bar{u} + d\bar{d} + s\bar{s})/\sqrt{3}, \\ \eta_8 &= (u\bar{u} + d\bar{d} - 2s\bar{s})/\sqrt{6}, \end{aligned} \tag{4}$$

θ being the mixing angle $\simeq -10.1^o$. Given the fact that the strange quark mass[8] is considerably different from zero, and from those of the up and down quarks, the chiral symmetry breaking effects, if any, could be larger in the eta meson case compared with that in the pion, in the context of the processes (1) and (2) respectively.

Two crucial differences between the reactions (1) and (2) that we wish to demonstrate here are the roles of the t-channel vector meson exchanges and the contributions of resonance excitations. For the pi-zero photoproduction near threshold the vector meson contributions are relatively small, as is the contribution from the $\Delta(1232)$ excitation. For eta photoproduction, the veter meson and the resonance excitation contributions are large, presenting a dynamically different situation from the pi-zero case. This also poses theoretical problems, as the number of free (or unknown) parameters increases, making it difficult to confidently predict the observables. Therefore. it is not surprising that theoretical treatments of the process (1) have not yet reached the sophistication that one finds in that of the pion photoproduction through the $\Delta(1232)$ resonance region[9]. The latter does not involve many channels, and the theoretical modelling of the process (2) can be designed to respect unitarity. In the case of the reaction (1), this crucial step is yet to be taken, including the results that we are going to report. Part of the problem is the paucity of data in the related channels, which does not allow unitary models to be critically tested. This problem of poor quality of data for the process (1), in particular, has plagued the understanding of the underlying physics. Thus far, most of the cross-section and polarization data are old[10], with two happy exceptions: one by the Tokyo group[11], the other that we are about to hear in this conference in the short report of Daehnick[12], also reported at the PANIC-XII and at the Illinois meeting of the American Physical Society, our first glimpse at results of the recent Bates experiment. This state of experimental affairs should change with the advent of new photon and electron factories - Mainz, Bonn, and others, and eventually CEBAF. Facilities such as COSY at Jülich, Germany, and the often-discussed possible extension of LAMPF, called PILAC, should make a difference on the understanding of the strong interaction of the η-nucleon system, which is half of the physics one is studying via the process (1).

THE EFFECTIVE LAGRANGIAN APPROACH TO ETA PHOTOPRODUCTION

The effective Lagrangian approach[1,13] helps us to sort out the tree-level structure (Fig.1) of the eta photoproduction amplitude. The procedure is exactly parallel to the pion photoproduction: we consider the leading s- and u-channel nucleon exchanges (nucleon Born terms) and add to that the leading veter meson exchangs in the t-channel; finally, we consider the s- and u-channel contributions including the excitations of nearby resonaces. Some similarities and differences with pi-zero photoproduction near threshold are obvious. There is one important difference that is subtle. In the case of pion photoproduction, the πNN coupling is preferred[14] to be pseudovector, in order to be in accord with the low-energy theorems and chiral symmetry. With mass of the eta meson being 548.8 MeV, there is less theoretical justification to demand the ηNN coupling to be pseudovector, and we prefer this question to be settled by the experimental observations in this particular instance. Thus, we choose the ηNN effective Lagrangian to be[15]

$$L_{\eta NN} = g_\eta[-i\epsilon\overline{N}\gamma_5 N\eta + (1-\epsilon)\frac{1}{2M}\overline{N}\gamma_\mu\gamma_5 N\partial^\mu\eta], \tag{5}$$

M being the nuleon mass. We determine the parameter ϵ from a fit to the data. The γNN coupling is described in the standard way:

$$L_{\gamma NN} = -e\overline{N}\gamma_\mu\frac{(1+\tau_3)}{2}NA^\mu + \frac{e}{4M}\overline{N}(k^s + k^v\tau_3)\sigma_{\mu\nu}NF^{\mu\nu}. \tag{6}$$

N, η, A^ν are nucleon, eta, photon field operators, k^s, k^v are isoscalar and isovector anomalous nucleon magnetic moments, $k^s = \frac{1}{2}(k_p + k_n)$, $k^v = \frac{1}{2}(k_p - k_n)$, with $k_p = 1.79$ and $k_n = -1.91$, $F^{\mu\nu} = \partial^\nu A^\mu - \partial^\mu A^\nu$, At threshold, only the E_{0+} multiple survives for the process (1). Using the above Lagrangians, one finds, for proton,

$$\begin{aligned} E_{0+}^{PV} &= -\frac{eg_\eta}{8\pi M}\frac{\beta}{(1+\beta)^{3/2}}(1-\frac{1}{2}\beta k_p), \\ E_{0+}^{PS} &= -\frac{eg_\eta}{8\pi M}\frac{\beta}{(1+\beta)^{3/2}}(1+k_p), \end{aligned} \tag{7}$$

with $\beta = \frac{\mu}{M}$, μ being the meson mass. Absent the nucleon anomalous magnetic moment ($k_p = 0$), the two expressions coincide - the famous equivalence theorem. Identical expressions are valid for the pi-zero meson case (2), with trivial sustitution of the appropriate coupling and mass.

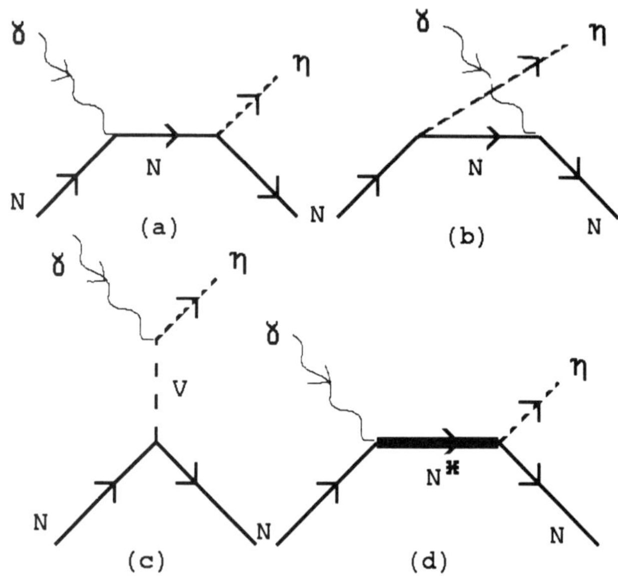

Fig 1: Leading contributions at the tree-level. N^*'s represent nearest I=1/2 nucleon resonances. V stands for vector meson exchange in the t-channel.

We next consider the role of the t-channel vecter meson exchanges. We have examined contributions from the t-channel exchange of ρ, ω, and ϕ mesons, and found the contribution due to ϕ to be unimportant. Hereafter, we shall omit it. The strong and electromagnetic vertices involving the vecter meson contribution in Fig.1c are described by the Lagrangian:

$$
\begin{aligned}
L_{VNN} &= -g_v \overline{N}\gamma_\mu N V^\mu + \frac{g_t}{4M}\overline{N}\sigma_{\mu\nu} N V^{\mu\nu}, \\
L_{V\eta\gamma} &= \frac{e\lambda}{4m_{\pi^+}}\epsilon_{\mu\nu\lambda\sigma} F^{\mu\nu} V^{\lambda\sigma}\eta,
\end{aligned}
\tag{8}
$$

with

$$
V_{\mu\nu} = \partial_\nu V_\mu - \partial_\mu V_\nu, \tag{9}
$$

analogous to the electromagnetic field tensor, g_v, g_t, λ are strong and electromagnetic couplings of the relevant vertices. Using the above interaction Lagrangians, we can show that the E_{0+} amplitude receives a contribution, near threshold, given by

$$E_{0+}^{V} = \frac{e\lambda}{8\pi m_{\pi^+}} \frac{\beta(2+\beta)\mu^2}{2(1+\beta)^{\frac{3}{2}}} \frac{g_v + g_t}{\mu^2 + M_V^2(1+\beta)}, \tag{10}$$

M_V being the mass of the vector meson involved in Fig.1c.

An important theoretical issue that we have investigated in the present context in the role of a form factor attached at the VNN vertex. Brown and coworkers have used a form factor of the type[16,17]

$$F(q^2) = \frac{\Lambda_1^2 - M_V^2}{\Lambda_1^2 + q^2}, \tag{11}$$

preferring a value of $\Lambda_1^2 \sim 2m_V^2$. We vary it between 1.2 and $2GeV^2$, motivated by the studies of nucleon-nucleon interaction[17]. On the other hand, electromagnetic structure of nucleons suggest a form factor of the type[18]

$$F(q^2) = \frac{\Lambda_2^2}{\Lambda_2^2 + q^2}, \tag{12}$$

with $\Lambda_2^2 \sim 0.8GeV^2$. The resultant reduction of E_{0+}^V from Eq.(10) is straightforward to compute, and is significant in either case.

We next turn our attention to the tree-level contributions of excitations of nucleon resonances in the photoproduction of eta mesons through contributions indicated in Fig.1d and corresponding amplitudes in the u-channel. While there is always a theoretical issue of duality and resultant overcounting, we shall take into account the leading s (and u-channel) nucleon resonance excitations, avoiding the former issue. Here there are two simplifications. First, since the eta meson has zero isospin, only I=1/2 nucleon resonances can contribute. Second, below 2 GeV, only two nucleon resonances have any significant decay branching ratio into the $N\eta$ channel: these are $S_{11}(1535)$ and $P_{11}(1710)$ resonances, with the respective branching ratios being 45 to 55% and $\sim$ 25% respectively. Next to come in with significant branching $N\eta$ ratio is $D_{13}(1700)$ with $\sim$ 4%, this resonance itself having only a (***) rating in the recent PDG tables[7]. To round it up , $P_{13}(1720)$ and $S_{11}(1650)$ have $\sim$ 3.5% and $\sim$ 1.5% as respective branching ratios to the $N\eta$ channel. Since we are primarily interested in the $S_{11}(1535)$ region, resonances above 2 GeV (such as $G_{17}(2190)$) will have small contribution mainly because they have large mass and small coupling to the $N\eta$ channel. For the present calculation shown in Figs.2-4, the resonances included are $S_{11}(1535), S_{11}(1650), P_{11}(1440)$ and $P_{11}(1710)$. The inclusion of the $D_{13}(1520)$ does not improve the fit considerably, but adds theoretical complications[19] specific to the spin-3/2 resonances.

The effective Lagrangians for the I=J=1/2 resonance excitations can be summarized as follows:

$$
\begin{aligned}
L_{\eta NR}^{PS} &= -ig_{\eta NR}\overline{N}\Gamma R\eta + h.c.,\\
L_{\eta NR}^{PV} &= \frac{f_{\eta NR}}{m_{\pi^+}}\overline{N}\Gamma_\mu R\partial^\mu\eta + h.c.,\\
L_{\gamma NR} &= \frac{e}{2(M_R+M)}\overline{R}(k_R^s + k_R^v\tau_3)\Gamma_{\mu\nu}NF^{\mu\nu} + h.c.,
\end{aligned} \tag{13}
$$

where R is the resonance field operator,M_R its mass, the transition magnetic couplings for the proton and neutron targets are respectively $k_R^p = k_R^s + k_R^v$ *and* $k_R^n = k_R^s - k_R^v$. The operator structure for the Γ, Γ_μ, and $\Gamma_{\mu\nu}$ are :

$$
\begin{aligned}
&\Gamma = 1,\ \Gamma = \gamma_\mu,\ \Gamma_{\mu\nu} = \gamma_5\sigma_{\mu\nu},\ (A)\\
&\Gamma = \gamma_5,\ \Gamma_\mu = \gamma_\mu\gamma_5,\ \Gamma_{\mu\nu} = \sigma_{\mu\nu},\ (B)
\end{aligned} \tag{14}
$$

where A and B correspond to nucleon resonances of odd and even parities respectively. In equ.(13), the ambiguity in the meson-nucleon-resonance coupling calls for the pseudoscalar (PS) or pseudovector (PV) options. The two couplings are related through

$$
\frac{f_{\eta NR}}{m_{\pi^+}} = \frac{g_{\eta NR}}{(M_R \pm M)}, \tag{15}
$$

the upper sign corresponding to even parity resonance. It turns out that the PS-PV ambiguity does not make a significant numerical difference in the fits obtained in this work, a contrast from the nucleon Born terms.

We shall complete the expression for the E_{0+} ,using the PS coupling of meson-nucleon-resonance, by including the contribution of the excitation of the S_{11} resonance. This is

$$
E_{0+}^{R,PS} = -\frac{eg_{\eta NR}k^p}{8\pi W(M_R+M)}\frac{ab_\eta(W-M)(W+M_R)}{W^2 - M_R^2 + iM_R\Gamma_T(W)}, \tag{16}
$$

with

$$
b_i^2(W) = \frac{(W+M)^2 - m_i^2}{2W},\quad a^2 = \frac{(W+M)^2}{2W}, \tag{17a}
$$

$$
\Gamma_T(W) = (\frac{W+M_R}{2W})\sum_i\left(\frac{b_i^2(W)}{b_i^2(M_R)}\frac{q_i}{q_i^R}\Gamma_i\right), \tag{17b}
$$

i=π,η, Γ_i is the partial decay width of S_{11} into (iN), R being S_{11} here, $q_i's$ are the cm momenta of the meson i. Finally, we give the relationship of the helicity amplitude $A_{1/2}^p$, defined by the PDG, to the transition moment k^p defined above:

$$
ek^p = \left(\frac{2M(M_R+M)}{(M_R-M)}\right)^{1/2} A_{1/2}^p. \tag{18a}
$$

Also, we have for S_{11},

$$
|g_{\eta NR}| = \left(\frac{4\pi M_R}{q_\eta^R b_\eta^2(M_R)}\Gamma_\eta\right)^{1/2}. \tag{18b}
$$

Expressions for other resonances are omitted here for brevity.

RESULTS OF OUR ANALYSIS

Since this a conference on particle production near threshold, we shall start with discussion on the meson photoproduction from the near threshold. Table I contrasts the eta photoproduction from the more familiar pi-zero photoproduction at the respective thresholds, by computing the E_{0+} multipole in the form

$$E_{0+} = a \times 10^{-3}/m_{\pi^+}, \tag{19}$$

where the units in equ.(19) is chosen in the conventional fashion[2] for the pion. The most important similarity is the importance of the choice of coupling of the nucleon Born term; the PS and PV couplings represent substantial difference if we keep the ηNN coupling fixed. Differences come from the contributions of the ρ and ω mesons, and the very large rôle of the $S_{11}(1535)$ excitation even near threshold for the eta meson. Thus, while the pi-zero photoproduction very near threshold probes only the nucleon Born terms, and, through it, the chiral symmetry effects, including breaking thereof , the eta photoproduction is additionally sensitive to the vector meson contribution and the excitation of the $S_{11}(1535)$ resonance, and even the $S_{11}(1650)$ resonance. Thus, the chiral symmetry breaking effects, if any, would be difficults to quantify in the eta case. Finally, as the vector meson contributions are important in the eta photoproduction, so are the rôles of the form factors at the meson-nucleon vertex, VNN. For Table I, We have used the couplings[21]

$$\begin{aligned} &g_\eta^2/4\pi = 0.51, \\ &\lambda_\rho G_v^\rho + \lambda_\omega G_v^\omega = 1.5, \;\; \lambda_\rho G_t^\rho + \lambda_\omega G_t^\omega = 4.5, \end{aligned} \tag{20}$$

for the eta photoproduction, and have taken[9]

$$\begin{aligned} &g_\pi^2/4\pi = 14.5, \\ &\lambda_\rho G_v^\rho + \lambda_\omega G_v^\omega = 3.2, \;\; \lambda_\rho G_t^\rho + \lambda_\omega G_t^\omega = -1.1, \end{aligned} \tag{21}$$

for the pion photoproduction. The form factor for the VNN vertex is taken to be (see equ.(11)) $\Lambda_1^2 = 1.2 GeV^2$.

Next we have displayed in Figs.2 and 3 a comparison of our best fit of the tree-level effective Lagrangian theory to the existing experimental data base[10], including the 1988 data set due to Homma <u>et. al.</u>[11] in the first figure. Note the importance of the s-channel contribution of the $S_{11}(1535)$ excitation. As expected, the curve excluding this contribution fails completely to describe either the differential cross-section or the recoil nucleon polarization. Clearly, the data on the recoil polarization are too poor to be of any quantitative value, except to give an indication of the interference of $S_{11}(1535)$ with other contributions. The lowest energy data point in Fig.2 is not reproduced. It is hard to make much out of the fit in Fig.3 except a qualitative claim of agreement.

Table I: The E_{0+} multipole contributions for the π^0 and η photoproduction at threshold. Leading contributions to "a" (see equ.(19)) are shown in the table. Note the differences of the vector meson and resonance contributions between the two cases.

Meson	Nucleon Born		Vector mesons ($\rho+\omega$)		Leading
	PS	PV	Without	With	Resonance
			Form Factor		Contribution
π^0	-7.93	-2.48	0.08	0.04	00.36
η	-3.72	-0.64	6.65	2.90	13.36

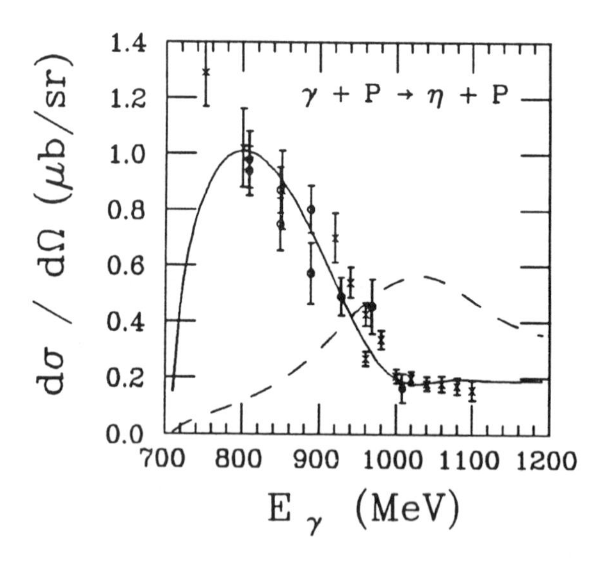

Fig.2 : Differential cross-section of eta photoproduction off proton target at cm angle of 90° as a function of the photon lab energy (in MeV). Recent data set of Homma <u>et al.</u> are marked with circles . The dotted curve excludes the $S_{11}(1535)$ contribution.

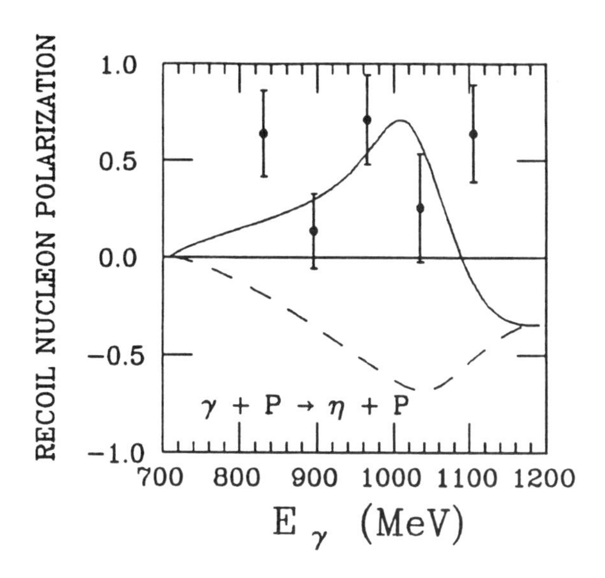

Fig.3 : Recoil nucleon polarization (see Ref. 10) for the eta photoproduction from proton target at cm angle of 90° as a function of the photon lab energy (in MeV). The absence of $S_{11}(1535)$ contribution is indicated by the dotted curve.

This brings us to one of the highlights of this conference: the data of the Pittsburgh-Boston-LANL group taken at Bates, which have been shown at this conference by Daehnick[12]. This group have been able to measure the angular distribution for (γ, η) reaction at photon lab energies of ~ 725 and $\sim 750 MeV$ at six angles each. These data are preliminary and the authors of this pionnering experiment have shown here their measured differential cross-section at five different angles, while the sixth, at 105°, is being analysed. While the lowest energy of the photon is $725 MeV$, some $16 MeV$ above the threshold, this set of data are of considerable interest to us. First, it prefers the PS effective coupling of the eta-nucleons vertex for the nucleon Born terms, with the parameter $\epsilon = 0.77$ in contrast to the pi-zero case, where $\epsilon = 0$. Perhaps this is an important hint of the chiral symmetry breaking effects we have referred to earlier. Second, it does confirm the important roles of the vector meson and the $S_{11}(1535)$ contributions. Third, it produces a strange non-isotropic angular distribution at $725 MeV$, which is hard to understand so far in our effective Lagrangian theory at the tree-level. We look forward to further exploration of these points by more precise experiments, which would be undoubtedly forthcoming in the future, as we explore further theoretical refinements. A comparison of our theoretical fits to the combined data sets, with the recent Bates data, is shown in Fig.4.

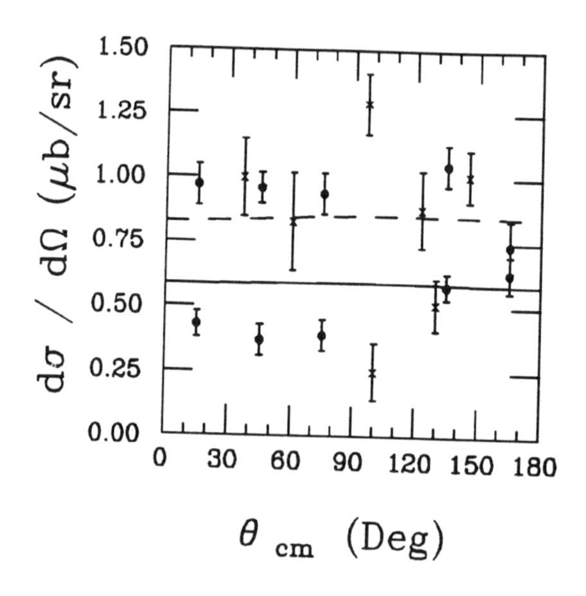

Fig.4 : Our fit of differential cross-section as function of the cm angle; included are the recent Bates data (circles) as reported at this conference by Dytman et al.[12]. The solid curve is for $725 MeV$ while the dashed one is for $750 MeV$.

The last topic we shall cover is the extraction of the helicity amplitude $A_{1/2}$ from the analysis of the existing data on the eta photoproduction. There is considerable theoretical interest in describing the structure of the $S_{11}(1535)$ resonance from the quark model, inspired by QCD. One of the intresting issues here is the mixing in the $(70, 1^-)$ multiplet of SU(6) of the $^2\underline{8}$ and $^4\underline{8}$ SU(3) states, as well as the mixing in the $S_{11}(1535)$ and $S_{11}(1650)$ wave functions. Connected with this, of course, is the nature of the nucleon wave function and the approximations involved in defining the electromagnetic transition operator. Amongst the different quark model calculations[20] the helicity amplitude $A_{1/2}$ is estimated to be 54 by Warns et al.,68 by Forsyth ,147 by Koniuk and Isgur, 156 by Feynman *et al.*,and 162 by Close *et al.* , for the proton target in the usual units. The extracted amplitude from the analysis of the (γ, π) data is quoted by PDG[7] to be 73±14, in poor agreement with the last two theoretical estimates. Our extracted amplitude has large theoretical uncertainties, with the best estimate falling in the range

$$A^p_{1/2} = (90 - 133) \times 10^{-3} GeV^{-1/2}, \quad (22)$$

We can conclude that the $S_{11} \rightarrow N\ \gamma$ transition amplitude extracted from the (γ, η) reaction, though not precisely estimated in this analysis, is in fair agreement with most of the quark model calculations[20].

CONCLUSIONS

Though the data base on the photoproduction of eta mesons near threshold is poor, it allows us to take a glimpse at the physics accessible by this process. Like pi-zero photoproduction near threshold, the nucleon Born terms for the eta photoproduction are sensitive to the chiral symmetry breaking effects and the nature of the effective meson-nucleon coupling. Unlike the former, the eta photoproduction receives large contributions from the t-channnel rho and omega exchanges, and the excitation of the nearby $S_{11}(1535)$ resonance. Contributions from other nucleon resonances must also be considered. Despite this complexity, the eta photoproduction process is a valuable tool to extract the $S_{11} \rightarrow N\ \gamma$ transition amplitude, of great interest to quark models and other realistic pictures of hadron structure. New Bates data[12] brings us experimentally lot closer to the eta production threshold, and presents some new puzzles, as well as a nice collaboration to our understanding of the photoproduction process at higher energy.

ACKNOWLEDGMENTS

We are grateful to the conference organizers for the their invitation. We thank S. Dytman and W. Daehnick for sharing with us the Bates data. This work is supported by the U.S. Dept. of Energy.

REFERENCES

1. M. Benmerrouche and N.C. Mukhopadhyay, in *Excited Baryons* 1988 (G.

Adams, N.C. Mukhopadhyay and P. Stoler, eds.), World scientific, Teaneck, N.J., 1989, See also the review by F. Tabakin, S.A. Dytman and A.S. Rosenthal in the same volume.

2. E. Mazzucato *et al.*, Phys. Rev. Lett. 57, 3144 (1986);*ibid.* 60, 749 (1988).

3. R. Beck *et al.*, Phys. Rev. Lett. 65, 1841 (1990).

4. R. Davidson and N.C. Mukhopadhyay, Phys. Rev. Lett. 60, 748 (1988).

5. B. Blankleider, this conf. and S. Nozawa, B. Blankleider and T. S.-H. Lee, Nucl. Phys. *A*513, 459 (1990).

6. L.M. Nath and S.K. Singh, Phys. Rev. *C*39,1207 (1989); T. S.-H. Lee and B.C. Pearce, Argonne Preprint (1990).

7. Particle Data Group, Phys. Lett. 239*B*, 1 (1990).

8. H. Leutwyler and J. Gasser, Phys. Rep. 87, 77 (1982).

9. R. Davidson, N.C. Mukhopadhyay and R. Wittman, Phys. Rev. *D* (in press).

10. H.R. Hicks *et al.*, Phys. Rev. *D*7, 2614 (1973) and refs. therein.

11. S. Homma *et al.*, J. Phys. Soc. (Japan) 57, 828 (1988).

12. W.W. Daehnick, this conf., and priv. comm. (1990); S.A. Dytman *et al.*, Contribution to Panic-XII (1990) and Bull. Am. Phys. Soc. 35, 1679 (1990).

13. See, for example, A. Donnachie and G. Shaw (eds.), *Electromagnetic Interactions of hadrons*, Plenum, New York (1978).

14. N. Dombey and B.J. Read, Nucl. Phys. *B*60, 65 (1973).

15. F. Gross, J.W. Van Orden and K. Holinde, Phys. Rev. *C*41, *R*1909 (1990).

16. G.E. Brown, in *Excited Baryons* 1988, Ref. 1.

17. R. Machleidt,K. Holinde and Ch. Elster, Phys. Rep. 149, 1 (1987).

18. M. Gari and W. Krümpelmann, Phys. Lett. 141*B*, 295 (1984) ;Z. Phys. *A*322, 689 (1985).

19. M. Benmerrouche, R. Davidson and N.C. Mukhopadhyay, Phys. Rev. *C*39, 2339 (1989).

20. R.P. Feynman *et al.*, Phys. Rev. *D*3, 2706 (1971); R. Koniuk and N. Isgur,*ibid.* *D*21, 1868 (1980);C.P. Forsyth, Ph.D. Thesis, Carnegie-Mellon Univ. (1981); F.E. Close and Z. Li,Phys. Rev. *D*42,2194 (1990), *ibid.*,*D*42,2207 (1990) and M. Warns, H. Schröder , W. Pfeil and H. Rollink,Phys. Rev. *D*42, 2215 (1990) .

21. The value of g_η is from our best fit to all data sets, while the couplings for the vector meson sector are taken from S.I. Dolinsky *et al.*,Z. Phys. *C*42, 511 (1989);G. Höhler and E. Pietarinen, Nucl. Phys. *B*95, 210 (1975) and W. Grein and P. Kroll, Nucl. Phys. *A*338, 332 (1980).

Neutral Pion - Photoproduction from the Proton near Threshold

R.Beck
University of Illinois

1. Introduction

The electromagnetic production of pions from the nucleon has been studied for more than thirty years. Today with the new c.w.– machines and powerful photon sources a luminosity of 10^{30} can be reached, which open a new class of real photon experiments. Now, even small cross sections at the production threshold, can be measured in reasonable experimental time and makes it possible to separate the non resonant s– wave production amplitude E_{0^+} from the dominant p– wave amplitudes.

2. Why is the threshold region interesting ?

At threshold the total cross section and the differential cross section is given by the s-wave amplitude E_{0+}

$$\frac{d\sigma}{d\Omega} = \frac{q}{k} \mid E_{0+} \mid^2$$

with $\quad$ q = pion momentum in the CMS
$\quad\quad\quad$ k = photon momentum in the CMS

The value of this s- wave amplitude is fixed by the low energy theorems. There are two low energy theorems, the Kroll- Rudermann theorem [1] and the Fubini, Furlan and Rossetti theorem [2] which calculate the amplitude E_{0+} as a function of the global properties of the nucleon and pion. The first one uses gauge invariance and the second one the partial conservation of the axial vector current (PCAC) to express the amplitude E_{0+} in terms of the pion and nucleon mass, magnetic moments and coupling constants. In the Table 1 we show the contribution from the theorems to the four physical channels. From this we see that the model dependence in the charge channels come in at the second order of x^2 ($x = m_\pi/m_N$ the pion to nucleon mass ratio) and of the order x^3 in the neutral channel. Also we find that neutral pionproduction is more sensitive to the partially conserved axial vector argument, than the charge channels which are dominated by the Kroll Rudermann theorem.

Table 1: Pion photoproduction amplitude E_{0+} in units of $10^{-3}/m_\pi$. The experimental data are compared to the LET (gauge invariance (CVC) and partially conserved axial vector current (PCAC)).

with $\mathrm{x} = x = m_\pi/m_N$ pion–nucleon mass ratio
$C = ef_\pi\ /4\pi$ coupling constants
μ_p, μ_n are the anomalous magnetic moments of the proton and neutron

channel	CVC	CVC + PCAC	LET	experiment
$\gamma p \to n\pi^+$	$\sqrt{2}C + Or(x)$	$\sqrt{2}C(1 - \frac{3}{2}x + Or(x^2))$	27.5	28.3 ± 0.5
$\gamma n \to p\pi^-$	$-\sqrt{2}C + Or(x)$	$\sqrt{2}C(-1 + \frac{1}{2}x + Or(x^2))$	−32.0	-31.9 ± 0.5
$\gamma p \to p\pi^{\circ}$	$0 + Or(x^2)$	$C(-x +_{-}1 + \frac{\mu_p}{2}x^2 + Or(x^3))$	−2.4	-1.8 ± 0.6
$\gamma n \to n\pi^{\circ}$	$0 + Or(x^2)$	$C(-\frac{\mu_n}{2}x^2 + Or(x^3))$	0.4	unknown

The numerical values from the low energy theorems (LET) are compared to the experimental results until 1986 in table 1. Today there are two experiments in the threshold region that get lower values for the E_{0+}-amplitude.

3. The experiments at threshold

a) The Saclay experiment [3]

The experiment utilized photons from the tagged annihilation facility [4] at the Saclay linear electron accelerator ALS, the accelerated positrons annihilate on a 40 LiH radiator and the low-energy annihilation photon is used to tag the high-energy one. The tagging detector provided a 1 MeV energy resolution for the tagged photon. The size of the tagging detector defined the total 8-MeV energy width of the photon beam. The photon flux, mainly limited by counting rates in the target, was 8000 sec^{-1}. The recoil proton energy was measured using the scintillation light output in the active target, a stock of twenty 5mm-thick scintillators viewed by individual photomultipliers.

Neutral pions were detected through their two photon decay in a cylindrical lead-glass Cherenkov detector surrounding the target and covering a $0.95 * 4\pi$ solid angle. The detector was divded into twelve pieces of a size such that a π^0– decay in this energy region could not fire two adjacent blocks. The lead-glass thickness was 6 radiation lengths. The inner surface of the detector was covered with twelve 5mm thick scintillators which vetoed charged particles, mostly initiated by the photon beam in the target.

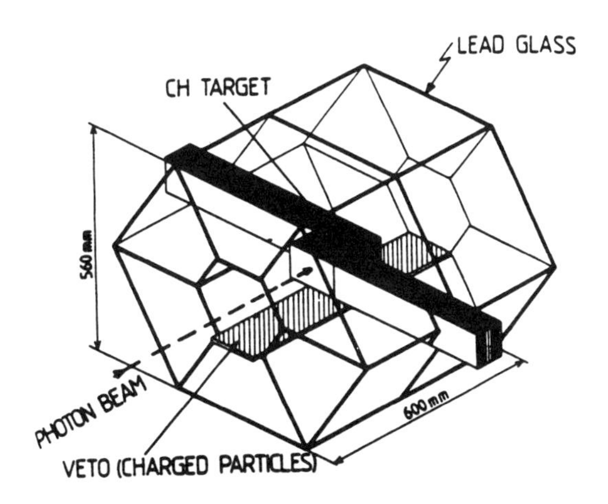

Figure 1: The Saclay π^0 – spectrometer

b) The Mainz experiment [5]

The c.w.– machine Mainz Microtron (MAMI A) [6] was used in combination with the Glasgow– Edinburgh– Mainz (GEM) tagging spectrometer [7] to study the photoproduction of neutral pions near threshold with monochromatic photons. In figure 2 the experimental set up is shown. A continues wave electron beam of 183.5 MeV, provided by MAMI A, hit a thin aluminium foil to produce bremsstrahlung photons. The photons are tagged by the electron, which are registered in 91 coincidence channels (overlapping scintillators) in the focal plane of the tagging spectrometer. This electron detector covers a energy range in one setup from 131.4 – 157.2 MeV with well defined energies (FWHM = 0.28 MeV). The combination of a tagged photon flux of $1.6 * 10^7$ and the 11 cm long liquid hydrogen target produced a luminosity of $4.6 * 10^{30}$.

The neutral pions were detected in a π^0 –spectrometer build out of 176 lead–glass detector arranged in two blocks of 8x11 detectors [8]. The individual detectors consists of an SF 5 lead glass block (dimensions $64x64x300mm^3$, the detector length corresponds to 12 radiation lengths). The two arrays of lead glass detectors were placed horizontally on either side of the beam at a distance of 50 cm from the liquid hydrogen target and the angle between the centroid of the arrays and the photon beam was chosen to be 80^0. Two thin scintillator were installed in front of each block to reduce charge particles.

Table 2: Comparison the Saclay and Mainz experimental set up.

	Saclay $\gamma + p \rightarrow p + \pi^0$	Mainz $\gamma + p \rightarrow p + \pi^0$
tagged photons	$8 * 10^3 sec^{-1}$	$1.6 * 10^7 sec^{-1}$
ΔE_γ energy range	$8MeV$	$26MeV$
δE_γ energy resolution	$1MeV$	$0.28MeV$
Target	active target CH $N_t = 4.8 * 10^{23}$	liquid hydrogen target $N_t = 4.6 * 10^{23} cm^{-2}$
π^0 –spectrometer	lead glass	lead glass
solid angle	$0.94 * 4\pi$	$0.17 * 4\pi$
total efficiency	0.55	0.09
- pions	700	15000

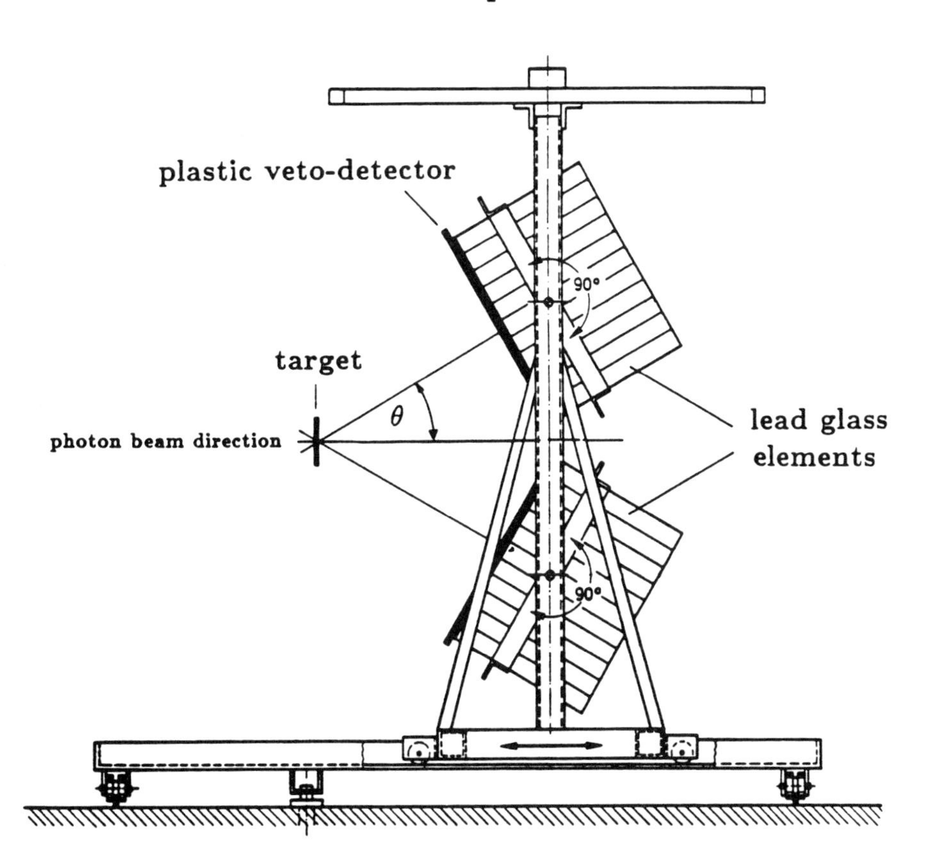

Figure 2: The Mainz- Giessen π^0 - spectrometer

4. The Results

(a) The total cross section

The figure 3 shows the total cross section $\gamma + p \rightarrow p + \pi^0$ as a function of the photon energy E_γ. There is agreement between the two data sets within the experimental errors bars. Because of the energy range $\Delta E_\gamma = 26$MeV the Mainz data were taken in one setting of the spectrometer and incoming electron energy.

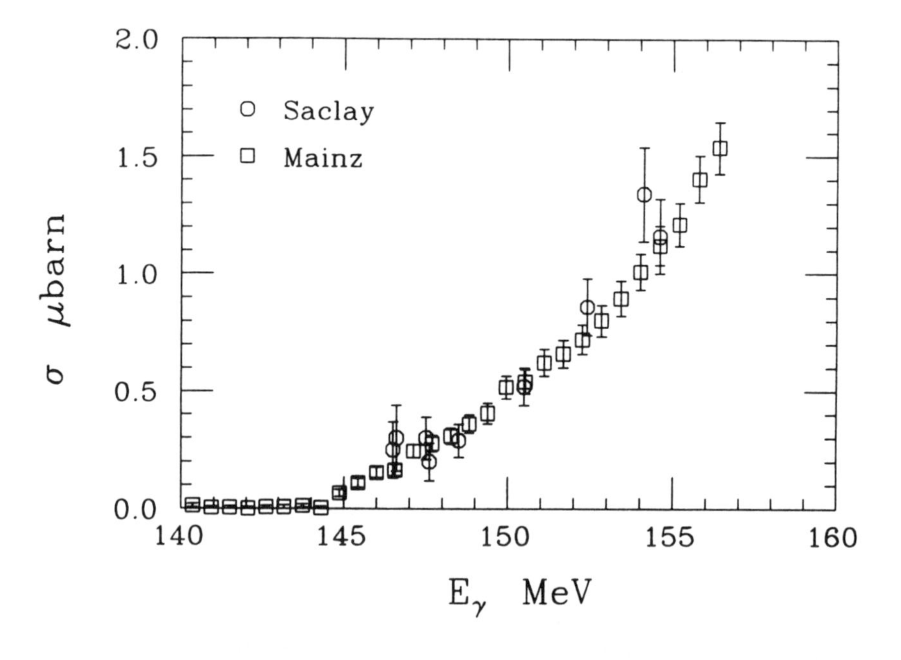

Figure 3: The total cross section $\gamma + p \rightarrow p + \pi^0$

(b) The differential cross section

The differential cross section for five energy intervals around photon energies $E_\gamma = 146.8$, 149.1, 151.4, 153.7 and 156.1 MeV are shown in figures 4 and 5. Since close to threshold only $l_\pi = 0{,}1$ partial waves are important and the differential cross section can be parametrized according to

$$\frac{d\sigma}{d\Omega} = \frac{q}{k}\,(A\cos\Theta_\pi + B\cos\Theta_\pi + C\cos^2\Theta_\pi)$$

Here q and k are the momentum of the pion and the photon, respectively, and Θ_π is the emission angle of the π^0 in the center of mass system relative to the photon momentum. The coefficients A, B and C are functions of the complex multipoles E_{0+}, M_{1+}, M_{1-} and E_{1+}. The s– wave amplitude E_{0+} can be determined from the forward backward– asymmetry in the angular distribution.

The relations

$$A + C = |E_{0+}|^2 + |3E_{1+} + M_{1+} - M_{1-}|^2$$

and

$$B = 2Re(E_{0+}(3E_{1+} + M_{1+} - M_{1-})^*)$$

show that E_{0+} can be determined as one solution of a quadratic equation assuming the imaginary part of the amplitude can be neglected compared to the real one. The fitted A, B and C– coefficients are shown in figure 6 (a)–(c).

Figure 7 shows the A– coefficient plotted against the energy dependence$(q \cdot k)$ of the p-wave, it should behave as a straight line. The solid line is a fit with the parametrisation

$$A = |E_{0+}|^2 + const(qk)^2$$

This gives the result

$$E_{0+} = (1.5 \pm 0.2)\,10^{-3}\,\frac{1}{m_\pi}$$

at threshold assuming no pion momentum dependence in the s– wave amplitude.

The solution from the quadratic equation (A + C and B) for the ReE_{0+} is shown in figure 8 for the five angular distributions in comparison with the result out of the A– coefficient and the result from Saclay. Also the predictions of the low energy theorems is shown as a dash dotted line. In order to compare the measured results with the predictions of the low energy theorem (LET) the measured values have to be corrected for the rescattering from the charge channel ($\gamma p \rightarrow n\pi^+ \rightarrow p\pi^0$). This correction was done in a K– matrix approach [9], which

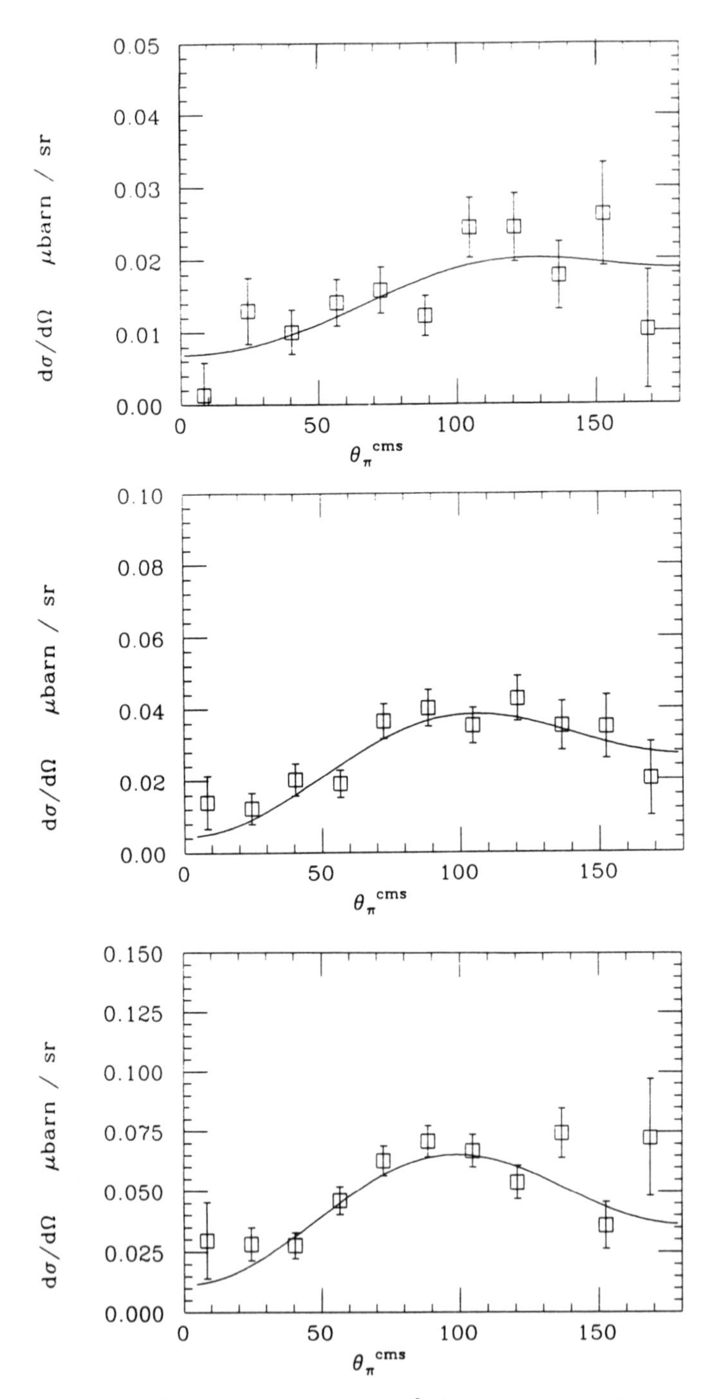

Figure 4: The differential cross section $\frac{d\sigma}{d\Omega}(\gamma + p \rightarrow p + \pi^0)$ for the energy intervals (a) $E_\gamma = 146.8$, (b) $E_\gamma = 149.1$, (c) $E_\gamma = 151.4 MeV$

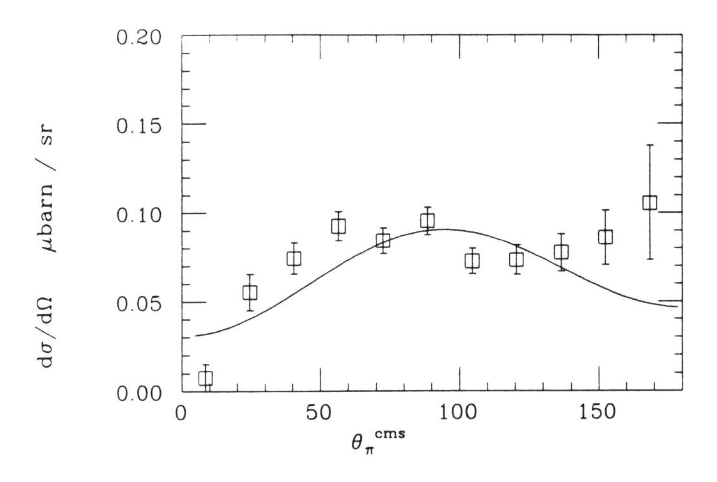

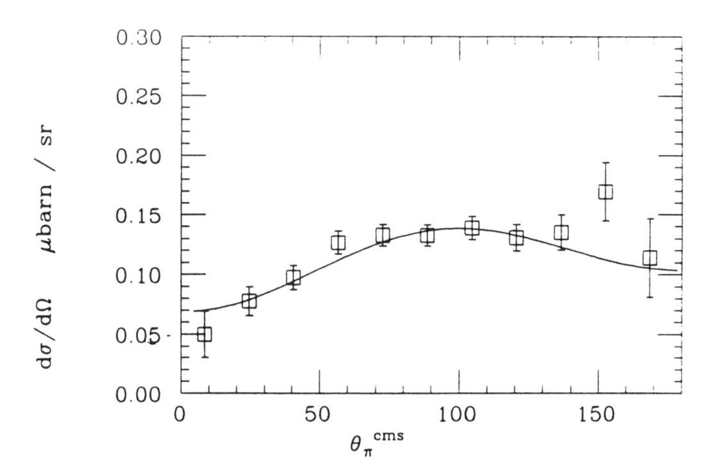

Figure 5: The differential cross section $\frac{d\sigma}{d\Omega}(\gamma + p \rightarrow p + \pi^0)$ for the energy intervals (d)$E_\gamma = 153.7$, (e) $E_\gamma = 156.1 MeV$

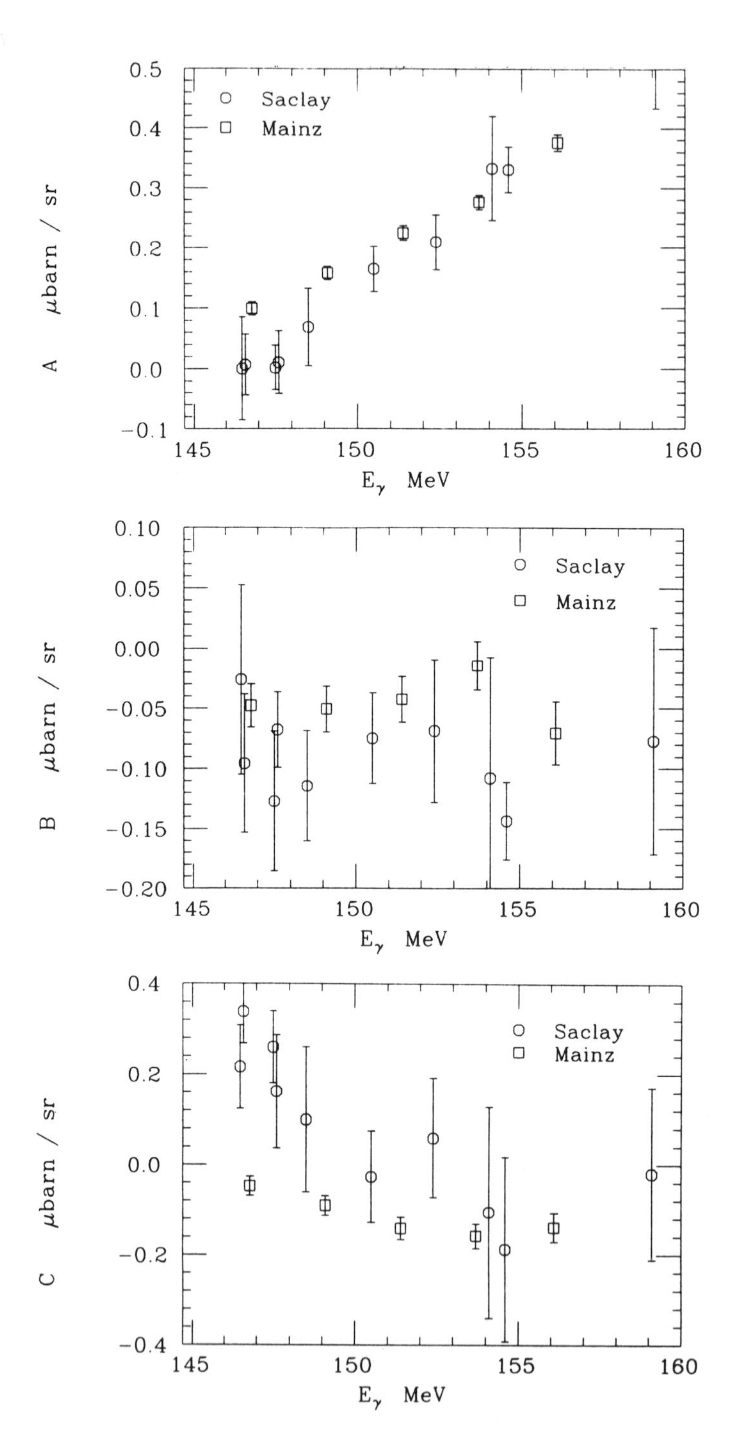

Figure 6: The angular distribution coefficients A, B and C

gives a pure real part below the charge threshold $E_\gamma(\gamma p \to n\pi^+) = 151.4\text{MeV}$ and a pure imaginary part above.

The solid line in figure 8 is a fit with the rescattering correction to the experimental data points for the E_{0+}– amplitude. With the rescattering correction the result for the s– wave amplitude is

$$\text{Saclay}: \quad \text{ReE}_{0+} = -(0.5 \pm 0.3)10^{-3}\frac{1}{m_\pi}$$

$$\text{Mainz}: \quad \text{ReE}_{0+} = -(0.35 \pm 0.1)10^{-3}\frac{1}{m_\pi}$$

5. Conclusions

A direct comparison between the measured s– wave amplitude and the predictions of the low energy theorems is not possible, because of the rescattering effect from the charge channel. However, at the threshold of the charge pion production there is no rescattering contribution and the extracted E_{0+} value is a factor 8 smaller than the low energy theorems predict.

To understand the difference between theory and experiment, measurements with polarized photons and polarized targets on all four production channels are necessary.

Acknowledgement: The author thanks A. Bernstein, D. Drechsel, V. Metag, B. Schoch, H. Stroeher and L. Tiator for many stimulating discussions. This work has been supported by the Deutsche Forschungsgemeinschaft, SFB 201.

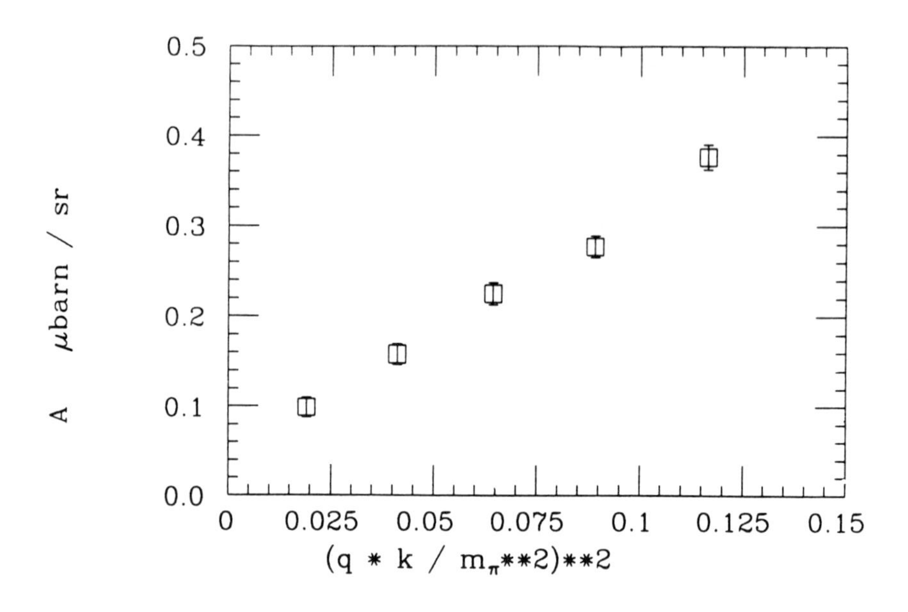

Figure 7: The A- coefficient

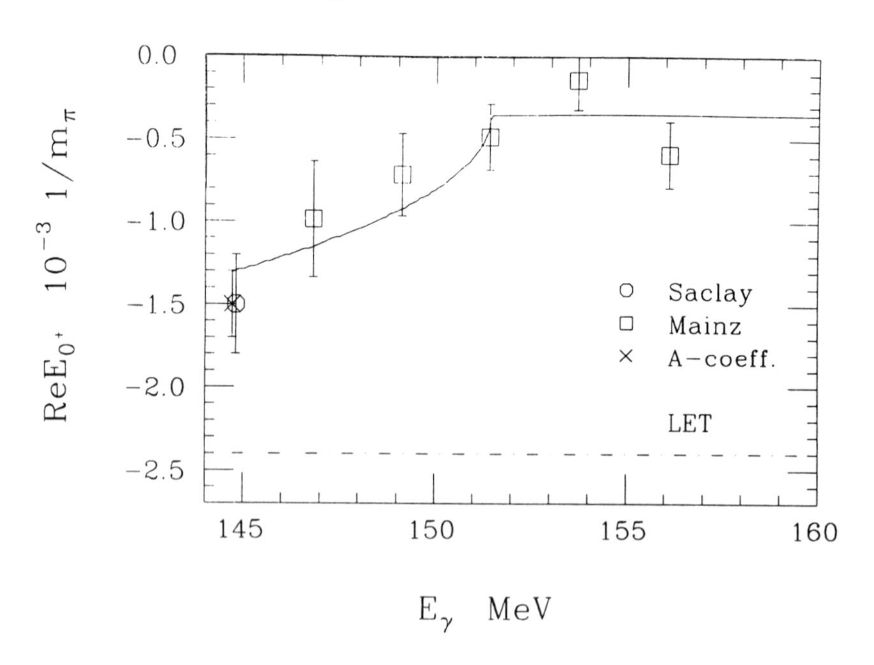

Figure 8: The energy dependence for ReE_{0+}

References

[1] N.M Kroll and M.A. Rudermann, Phys. ReV 93 (1954)233

[2] S. Fubini et.al., Nuovo Cimento 40 A (1965) 1171

[3] E. Mazzucato et.al., Phys. ReV. Lett. 57 (1986) 3144

[4] P. Argon et.al., Nucl.Inst. and Meth. A 228 (1984) 20

[5] R. Beck et. al., Phys. ReV. Lett.
R. Beck Ph.D. thesis (Mainz 1989)

[6] H. Herminghaus et.al., Nucl. Inst. and Meth. 138 (1976) 1

[7] J.D. Kellie et. al., Nucl. Instr. and Meth. A241 (1985) 153

[8] H. Stroeher et. al., Nucl. Instr. and Meth. A269 (1988) 568

[9] G. Fäldt, Nucl. Phys. A 333 (1980) 357
J.M. Laget, Physics Reports 69 (1981) 1

PION PHOTOPRODUCTION AT THRESHOLD*)

Dieter Drechsel
Institut für Kernphysik, Universität Mainz
D-6500 Mainz, Federal Republic of Germany

ABSTRACT

The predictions of low energy theorems on pion photoproduction are confronted with the exciting new experimental data near threshold. Isobars in the s- and u-channels and heavier mesons exchanged in the t-channel give only small corrections. The role of rescattering is discussed in some detail. A calculation of the threshold amplitude based on current algebra is able to describe the data but has serious shortcomings concerning gauge and Lorentz invariance. Following a derivation of the low energy theorems from field theory, we find that the discrepancy between experiment and theory could be due to a possible induced pseudoscalar of non-pionic origin in the axial vector current.

1. INTRODUCTION

The theory of photoproduction of pions was written in the 50's. Kroll and Ruderman[1] were the first to derive model-independent predictions in the threshold region, so-called low energy theorems (LET), by applying gauge and Lorentz invariance to the reaction $\gamma+N\rightarrow\pi+N$. The general formalism for this process was developped by Chew, Goldberger, Low and Nambu[2] (CGLN amplitudes). In 1965 Fubini et al.[3] extended the earlier predictions of LET by including also the hypothesis of a partially conserved axial current (PCAC). In this way they succeeded in describing the threshold amplitude as a power series in the ratio $\mu = m_\pi/m_N$ up to term of order μ^2. In 1967, Berends et al.[4] analysed the existing data in terms of a multipole de-composition and presented tables of the various multipole amplitudes contributing in the region up to excitation energies of 500 MeV.

* This work has been supported by the Deutsche Forschungsgemeinschaft (SFB 201)

Due to electromagnetic interactions the isospin symmetry between charged and neutral pions is broken, giving rise to a mass splitting of about 5 MeV and a very short life-time of neutral pions because of the decay $\pi^0 \to \gamma + \gamma$ ($\tau = 8.4 \cdot 10^{-17}$ sec, Ref. 5). The weak decay of charged pions ($\tau = 2.6 \cdot 10^{-8}$ sec, Ref. 5) is evidence for a pionic axial current,

$$J^{\alpha}_{5\mu}(\pi) = f_\pi \, \partial_\mu \, \Phi^{\alpha}_{\pi}, \tag{1}$$

where $f_\pi = 93$ MeV is the pion decay constant.

The pion (and its axial current!) is an essential ingredient for models of the nucleon. The original MIT model described the nucleon by 3 valence quarks confined to the volume of a bag by the "pressure" of the surrounding vacuum. It was soon realised that this model does not conserve the chirality. As long as the forces at the surface do not depend on the spin of the quarks, the helicity $\vec{\sigma} \cdot \hat{p}$ will change when the quarks are reflected at the surface. The chirality may be restored when the rigid surface is replaced by a pion cloud or, in so-called σ-models, by a "chiral combination" of the pion and a hypothetical sigma-meson.

We know from the theory of neutrinos that, a priori, chirality is only conserved for massless particles. However, QCD or QCD inspired theories have to describe massive nucleons. Such particles may be obtained, in a chirally invariant theory with massless "current" quarks, as nonperturbative solutions. When such massive nucleons are produced, there also appears a pion-like object at mass zero ("Goldstone boson") in such a way that overall chirality remains conserved. While the conservation of the (electric) charge may be derived as a consequence of a conserved vector current J_μ, a conserved chirality is related to a vanishing four-divergence of an axial current, $\partial^\mu J_{5\mu} = 0$.

If we allow for small but finite "current" quark masses, the pion obtains its finite mass and, at the same time, also the axial current is no longer conserved. Instead we obtain the PCAC relation,

$$\partial^\mu J^{\alpha}_{5\mu} = - f_\pi \, m^2_\pi \, \Phi^{\alpha}_{\pi}, \tag{2}$$

i.e. the axial current would be conserved if the pion was stable (pion decay constant $f_\pi \to 0$), for vanishing pion mass ($m_\pi \to 0$, as would be the case for m_u, $m_d \to 0$) and if there was no pion field ($\Phi^\pi \to 0$).

By simply counting the masses of the quark-antiquark pair in the pion and the 3 valence quarks in the proton, constituent quark models predict the mass ratio $\mu \approx 2/3$. In a model with vanishing current quark masses, on the other hand, the pion appears as a Goldstone particle leading to $\mu = 0$. The actual mass ratio $\mu \approx 1/7$ indicates the hybrid nature of the physical pion between Goldstone boson and $q\bar{q}$-state.

In view of the basic importance of the pion for our understanding of nucleons and nuclear forces, pion photoproduction has been and will be of considerable interest. Since this reaction is sensitive to the off-shell properties of the pion-nucleon interaction, i.e. to the short-range behaviour of the pion wave function, it may serve as a critical test of models of hadrons. Some of the contributions to this process are shown in Fig. 1, in particular the Born terms with nucleon and pion pole terms (singularity for $m_\pi \to 0$), the seagull or Kroll-Ruderman term as well as resonance contributions in the s-channel (nucleon resonances N^* and Δ^*) and the exchange of heavier mesons in the t-channel (ω, ρ, etc.). It is the essence of LET that the threshold amplitude is determined by the Born terms (Fig. 1a-d) and that all higher order terms should be small near threshold.

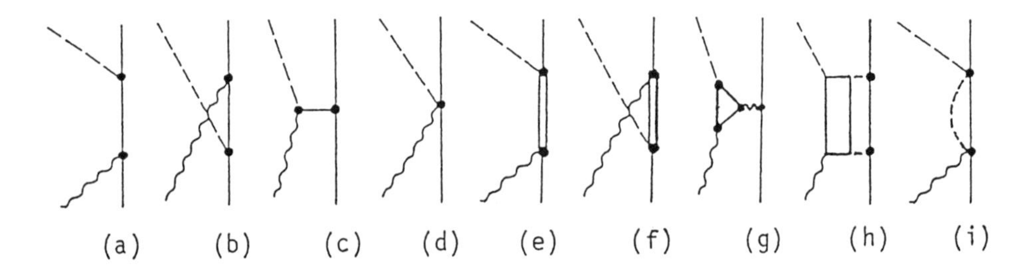

Fig. 1 Diagrams contributing to pion photoproduction: (a) direct and (b) crossed nucleon pole term, (c) pion pole, (d) Kroll-Ruderman term, (e) and (f) isobar excitation, (g) triangle anomaly (photon pole, vector meson exchange), (h) square anomaly, (i) rescattering.

In sect. 2 we will confront the predictions of LET with the new experimental data[6-8]. The leading corrections to LET are discussed in sect. 3. A different approach to the threshold amplitude, based on current algebra and proposed by Furlan et al.[9] in the 70's, is presented in sect. 4. It results in an additional "sigma term" related to the quark current masses. Though this new term gives rise to a better agreement with the data, the approach is neither gauge nor Lorentz invariant. Moreover, it will be shown in sect. 5 that there is no justification for such a term in a field-theoretical derivation of LET. Instead we find that the existing discrepancy between the data and LET could be due to an induced pseudoscalar of non-pionic origin in the axial current. Finally, a short summary and outlook is given in sect. 6.

2. LOW ENERGY THEOREMS

Low energy theorems (LET) have been a very powerful tool to determine the threshold amplitudes for reactions involving photons and pions. Covariance, gauge invariance and the (partial) conservation of the axial current (PCAC) allow to express these amplitudes in the low energy limit in terms of the global properties of the hadrons, such as their masses, magnetic moments, and coupling constants. As an example the Thomson limit for the scattering of photons at threshold (energy $\omega \to 0$) is

$$\frac{d\sigma}{d\Omega} \to \frac{2}{3}\left(\frac{Q^2}{M}\right)^2,$$

where Q and M are the total charge and mass of the struck object. The first contributions depending on the internal structure are $O(\omega^2)$ and proportional to the electric polarisability and magnetic susceptibility of the target[10,11].

While these theorems are exact in the case of photons, their application to pions involves an expansion about the unphysical "soft-pion limit" of massless pions, as a power series in $\mu = m_\pi/m_N$. Weinberg[12] and Tomozawa[13] were able to explain the surprisingly small s-wave phase shifts for the scattering of low energy pions within the framework of LET.

Near threshold the differential cross section for pion photoproduction should be dominated by the S-wave amplitude,

$$\frac{d\sigma}{d\Omega} \to \frac{|\vec{k}_\pi|}{|\vec{k}_\gamma|}\,|E_{0+}|^2, \tag{3}$$

where $\vec{k}_\pi$ and $\vec{k}_\gamma$ are the momenta of pion and photon, respectively.

The S-wave amplitude has the isospin decomposition

$$E_{0^+} = \frac{ef}{4\pi m_\pi}\,\frac{m}{W}\,\chi_f^+\left(\frac{1}{2}[\tau_\alpha,\tau_0]A^{(-)} + \tau_\alpha A^{(0)} + \delta_{\alpha 0} A^{(+)}\right)\chi_i, \tag{4}$$

with the pion-nucleon coupling constant f ($f^2/4\pi \simeq 0.08$), $W = m + m_\pi$ the total c.m. threshold energy, and χ_f and χ_i the isospinors of the final and initial nucleon, respectively. The isospin structure of Eq. (4) is due to the fact that the emission of a pion with isospin index $(-\alpha)$ has to be accompanied by a Pauli matrix τ_α in the nucleon's isospace while the absorbed photon is either isoscalar or isovector, leading to a factor 1 or τ_0, respectively. Hence, the amplitude $A^{(0)}$ is connected with the isoscalar part of the photon. The amplitudes $A^{(-)}$ and $A^{(+)}$ are related with the isovector part of the photon, the two time-orderings of absorption and emission appearing in symmetrised form as commutator and anticommutator of τ_α and τ_0.

The explicit evaluation of the Born terms in Fig. 1a-d in a chirally invariant model yields the isospin amplitudes

$$A^{(-)} = 1 + O(\mu^2), \qquad A^{(+,0)} = -\frac{1}{2}\mu + \frac{1}{4}\mu^{(v,s)}\mu^2 + O(\mu^3), \tag{5}$$

where $\mu^{(v,s)} = \mu_p \mp \mu_n$ are the isovector and isoscalar magnetic moments of the nucleon. The leading term of order 1, the "Kroll-Ruderman" amplitude[1] of Fig. 1d, is fixed by the gauge invariance of the electromagnetic current. The higher order terms in the mass ratio μ are determined by chiral invariance, i.e. by the partial conservation of the axial current (PCAC)[3,14]. Typical model-dependent corrections are of relative order μ^2.

The four physical amplitudes are

$$A(\gamma p \to n\pi^+) = \sqrt{2}\,(1 - \frac{1}{2}\mu + O(\mu^2)) \qquad A(\gamma n \to p\pi^-) = -\sqrt{2}\,(1 + \frac{1}{2}\mu + O(\mu^2))$$
$$A(\gamma p \to p\pi^0) = -\mu + \frac{1}{2}\mu_p\mu^2 + O(\mu^3) \qquad A(\gamma n \to n\pi^0) = -\frac{1}{2}\mu_n\mu^2 + O(\mu^3). \tag{6}$$

The ratios of these amplitudes can be obtained by evaluating the dipole moments of the pion-nucleon system in its c.m. frame.

The theoretical predictions of LET for the E_{0+} amplitude are shown in Tab. I in comparison with some model calculations and the experimental data. These predictions were assumed to be particularly powerful for neutral pion production with amplitudes of order μ and μ^2 for protons and neutrons, respectively.

Table I Pion photoproduction amplitude E_{0+} in units of $10^{-3}/m_\pi$. The experimental data are compared to various theoretical predictions. LET: soft pion limit, Eq. (6), CQM: constituent quark model[15], PV: Born terms (Fig. 1a-d) in pseudovector pion-nucleon coupling, CSB/ISB: current algebra including chiral symmetry breaking by an average quark mass m_0, Eq. (28), and an isospin symmetry breaking δm, Eq. (29).

channel	LET	CQM	PV	CSB/ISB	experiment
$\gamma p \to n\pi^+$	27.5	29.0	28.0	28.7±0.3	28.3 ± 0.5[16]
$\gamma n \to p\pi^-$	-32.0	-33.3	-32.2	-30.8±0.3	-31.9 ± 0.5[17]
$\gamma p \to p\pi^0$	- 2.4	- 2.7	- 2.5	- 0.8±0.5	- 0.5 ± 0.3[6]
					- 0.35 ± 0.1[8]
$\gamma n \to n\pi^0$	0.4	0.4	0.4	+ 1.6±0.4	unknown

In view of Eq. (6), a very small threshold cross section for neutral pion production had been expected. It came as a great surprise that the Saclay data[6] even fell below the predicted values. However, this puzzling result has been corroborated by a recent Mainz experiment[7,8]. As may be seen in Fig. 2, the threshold cross section does not show the typical S-wave behaviour predicted by the theory. Instead the data essentially follow the energy dependence given by the P_{33} resonance. Similarly there is little indication of the expected "cusp effect" in the cross section at E_γ = 151 MeV, where the threshold for the competing $n\pi^+$ channel opens.

The suppression of the E_{0+} amplitude can also be seen by inspecting the angular distribution shown in Fig. 3. Near threshold the differential cross section is given by

$$\frac{d\sigma}{d\Omega} = \frac{q}{k}\left(A + B\cos\Theta_\pi + C\cos^2\Theta_\pi\right)$$

$$A = \frac{1}{2}|2M_{1+} + M_{1-}|^2 + \frac{1}{2}|3E_{1+} - M_{1+} + M_{1-}|^2 + |E_{0+}|^2$$

$$B = 2\mathrm{Re}\left(E_{0+}\ (3E_{1+} + M_{1+} - M_{1-})^*\right) \qquad (7)$$

$$C = |3E_{1+} + M_{1+} - M_{1-}|^2 - \frac{1}{2}\ |2M_{1+} + M_{1-}|^2 - \frac{1}{2}\ |3E_{1+} - M_{1+} + M_{1-}|^2.$$

While LET predicts a cross section peaked in the backward region due to E_{0+}/M_{1+} interference in the B-term, the experimental data are much lower and nearly symmetrical about 90° due to the lack of E_{0+} strength.

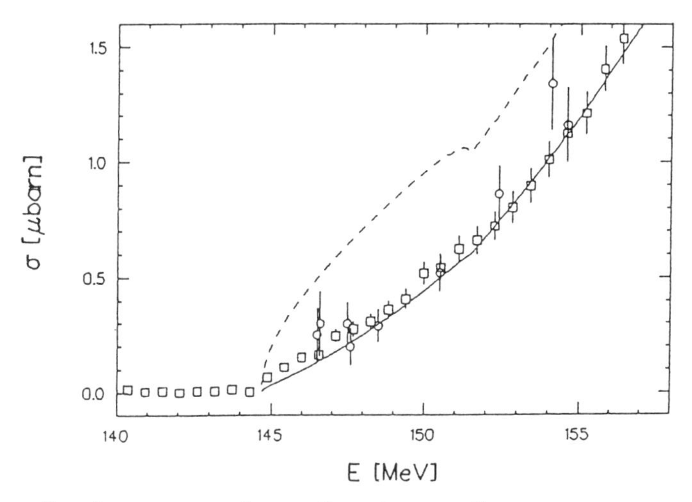

Fig. 2 Total cross section σ for the reaction $\gamma + p \to \pi + p$ as function of photon energy. Mainz data (□) according to Beck et al.[7,8], Saclay data (o) of Mazzucato et al.[6]. The dashed and solid curves are obtained with $E_{0+} = -2.5 \cdot 10^{-3}/m_\pi$ and $E_{0+} = -0.35 \cdot 10^{-3}/m_\pi$, respectively. The P-wave multipoles determined in the Mainz experiment are $M_{1+} = 7.95\ qk \cdot 10^{-3}/m_\pi^3$, $M_{1-} = -0.82\ qk \cdot 10^{-3}/m_\pi^3$ and $E_{1+} = 0$.

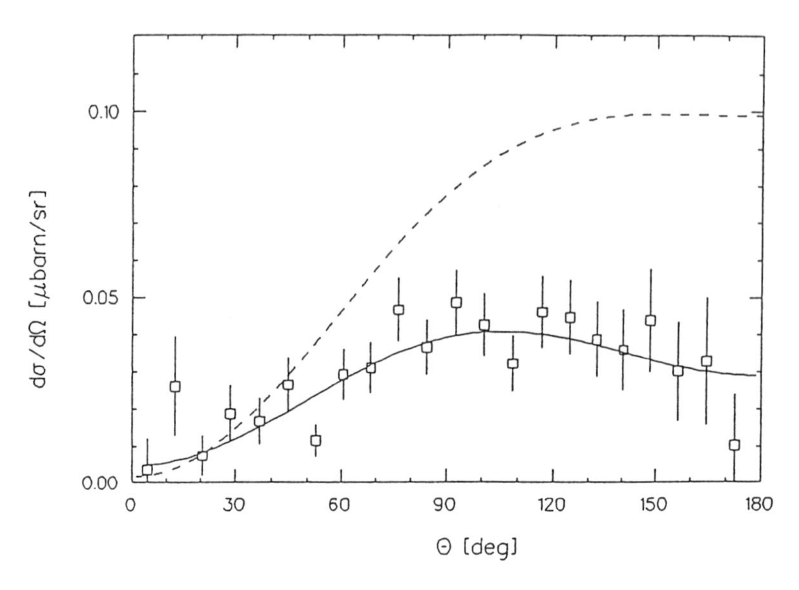

Fig. 3 Angular distribution $d\sigma/d\Omega$ for $\gamma + p \rightarrow \pi^0 + p$ at fixed photon energy E_γ = 149 MeV. The dashed and solid curves are obtained with multipoles as in Fig. 2. The data is from Beck et al.[7,8].

3. CORRECTIONS TO LET

The predictions of LET depend on gauge and Lorentz invariance as well as the assumption of a partially conserved axial current (PCAC), Eq. (2). Due to the appearance of m_π^2 on the rhs of Eq. (2), the divergence of the axial current does not influence the lowest order terms of the expansion of the threshold amplitude given in Eq. (6). However, there will be additional contributions to the divergence due to anomalies[18,19], e.g. the anomaly related to the decay $\pi^0 \rightarrow 2\gamma$ (photon exchange in Fig. 1g) with the divergence

$$D^{(\alpha)}_{\text{anom.}} = \delta_{\alpha 3} \frac{e^2}{32\pi^2} \varepsilon_{\mu\nu\rho\sigma} F^{\mu\nu} F^{\rho\sigma}, \tag{8}$$

where ε is an antisymmetrical tensor and F the electromagnetic field tensor. This anomaly gives a contribution of "leading order" for the production of neutral pions. However, its numerical value is only of the order of 1 %. Larger contributions arise if the exchanged photon is replaced by vector mesons. The exchange of ω mesons in the t channel decreases the amplitude for $\gamma + p \rightarrow \pi^0 + p$ by about 10 %, the ρ meson leads to a further reduction by about 3 %. Clearly, contributions of higher mesons cannot be excluded, but there is no indication for such effects at present.

Using an exchange mechanism as in Fig. 1g, with a structureless proton in the intermediate triangle, we obtain the following lowest order terms beyond the LET value of Eq. (5):

$$A^{(-)} = 1 + \frac{g_A^2}{4\pi}\,\frac{8}{3\pi}\,\frac{m_\pi^2}{m_A^2 + m_\pi^2} \tag{9}$$

$$A^{(+,0)} = -\frac{1}{2}\,\mu + \frac{1}{4}\,\mu^{(v,s)}\,\mu^2 + \frac{e^2}{4\pi}\,\mu^{(v,s)}\,\frac{\mu}{2\pi} + \frac{g_V(g_V + f_V)}{4\pi}\,\frac{m_\pi^2}{m_V^2 + m_\pi^2}\,\frac{\mu}{\pi} + \frac{g_T^2}{4\pi}\,\frac{m_\pi^2}{m_T^2 + m_\pi^2}\,\frac{16\mu}{3\pi}\,. \tag{10}$$

In these equations, g_A and m_A are coupling constants and masses, respectively, of axial vector-isovector mesons contributing to $A^{(-)}$. Similarly V and T refer to vector and tensor mesons, of isoscalar and isovector nature, appearing in $A^{(+)}$ and $A^{(0)}$, respectively. The constant f_V describes the induced tensor coupling appearing, e.g., for the ρNN-vertex. It is interesting to note that all higher order t-channel corrections are positive.

Further contributions come from intermediate Δ states and higher resonances in the s and u channels (Fig. 1e,f). Explicit calculations[15,20,21] show large model dependencies but relatively small effects ($\lesssim$ 10 %). Even the sign of the Δ contribution depends on the particular coupling and propagator used in the model[20].

Due to the suppression of neutral pion production at threshold by a factor μ, the rescattering graph of Fig. 1i may become important[22,23]. It involves the production of a positive pion followed by charge exchange. The rapid cross-over from complex to real values of the amplitude might give rise to a cusp effect at threshold ("Wigner cusp"), see Fig. 2. Since the loop diagram diverges, the results are strongly dependent on form factors and/or renormalisation procedures. However, the lack of a pronounced cusp in the data is an indication that rescattering effects are not a likely explanation of the discrepancy.

Earlier estimates of rescattering were based on an extrapolation of the R-matrix to momenta below π^+n-threshold[23]. The relevant matrix element is related to the imaginary part of the rescattering diagram. Being on-shell, it can be expressed by the photoproduction of a π^+ followed by the charge exchange scattering $\pi^+ n \rightarrow \pi^0 p$. While the photoproduction of charged pions is much more likely than that one of neutral pions, the process is reduced by the smallness of the scattering length for charge exchange near threshold,

$$E_{0+} = E_{0+}(p\pi^0) + ik_{\pi^+}\,a(n\pi^+ \rightarrow p\pi^0)\,E_{0+}(n\pi^+), \tag{11}$$

where k_{π^+} is the momentum of the π^+. Below threshold, the correction is obtained by $k_\pi \rightarrow ik_\pi$, resulting in a real value of the S-wave amplitude. Unfortunately, this correction increases the discrepancy between LET and the data.

Final state interactions have recently been investigated in a dynamical model by Yang[24] and by Nozawa et al.[25] in a coupled channels calculation. As an important result these authors find large contributions from the real part of the rescattering amplitude, which was previously neglected. As a result Nozawa et al.[25] obtain a threshold value of $E_{0+} = -1.92 \cdot 10^{-3}/m_\pi$. Since this value contains corrections derived from vector meson exchange, the total contribution of rescattering is only about 15 %. It has the right sign but is too small to explain the Mainz data. It is interesting to note that the calculations of Yang[24] lead to quite different predictions depending on the pion-nucleon interaction. Since these potentials are phenomenological ones, they generally do not obey the PCAC relation and, hence, there is no reason to expect the results of LET. A comparison of the experimental values with the predictions of LET and of Ref. 25 has been given in Fig. 4. In a new evaluation of the Mainz data[7] the authors find a second solution with threshold values of E_{0+} closer to LET. However, this solution also implies an unusual threshold behaviour of M_{1+} and a very discontinuous cross-over from solution I to II near charged pion threshold.

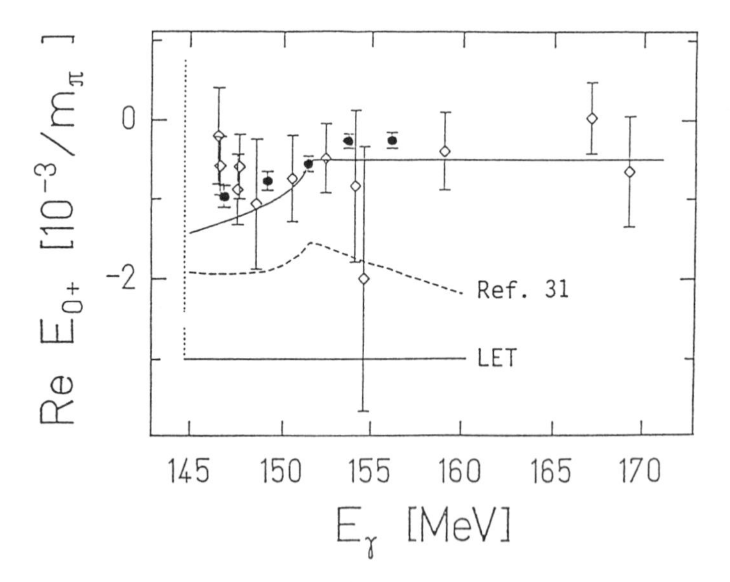

Fig. 4 Experimental values for E_{0+} as function of photon energy according to Ströher[26] in comparison with the predictions of LET and Nozawa et al.[5].

Including the off-shell contributions of rescattering, Nozawa et al.[25] find the following structure for the production amplitude

$$A_{\gamma p\to\pi^0 p} = e^{i\delta}\cos\delta\left\{A^B_{\gamma p\to\pi^0 p} - i\Theta(E-m_n-m_{\pi^+})\,F_{\pi^+ n\to\pi^0 p}\,A^B_{\gamma p\to\pi^+ n} + \sum_{\pi N} P\int_0^\infty dk\,k^2\,K_{\pi N\to\pi^0 p}\,A^B_{\gamma p\to\pi N}\,(E-E_N-E_\pi)^{-1}\right\}. \tag{12}$$

The amplitude has the overall factor $\exp(i\delta)$, where δ is the phase shift for elastic $\pi^0\pi$ scattering, as required by the Fermi-Watson theorem. The three terms in curly brackets on the rhs of Eq. (12) are

- the Born amplitude for $\gamma p\to\pi^0 p$,
- the imaginary contribution of π^+ production followed by charge exchange scattering. It appears only above $\pi^+ n$ threshold and, in this case, reduces to the correction term of Eq. (11) in lowest order perturbation theory,
- the real contributions of rescattering expressed by the principal value integrals, P.

The parameters of Ref. 25 have been fitted to pion-nucleon scattering and pion photoproduction in the Δ region (P-wave amplitudes M_{1+} and E_{1+}). With a reasonable cut-off Λ = 650 MeV they obtain a good overall agreement with the data at higher energies. In the threshold region the calculation is characterised by strong cancellations of the effects of final state interactions for $\pi^+ n$ and $\pi^0 p$ channels as well as pion pole and Kroll-Ruderman terms (see tab. II).

Table II Contributions of Born diagrams and final state interactions to the threshold s-wave amplitude $E_{0+}(\gamma p\to\pi^0 p)$ in units of $10^{-3}/m_\pi$. B: Born terms for the diagrams (a)-(d) of Fig. 1 and ω/ρ exchange (Fig. 1g). FSI(π^0): corrections for rescattering of π^0. FSI(π^+): production of π^+ followed by charge exchange scattering. Results from Ref. 25.

	(a)	(b)	(c)	(d)	ω/ρ	sum
B	-1.26	-1.25	0	0	.22	-2.29
FSI(π^0)	- .10	.57	0	0	.05	.52
FSI(π^+)	- .53	- .24	-3.07	+3.57	.10	- .15
sum	-1.89	- .92	-3.07	+3.57	.37	-1.92

The final result shows that higher order contributions cancel to a large degree. This is also borne out in an analytical calculation by Naus et al.[27]. In a consistent expansion of the threshold amplitude about the soft-pion point, the authors precisely obtain LET. All off-shell behaviour of the vertices disappears in the final answer, and rescattering corrections only contribute in higher order terms not specified by LET.

4. CURRENT ALGEBRA

On the usual energy scale of hadrons, the mass terms of the 3 lightest quarks (u,d,s) in the QCD Lagrangian can be neglected. In this limit of vanishing "current" masses, we can define 8 conserved vector and axial currents,

$$J^a_\mu = \bar{q}\, \gamma_\mu \frac{\lambda^a}{2}\, q \tag{13}$$

$$J^a_{5\mu} = \bar{q}\, \gamma_\mu \gamma_5 \frac{\lambda^a}{2}\, q\,, \tag{14}$$

where the Gell-Mann matrices refer to the flavours of the 3 lightest quarks. The corresponding charges,

$$Q^a = \int d^3\vec{x}\, J^a_0\,(x), \qquad Q^a_5 = \int d^3\vec{x}\, J^a_{50}\,(x) \tag{15}$$

are constants of motion.

As postulated first by Gell-Mann, the equal-time commutators (ETC) of these charges fulfil the SU(3) × SU(3) algebra[28]

$$[Q^a, Q^b] = i f^{abc}\, Q^c\,, \quad [Q^a_5, Q^b_5] = i f^{abc}\, Q^c, \quad [Q^a, Q^b_5] = i f^{abc}\, Q^c_5\,. \tag{16}$$

Often we also assume such relations for commutators of charges and currents, i.e.

$$[Q^a, J^b_\mu(x)] = i f^{abc}\, J^c_\mu(c) \text{ etc.} \tag{17}$$

These symmetries are broken by the mass term in the QCD Lagrangian,

$$M = \begin{pmatrix} m_u & 0 & 0 \\ 0 & m_d & 0 \\ 0 & 0 & m_s \end{pmatrix}, \tag{18}$$

giving rise to the relations

$$\partial^\mu J^a_\mu = i\bar{q}\left[M, \frac{\lambda^a}{2}\right] q\,, \qquad \partial^\mu J^a_{5\mu} = i\bar{q}\left\{M, \frac{\lambda^a}{2}\right\} \gamma_5\, q\,, \tag{19}$$

i.e. the 4-divergences are given by the commutator or anticommutator of the mass matrix with the Gell-Mann matrix describing the flavour of the vector or axial current, respectively. We note that the electromagnetic current,

$$J_\mu^{em} = J_\mu^3 + \frac{1}{\sqrt{3}} J_\mu^8, \tag{20}$$

contains the diagonal matrices λ^3 and λ^8. Therefore, it commutes with the diagonal mass matrix, leading to a conserved electromagnetic current.

If we allow for finite "current" quark masses m_u and m_d, also the pion obtains a finite mass m_π according to the relation[29]

$$m_\pi^2 f_\pi^2 = -(m_u + m_d)\langle\bar{q}q\rangle, \tag{21}$$

where f_π = 93 MeV is the pion decay constant. At the same time, chiral symmetry is broken. However, it is broken at the scale of the small current quark masses of a few MeV, not at the scale of the much larger effective masses. On the phenomenological level, the divergence of the axial current of Eq. (19) may be expressed in terms of the effective hadronic degrees of freedom by the PCAC relation, Eq. (2).

Following the work of Furlan et al.[9] we now introduce the operator

$$Q_{5L}^\alpha = Q_5^\alpha + \frac{i}{m_\pi} \dot{Q}_5^\alpha, \tag{22}$$

having the matrix elements

$$\langle\pi^\alpha|Q_{5L}^\alpha|0\rangle = 0, \qquad \langle 0|Q_{5L}^\alpha|\pi^\alpha\rangle = 2\langle 0|Q_5^\alpha|\pi^\alpha\rangle = 2if_\pi. \tag{23}$$

Using Eq. (19), the symmetry breaking may be expressed at the microscopical level by the quark mass matrix,

$$\dot{Q}_5^\alpha = i[H, Q_5^\alpha] = \int d^3\vec{r}\, \partial^\mu J_{5\mu}^\alpha = i \int d^3\vec{r}\, \bar{q}\gamma_5 \left\{M, \frac{\lambda^\alpha}{2}\right\} q. \tag{24}$$

Next, Furlan et al. evaluate the commutator of Q_{5L}^α and J_μ^{em}, Eqs. (22) and (20), respectively, first by a direct calculation at the quark level and second by saturating the commutator with a complete set of intermediate states (π,N,πN,...). As a result they obtain the following equation:

$$\begin{aligned} &\langle b|[Q_{5L}^\alpha, \varepsilon\cdot J^{em}]|a\rangle \\ &= \sum_\pi \left(\langle 0|Q_{5L}^\alpha|\pi^\alpha\rangle\langle b\pi^\alpha|\varepsilon\cdot J^{em}|a\rangle - \langle b|\varepsilon\cdot J^{em}|a\pi^\alpha\rangle\langle\pi^\alpha|Q_{5L}^\alpha|0\rangle\right) \\ &+ \sum_N \left(\langle b|Q_{5L}^\alpha|N\rangle\langle N|\varepsilon\cdot J^{em}|a\rangle - \text{c.t.}\right) + \sum_{n\neq N,\pi}\left(\langle b|Q_{5L}^\alpha|n\rangle\langle n|\varepsilon\cdot J^{em}|a\rangle - \text{c.t.}\right). \end{aligned} \tag{25}$$

In the first sum on the rhs we find the matrix elements of the electromagnetic current for both photoproduction ($a \to b\pi^\alpha$) and radiative capture ($a\pi^\alpha \to b$). Due to the definition of Q^α_{5L}, however, the pionic matrix element accompanying radiative capture vanishes, and the two reactions are decoupled. With $T_{\gamma a\to\pi b} = i\langle b\pi^\alpha|\varepsilon\cdot J^{em}|a\rangle$, Eq. (25) can be cast into the form

$$T_{\gamma a\to\pi b} = \frac{1}{f_\pi}\langle b|[Q^\alpha_5,\varepsilon\cdot J^{el}]|a\rangle + \frac{i}{m_\pi f_\pi}\langle b|[\dot{Q}^\alpha_5,\varepsilon\cdot J^{el}]|a\rangle$$
$$- \frac{1}{f_\pi}\sum_N\left(\langle b|Q^\alpha_{5L}|N\rangle\langle N|\varepsilon\cdot J^{el}|a\rangle - \text{c.t.}\right) + \text{h.o.t.} \tag{26}$$

On the rhs of this equation we find, in the order of their appearance, the familiar Kroll-Ruderman term[1], the additional term describing chiral symmetry breaking, the Born terms (s and u channels) and higher order terms involving at least two pions in intermediate states. In a single quark model of the nucleon, with symmetrical S-wave functions, $q = N(f, ig\vec{\sigma}\cdot\hat{r})^T$, and using the Goldberger-Treiman relation, Nath and Singh[20] have obtained the following contributions for the chiral symmetry breaking ("CSB") terms to the isospin amplitudes

$$A^{(-)}_{(CSB)} = 0,\quad A^{(+)}_{(CSB)} = 2\eta\,\frac{m_0}{m_\pi},\quad A^{(0)}_{(CSB)} = \frac{10}{9}\,\eta\,\frac{m_0}{m_\pi}, \tag{27}$$

where

$$\eta = \frac{3}{5}\int\left(f^2 + \frac{1}{3}g^2\right)r^2dr \Big/ \int\left(f^2 - \frac{1}{3}g^2\right)r^2dr = \left(1 + \frac{3}{5}G_A\right)\Big/2G_A \sim 0.7. \tag{28}$$

Up to this point we have assumed that the two light quarks have the same current mass m_0. Obviously, this is not the case and the mass splitting δm is of the same order as the masses themselves. We expect, therefore, that the isospin symmetry expressed in Eq. (4) will be broken. A calculation including δm shows that the amplitude $A^{(+)}$ in Eq. (27) has to be replaced by[31]

$$A^{(+)}_{(ISB)} = 2\eta\,\frac{m_0}{m_\pi}\left(1 + \frac{\delta m}{2m_0}\,\frac{1+5\tau_3}{3}\right). \tag{29}$$

The resulting predictions for the various physical production channels are shown in column CSB/ISB of tab. I. The errors correspond to the uncertainties in the quark masses[30]. It is obvious that the effects of the current quark masses are quite substantial, at least for the production of neutral pions.

In a recent extension of this work, Schäfer and Weise[32] have found a strong model-dependence of this effect in a chiral model with a pion cloud surrounding the quark bag. In particular, the additional term decreases strongly with the radius of the quark bag. It is interesting to note that the size of the effect is related, in the model of Ref. 32, to the strength of the singlet axial coupling constant responsible for the "spin crisis" in deep inelastic lepton scattering.

While the idea of Furlan et al. may explain all or part of the discrepancy between LET and the data, there are some serious flaws connected with it. In particular,

- the ansatz Eq. (22) is not Lorentz invariant,
- the additional term in Eq. (27) is not gauge invariant,
- there are some problems to recover the usual Born terms from Eq. (26), and
- there is no motivation for the additional term in a conventional derivation of LET (see sect. 5).

5. LET REVISITED

While the applications of LET to pion photoproduction have followed a somewhat different path[3,14], we will now derive these theorems by a field-theoretical approach.

To lowest order in the electromagnetic current, the S matrix element for the reaction $\gamma(q) + N(p) \rightarrow \pi^{\alpha}(k) + N(p')$ is given by

$$S = -ie \int d^4x\, e^{-iqx}\, \varepsilon^{\mu}(q\lambda)\, N_{\gamma}^{1/2} \langle N'\pi^{\alpha}|J_{\mu}(x)|N\rangle, \tag{30}$$

where ε^{μ} is the polarisation vector of the electromagnetic field and $N_{\gamma} = (2\pi)^{-3}(2q_0)^{-1}$ its normalisation. Note that the hadronic systems are described by exact solutions of the Lagrangian with appropriate "in" and "out" boundary conditions.

If we neglect strange components in nucleons and pions, the current operator at the quark level can be expressed by its usual isoscalar and isovector components,

$$J_{\mu} = \sum_i \frac{\frac{1}{3} + \tau_0(i)}{2} \gamma_{\mu}(i). \tag{31}$$

On a phenomenological level, in a model with nucleons and pions, we have

$$J_{\mu} = J_{\mu}^{N} + J_{\mu}^{\pi} + J_{\mu}^{int}. \tag{32}$$

The current operators of free nucleons and pions are

$$J_\mu^N = \bar{\psi}\frac{1+\tau_0}{2}\gamma_\mu\psi + \frac{\partial^\nu}{2m}\left(\bar{\psi}\frac{\varkappa_s+\varkappa_v\tau_0}{2}\sigma_{\mu\nu}\psi\right) \tag{33}$$

$$J_\mu^\pi = (\vec{\Phi}\times\partial_\mu\vec{\Phi})_3 = \varepsilon_{\beta\gamma 3}\,\Phi^\beta\partial_\mu\Phi^\gamma . \tag{34}$$

Note in particular that the nucleon current includes the familiar Pauli term. The derivative in this term can be removed, $\partial^\nu \to iq^\nu$, by a partial differentiation in Eq. (30). While the first two currents on the rhs of Eq. (32) are completely determined by the well-known properties (Lagrangian) of free nucleons and pions, the interaction current is strongly model-dependent. In the Weinberg model, e.g., there appear many additional terms containing both nucleon and pion field operators.

Eq. (30) can be simplified using translation invariance

$$\langle N'\pi|J_\mu(x)|N\rangle = e^{-i(p-p'-k)\cdot x}\langle N'\pi|J_\mu(0)|N\rangle$$

and integrating over the space variable,

$$S = -i(2\pi)^4\delta^4(q+p-p'-k)\;N_\gamma^{1/2}\varepsilon^\mu\langle N'\pi^\alpha|J_\mu(0)|N\rangle . \tag{35}$$

Next we apply the reduction technique of Lehmann, Symanzik and Zimmerman[33]. With the assumptions of a local field theory and microcausality the S matrix can be cast into the form

$$S = (2\pi)^4\delta^4(q+p-p'-k)\,N_\gamma^{1/2}N_\pi^{1/2}\int d^4x\,(-k^2+m_\pi^2)e^{ikx}$$

$$\langle N'|T(\Phi^\alpha(x)\,\varepsilon\cdot J(0))|N\rangle , \tag{36}$$

where $\Phi^\alpha(x)$ is the local pion field, an exact solution of the hadronic Lagrangian, and T the time-ordering operator.

Redefining the nucleon states such that the normalisation factors $N_N = (2\pi)^{-3}(E+m)/2m$ appear explicitly, the integral in Eq. (36) is exactly the invariant amplitude M in the notation of Bjorken and Drell[34]. It would be straightforward to calculate this amplitude within the framework of a specific Lagrangian, e.g. pseudoscalar πN coupling, by evaluating "all Feynman graphs" with the familiar rules for vertices and propagators. However, we cannot be assured that this result makes any sense because of the large value of the coupling constant.

The physical assumption to bring us closer to a reliable numerical prediction is the PCAC hypothesis, Eq. (2), relating the divergence of the axial current to the local pion field. The hypothesis is, more precisely, that all higher order terms (heavier mesons, Φ^3 terms in the Weinberg model etc.) in this equation may be neglected at low energies. This enables us to rewrite the invariant amplitude as

$$M = \frac{k^2 - m_\pi^2}{f_\pi m_\pi^2} \int d^4x\, e^{ikx} \langle N'|T(\partial^\nu J_{5\nu}^\alpha(x)\, \varepsilon \cdot J(0))|N\rangle . \tag{37}$$

Eliminating the differential operator by a partial integration we obtain two contributions, the first by differentiation on the plane wave, the second by differentiation on the step functions $\Theta(x_0)$ and $\Theta(-x_0)$ implicitly contained in the time ordering operator,

$$M = \frac{-k^2 + m_\pi^2}{f_\pi m_\pi^2} \int d^4x\, e^{ikx} \Big\{ ik^\nu \langle N'|T(J_{5\nu}^\alpha(x),\ \varepsilon \cdot J(0))|N\rangle$$
$$+\, \delta(x_0) \langle N'|[J_{50}^\alpha(x), \varepsilon \cdot J(0)]|N\rangle \Big\} . \tag{38}$$

The second term in this equation contains the equal-time commutator (ETC) of the axial charge and the electromagnetic current. Due to Gell-Mann's hypothesis on ETC's (see Eqs. (16) and (17))

$$\int d^3x \left[J_{50}^\alpha(x), J_\mu^{em}(0) \right]_{x=0} = i\varepsilon_{\alpha 3 \beta} J_{5\mu}^\beta , \tag{39}$$

and this term will yield the familiar Kroll-Ruderman term[1], dominating the threshold production of charged pions but vanishing for neutral pions.

Unfortunately, there is still a flaw in Eq. (38). For physical pions $k^2 \to m_\pi^2$, and the matrix element would vanish if it were not about the pion pole terms hidden in the time-ordered product. This problem is usually circumvented, more or less carefully, by evaluating the rhs of the equation in the "soft-pion limit", $k^2 \to 0$, as a power series in m_π^2. Instead, we follow a procedure suggested by Weisberger[35] to first evaluate the contribution of the axial current of the pion. Similar as in the case of the electromagnetic current, the axial current of the hadronic system should have the (phenomenological) form

$$J_{5\nu}^\alpha = J_{5\nu}^{N,\alpha} + f_\pi \partial_\nu \Phi^\alpha + J_{5\nu}^{int,\alpha} \equiv f_\pi \partial_\nu \Phi^\alpha + \tilde{J}_{5\nu}^\alpha . \tag{40}$$

Inserting this expression into Eq. (38) and integrating by parts once more, we find

$$\frac{f_\pi m_\pi^2}{-k^2 + m_\pi^2} M = \int d^4x\, e^{ikx} \Big\{ ik^\nu \langle N'|T(\tilde{J}_{5\nu}^\alpha\, \varepsilon \cdot J)|N\rangle + \delta(x_0) \langle N'|[J_{50}^\alpha, \varepsilon \cdot J]|N\rangle$$
$$-\, i m_\pi f_\pi\, \delta(x_0) \langle N'|[\Phi^\alpha, \varepsilon \cdot J]|N\rangle + f_\pi k^2 \langle N'|T(\Phi^\alpha, \varepsilon \cdot J)\|N\rangle \Big\} . \tag{41}$$

According to Eq. (36) we can identify the last term on the rhs with the invariant amplitude M multiplied by $f_\pi k^2/(-k^2 + m_\pi^2)$. If we combine this term with the lhs of the equation, the pole structure vanishes. The threshold amplitude in the cm frame is simply obtained by $k \to (m_\pi, 0)$, and our final result is

$$M = \frac{\varepsilon^\mu}{f_\pi} \int d^4x\, e^{im_\pi x_0} \Big\{ im_\pi \langle N'|T(\tilde{J}^\alpha_{50}(x), J_\mu(0))|N\rangle + im_\pi f_\pi\, \delta(x_0)$$

$$\cdot \langle N'|[\Phi^\alpha(x), J_\mu(0)]|N\rangle + \delta(x_0)\langle N'|[J^\alpha_{50}(x), J_\mu(0)]|N\rangle \Big\}. \tag{42}$$

As has been stated before, the last term on the rhs is the Kroll-Ruderman term. The other two terms are of order m_π. The time-ordered product yields the s- and u-channel Born terms to lowest order in pseudovector (PV) πN coupling. The second term gives the t-channel Born term to lowest order in the pion source function. It does not contribute to pion photoproduction at threshold.

In stating that Eq. (42) is equivalent to PV coupling, we have tacitly assumed that

$$\tilde{J}^\alpha_{5\mu} = g_A \bar{\psi}\, \gamma_\mu \gamma_5 \frac{\tau^\alpha}{2}\, \psi. \tag{43}$$

The matrix element of the full axial current, Eq. (40), between nucleon states has the structure

$$\langle p'|J^\alpha_{5\mu}|p\rangle = G_A\, \bar{u}_{p'} \gamma_\mu \gamma_5 \frac{\tau^\alpha}{2}\, u_p + \frac{G_P}{2m} \bar{u}_{p'} (p'-p)_\mu\, \gamma_5 \frac{\tau^\alpha}{2}\, u_p. \tag{44}$$

The renormalisation of the axial coupling, $g_A = G_A(0) \approx 5/4$, should be explained by the pion cloud surrounding the bare nucleon. In addition, the pion cloud and its axial current lead to the induced pseudoscalar current proportional to G_P. Since the pion cloud is related to the pion source function,

$$(\Box + m_\pi^2)\, \Phi^\alpha = -ig\, \bar{\psi}\, \gamma_5\, \tau^\alpha\, \psi, \tag{45}$$

the axial coupling constant g_A is connected with the pion decay constant f_π by the Goldberger-Treiman relation,

$$m g_A = f_\pi g. \tag{46}$$

Suppose now that there is an additional induced pseudoscalar at the level of the current operator, Eq. (43):

$$\tilde{J}^\alpha_{5\mu} = g_A \bar{\psi}\, \gamma_\mu \gamma_5 \frac{\tau^\alpha}{2}\, \psi - \frac{g_P}{2m}\, i\, \partial_\mu \left(\bar{\psi} \gamma_5 \frac{\tau^\alpha}{2}\, \psi \right). \tag{47}$$

This additional term should be of non-pionic origin, e.g. due to other degrees of freedom of quarks and gluons.

Similar as in the case of the Pauli coupling for the electromagnetic current operator in Eq. (33), we have assumed that there is a derivative coupling at the operator level. While parts of the magnetic moments may be of pionic origin and have to be renormalised to appropriate order in a power series expansion in g, it seems obvious that the internal structure of the nucleon cannot be fully explained by pionic degrees of freedom.

The immediate consequence of the induced pseudoscalar in Eq. (47) is the appearance of a derivative under the time ordering operator in Eq. (42). Another partial integration leads to an additional commutator resembling very closely the structure of the sigma-term of Furlan et al. If we choose, e.g., $g_p \approx 1$, the Goldberger-Treiman relation is changed only at the percent level. This small change suffices, however, to cancel the terms of order m_π and to "obtain agreement" with the experimental data.

The current of Eq. (47) could be easily realised in an extended PV-model with an interaction Lagrangian

$$L^{int} = \frac{f}{m_\pi}\left(\bar{\psi}\,\gamma_\mu\gamma_5\,\vec{\tau}\,\psi - \frac{g_p}{2mg_A}\,i\,\partial_\mu\,(\bar{\psi}\,\gamma_5\,\vec{\tau}\,\psi)\right)\cdot\partial^\mu\vec{\Phi}. \tag{48}$$

It is obvious that minimal coupling of the electromagnetic current will lead to an additional contact term, as required by gauge invariance of the theory.

Another interesting aspect of the model is that it has no effect on the reaction $\gamma+n\to\pi^0+n$, quite contrary to the model of Furlan et al. (see tab. I). Therefore, the leading contribution for the neutron will still be due to its magnetic moment, which is certainly in accordance with the general spirit of LET.

6. SUMMARY AND OUTLOOK

Pion photo- and electroproduction at threshold should be described by the predictions of LET. These theorems are based on Lorentz invariance, current conservation and the partial conservation of the axial current (PCAC). The data for charged pion production agree with LET within the experimental (and theoretical) error bars. In order to test the isospin invariance of the 4 physical production channels, more precise data would be required, in particular for the neutron (deuteron target!).

It came as a large surprise when the new Saclay data and, more recently, the Mainz data for neutral pion production showed large deviations from LET. Various higher order corrections, e.g. isobar contributions, vector meson exchange and final state interactions only give rise to relatively small corrections of the order of 10 %.

The data may be described within a current algebra approach of Furlan et al., resulting in a pion photoproduction "sigma-term". However, there are several serious flaws in its derivation, e.g. the additional sigma-term is not gauge invariant.

A derivation of LET based on the LSZ reduction technique shows no justification for a sigma-term. However, a contribution of similar structure will arise if the axial current operator of the nucleon would contain an induced pseudoscalar of non-pionic origin. In that case there appears an additional commutator term giving rise to a modification of LET. Without changing the Goldberger-Treiman relation by more than a few percent, such an induced pseudoscalar could cancel the terms of order m_π in LET, leading to good agreement with the data.

Experimental data for the reaction $n(\gamma,\pi^0)n$ would be of extreme interest. The predictions for this process differ by up to one order of magnitude. Independent of any theoretical interpretation, such an experiment could test a possible breakdown of isospin invariance at threshold.

Future polarisation degrees of freedom will allow a cleaner separation of multipoles. In particular, the photon asymmetry could be used to determine the small "background" multipoles E_{1+} and M_{1-} whose magnitudes at threshold are somewhat disputed. Experiments with virtual photons, particularly the reaction $(e,e'\pi^0)$, will serve to measure the longitudinal multipoles, which are sensitive to the pion pole term and induced pseudoscalar contributions.

In conclusion the discrepancy between LET and the data for the reaction $\gamma p \rightarrow p\pi^0$ persists. The origin of this puzzle is still under discussion. Since this discrepancy could affect very basic notions about the structure of hadronic matter, such as the validity of the PCAC relation or the structure of the axial current, new and more precise data will be required. In particular the proposed theoretical models differ quite strongly in their predictions for the reaction $n(\gamma,\pi^0)n$ and $(e,e'\pi^0)$, and experiments with polarised photons or targets will allow a more careful measurement of the small background multipoles.

REFERENCES

1. N.M. Kroll and M.A. Ruderman, Phys. Rev. 93, 233 (1954).
2. G.F. Chew, M.L. Goldberger, F.E. Low and Y. Nambu, Phys. Rev. 106, 1345 (1957).
3. S. Fubini, G. Furlan and C. Rossetti, Nuovo Cim. 40, 1171 (1967).
4. F.A. Berends, A. Donnachie and D.L. Weaver, Nucl. Phys. B4, 1 (1967).
5. M. Aguilar-Benitez et al. (Particle Data Group), Phys. Lett. B204, 1 (1988).
6. E. Mazzucato et al., Phys. Rev. Lett. 57, 3144 (1986).
7. R. Beck, F. Kalleicher, B. Schoch, J. Vogt, G. Koch, H. Ströher, V. Metag, J.C. McGeorge, J.D. Kellie and S.J. Hall, to be published, and B. Schoch in Proc. Elba workshop (1987).
8. R. Beck in: Proceedings of the topical conference on Particle Production Near Threshold

9. G. Furlan, N. Paver and C. Verzegnassi, Nuovo Cim. 20, 118 and 295 (1974).
10. F.E. Low, Phys. Rev. 96, 1428 (1954).
11. M. Gell-Mann and M.L. Goldberger, Phys. Rev. 96, 1433 (1954).
12. S. Weinberg, Phys. Rev. Lett. 17, 616 (1966).
13. Y. Tomozawa, Nuovo Cim. 46A, 707 (1966).
14. P. de Baenst, Nucl. Phys. B24, 633 (1970).
15. D. Drechsel and L. Tiator, Phys. Lett. B148, 413 (1984).
16. M.I. Adamovitch, Proc. P.N. Lebedev Physics Institute 71, 119 (1976).
17. J. Spuller et al., Phys. Lett. B67, 479 (1977).
18. S. Adler in: Lectures on Elementary Particles and Quantum Field Theory, Cambridge 1970.
19. R. Jackiw in: B. de Witt and R. Stora (eds.), Relativity, Groups and Topology, Amsterdam 1984.
20. L.M. Nath and S.K. Singh, Phys. Rev. C39, 1207 (1989).
21. R.D. Peccei, Phys. Rev. 176, 1812 (1968).
22. G.L. Shaw and M.H. Ross, Phys. Rev. 126, 806 (1962).
23. J.M. Laget, Phys. Rep. 69, 1 (1981).
24. S.N. Yang, Phys. Rev. C40, 1810 (1989).
25. S. Nozawa, B. Blankleider and T.-S.H. Lee, Phys. Rev. C41, 213 (1990).
26. H. Ströher, to be published.
27. H.W.L. Naus, J.H. Koch and J.L. Friar, Phys. Rev. C41, 2852 (1990).
28. S.L. Adler and R.F. Dashen, Current algebras and applications to particle physics (Benjamin, New York and Amsterdam) (1968).
29. M. Gell-Mann, R.J. Oakes and B. Renner, Phys. Rev. 175, 2195 (1968).
30. J. Gasser and H. Leutwyler, Phys. Rep. 87, 1 (1982).
31. L. Tiator and D. Drechsel, Proc. 12th Int. Conf. on the Few Body Problem in Physics, Vancouver, Canada, 1989.
32. T. Schäfer and W. Weise, to be published in Phys. Lett.
33. H. Lehmann, K. Symanzik and W. Zimmermann, Nuovo Cim. 1, 205 (1955).
34. J.D. Bjorken and S.D. Drell, Relativistic Quantum Mechanics (McGraw-Hill, N.Y., 1964).
35. W.I. Weisberger in: M. Chrètien and S.S. Schweber (eds.), Elementary Particle Physics and Scattering Theory, vol. 1 (1967).

(γ, π) REACTIONS IN NUCLEI

R.C. Carrasco, E. Oset

Departamento de Física Teòrica and IFIC, Universidad de Valencia–CSIC, 46100 Burjassot (Valencia) Spain.

L.L. Salcedo,

Departamento de Fisica Moderna, Universidad de Granada, 18001 Granada, Spain.

In a recent paper we have developed a microscopic theory to describe the interaction of photons with nuclei, by looking in detail at the different excitation mechanisms of the nucleus[1]. The theory allows one to evaluate the total photonuclear cross section in any nucleus and it also allows to separate the contribution to the cross section from the (γ, π) events and from direct photon absorption. There is another source of photon absorption, which we call indirect, which comes from (γ, π) events followed by the absorption of the pions through the nucleus. This second part requires a detailed analysis of the final state pion–nuclear interaction . For this purpose, we follow a computer simulation procedure. The probability of producing a π^+, π^o, π^- in a (γ, π) step at any point of the nucleus and with a certain momentum is calculated in ref.[1] Hence pions are ramdomly generated but weighted by this probability. Afterwards, they are allowed to propagate in the nucleus. In this second step we use information of ref. [2],where the reaction probabilities for the pion, quasielastic, charge exchange and absorption were studied. In that work a simultaneous study of all these reactions, using as a theoretical input the reaction probabilities and a computer simulation procedure to follow all the ramifications of the pions, was done. The agreement of the theoretical results with experiment for different reactions and different nuclei was very good. We follow here the same steps and hence there are some of the pions which undergo quasielastic collisions or charge exchange or absorption. In this way, we know the amount of indirect absorption plus the cross sections $d^2\sigma/d\Omega\, dE$ for the π^+, π^o, π^-. This indirect absorption plus the direct one from ref. [1] is the total γ absorption cross section (no pions in the final state).

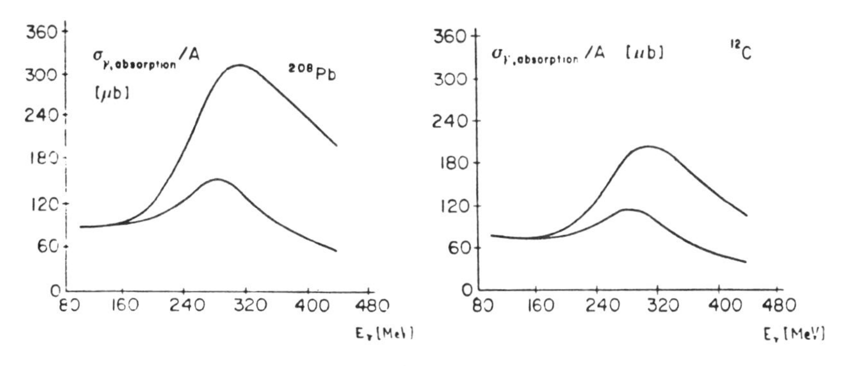

fig. 1.-

In fig. 1 we plot the amount of direct and total photon absorption per nucleon in ^{12}C and in ^{208}Pb. Of course, below the pion production threshold all the absorption is direct. The other interesting feature is that while the direct photon absorption per nucleon in ^{208}Pb is only 20% bigger than in ^{12}C, the indirect absorption per nucleon is much bigger , as one could easily gess, since the pions have to travel a larger distance through ^{208}Pb.

In fig. 2 we plot two selected differential cross sections for the (γ, π) channel. We have calculated cross sections for different nuclei, charge channels and kinematical regimes, and compared with the experiments of ref. [3)]. The agrement with the data is overall quite good, indicating that we have succeeded in developping a very accurate theoretical tool to describe the photonuclear reactions. It is our aim to proceed with the calculations and extend the present methods to the virtual photon regime, entering the realm of the rich physics in electron scattering.

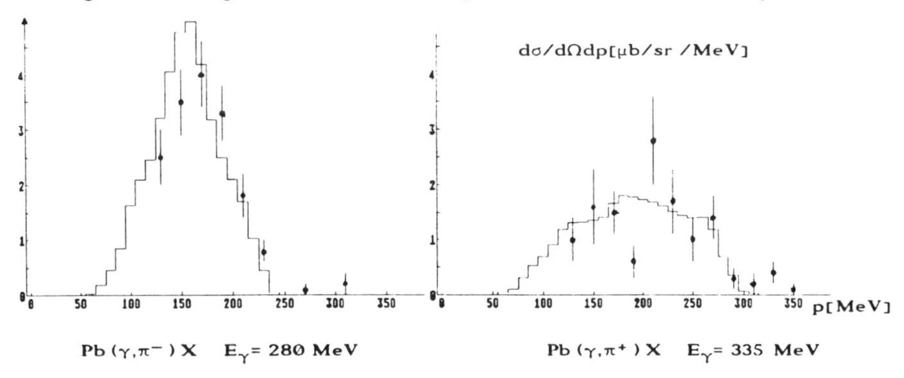

fig. 2.-

REFERENCES

1) R.C. Carrasco and E. Oset, University of Valencia preprint.
2) L.L. Salcedo, E. Oset, M.J. Vicente-Vacas and C. García-Recio, Nucl. Phys. A484 (1988) 557.
3) R. Stenz, PhD. Thesis, Bonn, 1986; J. Arends et al. Nuc.Phys. A , in print.

ETA PHOTOPRODUCTION FROM THE PROTON NEAR THRESHOLD

W.W. Daehnick

University of Pittsburgh, Pittsburgh, PA 15260

ABSTRACT

Differential cross sections are presented for eta production on the proton with photon beam energies of 725 and 750 MeV. Experiments of 20 years ago established the dominance of the S11(1535)N* resonance for higher photon energies. The purpose of this experiment was to investigate the coupling to other intermediate states in the H(γ,η) reaction near threshold.

EXPERIMENT

This is a preliminary report of a recently completed experiment at Bates, which aimed at a better determination of the H(γ,η) reaction near threshold. The data are the result of a collaboration of the University of Pittsburgh (S.A. Dytman, C.W. Alcorn, W.W. Daehnick, J.G.Hardie, and M.Yamazaki), of Boston University (E. Booth and J. Miller), Los Alamos National Laboratory (M.J. Leitch, S. Mishra, and J.C.Peng) and the MIT Bates Laboratory (D. Tieger and K.F. von Reden). The spokesman of the experiment was S. A. Dytman.

This experiment was the first measurement of η mesons at Bates. We produced η mesons with a highly collimated Bremsstrahlung beam of photons in a 10 cm long liquid hydrogen target. The etas were detected via the 39% decay branch to two photons. Coincident γ detection was accomplished by two large hodoscopes, each consisting of 24 Pb-glass scintillators typically arranged as shown in Fig. 1.

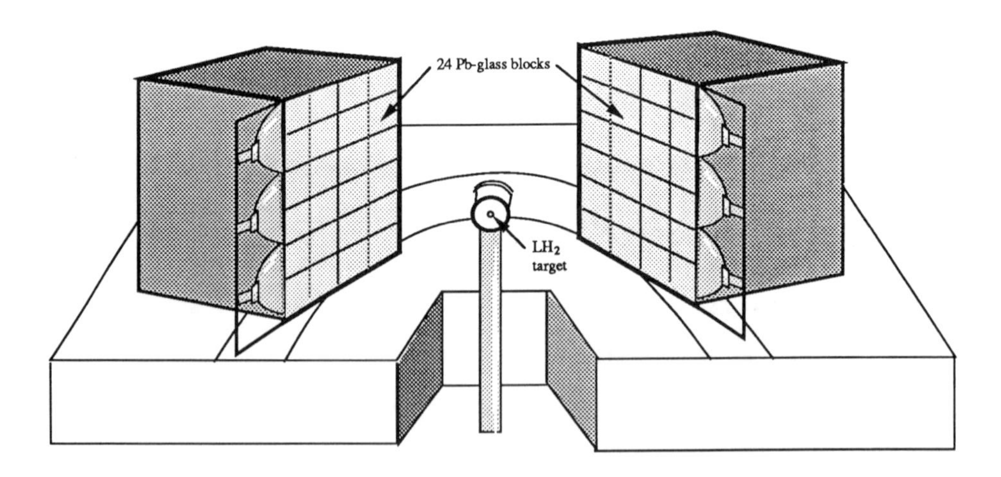

Fig. 1. View of (γ,η) experiment along photon beam line. Each tower holds 24 Pb-glass blocks, with block dimensions of 19 cm x 19 cm x 30 cm, in a 4 x 6 array. The "towers" were positioned at various angles around the target.

Energy calibrations were obtained by placing each hodoscope in a low flux electron beam. For normalization and testing of the Monte Carlo acceptance calculation we detected π^o mesons at $\theta_{CM} = 90°$ with an endpoint energy of 328.3 MeV. Since η production on the proton requires a minimum photon energy of 709.3 MeV, the beam end point energy used for the η runs was 768±10 MeV. It was measured using elastic electron scattering with the ELSSY spectrometer. With the hodoscopes 1.25 m from the target the acceptance for etas was broad, about 70 MeV in beam photon energy and 60° in θ_{CM}. Gammas from the η decay were identified through their kinematic and energy correlation. Fig. 2 shows preliminary center-of-mass angular distributions for beam photon energy slices centered at 725 and 750 MeV.

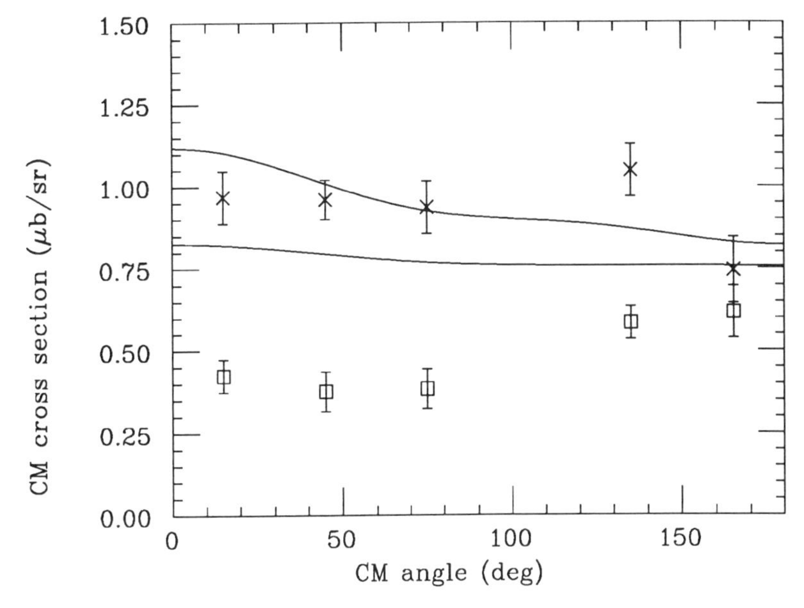

Fig. 2. Preliminary angular distributions for Bates experiment. The x symbol is used for 750 MeV beam energy and the box is used for 725 MeV.

The solid lines in Fig. 2 are calculations based on our fits to all previous $\gamma p \rightarrow \eta p$ data. The model includes s-channel resonances with fixed non-resonant Born amplitudes. This is an updated version of the Hicks analysis of 1973. (Both versions suffer from a very small data base and a lack of theoretical constraints.) We note that the cross-sections for 750 MeV agree reasonably well with earlier data (upper curve). 725 MeV cross sections have not been measured before, but the points are significantly below expectations from the previous fit. The 725 MeV points also exhibit a deviation from isotropy. Such anisotropy might be expected if in addition to the $S_{11}(1535)$ resonance a second resonance contributes. Possible candidates are the $P_{11}(1440)$ or $D_{13}(1520)$ resonances. The final report on this experiment will include some additional data, in particular a point at 105°.

SESSION C

Meson Production with Hadronic Beams

MEASUREMENTS OF $NN \rightarrow d\pi$ NEAR THRESHOLD

D.A. Hutcheon
TRIUMF, 4004 Wesbrook Mall, Vancouver, B.C., Canada, V6T 2A3

ABSTRACT

New, precise measurements of the differential cross sections for $np \rightarrow d\pi^0$ and $\pi^+ d \rightarrow pp$ and of analyzing powers for $pp \rightarrow d\pi^+$ have been made at energies within 10 MeV (c.m.) of threshold. They allow the pion s-wave and p-wave parts of the production strength to be distinguished unambiguously, yielding an s-wave strength at threshold which is significantly smaller than the previously accepted value. There is no evidence for Charge Independence Breaking nor for πNN resonances near threshold.

INTRODUCTION

Pions are, by far, the lightest of the strongly-interacting particles and consequently are of especial importance in studying the nuclear force. In models based on nucleons interacting through meson exchange, the long-range part of the internucleon force is due to one-pion exchange (OPE). In a quark picture the non-zero pion mass reflects the non-zero masses of the u and d quarks; in the limit of zero pion mass, hadronic systems of u and d quarks would exhibit Chiral Symmetry. This circumstance leads to a number of relationships between the pion decay constant, the pion mass, the πNN coupling constant, and the s-wave amplitudes for elastic scattering, charge exhange and radiative capture in πN systems. Measurements of pion production near threshold emphasize the s-wave part of πN interaction; the N-Δ intermediate states, which dominate pion production at higher energies, are suppressed.

Study of pion production in the 2-nucleon system has been heavily concentrated on the $pp \rightarrow d\pi^+$ reaction. This was due to the advantages of a 2-body final state: (1) there is a single independent kinematical variable (e.g. pion c.m. angle), so results can be presented in concise form, (2) backgrounds can be suppressed by measuring redundant kinematical variables (e.g. pion momentum, deuteron angle or momentum), and (3) a unique final-state NN configuration (the deuteron) limits the number of partial-wave amplitudes coming into play. For $pp \rightarrow d\pi^+$, all particles are charged and it is also possible to study the inverse reaction $\pi^+ d \rightarrow pp$.

THE PAST: EXPERIMENT AND THEORY

The Phenomenological Model

Dating from the early 1950's, the phenomenological model of Watson and Brueckner[1] sought to explain the energy dependence of $pp \rightarrow d\pi^+$ cross sections in terms of angular momentum barriers, assuming short-range nuclear forces. The model predicted

$$\sigma_{\text{tot}}(pp \rightarrow d\pi^+) = \alpha\eta + \beta\eta^3 \tag{1}$$

where η is the pion c.m. momentum in units of $m_\pi c$ and α,β are constants which give the strength for s-wave and p-wave pion production, respectively. A subsequent refinement of the model included factors for Coulomb repulsion between the π^+ and the d.

Considerable effort during the 1950's and early 1960's went into attempts to determine α because it could be related, through a chain of measured and calculated ratios, to $\pi N \rightarrow \pi N$ and $\gamma p \rightarrow \pi^+ n$ cross sections at low pion energy. This effort was handicapped by inconsistent measurements and lack of data close to threshold.

The $\pi^+ d \rightarrow pp$ Experiment of Rose

The first and, until very recently, only experiment to measure cross sections near threshold was that of Rose[2] in 1967. He studied the inverse reaction, $\pi^+ d \rightarrow pp$, by photographing tracks of low-energy pions stopping in a liquid deuterium bubble chamber. In all, over 2600 reactions took place within the fiducial volume at interaction energies from 2 to 16 MeV, as shown in Fig. 1. The measured absorption cross sections are shown in Fig. 2, along with the best fit to

$$\sigma_{tot}(\pi^+ d \rightarrow pp) = (2/3)(\mathrm{k}/\mathrm{m}_\pi c\eta)^2(\mathrm{C}_1^2\alpha\eta + \mathrm{C}_2^2\beta\eta^3) \qquad (2)$$

C_1 and C_2 being Coulomb barrier factors and k the c.m. momentum of the protons. Rose obtained $\alpha = 240\pm20$ μb and $\beta = 520\pm200\mu$b from the fit. At his lowest energy, the Coulomb factor was 0.78.

Theories of $pp \rightarrow d\pi^+$

Calculations of the amplitude for emission of a pion by one of the initial nucleons (Fig. 3a) came up with α much smaller than that measured (see, for example, Koltun and Reitan[3]). This was due to: (1) near cancellation of terms from S- and D-wave parts of the deuteron wave function, and (2) the "momentum mismatch" between the initial-state nucleon pair and the final-state pair. A much larger contribution came from diagrams in which the pion rescatters from the second nucleon (Fig. 3b,c). The relative contributions of the direct term, of πN P-wave rescattering, and of S-wave rescattering may be seen in Fig. 4, taken from work of Vogelzang, Bakker and Boersma[4].

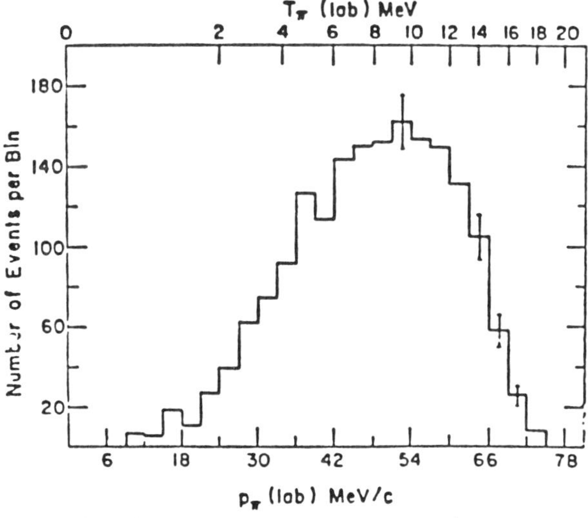

Fig. 1. Momentum distribution of pions undergoing a $\pi^+ d \rightarrow pp$ reaction in the experiment of Rose (ref.2).

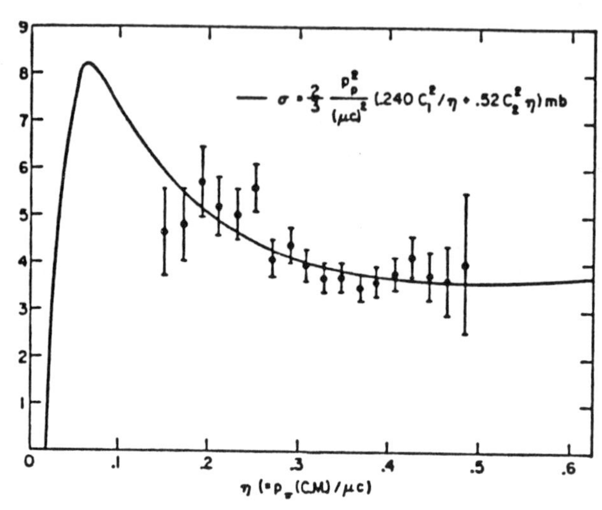

Fig. 2. Total absorption cross sections, $\sigma_{\mathrm{tot}}(\pi^+ d \rightarrow pp)$, measured in the work of ref.2. The line is a fit to equation 2.

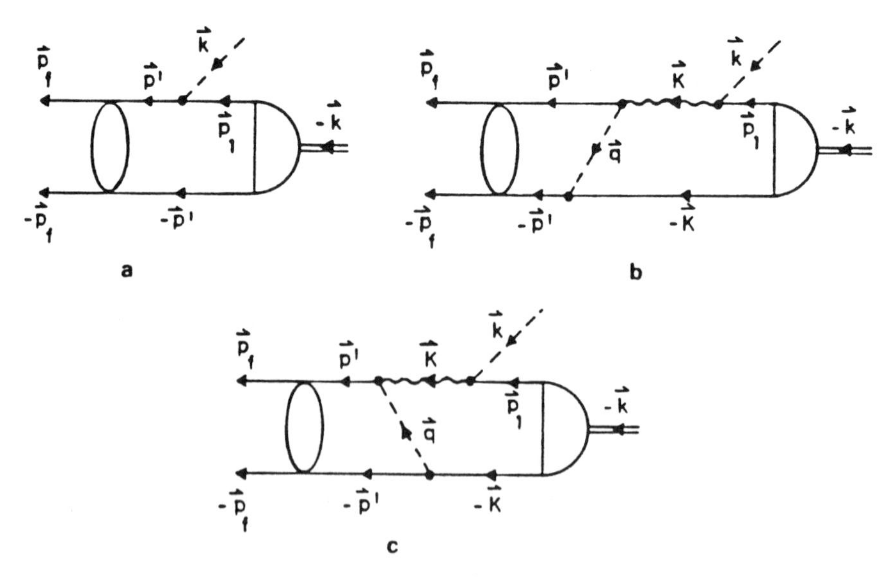

Fig. 3. Contributions to the $pp \rightarrow d\pi^+$ amplitude: (a)one-nucleon emission, (b)rescattering by the second nucleon, (c)rescattering of a "backward-going" pion.

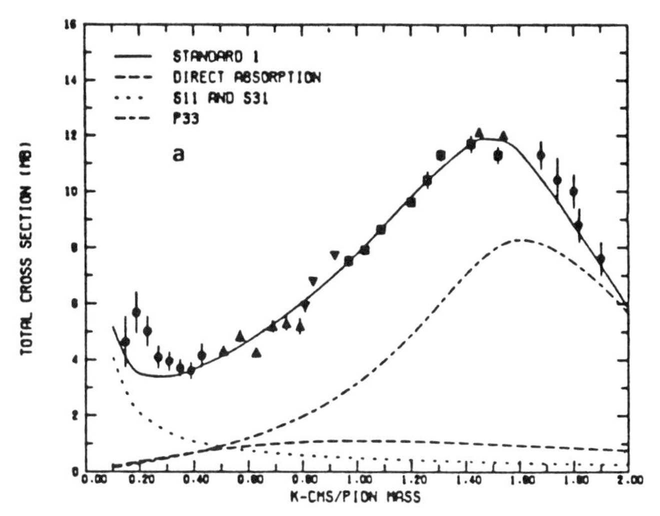

Fig. 4. Contributions of direct emission, S-wave pion rescattering, and P-wave rescattering to $\sigma_{\text{tot}}(\pi^+ d \to pp)$ as calculated in ref. 4.

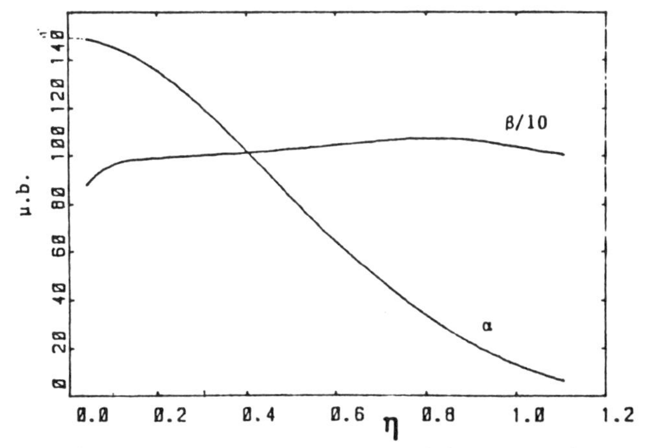

Fig. 5. The dependence upon pion momentum, η, of the $pp \to d\pi^+$ s-wave strength, $\alpha\eta$, and the p-wave strength, $\beta\eta^3$, as calculated in ref. 6.

A different approach, followed by groups at Flinders, Lyons, and Rehovot, sums up contributions from diagrams 3a and 3b plus higher terms in a rescattering series. Their Faddeev method treats NN→NN, $NN \to d\pi$, and πd→ πd consistently, respecting 3-body unitarity. It does not, however, include diagrams such as 3c having four particles in an intermediate state. The calculations of Afnan and Thomas[5] and of Blankleider[6] concluded that low-energy s-wave pion production could not be described by a constant value of α (Fig. 5).

This result from Faddeev calculations prompted Spuller and Measday[7] to re-fit the data of Rose and others, allowing $\alpha = \alpha_0 + \alpha_1\eta$ and adding a Breit- Wigner peak at the Δ resonance. The new functional form resulted in higher values of predicted threshold yield, for example their often-cited solution F having $\alpha_0 = 270 \pm 40 \mu$b.

RECENT MEASUREMENTS NEAR THRESHOLD

$np \to d\pi^0$ Cross Sections

An $np \to d\pi^0$ cross section should be the same as that for $pp \to d\pi^+$ within an iso-spin coupling factor of 1/2, after correcting the latter reaction for Coulomb barrier effects. A group based at the University of Alberta and TRIUMF has used $np \to d\pi^0$ to make precision measurements[8] much closer to $\eta = 0$ than Rose was able to reach by $\pi^+ d \to pp$. The great advantage of this reaction is that there is no Coulomb repulsion. (In fact, all the model calculations cited have ignored Coulomb effects!) Deuterons were detected in singles in a magnetic spectrometer, avoiding problems of π^0 detection. Deuterons were confined to a small cone of forward angles (Fig. 6), and the position of the p-θ locus permitted a precise determination of η without measuring the beam energy. Simultaneous acquisition of $np \to pn$ data allowed yields to be normalized to the well-known elastic scattering cross section.

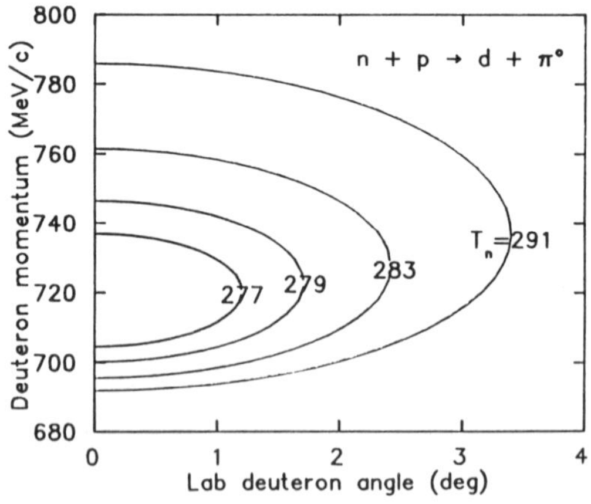

Fig. 6. Relationship of lab angle and momentum of deuterons produced by the $np \to d\pi^0$ reaction near threshold.

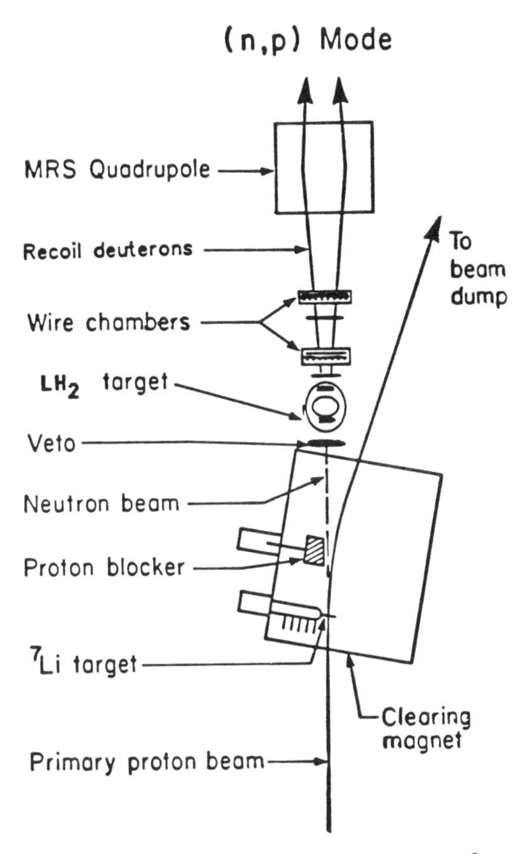

Fig. 7. Schematic view of the layout of the TRIUMF $np \to d\pi^0$ experiment (ref. 8).

The TRIUMF Chargex facility[9] produced a secondary neutron beam of about 0.75 MeV FWHM via the ^{7}Li(p,n) reaction, the 0° beam of neutrons struck a liquid hydrogen target, and recoiling deuterons were detected in the Medium Resolution Spectrometer (MRS), as shown schematically in Fig. 7. For a primary target of Pb, a measurable fraction of the primary beam left the target as neutral hydrogen. The H°'s formed a pencil beam which defined true zero degrees through the LH_2 target and the MRS. The facility gave a very clean locus of deuterons (Fig. 8), with negligible background even for cross sections as low as 1μb.

At each beam energy the deuteron data were binned in a p-θ array (after correction for MRS acceptance) and fitted to a calculated yield. Proton data were fitted at the same time to provide the normalization factor. The calculated yield was obtained from the following ingredients:

1. the neutron beam energy distribution, based on known bound states of ^{7}Be plus a continuum inferred from proton spectra,

2. a c.m. differential cross section

$$4\pi \cdot d\sigma/d\Omega^* = [\sigma(T_0) + (d\sigma/dT_n)(T_n - T_0) + \cdots] \times [1 + \{(A_2/A_0) + (d(A_2/A_0)/dT_n)(T_n - T_0)\} \cdot P_2(\theta)] \quad (3)$$

3. transformation from c.m. to lab variables,

4. smearing of the deuteron locus due to multiple scattering and differential energy loss in the LH_2 target.

Parameters of the fit included the nominal energy T_0, the energy spread of the proton beam, σ, $d\sigma/dT_n$, $d^2\sigma/dT_n^2$, A_2/A_0, and $d(A_2/A_0)/dT_n$. Good values of χ^2 were obtained; some of the distributions projected in p or θ are shown in Fig. 9. As a by-product, the experiment gave an improved calibration of beam energy *vs* cyclotron extraction foil radius (Fig. 10).

$\pi^+ d \to pp$ Cross Sections

Less than a month old are results from measuring $\pi^+ d \to pp$, obtained at LAMPF by an Arizona State University/University of Virginia/ University of South Carolina

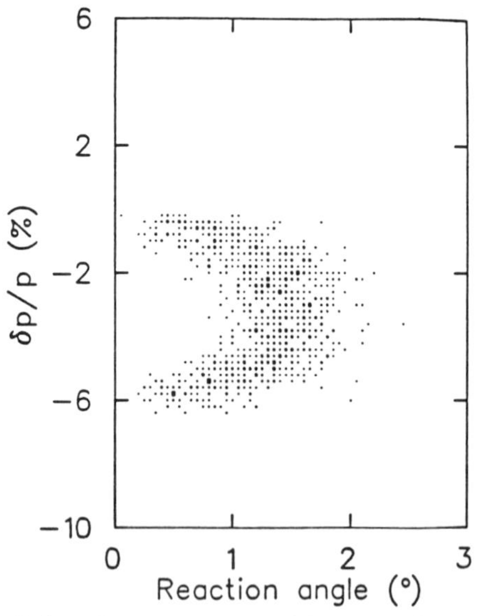

Fig. 8. Correlation of lab angle *vs* momentum of deuterons from $np \to d\pi^0$ at nominal beam energy 278 MeV (ref. 8).

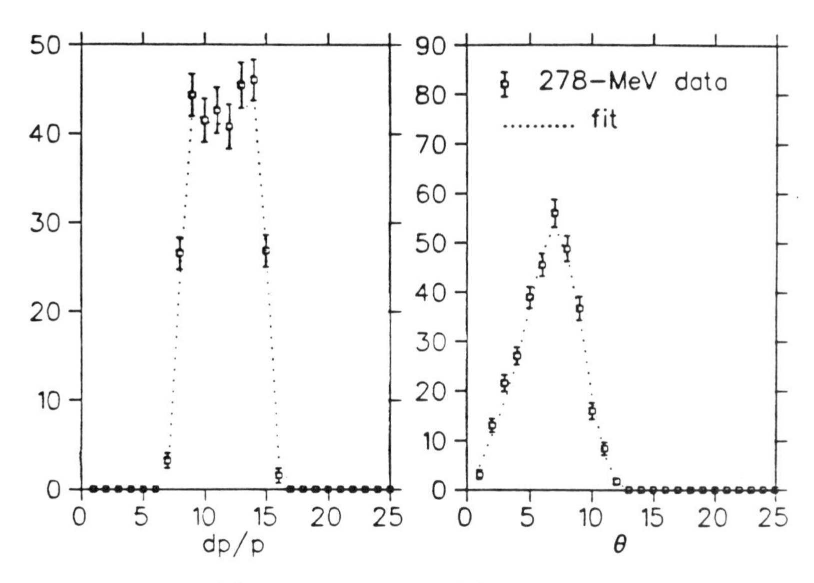

Fig. 9. Distributions in (a)deuteron angle, and (b)deuteron momentum for $np \rightarrow d\pi^0$ (ref. 8) at nominal beam energy 278 MeV. The dotted lines are the result of a fitting procedure described in the text.

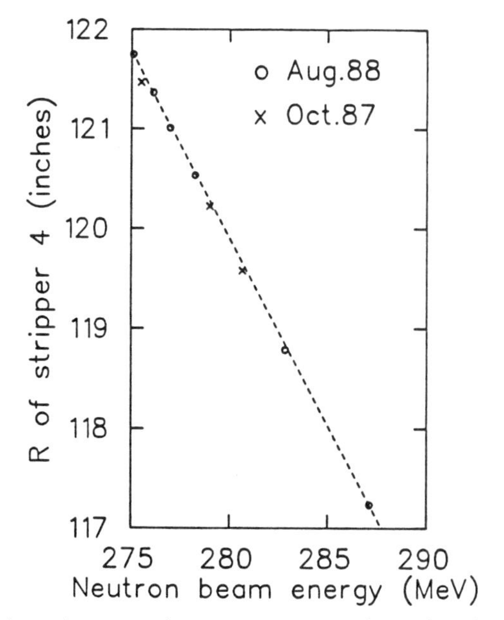

Fig. 10. Fitted value of neutron beam energy *vs* radius of cyclotron extraction foil in the TRIUMF $np \rightarrow d\pi^0$ experiment (ref. 8).

collaboration[10]. A pion beam from the Low Energy Pion Channel was incident upon an active target of deuterated scintillator. The purity of the pion beam was enhanced by use of a crossed E and B field velocity filter. Protons from the reaction were detected in coincidence in E-ΔE scintillator telescopes. The setup is shown schematically in the inset to Fig. 13.

Differential cross sections were obtained for three angles at pion energies 3.7, 5.0, 9.6, 15.2, and 20.5 MeV. The two lowest energies were achieved by placing degraders of graphite and aluminum directly in front of the target. Careful monitoring of the muon contamination in the beam (up to 50%) and pions stopped in the target (about 6% at 3.7 MeV) was necessary.

Analyzing Power in $pp \rightarrow d\pi^+$

The University of Alberta/TRIUMF group have also measured the vector analyzing power, A_y, for polarized beams 3 and 8 MeV above threshold. As in the $np \rightarrow d\pi^0$ experiment, deuterons were detected in singles by the MRS spectrometer. Since the beam passed through the front-end scintillators, its intensity was limited to about 10^6/sec by 2-stage collimation (Fig. 11). A miniature beam dump, 1 cm in diameter, stopped most of the primary beam in front of the spectrometer; it limited the acceptance for recoil deuterons to approximately 20° – 160° c.m.

Analyzing powers were extracted by comparing the beam spin-up *vs* the spin-down yields as a function of deuteron azimuthal angle, ϕ. It was not necessary (nor was it possible) to make an accurate determination of absolute cross sections.

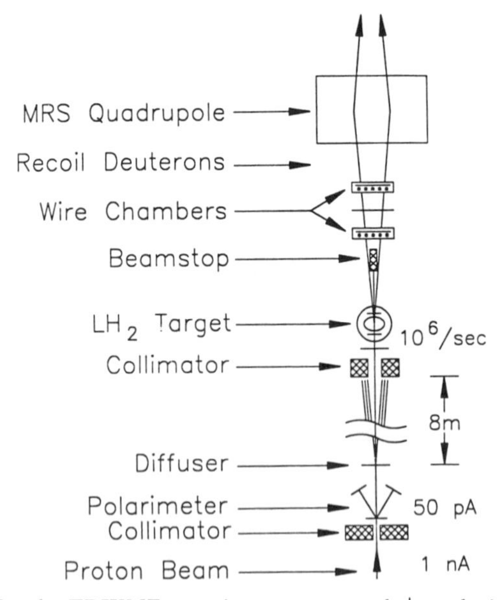

Fig. 11. Layout for the TRIUMF experiment on $pp \rightarrow d\pi^+$ analyzing powers.

RESULTS

A New Value for α at Threshold

The total cross sections measured in the Alberta/TRIUMF $np \to d\pi^0$ experiment are shown in Fig. 12. They are fitted very well by

$$\sigma_{tot}(np \to d\pi^o = (1/2)[(184 \pm 5)\eta \ + \ (781 \pm 79)\eta^3]\mu b \ . \quad (4)$$

The value of α is significantly lower than the 240±20 μb which Rose[2] obtained from his data and the 270±40 μb of Spuller and Measday[7] solution F. Rose's data, transformed assuming Detailed Balance and Charge Independence, and with the Coulomb factors of Reitan[11], are also plotted on Fig. 12. They agree with the $np \to d\pi^0$ data at high η, but lie above them near threshold. In contrast, the LAMPF data agree with the $np \to d\pi^0$ cross sections in the region of overlap (Fig. 13) and also match up well with data taken earlier by that group at higher energies. Figure 13 also illustrates the level of uncertainty in calculations: the VBB "Standard-1" and "Standard-3" calculations differ mainly in that the latter includes the contribution of backward-going pions.

The recent cross section data do not suggest any πNN resonance near threshold.

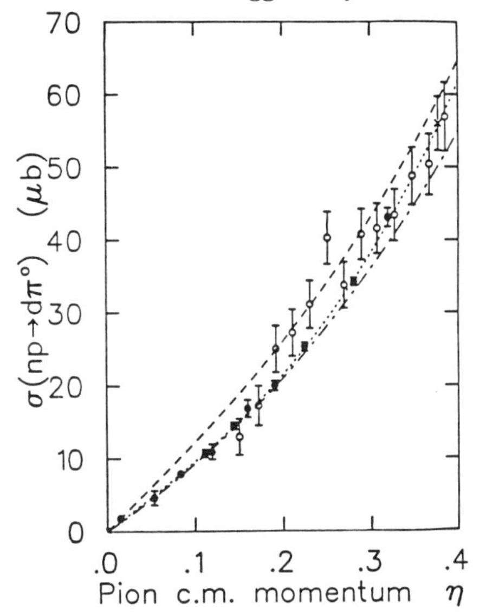

Fig. 12. Total cross sections $\sigma_{tot}(np \to d\pi^0)$ as a function of pion c.m. momentum, η. The solid points are from ref. 8 and the open circles are the $\pi^+ d \to pp$ data of ref. 2, converted by Detailed Balance. The dotted line is equation 1 fitted to the data of ref. 8. The dash-dot curve is the "Standard-3" calculation of ref. 4.

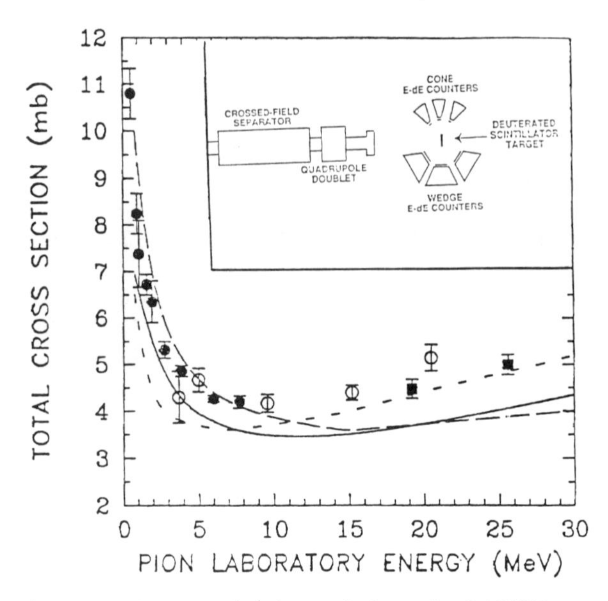

Fig. 13. Total cross sections, $\sigma_{tot}(\pi^+d \to pp)$, from the LAMPF experiment (ref. 10) adjusted by the Coulomb barrier factor of Reitan[11] (open circles). The inset is a schematic view of the apparatus. Earlier measurements by the same group[17] (squares) and results of ref.8 converted by Detailed Balance (solid circles) are also presented. The solid line is the calculation of ref. 6, and the long (short) dashed lines are the Standard-1 (Standard-3) results of ref.4.

It is interesting to compare the invariant matrix elements (IME's) for $np \to d\pi^0$ with those of pd$\to{}^3$Heπ^0. The IME's are defined by

$$| \mathrm{M} |^2 = (8\pi/\hbar c)^2 \mathrm{g} \cdot s(\mathrm{p}_i/\mathrm{p}_f) \cdot \mathrm{d}\sigma/\mathrm{d}\Omega \qquad (5)$$

where g is a spin-multiplicity factor, s is the Mandelstam variable, and $\mathrm{p}_i,\mathrm{p}_f$ the initial and final c.m. momenta. In the limit $\eta \to 0$, $|\mathrm{M}(\mathrm{pd}\to{}^3\mathrm{He}\pi^0)|^2 \;/\; |\mathrm{M}(np \to d\pi^0)|^2 = 1.1$, based on a near-threshold yield of $\sigma_{Tot} = 12.9\ \eta\ \mu$b for the pd$\to{}^3He\pi^0$ reaction[12].

The Pion p-wave Amplitudes

If only s-wave and p-wave pions are produced, the $NN \to d\pi$ reaction is completely described by just three partial-wave amplitudes characterized by the NN orbital and total angular momenta (L,J) and the pion angular momentum (l). They are listed in Table 1, along with the lowest permitted $\mathcal{L}$ of an intermediate N-Δ pair. The amplitude a_2, which allows $\mathcal{L}$ =0, becomes the dominant amplitude at slightly higher energies. In contrast, the other p-wave amplitude, a_0, requires $\mathcal{L}$=2 and probably is

Table 1. Partial wave amplitudes for s-wave and p-wave π production in $pp \to d\pi^+$.

Amplitude	$(^{2S+1}L_J)pp$	$(\ell)_{\pi d}$	$(\mathcal{L})_{N\Delta}$
a_0	$^3S_0$	1	2
a_1	$^3P_1$	0	1
a_2	$^1D_2$	1	0

due to πN P_{11} rescattering. For the moment, assume a_0 is negligible compared to a_2. Then

$$4\pi \cdot (d\sigma/d\Omega) = (1/4)[|a_1|^2 + |a_2|^2] + (1/4)|a_2|^2 \cdot P_2(\theta) . \tag{6}$$

That is, the ratio of coefficients of Legendre polynomials, A_2/A_0, gives the fraction of production strength which is p-wave. Figure 14 shows A_2/A_0 data from the TRIUMF experiment[8], the LAMPF experiment[10], and a $np \to d\pi^0$ experiment of the Freiburg group[13]. The different data sets are in good accord. Both the "Standard-3" perturbative calculation of Vogelzang, Bakker, and Boersma[4] (dash-dot curve) and the Faddeev calculation of Blankleider[6] (solid curve) reproduce the data. The Watson-Brueckner phenomenological model[1] (dotted curve) fails, however.

If the amplitude a_0 is present, the cross section becomes

$$\begin{aligned} 4\pi \cdot (d\sigma/d\Omega) &= (1/4)[|a_0|^2 + |a_1|^2 + |a_2|^2] \\ &\quad + (1/4)[|a_2|^2 - 2\sqrt{2}\mathrm{Re}(a_0 a_2^*)] \cdot P_2(\theta) \end{aligned} \tag{7}$$

and the analyzing power is

$$4\pi \cdot (d\sigma/d\Omega)A_y = (1/4)[-\sqrt{2}\mathrm{Im}(a_0 a_1^*) + \mathrm{Im}(a_1 a_2^*)] \cdot P_1^1(\theta) \tag{8}$$

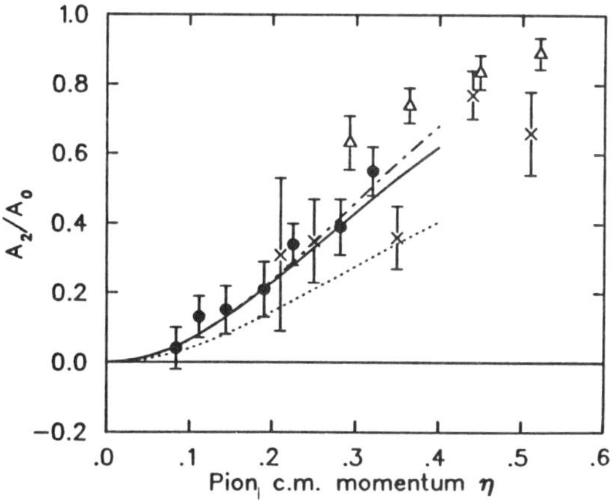

Fig. 14. Ratio of Legendre polynomial coefficients A_2/A_0 *vs* pion c.m. momentum for $np \to d\pi^0$ experiments at TRIUMF (solid dots) and Freiberg (triangles), and for $\pi^+ d \to pp$ at LAMPF (crosses). The curves are model predictions described in the text.

There is evidence[14] that the phases of the partial-wave amplitudes near threshold are given by Watson's Theorem: $\delta(NN \to d\pi) = \delta(\mathrm{NN}{\to}\mathrm{NN}) + \delta(\pi\mathrm{d}{\to}\pi\mathrm{d})$. With this assumption, one can predict A_y as a function of $| a_0 |$. The A_y data from 3 and 8 MeV above threshold beam energy are presented in Figs. 15 and 16; the dashed lines are the 1-σ error band from a 1-term Legendre polynomial fit. A simultaneous fit of these data plus the $np \to d\pi^0$ cross sections at the same η gave a slightly negative best-fit value for $| a_0 |$, suggesting the phases may not be exactly as predicted by Watson's Theorem. However, if a_0 is taken to be zero, the A_y (vertical line at 90° in Figs. 15 and 16) falls only slightly below the data. If $| a_0 |$ is set to 10% of $| a_2 |$, the Watson's Theorem phases give $A_y(90°) = -0.28$ at 290.7 MeV; in this case the

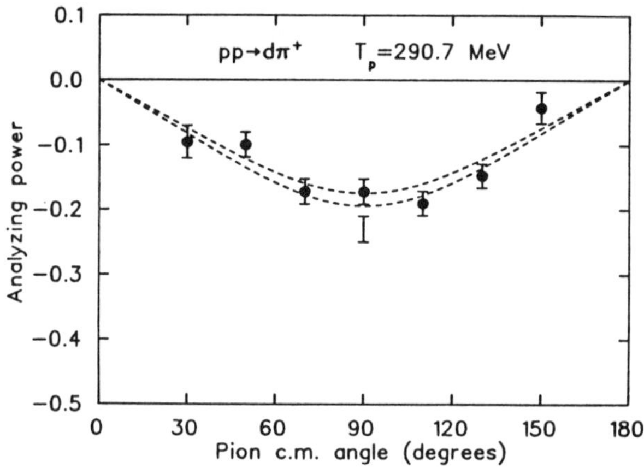

Fig. 15. Polarized beam analyzing powers measured in $pp \to d\pi^+$ at 290.7 MeV ($\eta = 0.14$). The error band for a 1-term Legendre polynomial fit is shown as dashed lines.

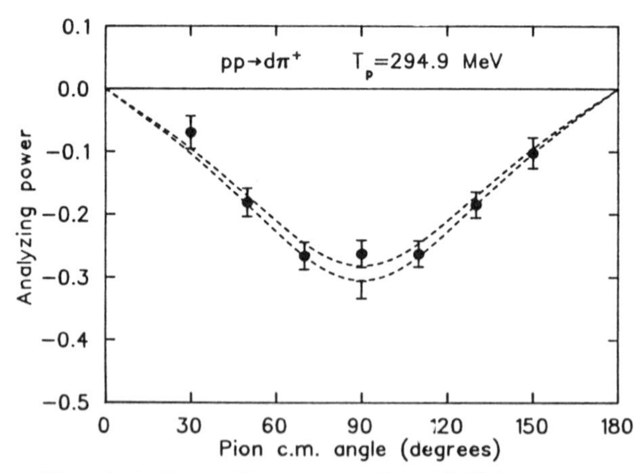

Fig. 16. As figure 15, except at 294.9 MeV ($\eta = 0.21$).

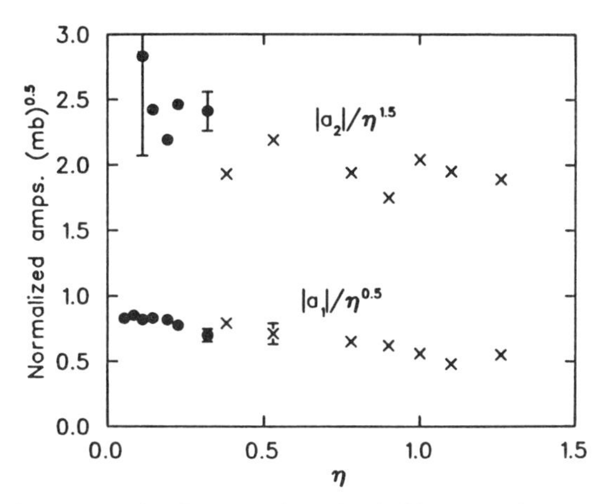

Fig. 17. Partial-wave amplitudes a_1 and a_2 divided by centrifugal barrier factors *vs* pion c.m. momentum. The crosses are amplitudes from a partial-wave fit by Bugg[14], converted to the amplitude normalization of Blankleider[6].

best-fit $|a_2|$ increases by 10% with respect to its value with vanishing a_0. At 400 MeV ($\eta = 0.9$), Bugg *et al.*[15] found $|a_0| / |a_2| = 0.02$.

We are now in a position to display the energy dependence of the amplitudes a_1 and a_2, with the caveat that there is an uncertainty in a_2, for the reason just discussed. In order to remove the centrifugal barrier effects, we plot $a_1/\eta^{1/2}$ and $a_2/\eta^{3/2}$. As well as the new near-threshold results, the amplitudes from the partial-wave fit of Bugg[14,15] (converted to Blankleider's normalization) are shown in Fig. 17. There is very modest variation with energy from threshold to 500 MeV, after removal of the centrifugal barrier terms. Bugg[14] has attributed the decrease in normalized a_1 to energy dependence of the πN S_{31} and S_{11} rescattering.

Charge Symmetry and Charge Independence

An asymmetry in $np \rightarrow d\pi^0$cross section at c.m. angle θ compared to that at angle $(\pi - \theta)$, i.e. the presence of odd terms in a Legendre polynomial expansion, requires Charge Symmetry Breaking. A term $A_1P_1(\theta)$ was included in the fit to $np \rightarrow d\pi^0$ data at 277 MeV, where the full range of deuteron momenta was comfortably within the acceptance of the spectrometer. The result was A_1/A_0 = -0.01±0.07, the large uncertainty reflecting the statistical error on approximately 1000 events. Since, in any case, the cross section at 277 MeV is dominated by s-wave production, a large CSB effect is unlikely at this energy.

The Nijmegen group have analyzed low-energy nucleon-nucleon elastic scattering data and concluded[16] that the πNN coupling constant for π^0 is about 8% lower than that for charged pions. The ${}^3P_J$ amplitudes showed the greatest sensitivity to this

Charge Independence Breaking, and we may ask whether near-threshold $NN \to d\pi$, where the dominant amplitude is $^3P_1 \to s$, is affected. Charge Independence requires that $2\ \sigma_{tot}(np \to d\pi^0) = \sigma_{tot}(pp \to d\pi^+)$, after correction for "purely electromagnetic" effects, such as Coulomb repulsion. The proper incorporation of the long-range Coulomb interaction in the three-hadron system is a very hard problem, and it is by no means clear that simple Coulomb barrier corrections are sufficient at energies very close to threshold. Another question is how to account for the mass difference of π^0 and π^+ and of NN energy at threshold; we have compared cross sections at equal values of pion c.m. velocity, but perhaps some other variable is more appropriate. If the LAMPF $\pi^+ d \to pp$ data are corrected using the Reitan[11] Coulomb barrier expression, the lowest-energy points agree with the $np \to d\pi^0$result (within statistical uncertainties of 15% and 7%), but the data of Rose[2] are of order 20% higher. In summary, there is no compelling evidence for CIB but the uncertainties are rather large, both for theory and experiment.

FUTURE DIRECTIONS

Partial-wave Amplitudes

We now have a good determination of the s-wave and principal p-wave amplitudes in the threshold region. There is a slow variation with energy after centrifugal barrier factors are applied, with no evidence for πNN resonances. Watson's Theorem appears to provide a good, though perhaps not perfect, guide to the phases. However, there is only a fuzzy upper limit for the size of the lesser p-wave amplitude, a_0. Is there some other spin observable which can isolate this amplitude? The best candidate is the deuteron vector polarization, it_{11}, which depends upon interference between the p-wave amplitudes, $\text{Im}(a_0 a_2^*)$. The bad news is that interference terms between the s-wave amplitude a_1 and the d-wave amplitudes may be larger — p-p interference should have the same η dependence as s-d interference according to centrifugal barrier arguments. There appears to be no prospect for model-independent determination of this amplitude by near-threshold measurements.

Symmetries

The TRIUMF result for front-back asymmetry in $np \to d\pi^0$ cross sections came as a by-product, and would better have been done at a higher beam energy and with higher statistics. The SASP spectrometer, now under construction at TRIUMF, will offer larger acceptance than the MRS, plus a 25% momentum bite, and would be an attractive instrument for an experiment dedicated to looking for Charge Symmetry Breaking. Calculations of CSB have concentrated on higher energies, but there is no *a priori* reason effects can't be seen close to threshold, provided the p-wave production amplitude is of reasonable size.

A better measurement of the $pp \rightarrow d\pi^+$ total cross section would permit a more stringent test of the Charge Independence relation between that reaction and $np \rightarrow d\pi^0$. Again, the SASP spectrometer offers prospects of improving on the previous TRIUMF results by an experiment dedicated to precision comparison of the cross sections. Another possibility is to take advantage of the thin targets of a proton cooler ring and detect the pions, rather than the deuterons, from $pp \rightarrow d\pi^+$.

Theory

Many calculations of $pp \rightarrow d\pi^+$observables around resonance energy (580 MeV) have been made. Perturbative, coupled channels, and Faddeev models have had a fair degree of success, although none has given detailed agreement with all observables. Underestimate of spin-triplet strength has been identified as a common failure mode of the models. The new data near threshold, with both triplet and singlet strengths determined, provide a new playing field for testing models. The results are easily summarized: there are only two important partial waves, whose magnitudes vary slowly apart from centrifugal barrier effects, and whose phases appear to be near the predictions of Watson's Theorem. In this simple world it should be possible for theorists to focus their attention on the essentials of the reaction. One feature of the new playing field is that one is well removed from the peak of the Δ resonance, and the contribution of S-wave πN rescattering should be more easily seen.

The omission of backward-going pions from Faddeev models and the truncation of the multiple-scattering series in perturbative models both appear to be significant shortcomings. These and other uncertainties in the models are much greater than the errors in near-threshold data. For now, the greatest opportunities for progress in understanding near-threshold $NN \rightarrow d\pi$ appear to lie with the theorists.

ACKNOWLEDGEMENTS

It is a pleasure to acknowledge many helpful discusions with my colleagues at TRIUMF and the University of Alberta, and to thank Dr. Barry Ritchie for communicating the LAMPF results prior to publication.

REFERENCES

1. K. Watson and K. Brueckner, Phys. Rev. **83**, 1 (1951).
2. C.M. Rose, Phys. Rev. **154**, 1305 (1967).
3. D.S. Koltun and A.R. Reitan, Phys. Rev. **141**, 1413 (1966).
4. J. Vogelzang, B.L.G. Bakker and H.J. Boersma, Nucl. Phys. **A452**, 644 (1986).
5. I.R. Afnan and A.W. Thomas, Phys. Rev. C **10**, 109 (1974).
6. B. Blankleider, Ph.D. thesis, Flinders University, 1980.
7. J. Spuller and D.F. Measday, Phys. Rev. D **12**, 3550 (1975).

8. D.A. Hutcheon *et al.*, Phys. Rev. Lett. **64**, 176 (1990).
9. R. Helmer, Can. J. Phys., **65**, 588 (1987).
10. B.G. Ritchie *et al.*, preprint 1990, submitted to Phys.Rev.Lett.
11. A. Reitan, Nucl. Phys., **B11**, 170 (1969).
12. M. Pickar, A.I.P. Conf. Proc. **79**, 143 (1981).
13. G. Jones, A.I.P. Conf. Proc. **79**, 15 (1981).
14. D.V. Bugg, J. Phys. G **10**, 47 (1984).
15. D.V. Bugg, A. Hasan and R.L. Shypit, Nucl. Phys. **A477**, 546 (1988).
16. J.R. Bergenvoet *et al.*, Phys. Rev. Lett. **59**, 2255 (1987).
17. B.G. Ritchie *et al.*, Phys. Rev. C **24**, 552 (1981).

TOTAL CROSS SECTION FOR $p+p \rightarrow p+p+\pi^0$ CLOSE TO THRESHOLD

H.O. Meyer

Department of Physics, Indiana University and
Indiana University Cyclotron Facility, Bloomington, IN 47405

ABSTRACT

The total cross section for the reaction $p+p \rightarrow p+p+\pi^0$ was measured at nine center-of-mass energies from 1.5 MeV to 23 MeV above threshold. The experiment was carried out with the Indiana Cooler, making use of the experimental advantages of an stored, electron-cooled beam. Over most of the covered energy range only the lowest possible angular momentum state contributes to the exit channel. The energy dependence of the total cross section close to threshold is discussed.

1. CHARACTERISTICS OF $pp \rightarrow pp\pi^0$ NEAR THRESHOLD

The reaction $NN \rightarrow NN\pi$ represents the dominant inelasticity in NN collisions and is therefore fundamental to an understanding of the nucleon-nucleon (NN) interaction. Close to threshold, as we shall see, the description of this process is simplified considerably, yet theoretical studies of the near-threshold pion production in the NN system have stagnated 25 years ago, presumably due to the lack of relevant experimental information.

Customarily, the partial waves in the exit channel of $NN \rightarrow NN\pi$ reactions are labelled[1)] by $L\ell$, where L is the angular momentum of the nucleon pair and ℓ is the angular momentum of the pion with respect to the nucleon pair. Within 100 MeV (cms) of threshold, because of the short range of the strong interaction, only Ss, Sp, Ps, and Pp final states contribute significantly; this prediction is strengthened for the reaction $pp \to pp\pi^0$ by the fact that the Sp final state is forbidden in reactions with isospin T=1 in initial and final NN states.

In the $pp \to pp\pi^0$ reaction, the normally dominant pion production via an intermediate ΔN state, containing an excited nucleon, is suppressed by the fact that the formation of the N and the Δ in a relative S state is forbidden. The rescattering contribution is also small since the dominant isovector component of low-energy πN scattering cannot contribute. Thus, near threshold, the $pp \to pp\pi^0$ reaction is expected to be strongly dominated by the direct-production Born term.

The energy dependence of the cross section is customarily expressed in terms of η, the largest possible center-of-mass pion momentum (with nucleons at rest relative to each other) divided by the pion mass. Based on phase space arguments, the assumption of $\eta << 1$, and a simple treatment of the final state interaction, it has been predicted[2,3] that the total cross section close to threshold can be written as

$$\sigma_{tot} = B_0 \eta^2 + B_1 \eta^6 + B_2 \eta^8 \tag{1}$$

where the three terms correspond to the Ss, Ps, and Pp exit channels, respectively. The expansion, Eq.1, usually serves to decompose the total cross section into different angular momentum states. However, as shall be demonstrated in Sect.4, some of the assumptions made in treating the final-state interaction between the nucleons in the exit channel are not valid close to threshold. Nevertheless, it is obvious that sufficiently close to threshold, only the exit channel of type Ss contributes to $pp \to pp\pi^0$. In this case, the two final nucleons are in a $^1S_0$ state, and the entrance channel is limited to a single partial wave ($^3P_0$). Based on data at higher energies, where Ps and Pp states do contribute, the dominance of the Ss configuration is expected within about 20 MeV (cms) of threshold.

2. THE EXPERIMENT

2.1. Introduction

Intermediate-energy light ion storage rings, equipped with an intense, co-moving electron beam for phase space cooling of the stored proton beam[4], are new tools in the hands of the nuclear physicist and a number of facilities are presently commissioning their machines for nuclear research with stored beams and internal targets. The experiment described here has been carried out with the Indiana Cooler[5] and represents the first use of such a machine in nuclear physics. The participants include A. Berdoz, F. Dohrmann, J.E. Goodwin, M.G. Minty, H. Nann, P.V. Pancella, S.F. Pate, R.E. Pollock, B. v. Przewoski, T. Rinckel, M.A. Ross, F. Sperisen, and the author, all from IUCF.

We have measured the total cross section for the process $pp \rightarrow pp\pi^0$ at a number of bombarding energies close to threshold. Previous experiments were hampered by the fact that the cross sections are small (because resonant production is suppressed in $pp \rightarrow pp\pi^0$) and by the fact that the process of interest is in competition with background pion production from nuclei other than hydrogen in the beam (e.g., the windows of target cells). The experiment described here uses a windowless internal gas target, and is the first to cover an energy range low enough that one might expect only the Ss final state to contribute. Due to the properties of the novel experimental environment, it also becomes feasible to measure the four-momenta of both outgoing protons, making it possible to completely reconstruct the three-body final state.

2.2. The Experimental Setup

Of the six straight sections of the storage ring, three are for injection, electron cooling and beam manipulation, while the others are provided for experiments. Fig.1 shows the setup for this experiment which has been mounted in the center of the

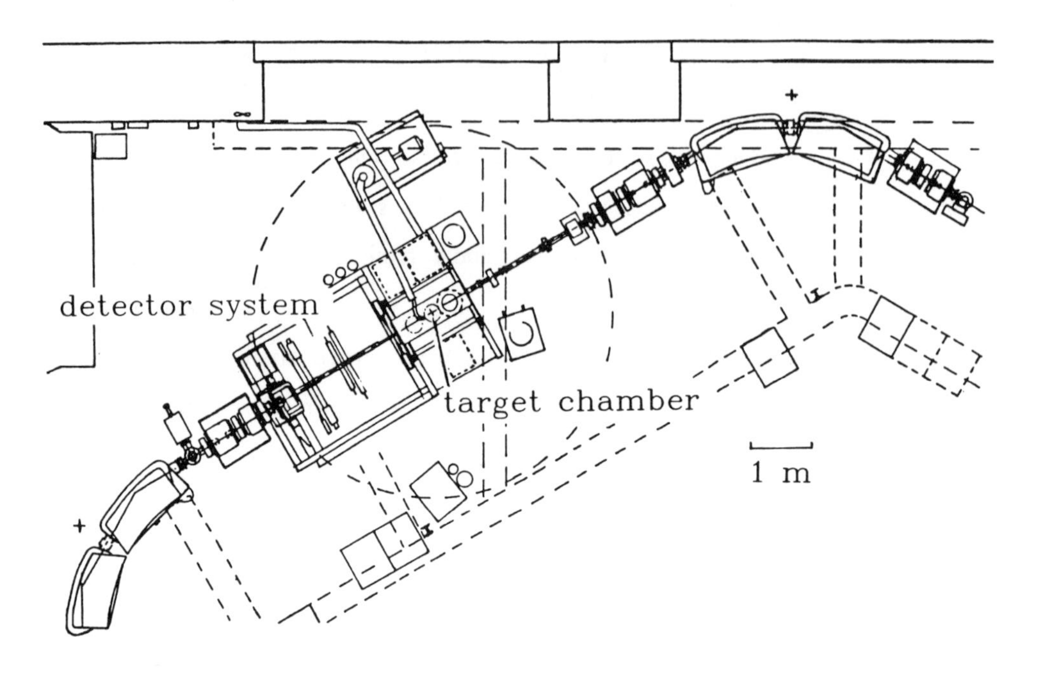

Fig.1. View of the experimental straight section preceding the electron cooler.

section preceding the electron cooling device. Aperture functions at the target are β_x=0.51 m, β_y=1.03 m, and the dispersion is 4.1 m.

A close-up of the target chamber which houses an internal hydrogen target is shown in Fig.2. A gas jet, emerging from a 0.11 mm diameter nozzle (N), cooled to 40 K, crosses the stored beam. Half of the gas flow (typically $6 \cdot 10^{19}$ molecules per s) is pumped away through a catcher opposite the jet (C). The remainder of the gas is removed in a three-stage differential pumping arrangement (1-3 in Fig.2). This results in a hydrogen target of $2 \cdot 10^{15}$ to $5 \cdot 10^{15}$ atoms/cm^2 total thickness. About 80% of the target thickness is concentrated within ± 1 cm of the jet; the jet itself, measured at beam height, is 3 mm (FWHM) wide.[6] At the target, the unobstructed area through which the beam passes is 12 mm high and 24 mm wide. The pumping apertures on the downstream side are holes in 24 μm thick aluminum foils and the end of the target box consists of a 127 μm thick steel foil. This minimizes the amount of material the exiting charged particles have to traverse before being detected.

The detector arrangement (see Fig.3) downstream of the target chamber has cylindrical symmetry around the beam axis. A 1.5 mm thick scintillator (F), segmented into quadrants, is followed by two pairs of orthogonal wire planes, and by a 102 mm thick (E) and a 6.4 mm thick (V) scintillator, both segmented into octants. The wire chambers feature a central hole for the beam pipe to pass through and have been specially designed[7] for this experiment. The direction of each outgoing charged particle is determined by the wire chambers; two pairs of planes, oriented as shown in Fig.3, are needed to resolve ambiguities when more than one charged particle is detected.

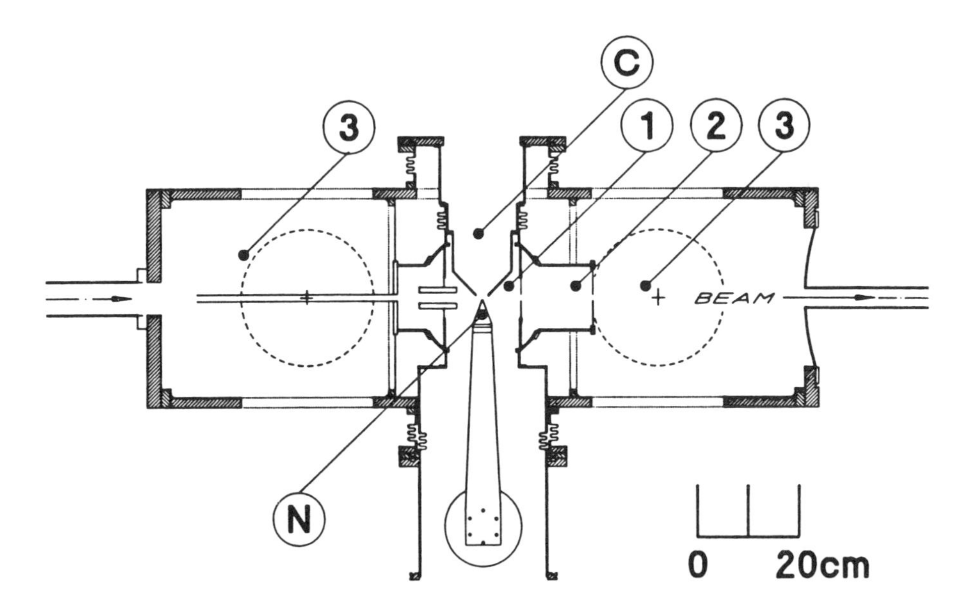

Fig.2. Side view of the target chamber and the internal hydrogen gas jet target. Note the thin exit foil on the downstream end. The labels are described in the text.

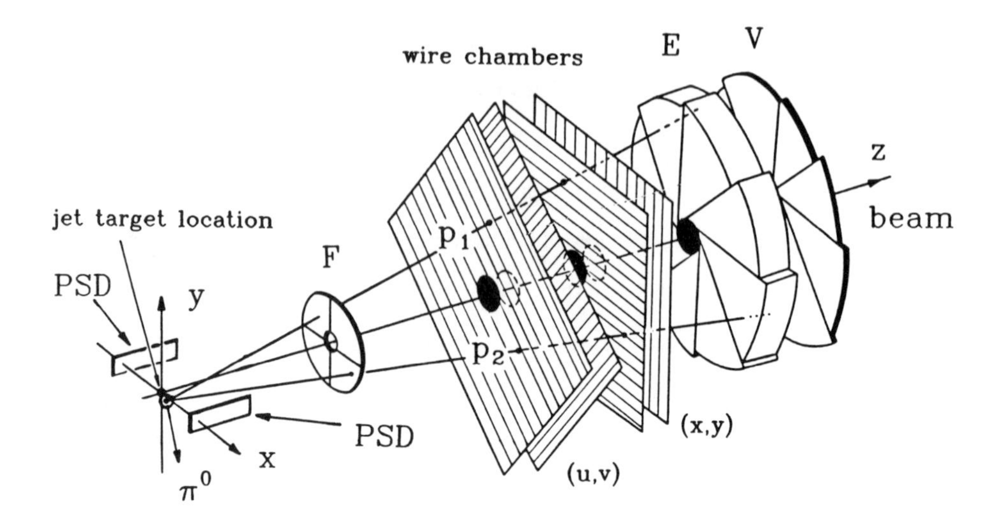

Fig.3. Schematic view of the detector arrangement. The components are described in the text.

Near threshold, the laboratory angle of protons from the $pp \to pp\pi^0$ reaction is limited: the detector was built to completely cover this angular region with the exception of the central hole for the beam pipe. As the bombarding energy is changed, the positions of the detector elements are adjusted in order to accommodate the corresponding changes in the kinematically allowed cone. In addition, for the measurements above 290 MeV, the angular acceptance of the detector has been increased by relocating the jet target from the middle position in the target box to a position closer to the exit foil. The angular acceptance of the detector system is from 4° to 20° in the most upstream position, and 2° to 10° in the most downstream position. The thickness of the E detector is chosen to stop up to 120 MeV protons. The impact position of the protons on the E detector is deduced from the wire chamber information and used to correct for a position dependence of the light collection efficiency in the elements of the E detector.

The trigger for a ppπ^0 event is a coincidence between E and F, but no pulse in V, and, in addition, a response from more than one segment of the E detector, signalling a multi-particle final state. In a separate event stream, elastically scattered protons are acquired (see Sect.2.3). For this, the trigger is a coincidence between all three scintillator planes and no condition on the multiplicity. In order to evaluate the importance of accidental coincidences, the experiment is also triggered at random times during data taking. By this method, the accidental rate was found to be low enough to be negligible.

2.3. Determining the Integrated Luminosity

For an internal target in a stored beam it is impractical and difficult to measure the beam current and the target thickness with sufficient accuracy to derive the luminosity. Instead, the integrated luminosity is obtained by comparing the number of p+p elastic scattering events, observed concurrently with pion production, with the tabulated[8] value of the p+p elastic scattering cross section.

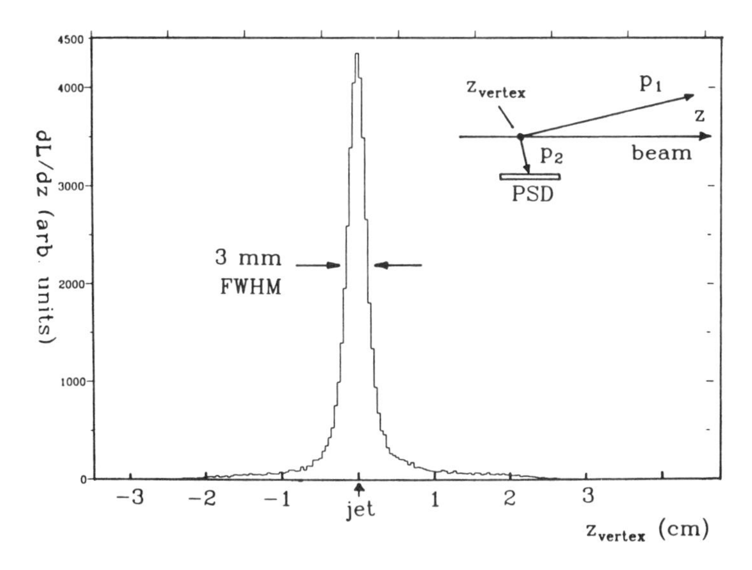

Fig.4. Differential luminosity along the beam in the target region. Since the beam size is smaller than the jet diameter, this is proportional to the gas density.

When detecting elastically scattered protons at forward angles, one finds that a large fraction of the observed particles are beam protons outside the ring acceptance which have been scattered from material close to the beam. In order to reject such events, low-energy recoil protons from p+p elastic scattering are observed in coincidence by two 7 mm by 45 mm position-sensitive silicon detectors ('PSD' in Fig.3). They are mounted 11 cm from the beam in a horizontal plane on either side of the target with their position-sensitive direction along the beam axis. Practically all such coincidences satisfy the kinematics of p+p elastic scattering. The position of the recoils in the silicon detector and the angle of the forward protons determine the location of each scattering center, and allow the reconstruction of the luminosity along the beam. This yields the density profile of the gas jet (see Fig.4). In addition, the vertex reconstruction yields information on the uniform gas density outside the jet region; this is needed since the active luminous volume for pion production is somewhat longer than that for coincident elastic scattering.

2.4. Measurements

Data are acquired in so-called cycles. Each cycle consists of injection of 45 MeV protons from dissociating a 90 MeV H_2^+ beam on a thin carbon stripper foil (2 s), acceleration to the final energy (4 s), data acquisition while cooling (10 s), and restoration of initial conditions (4 s). Because of the high radiation background during injection and acceleration, the high voltage is applied to the wire chambers only during the data acquisition phase. The flow of gas through the target nozzle is also limited to this phase, in order to lower the gas load to the ring vacuum.
During data taking, the RF cavity was run at a constant frequency; the harmonic number was 6. Stored beam currents ranged from 5 to 25 μA, resulting in a typical average luminosity of $5 \cdot 10^{28}$ $cm^{-2}s^{-1}$. During a total time for production runs of 110 h an integrated luminosity of about 16 nb^{-1} was achieved.

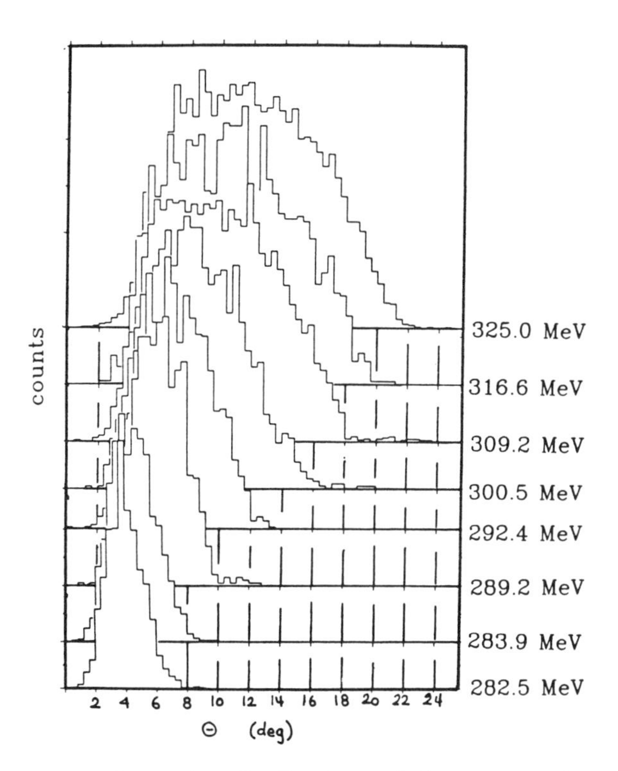

Fig.5. Histograms of angular distributions of the protons from the $pp \rightarrow pp\pi^0$ reaction, illustrating the increase of the kinematically allowed cone as the bombarding energy is increased.

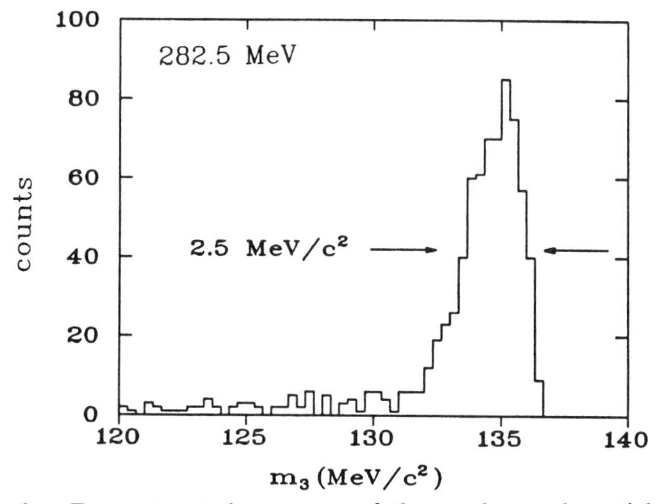

Fig.6. Reconstructed mass m_3 of the unobserved particle. The bombarding energy in this case is 282.5 MeV.

Measurements have been carried out at nine bombarding energies ranging from 282.5 MeV to 325 MeV. At energies below 290 MeV, the absolute value of the bombarding energy is determined with a precision of ± 200 keV from the strongly energy-dependent maximum laboratory reaction angle of the $pp\pi^0$ protons (illustrated in Fig.5). From the known RF frequencies, a ring circumference of 86.77±0.01 m is deduced from these data. At energies above 290 MeV, where the maximum proton angle varies less rapidly with energy, the bombarding energies were determined from the circumference and the respective RF frequencies.

3. DATA ANALYSIS

3.1. Definition of $pp\pi^0$ Events

Charged particles in the E detector are identified as protons by a condition on the relation between the time of flight from the F to the E detector and the energy deposited in the E detector. The particle tracks reconstructed from the wire chambers are required to be consistent with the individual elements of the F and E detector which have registered a hit. From the energy in the E detector and the direction obtained from the wire chambers the mass m_3 of the unobserved particle is reconstructed: as can be seen from Fig.6, a clean peak at the π^0 mass is observed. The area under this peak determines the number N_π of observed $pp\pi^0$ events.

A significant number of two-prong events deviate from the expected pattern of two hits in all of the four wire chamber planes. Spurious extra hits are eliminated by the condition that the real event is the one for which the shortest distance between the two reconstructed tracks is the smallest. Missing hits are either due to the two prongs lining up along the wire direction in one of the planes or due to a chamber inefficiency which is determined independently from single-prong events. Thus, events with a hit missing in one of the planes, or with one additional hit in up to two of the planes, are also included in the data sample.

3.2. Corrections

A Monte Carlo program is used to simulate the response of the detector system, including the effects of multiple scattering, the wire separation, energy and time resolution of the scintillators, and the finite target volume. Excellent agreement with the measured spectra is obtained. The three-body final state is taken to be phase-space distributed with the option to include the final-state interaction between the two outgoing protons via an effective range expansion.[9] At bombarding energies beyond 300 MeV, inclusion of the final-state interaction has a noticeable effect on the proton angular distribution. This is demonstrated in Fig.7 where simulations with and without final-state interaction are compared to the data. The accepted value of a = -7.82 fm for the scattering length[10] was used when calculating the solid curve in

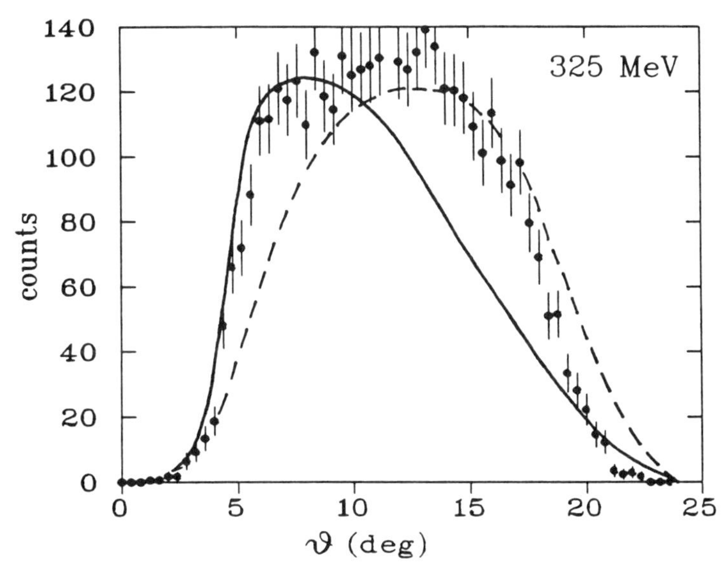

Fig.7. Histogram of the angular distribution of the protons from $pp \rightarrow pp\pi^0$ at 325 MeV. The dashed curve is a simulation assuming phase space distribution of the three outgoing particles. The solid curve represents a calculation where the effective range expansion has been used to take into account the final state interaction between the two protons. The normalization of both curves is arbitrary.

Fig.7. The angular distributions can be reproduced by choosing an effective range about four times smaller than the actual value. This means that the final-state interaction is obviously important in describing this reaction, but that the effective range parametrization of the NN interaction is inadequate.

Special care is taken to determine the fraction of $pp\pi^0$ events that are not detected because of the dead zone in the center of the detector. The phenomenological extrapolation into the dead region is based on the shape of the angular distribution predicted by the Monte Carlo simulation. The measured number of $pp\pi^0$ events is then corrected accordingly. The procedure is model-independent since, close to threshold, where the corrections are relatively large, the final state interaction, if included, has very little effect. Fig.8 illustrates the size of this correction at a bombarding energy of 286.5 MeV: the data (dots) are compared (a) to a full simulation and (b) to one without events with at least one prong in the

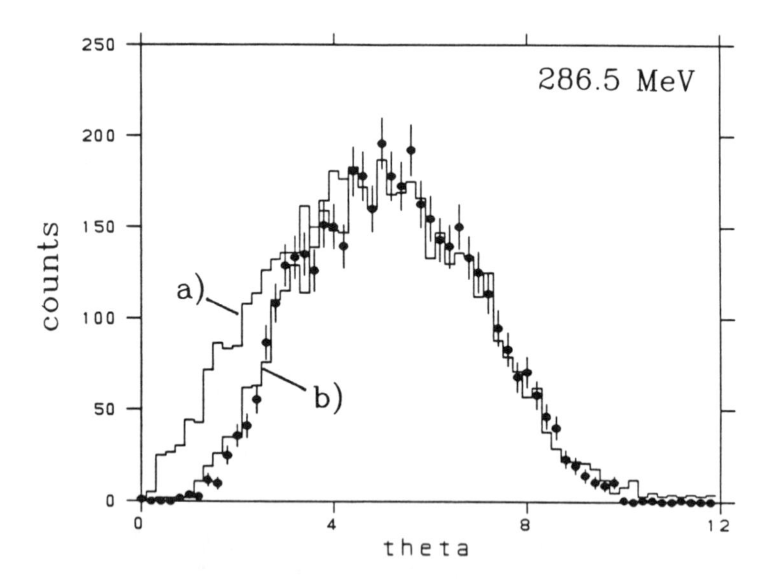

Fig.8. Angular distribution of outgoing $pp\pi^0$ protons at 286.5 MeV bombarding energy. Histogram (b) is a Monte Carlo simulation of the experiment. In calculation (a) the effect of the hole is ignored.

central hole. The correction factor due to loss in the central hole ranges from 1.69 at the lowest, to 1.24 at the highest energy.

Additional corrections arise from the efficiencies of the scintillators and the wire chambers. The scintillator efficiencies are estimated from measured total reaction cross sections and total (p,n) cross sections. The wire chamber performance characteristics, i.e., the role of accidental hits, missing wires, the distribution of wire cluster sizes, and the fusion of individual clusters into one, has also been simulated by a Monte Carlo calculation. The calculation is adjusted to reproduce the experimental distribution of multiplicities and then used to determine the wire chamber efficiency for two-prong events. The simulation is also used to determine the fraction of events with both protons in the same segment of the E detector (which would fail to trigger the experiment). The combined correction due to these three effects is at most 5%.

3.3. Tests

A number of test runs were carried out with the intention to check the experimental procedure. Measurements at the same energy but with different detector and target positions, and separated in time by two months were found to yield compatible values for the cross section. This supports the validity of the hole correction and the reproducibility of the procedure. At most energies, runs were also taken with gas bleeding into the target region instead of entering through the cold nozzle. This results in a uniform gas density, defined by the differential pumping impedances. Compatible results verify the procedure used to determine the absolute normalization.

3.4. Results

The final results for the total cross section σ_{tot} are computed from the observed number of pions, N_π, the absolute, integrated luminosity L deduced from p+p elastic scattering, and a correction factor g. This factor is mainly due to the loss of events in the dead zone of the central hole; smaller contributions include the measured wire chamber efficiency and the scintillator efficiency, as described above. The resulting cross sections, divided by η^2, are shown as solid dots in Fig.9, together with previous measurements.[11-14] The errors shown contain the contribution from counting statistics as well as estimated uncertainties in the luminosity L and the correction factor g. The errors do not include an estimated 5% error of the overall normalization of the data which is due to the uncertainty of the p+p elastic scattering cross section obtained from the literature.[8]

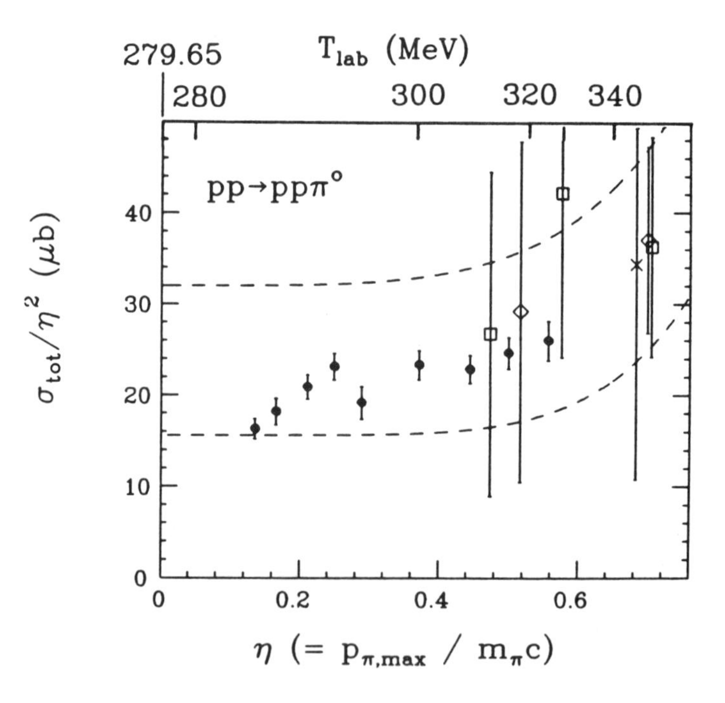

Fig.9. The total cross section for $pp \to pp\pi^0$ versus the dimensionless pion momentum. The data are from present work (solid dots), Ref.11 (cross), Ref.12 (squares), and Ref.14 (diamonds). The curves represent Eq.1 and are discussed in the text.

4. DISCUSSION

4.1. Energy Dependence near Threshold

In order to discuss the present data it is necessary to reexamine the energy dependence of the $pp \rightarrow pp\pi^0$ cross section near threshold. Since we are interested in the energy dependence only, we ignore constant factors. Then, the total cross section is given by

$$\sigma_{tot} \propto \sum_{Ll} \int |M_{Ll}|^2 \, d\rho_{Ll} \qquad (2)$$

where the sum extends over the angular momenta, $L\ell$, in the exit channel, M is the reaction amplitude, and $d\rho$ the phase space element.

The kinetic energy T_{cm} available in the exit channel is shared by three particles. Let **q** be the pion momentum in the three-body center-of-mass system and **p** the nucleon momentum relative to the NN center of mass, as shown in Fig.10. The largest possible pion momentum is

$$q_{max} = \sqrt{\frac{4m_p m_\pi c}{m_\pi + 2m_p} T_{cm}} \qquad (3)$$

It is related to the η parameter introduced in Sect.1 by $q_{max} = \eta \, m_\pi \, c$. Defining the kinematical variable x ($0<x<1$), we get for **q** and **p**

$$q = x \, q_{max} \qquad (4)$$

$$p = \sqrt{1-x^2} \, \frac{q_{max}}{2} \sqrt{1+2\frac{m_p}{m_\pi}} \qquad (5)$$

As mentioned in Sect.1, the partial waves in the $NN\pi$ system are labelled by the angular momentum L in the NN system and by the angular momentum ℓ of the pion with respect to the NN center of mass. The phase space element corresponding to a given $L\ell$ is then given by

$$d\rho_{Ll} \propto q^{2l+2}\, p^{2L+1}\, dq \qquad (6)$$

If the amplitude M is assumed to be constant over the small energy range of the threshold region, the energy dependence arises from phase space only. Inserting (4) and (5) into (6) and integrating from x=0 to x=1 yields for the partial cross section

$$\sigma(Ll) \propto q_{max}^{2L+2l+4} \quad . \qquad (7)$$

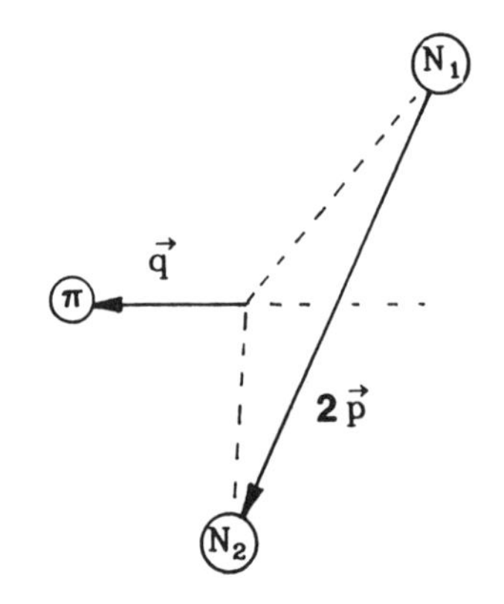

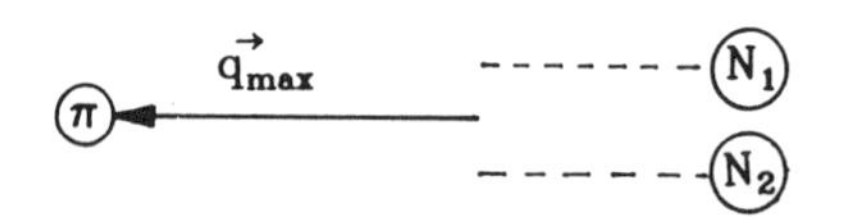

Fig.10. Definition of the kinematical variables in the NNπ center-of-mass system.

If the two nucleons in NNπ are in an L=0 state, the strong final-state interaction between them causes a significant energy dependence of the amplitude. Following Watson's treatment[9)] to factor out this energy dependence, one gets

$$|M_{Ss}|^2 = \frac{|M_{Ss}^0|^2}{k^2(1+\cot\delta_0)^2} \qquad (8)$$

where $k=p/\hbar$ and δ_0 is the pp S-wave phase shift. The latter can be replaced by an effective range expansion

$$k\cot\delta_0 = -\frac{1}{a} + \frac{1}{2}bk^2 + \cdots \quad , \tag{9}$$

where $a=-7.82$ fm is the pp scattering length and $b=2.7$ fm the effective range[10]. Inserting (9) into (8) leads to the following energy dependence of the Ss partial wave cross section

$$\sigma_{Ss} \propto \int \frac{q^2\ p\ dq}{a^2p^2+\hbar^2c^2\left[abp^2/(2\hbar^2c^2)-1\right]^2} \quad . \tag{10}$$

In the past, the usual procedure[2,3] has been to omit the second term in the denominator of Eq.(10) in order to carry out the integral analytically. This leads to

$$\sigma_{Ss}(approx.) \propto \int \frac{q^2dq}{p} = q^2_{max}\int_0^1 \frac{x^2dx}{\sqrt{1-x^2}} \propto q^2_{max} \quad , \tag{11}$$

i.e., the well-known and accepted η^2 dependence. However, the above approximation is equivalent to requiring that $p \gg 25$ MeV/c. This condition never holds for the whole range of the integration (which includes $p=0$) and it is clearly invalid below $\eta \approx 0.2$. In order to demonstrate the effect of this approximation, Eq.(10) has been integrated numerically. The results are illustrated in Fig.10. The dashed curve corresponds to $\sigma(Ss)/\eta^2$ as given by Eq.(10) without any approximations. The dot-dash curve results if the effective range parameter in (10) is set to $b=0$. The dotted line is obtained when all of the second term in the denominator of (10) is omitted; it shows $\sigma(Ss)/\eta^2=$const, as expected. All curves shown in Fig.11 have been scaled together to correspond to the measurement, but the relative normalization between them has been preserved.

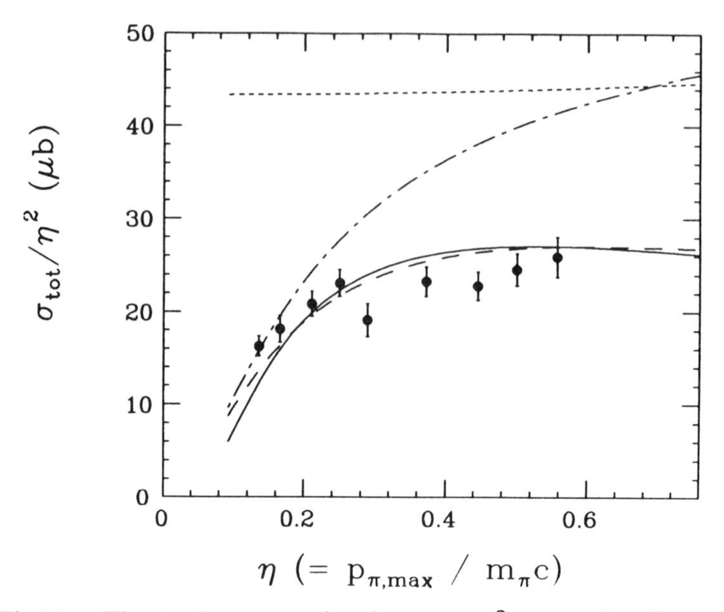

Fig.11. The total cross section for $pp \to pp\pi^0$ versus the dimensionless pion momentum. The data are from the present work. The curves show the expected energy dependence based on phase space and the pp final-state interaction. The curves differ in the assumptions made in evaluating the final-state interaction and are discussed in Sect.4.1.

In the preceding discussion the Coulomb force between the two protons has been neglected. It turns out, however, that the effect of this Coulomb repulsion is, contrary to intuition, rather weak. This has been demonstrated by replacing the effective range expansion (9) by a form that takes the Coulomb force into account[15,16]. The result is shown as solid line in Fig.11, which is found to differ very little from the (uncharged) Eq.(10), when evaluated properly (dashed line).

4.2 Comparison of the Data to Theory

In Fig.9, together with the data, are shown two previously published[12,14], polynomial fits according to Eq.1. These fits (dashed line[12], dot-dashed line[14]) yield the coefficients B_0=32 μb and B_0=15.6 μb for the contribution of the lowest angular momentum state. Presumably, this should be compared with theoretical values for B_0 ranging from 11 to 18 μb.[3,17] However, such a comparison is questionable since in the treatment of the energy dependence of the reaction amplitude in Refs. 3,17 approximation was used which is invalid in the energy region where the comparison is made, as demonstrated in the previous paragraph. As expected, it can be seen from Figs.9 and 11 that our measurement is not compatible with a near-threshold energy dependence of σ_{tot} proportional to η^2, in fact, the data support the more correct energy dependence discussed in Sect.4.1. The present measurement provides a strong incentive for theoretical efforts to reevaluate the energy dependence of the NN→NNπ process close to threshold, treating the final-state interaction consistently with the production reaction and with a realistic NN potential. The necessity to do so is also illustrated by our failure to account for the measured proton angular distributions by simply using an effective range description of the NN final state, as was mentioned in Sect.3.2. The theoretical effort should be facilitated by the fact that only one partial wave contributes and that the transition matrix element is dominated by the Born term.

5. OUTLOOK

The work reported here is the first nuclear physics experiment carried out with a storage ring with electron cooling. It clearly illustrates the potential of the new technology: aside from the fact that synchrotron acceleration gives IUCF access to the NNπ threshold, this experiment has benefitted from the small interaction energy spread, the low background due to the windowless target, the accessibility of small angles, and the observability of low-energy recoils.

A promising future application of this technology involves polarized beams and polarized targets. At present, considerable experience with storing and manipulating polarized proton beams has already been gained at IUCF, and the target thickness that can be achieved with a polarized atomic beam source injecting into a buffer cell[18] is reaching the order needed for internal target experiments. A measurement of spin correlation coefficients for the $pp \to pp\pi^0$ reaction, for example, would yield additional information on individual angular momentum states in the final state.

6. ACKNOWLEDGEMENTS

The author would like to thank D. Hutcheon, TRIUMF, for suggesting that the near-threshold energy dependence of $NN \to NN\pi$ should be reexamined. This work has been supported in part by the Deutscher Akademischer Austauschdienst and by the US National Science Foundation under grant NSF PHY 87-14406.

REFERENCES

1. A.H. Rosenfeld, Phys.Rev. 96, 139 (1954).

2. M. Gell-Mann and K.M. Watson, Ann.Rev.Nucl.Sci. 4, 219 (1954).

3. D.S. Koltun and A. Reitan, Phys.Rev. 141, 1413 (1966).

4. F.T. Cole and F.E. Mills, Ann.Rev.Nucl.Part.Sci. 31, 295 (1981), and references mentioned therein.

5. R.E. Pollock, Proc. IEEE Particle Accelerator Conf., Chicago 1989, ed. F. Bennett and J. Kopta, IEEE Conf. 1989, p. 17.

6. F. Sperisen, A. Berdoz, H.O. Meyer, R.S. Barbieri, and R.A. Bonham, Nucl. Instr. Meth. A274, 604 (1989).

7. K. Solberg, A. Eads, J.Goodwin, P. Pancella, H.O. Meyer, T. Rinckel, and A. Ross, Nucl. Instr. Meth. A281, 283 (1989).

8. R.A. Arndt, L.D. Roper, R.A. Bryan, R.B. Clark, B.J. VerWest, and P. Signell, Phys.Rev. D28, 97 (1983), and the data base SAID mentioned therein.

9. K.M. Watson, Phys. Rev. 88, 1163 (1952).

10. H.P. Noyes, Ann. Rev. Nucl. Sci. 22, 465 (1972).

11. R.A. Stallwood, R.B. Sutton, P.H. Fields, J.G. Fox, and J.A. Kane, Phys.Rev 109, 1716 (1958).

12. A.F. Dunaitsev and Yu.D. Prokoshkin, Zh.Eksp.Teor.Fiz. 36, 1656 (1959) (Sov.Phys.JETP 9, 1179 (1959).

13. F. Shimizu, Y. Kubota, H. Koiso, F. Sai, S. Sakamoto and S.S. Yamamoto, Nucl.Phys. A386, 571 (1982).

14. S. Stanislaus, D. Horvath, D.F. Measday, and A.J. Noble, Phys. Rev. C41, R1913 (1990).

15. T. Wu and T. Ohmura, Quantum Theory of Scattering, Prentice Hall, Englewood Cliffs, NJ, 1962, p.88 ff.

16. B.J. Morton, E.E. Gross, E.V. Hungerford J.J. Malanify, and A. Zucker, Phys.Rev. 169, 825 (1968).

17. V.P. Efrosinin, D.A. Zaikin, and I.I. Osipchuk, Z.Phys. A322, 573 (1985).

18. W. Haeberli, Proc.Int.Workshop on Polarized Sources and Targets, Helv.Phys.Acta 59, 597 (1986).

PION PRODUCTION IN *NN* COLLISIONS

B. Blankleider
Institut für Physik, University of Basel, CH-4056 Basel, Switzerland
and Paul Scherrer Institute, CH-5232 Villigen PSI, Switzerland

ABSTRACT

We present a brief overview of modern $NN \rightarrow \pi d$ and $NN \rightarrow NN\pi$ calculations. Our aim is to make a comparative study, emphasizing the relationship between a number of recent models. In view of current experimental interest, we complement this survey by discussing some results of a $NN-\pi NN$ model for low energy pion production. We conclude by summarizing the present theoretical difficulties and discuss possible directions for the future.

INTRODUCTION

Single pion production in *NN* collisions has been a subject of intense study for more than forty years. Although we do not intend to present a historical survey (one was presented here last time[1]), it cannot be left unmentioned that our early understanding in terms of s and p wave pion rescattering mechanisms is in part due to work done here at Indiana University in the 1950's.[2] However, progress toward a *quantitative* understanding has been much slower. Indeed many of the problems summarized nine years ago[1] remain with us today, despite an on-going research effort during this period.

Here we shall discuss a variety of models used to investigate pion production. As some of the models differ markedly in their technical aspects, the relationship between them may not be readily transparent to a non-specialist. Here we therefore concentrate on the physical content of each calculation, relating as much as possible one model to the other. As this is only in the nature of a summary, we note that a much more exhaustive comparison of different models and data may be found in a recent review.[3]

It is probably true to say that most of the more recent models have been developed mainly with the aim of describing the intermediate energy sector dominated by the πN P_{33} (1232) resonance. This is perhaps not surprising since the meson factories were designed largely to explore this energy region. As a result, very few low energy results have been reported using recent pion production models. To help offset this imbalance, we shall discuss the recent low energy measurements[4,5] of $np \rightarrow d\pi^0$ and $pp \rightarrow pp\pi^0$ in the context of the $NN-\pi NN$ model.[24] Some surprising effects of multiple scattering are found. In particular, πN rescattering in the S_{31} partial wave is especially prominent in our model, and acts as a doorway state to other channels. We therefore expect a close link between the understanding of low energy pion production, and the understanding of the off-shell behavior of s-wave πN amplitudes.

DISTORTED WAVE BORN APPROXIMATION

In the early days of pion physics, attempts were made at understanding pion production in NN collisions by appealing to field theoretic models. Typically one started from a Lagrangian describing the interactions between pion and nucleon fields, and after perhaps some approximations needed for practical calculations, a few diagrams were calculated. Initial and final state interactions between nucleons were taken into account phenomenologically by sandwiching these diagrams between NN wave functions.

Such a picture was rigorously derived by Woodruff[6] using a method of major significance for later theoretical developments. His argument took only three lines of algebra and goes as follows. The pion production amplitude F, represented by the first diagram of Fig. 1(a), is assumed to be expandable into an infinite sum of perturbation diagrams involving any number of pions and two nucleons. In this sum one groups together those diagrams that have *at least one pion in every intermediate state*; this partial sum, denoted by $F^{(1)}$, is represented by the second diagram of Fig. 1(a). All the rest of the diagrams are then of the form given by the third diagram of Fig. 1(a) where an intermediate state with two nucleons but no pions is explicitly exposed. Thus altogether one has the equation $F = F^{(1)} + F^{(1)}GT$ with G being the two nucleon propagator and T the off-shell NN amplitude. The next step groups together those diagrams of $F^{(1)}$ that have at least one *off-shell* pion in every intermediate state - denoted by $\tilde{F}^{(1)}$. $F^{(1)}$ can therefore be written as $\tilde{F}^{(1)}$ plus a term which explicitly has an intermediate state with no off-shell pions. This is diagrammatically illustrated in Fig. 1(b) or in the equivalent algebraic notation as $F^{(1)} = \tilde{F}^{(1)} + TG\tilde{F}^{(1)}$. Finally one substitutes the equation of Fig. 1(b) into the one of Fig. 1(a), resulting

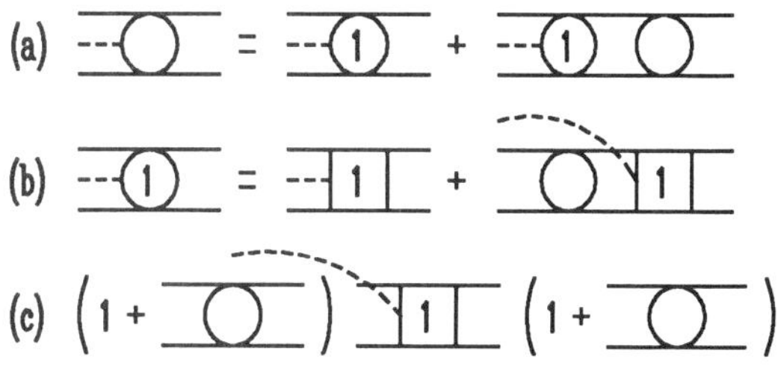

Fig. 1 Representing the amplitude for $NN \rightarrow NN\pi$ in terms of a pion production mechanism plus distorted waves. Dashed lines represent pions, solid lines are nucleons. An empty circle represents a full amplitude, a circle containing 1 denotes in states with at least one pion in every intermediate state, and the square containing a 1 denotes states with at least one *virtual* pion in every intermediate state. Substituting (b) into (a) gives (c) as the final result.

the full pion production amplitude being expressed as in Fig. 1(c), or algebraically as

$$F = (1 + TG)\tilde{F}^{(1)}(1 + GT). \tag{1}$$

This is an exact result expressing the pion production amplitude in terms of a pion production mechanism, $\tilde{F}^{(1)}$, and *NN* distorted waves $(1 + TG)$ and $(1 + GT)$.

In practice, $\tilde{F}^{(1)}$ is often replaced by the lowest order contributions shown in Fig. 2, and T is taken to arise from a phenomenological *NN* potential. An example is the model of Koltun and Reitan[7] for reactions $pp \to d\pi^+$ and $pp \to pp\pi^0$ at threshold. Their model is summarized by the pion-nucleon interaction density $H = H_0 + H_1 + H_2$ where

$$H_0 = (4\pi)^{1/2}(f/\mu)i\boldsymbol{\sigma} \cdot \{\boldsymbol{\nabla}_\pi [\tau \cdot \phi(x)] + [\mathbf{p}\tau \cdot \pi(x) + \tau \cdot \pi(x)\mathbf{p}]/2M\} \tag{2}$$
$$H_1 = 4\pi\lambda_1\mu^{-1}\phi^2(x), \qquad H_2 = 4\pi\lambda_2\mu^{-2}\tau \cdot \phi(x) \times \pi(x) \tag{3}$$

which was used in calculating the first three diagrams of Fig. 2. The term H_0 results from a non-relativistic reduction of the pseudovector πNN vertex. The first term of Eq. (2) is the static p-wave pion production vertex, the second term models the recoil of the system and corresponds to s-wave pion production. Terms H_1 and H_2 model s-wave πN rescattering with effective zero range interactions. Demanding that H_1 and H_2 reproduce zero energy πN data in Born term, relates the constants λ_1 and λ_2 to the s-wave isoscalar and isovector πN scattering length combinations $a_1 + 2a_3$ and $a_1 - a_3$ respectively. Note that although the exchanged pions in Figs. 2(b) and (c) are significantly off-mass-shell, the πN amplitudes used in this model are purely on-shell. Following Gell-Mann and Watson,[8] it is usual to write threshold pion production total cross sections in the form

$$\sigma = \alpha\eta + \beta\eta^3, \qquad \sigma' = \alpha'\eta^2 + \alpha''\eta^6 + \beta'\eta^8 \tag{4}$$

for $NN \to \pi d$ and $NN \to NN\pi$ reactions respectively. Here η is the c.m. pion momentum in units of pion mass, and it is often assumed that the coefficients α and β (corresponding to s and p-wave pion production) are largely independent of η. In spite of its simplicity, the Koltun and Reitan model gave values $\alpha = 146\mu$b and $\alpha' = 17\mu$b in fair agreement with current experiments.[4,5]

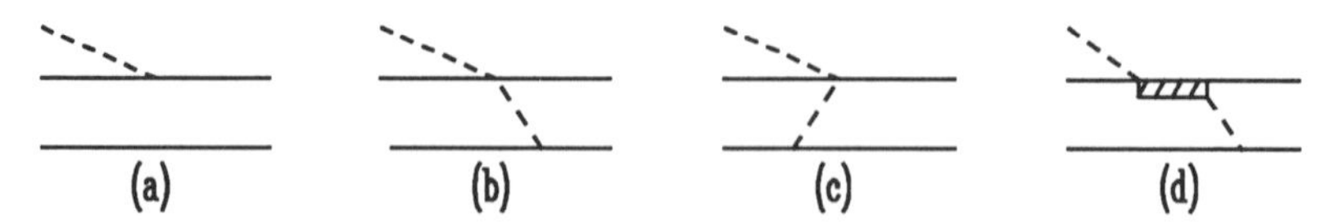

Fig. 2 Pion production mechanisms in DWBA models: (a) direct, (b) forward rescattering, (c) backward rescattering, (d) Δ resonance.

In a number of articles,[9] the model of Koltun and Reitan was extended to include p-wave pion production by inclusion of the diagram of Fig. 2(d). Here pion rescattering takes place via the formation of the Δ resonance. With other refinements like inclusion of off-shell effects and vector meson exchange, the qualitative features of $pp \to d\pi^+$ observables were reproduced across the Δ resonance region. It was however generally concluded that a more accurate multiple scattering description would be needed for a more precise description of pion production.

RELATIVISTIC MODELS

Rather than rely on one or another Lagrangian to generate the elementary interactions, one can instead write down the most general form of the interaction in terms of its Lorentz structure. Thus König and Kroll[10] evaluated Fig. 3(a) as the invariant amplitude

$$M = \frac{g}{\sqrt{4\pi}} \bar{u}(p_1')[A + \not{q}B]u(p_1) \frac{i}{(p_2 - p_2')^2 - m_\pi^2} \bar{u}(p_2')\gamma_5 u(p_2) \tag{5}$$

where $[A + \not{q}B]$ is the general form of the πN amplitude. After expressing the invariant amplitudes A and B in terms of partial wave amplitudes $T_{l\pm}$, a way was chosen to continue off-shell by allowing all kinematical factors to vary away from their on-shell value, but with the partial wave amplitudes themselves fixed at their on-shell value. With this prescription, πN phase shift data could be used directly to generate a version of the off-shell πN invariant amplitude. With a simple phase prescription for including initial and final state interactions (discussed shortly) the model was quite successful at describing scattering data for a variety of different charge states of $NN \to NN\pi$. However the total cross sections for $pp \to pp\pi^0$ and $np \to np\pi^0$ were not well reproduced, particularly at low energies. Later this model was extended by Grein *et al.*[11] and further by Locher and Švarc[12] to calculate the reaction $pp \to \pi^+ d$. The model consisted of the diagrams in Figs. 3(b) and (c), and again involved the full complexity of the relativistic theory. Examples of results obtained with these models are given in Figs. 4 and 5.

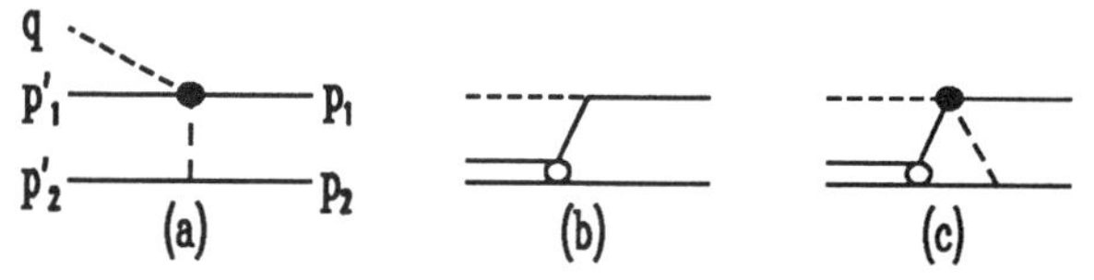

Fig. 3 Pion production mechanisms in relativistic models. All "blobs" represent invariant amplitudes. (a) $NN \to NN\pi$ (b) direct production in $NN \to \pi d$ (c) pion rescattering in $NN \to \pi d$.

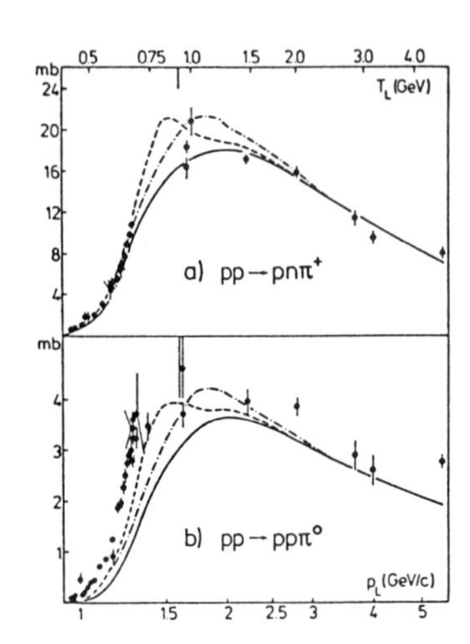

Fig. 4 Reaction cross sections in the relativistic model of König and Kroll[10] - solid lines. Chained lines result from additional dibaryon contributions.

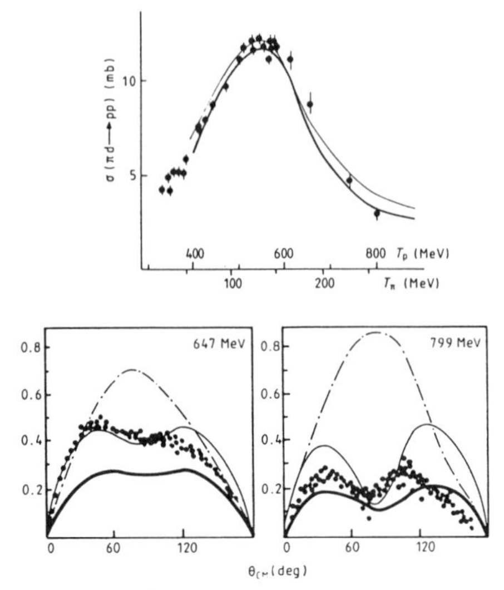

Fig. 5 (a) Total cross section and (b) asymmetry A_{y0} in the relativistic model of Locher and Švarc [12] - thick solid curves (thin solid curves are *ad hoc* modifications of triplet waves). The chain curves are from $NN-\pi NN$ calculations of the Lyon group.[26]

Here initial (ISI) and final (FSI) state interactions were included using prescriptions based on the so called Sopkovich formula. This expresses the full $NN \rightarrow \pi d$ amplitude $T_{\pi d,NN}$ in terms of the Born term $B_{\pi d,NN}$ (in this case, sum of Figs. 3(b) and (c)) as

$$T_{\pi d,NN} = S^{\frac{1}{2}}_{\pi d,\pi d}\, B_{\pi d,NN}\, S^{\frac{1}{2}}_{NN,NN} \tag{6}$$

where the square roots of S-matrices are used to include the elastic channel distortions. Note that this is an on-shell prescription with no integrals implied connecting the S-matrices and the Born term. Interestingly, it was found that the distortions are dominated by the πd channel; distortions in this channel were *not* taken into account in the DWBA approaches discussed above.

Considering the very careful treatment of the Born term, it is important to estimate the effect of treating ISI and FSI approximately with an on-shell prescription. To this end, Alfred Švarc and I[13] have examined the Sopkovich approximation in a simple two-body coupled channels calculation of the processes $NN \rightarrow \pi d$, $\pi d \rightarrow \pi d$, and $NN \rightarrow NN$. Separable potentials describing transitions between the channels NN, πd, and $N\Delta$ were constructed by solving the two-body Lippmann-Schwinger equation and fitting the resulting t-matrices to the amplitudes coming from a more sophisticated Faddeev calculation of the πNN system[24] (see below). This way we can directly compare the true $NN \rightarrow \pi d$

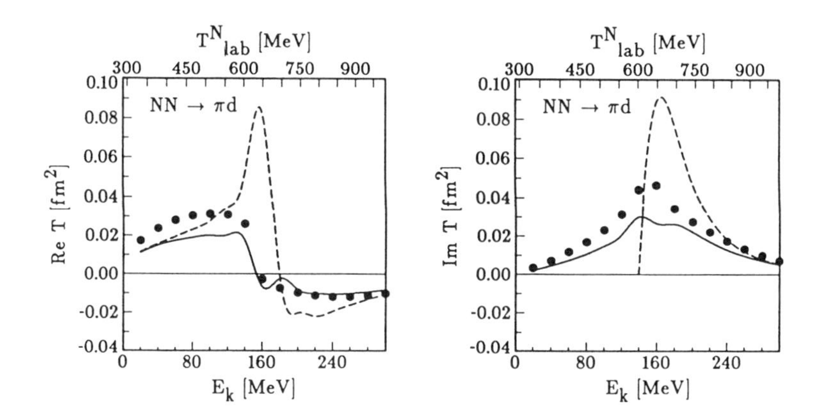

Fig. 6 Comparison of the exact t-matrix in the $J = 2^+$ channel for the process $NN \rightarrow \pi d$ (filled circles) with the Sopkovich modified Born term (solid line). Also shown is the unmodified Born term (dashed line).

amplitude $T_{\pi d,NN}$ with the prescription of Eq. (6). The results are shown in Fig. 6 for the dominant $J = 2^+$ channel. Comparing the Born term $B_{\pi d,NN}$ (dashed curve) with the Sopkovich prescription for $T_{\pi d,NN}$ (solid curve), we see that this correction for initial and final state interactions does indeed bring one closer to the exact $NN \rightarrow \pi d$ t-matrix (solid dots). Nevertheless, the Sopkovich approximation appears to be quantitatively accurate only above 200 MeV c.m. kinetic energy. Within this model, off-shell distortions play an important role at resonance as well as at low energies.

COUPLED CHANNELS MODELS

Because one is interested in energies above pion production threshold, and in view of the importance of distortions, attempts have been made to dynamically include pionic and resonance degrees of freedom into the initial and final state interactions. Here the word *dynamically* means that coupled scattering equations are solved for a number of explicitly included channels.

(a) Helsinki model

One of the first coupled channel models was worked out in detail by a Helsinki group - in particular by Green and Niskanen.[14,15] Their model can be viewed as a special case of an extended DWBA model that treats N and Δ on an equal footing. For the derivation of Fig. 1 such an extension means adding extra diagrams where N is replaced by Δ in intermediate states. Thus in the most general case Fig. 1(a) would be replaced by

B_1

$= \; + \; \sum_{B_1 B_2 = N,\Delta}$

B_2

where all 2-baryon states (NN, $N\Delta$ and $\Delta\Delta$) are explicitly taken into account. Clearly the derivation will follow through exactly as before if we now interpret all expressions as 2×2 matrices in the baryon space spanned by N and Δ. Thus the final pion production amplitude will be again given by Fig. 1(c), which we now interpret to have a sum over all intermediate 2-baryon states, thus

$$F_{\pi NN,NN} = \sum_{all B's = N\Delta} (\delta_{NN,B_1'B_2'} + T_{NN,B_1'B_2'} G_{B_1'B_2'}) \tilde{F}^{(1)}_{\pi B_1'B_2',B_1B_2} (\delta_{B_1B_2,NN} + G_{B_1B_2} T_{B_1B_2,NN}). \tag{7}$$

It can be seen that in addition to NN elastic amplitudes, one also needs amplitudes for $NN \to N\Delta$ and $NN \to \Delta\Delta$. For these we can again use the diagrammatic argument to obtain the coupled channels Lippmann-Schwinger equations

$$T_{B_1'B_2',B_1B_2} = T^{(1)}_{B_1'B_2',B_1B_2} + \sum_{B_1'',B_2''=N\Delta} T^{(1)}_{B_1'B_2',B_1''B_2''} G_{B_1''B_2''} T_{B_1''B_2'',B_1B_2} \tag{8}$$

which link all the amplitudes in the pure 2-baryon sector. Within this scheme, the Helsinki model corresponds to (i) replacing $T^{(1)}_{B_1'B_2',B_1B_2}$ with phenomenological potentials $V_{B_1'B_2',B_1B_2}$ in order to fit NN elastic experimental data, (ii) dropping $\Delta\Delta$ states and setting the potential $V_{N\Delta,N\Delta}$ to zero, and (iii) retaining only the pion production mechanisms illustrated in Fig. 7.

In the philosophy of this model, the potentials $V_{NN,NN}$ and $V_{NN,N\Delta}$ are essentially phenomenological and form one of the basic inputs to the model. In one of the latest calculations $V_{NN,NN}$ was taken to be the Reid potential modified to give the right phase shifts below the inelastic threshold, and $V_{NN,N\Delta}$ was based on a static meson exchange picture.[15] For the pion production mechanism itself, the usual form of the $\pi N\Delta$ vertex was used ($\nabla_\pi \cdot S\,T \cdot \phi$), while for direct and s-wave pion production essentially the model of Koltun and Reitan was used. This model has been one of the most successful at describing $pp \to \pi^+ d$ observables

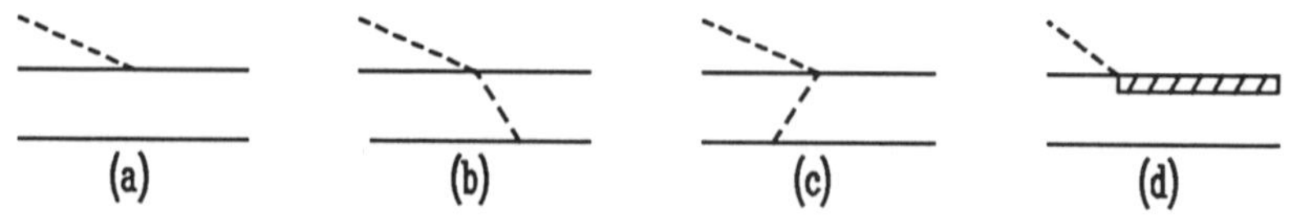

Fig. 7 Pion production mechanism in the Helsinki model. Diagrams (a), (b) and (c) give the model for $\tilde{F}^{(1)}_{\pi NN,NN}$ while (d) models $\tilde{F}^{(1)}_{\pi NN,N\Delta}$.

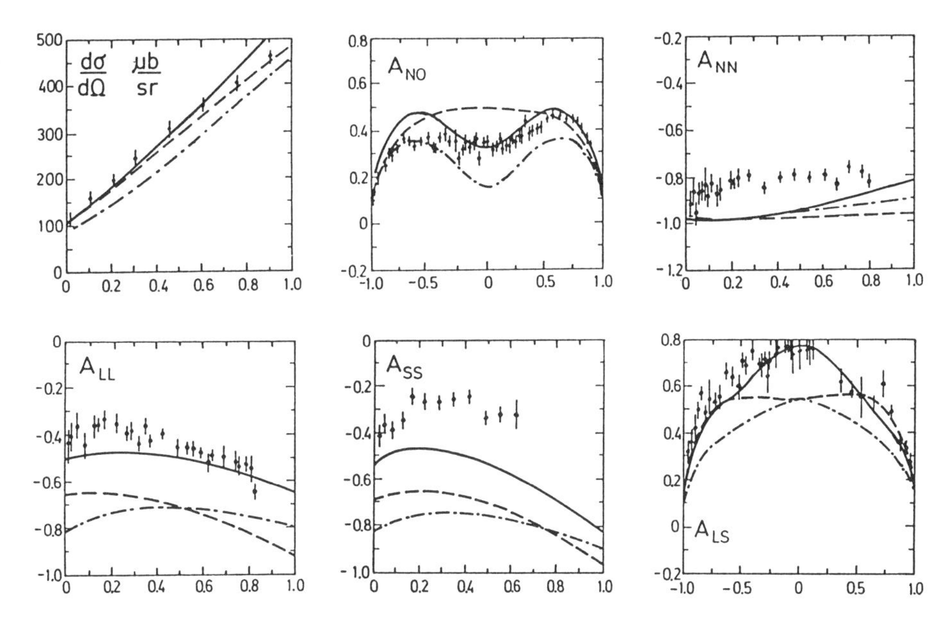

Fig. 8 The differential cross section and asymmetries for $pp \to \pi^+ d$ as a function of $\cos\theta_\pi$ or $\cos^2\theta_\pi$. Shown are results of the Helsinki model[15] (solid curve), and the $NN-\pi NN$ models of Ref. 24 (dashed curve) and Ref. 28 (dash-dot curve).

- see Fig. 8 where a comparison is made with data and with two calculations using the $NN-\pi NN$ model (discussed below).

(b) Argonne-Hannover model

It is evident that the treatment of NN distortions in the Helsinki model neglects all explicit reference to three (or more) body states. Because single pion production dominates the inelastic cross section at intermediate energies, it is expected that at least the explicit inclusion of πNN states may be necessary for a quantitative description of $NN \to NN\pi$.

A model which essentially extends the one of Helsinki to include πNN contributions has been developed by an Argonne group[16] with a number of recent calculations being carried out by Lee and Matsuyama.[17] The same model has also been pursued in Hannover.[18] As in the Helsinki model, amplitudes $T^{(1)}_{NN,NN}$ and $T^{(1)}_{NN,N\Delta}$ of Eq. (8) are replaced by phenomenological potentials $V_{NN,NN}$ and $V_{NN,N\Delta}$. On the other hand, the amplitude $T^{(1)}_{N\Delta,N\Delta}$ is retained, and assumed to arise from the three-body multiple scattering of πNN states through a $\pi N\Delta$ vertex $h_{\pi N\Delta}$ as well as through (separable) πN and NN potentials - $v_{\pi N,\pi N}$ and $v_{NN,NN}$ respectively - see Fig. 9. Note that $v_{NN,NN}$ is in effect *only* while there is a pion in flight, while $V_{NN,NN}$ is used when no pion is present. For simplicity,

Fig. 9 Representative diagrams contributing to $T^{(1)}_{N\Delta,N\Delta}$ in the Argonne-Hannover model.

the πNN vertex is excluded from this description, although it is implicitly taken into account through the potentials $V_{NN,NN}$ and $V_{NN,N\Delta}$.

As displayed in Fig. 9, $T^{(1)}_{N\Delta,N\Delta}$ is made up of an infinite series of three-body contributions. These may be effectively summed by solving an appropriate integral equation (for separable πN and NN potentials, this is the Alt-Grassberger-Sandhas form of the Faddeev equation).

The coupled equations for this model then take the form shown diagrammatically in Fig. 10. A feature of this model is that it simultaneously describes the reactions $NN \rightarrow NN$, $\pi d \rightarrow \pi d$, $\pi d \rightarrow \pi NN$, $NN \rightarrow \pi d$, and $NN \rightarrow NN\pi$. As described so far, this model allows pion production to take place explicitly only through the decay of the Δ resonance. An attempt was made to overcome this difficulty in the most recent work of Ref. 17 where an extra three-body $NN \rightarrow NN\pi$ operator was introduced.

With $V_{NN,NN}$ taken to be the Paris potential with Δ components subtracted (since this model includes them via $V_{NN,N\Delta}$) the results of Lee and Matsuyama[17] for NN scattering compare quite favorably with experimentally determined phase shifts and inelasticities up to 1 GeV. By contrast, we note that the input Paris potential is valid only below the pion production threshold ($\sim$ 280 MeV). Nevertheless, agreement with NN experiments is not so good for sensitive polarization quantities like $\Delta\sigma_L$ and $\Delta\sigma_T$.

For πd elastic scattering, this calculation gives results which are typical of a number of other models; namely, the forward angle differential cross sections are well predicted but there is a large overestimation at the backward angles.

For pion production, the only published result of Matsuyama and Lee is for the reaction $pp \rightarrow pn\pi^+$ at 800 MeV - Fig. 11. Although the general features of the Δ and final state interaction peaks are well reproduced, there is plenty of room for improvement.

(c) $NN-\pi NN$ model

If it were not for pion absorption, standard three-body theory could be used directly to describe the πNN system. Indeed, some early calculations[19,20] used Faddeev equations in a strictly three-body model where pion absorption was treated as a bound state of a pion and nucleon (bound by m_π MeV). The problem of course was that only one of the nucleons (the bound state) was allowed to emit a pion. Attempts to solve this inconsistency led to the $NN-\pi NN$ model.

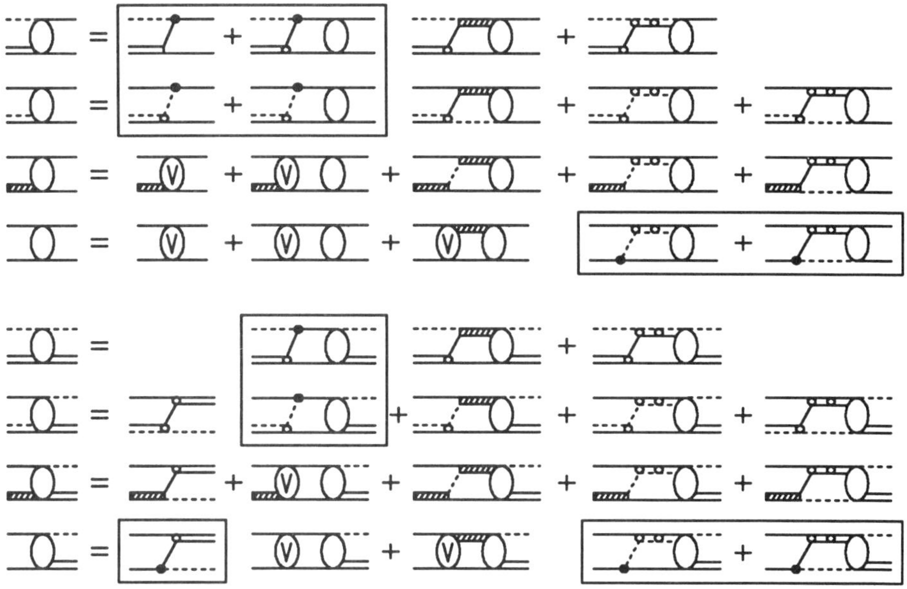

Fig. 10 Comparison of the Argonne-Hannover three-body equations with the ones of the $NN-\pi NN$ model. The first four equations couple pion production with NN elastic scattering, while the last four couple pion absorption with πd elastic scattering. In the Argonne-Hannover model the contributions enclosed in boxes are missing, while the diagrams marked with a "V" represent phenomenological potentials $V_{NN,NN}$ and $V_{N\Delta,NN}$. By contrast, the $NN-\pi NN$ model retains all contributions and the "V" amplitudes are given by non-static one pion exchange.

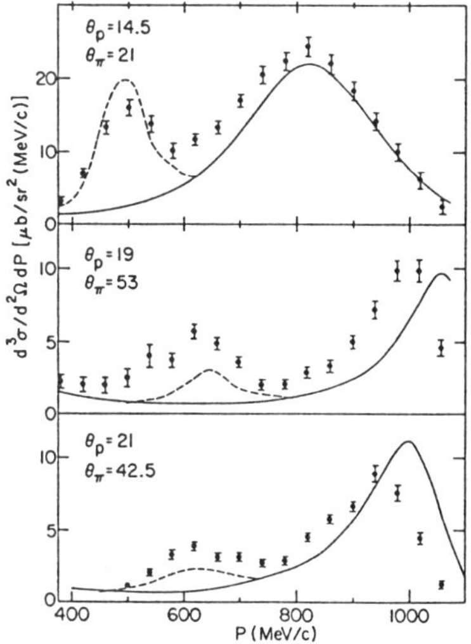

Fig. 11 Differential cross sections for $pp \rightarrow pn\pi^+$ at 800 MeV, calculated by Lee and Matsuyama[17] using the Argonne-Hannover model. The solid curves result when the very last interaction is a Δ decaying into a pion and nucleon; the dashed curves include the case when the very last interaction is elastic scattering of the nucleons through the $^3S_1$-$^3D_1$ channel.

This model, developed originally mainly by Afnan, Avishai, Blankleider, Mizutani, Rinat, and Thomas,[21–23] leads to Faddeev-like equations (we'll call them the $NN-\pi NN$ equations) and allows either nucleon to emit or absorb a pion as long as the total number of particles remains three or less.

In overall structure, the $NN-\pi NN$ model can be thought of as the Argonne-Hannover model where an explicit πNN vertex is introduced, and where the potentials $V_{NN,NN}$ and $V_{NN,N\Delta}$ are no longer phenomenological but are generated from the meson-baryon vertices (in the simplest case πNN and $\pi N\Delta$). The $NN-\pi NN$ equations are compared to the ones of the Argonne-Hannover model in Fig. 10. Evidently the ultimate goal is no less than a completely unified description of the reactions $NN \rightarrow NN$, $\pi d \rightarrow \pi d$, $\pi d \rightarrow \pi NN$, $NN \rightarrow \pi d$, and $NN \rightarrow NN\pi$ within the three-body sector.

Calculations using the $NN-\pi NN$ model or similar Faddeev-like approaches have been performed by a host of groups.[23–32] Examples of such calculations for $pp \rightarrow \pi^+ d$ were given in Figs. 5 and 8.

One of the fundamental hurdles faced by the $NN-\pi NN$ model is the modelling of the short range part of the NN interaction. This indeed is one advantage of the Argonne-Hannover model; there part of the range short interaction is put in "by hand" through the use of separately constructed potentials (e.g. Paris potential). On the other hand, there is still much room left in the $NN-\pi NN$ model to include explicit heavy meson exchanges. Indeed, any number of these could also be put in "by hand". More interesting would be, however, to treat the pion and heavier mesons on an equal footing. For example, including the ρ meson on an equal footing would result in coupled $NN-\pi NN-\rho NN$ equations describing simultaneously both π and ρ production. That this may give a realistic description is suggested by $NN-\pi NN$ calculations that generate a surprising amount of short range repulsion using pure πNN states through one-pion and one-nucleon exchange: see for example Ref. 24 where the $^1S_0$ phase shift changes sign with increasing energy even though no heavy mesons were included.

Although the $NN-\pi NN$ model represents significant progress over the original purely Faddeev description, there remain fundamental problems with the treatment of the πN P_{11} channel - the partial wave where pion absorption takes place. Within this model, the P_{11} amplitude is written in terms of nucleon pole and background parts. The background part contains all contributions apart from the pole term, and is usually modelled with a separable potential. In the two nucleon sector the pole part is illustrated in Fig. 12(a) while the underlying lowest order contribution to the background part is shown in Fig. 12(b). Figs. 12(c) and (d) show the antisymmetric partners of (a) and (b) respectively. Although Figs. 12(c) and (d) would not be differentiated in a covariant theory, practical calculations still rely on a 3-dimensional time ordered approach where the contribution of Fig. 12(d) is neglected in order to avoid the complications of four-body states.

In relation to πd elastic scattering, Jennings and Rinat[33] showed how retaining the diagram of Fig. 12(d) can significantly improve results for the tensor polarization t_{20}. Their result has been further supported by a recent measurement[34] of πd vector polarization iT_{11}. In a more sophisticated calculation, Mizutani *et al.* [27] came to a similar conclusion.

Perhaps an even more fundamental problem of neglecting Fig. 12(d) concerns the dressings of one and two-nucleon propagators in the NN–πNN model. There the bare one nucleon propagator g_0 is dressed by one-pion loops Σ as illustrated in Fig. 13(a) (here for simplicity we neglect rescattering via the background term). Thus the dressed propagator is given by the series $g = g_0 + g_0\Sigma g_0 + g_0\Sigma g_0\Sigma g_0 + \ldots$ Now using the same Σ to dress the two-nucleon propagator would give rise to a similar series which would contain, besides three-body states, four-body states as in Fig. 13(b) where the nucleons are being dressed "at the same time". Such diagrams arise from the mechanism in Fig. 12(d) and are therefore neglected. This results in each of the nucleons in the two-nucleon sector being underdressed with consequent problems of wave function normalizations. The numerical consequences of this problem are presently being investigated.[35]

Despite the variety of implementations of the $NN - \pi NN$ model and indeed despite the differences between all the coupled channels models, numerical results from them all have a (by now well reported) number of common deficiencies. They are as follows: (i) At the higher energies, the πd elastic differential cross section is significantly overestimated at backward angles. (ii) The $\vec{p}\vec{p} \rightarrow \pi^+ d$ polarization observables A_{xx}, A_{yy}, and A_{zz} all display a missing strength in the triplet channels. (iii) Perhaps correlated with (ii), the NN triplet partial waves are weaker than that indicated by partial wave phase shifts. (iv) Some coupled channels calculations underestimate $pp \rightarrow \pi^+ d$ cross sections by up to 30%. (v) In all reactions, most polarization observables are badly described.

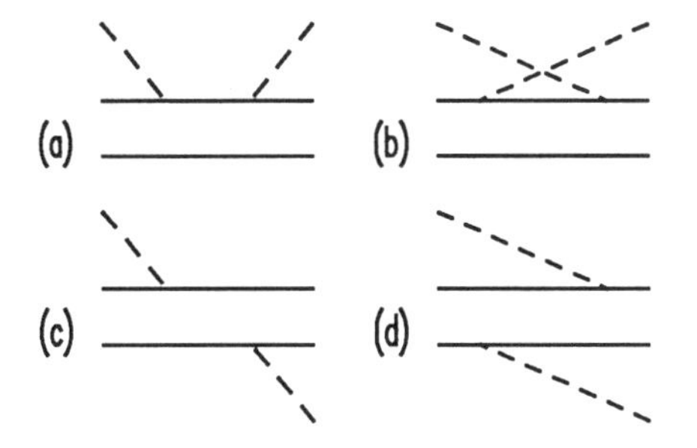

Fig. 12 (a) nucleon pole term, (b) lowest order contribution to background term, (c) antisymmetric partner of (a), (d) antisymmetric partner of (b).

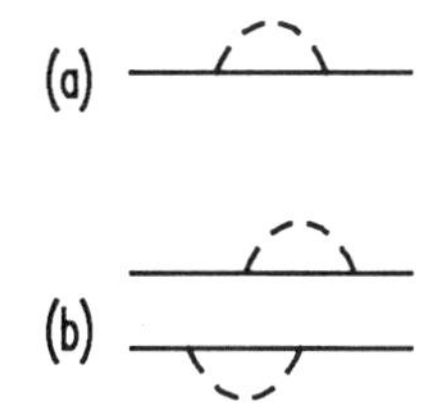

Fig. 13 (a) one-pion dressing Σ, (b) one-pion dressing of two nucleons "at the same time".

Some of these problems may be addressed by extending the $NN - \pi NN$ model to treat nucleons and Δ's on an equal footing, the so called $BB - \pi BB$ model. [36] Clearly much remains to be done before a satisfactory quantitative understanding of pion production is achieved.

PION PRODUCTION AT LOW ENERGIES

As mentioned previously, pion production in the low energy region has not been as well investigated as at intermediate energies. Here we would therefore like to discuss some results of the $NN - \pi NN$ model near pion production threshold.

It is convenient to discuss both the reactions $NN \rightarrow \pi NN$ and $NN \rightarrow \pi d$ in terms of partial wave amplitudes using the same coupling scheme. Thus in general the final state two-nucleon pair forms a 'quasi-deuteron' whose quantum numbers are given by the usual $^{2S+1}L_j$ symbol.

Close to threshold we expect the largest contribution to come from states where the two nucleons in the quasi-deuteron are in a relative s-state, while the pion may be in an s or p-state with respect to the centre of mass of the quasi-deuteron.[8] In this case there are just six possible partial waves as enumerated in Table I. For $^3S_1$ (coupled to $^3D_1$) quasi-deuterons (isospin zero) we denote the amplitudes by a_0, a_1, a_2, while for $^1S_0$ quasi-deuterons (isospin one) we write them as b_0, b_1, b_2. For deuterons or quasi-deuterons whose nucleons are strongly correlated, we may write effective two-body differential cross sections. In terms of the six amplitudes these are given (for a few cases) by

$$d\sigma/d\Omega\,[pp \rightarrow \pi^+(np)_{^3S_1}] = K\left\{|a_0|^2 + 3|a_1|^2 + 5|a_2|^2 + [5|a_2|^2 - \sqrt{40}Re(a_0a_2^*)]\,P_2(x)\right\} \quad (9)$$

$$d\sigma/d\Omega\,[np \rightarrow \pi^-(pp)_{^1S_0}] = K\left\{|b_0|^2/4 + |b_1|^2/2 + |b_2|^2/2 + [Re(b_0b_2^*) - Re(b_0b_1^*)/\sqrt{2}]\,P_1(x) + [\,|b_2|^2/2 - \sqrt{2}Re(b_1b_2^*)]\,P_2(x)\right\} \quad (10)$$

$$d\sigma/d\Omega\,[pp \rightarrow \pi^0(pp)_{^1S_0}] = K\,|b_0|^2/2 \quad (11)$$

$$K = \tfrac{1}{(2s_1+1)(2s_2+1)}\pi^2 k_f/k_i\,\mu_i\mu_f\,; \qquad x = cos\theta. \quad (12)$$

As written, these expressions hold both for pion production and absorption, only the spin average factor and momenta need be appropriately set in Eq. (12).

TABLE I. Partial wave channels for $NN \rightarrow \pi d^*$, where d^* is a 'quasi-deuteron'

I	J^π	T	NN	L_{NN}	S_{NN}	d^*	$L_{\pi d}$	$S_{\pi d}$	$L_{N\Delta}$	$S_{N\Delta}$
a_0	0^+	1	$^1S_0$	0	0	$^3S_1$	1	1	2	2
a_1	1^-	1	$^3P_1$	1	1	$^3S_1$	0	1	1	1
a_2	2^+	1	$^1D_2$	2	0	$^3S_1$	1	1	0	2
b_0	0^-	1	$^3P_0$	1	1	$^1S_0$	0	0	1	1
b_1	1^+	0	$^3S_1$	0	1	$^1S_0$	1	0	–	–
b_2	1^+	0	$^3D_1$	2	1	$^1S_0$	1	0	–	–

(a) $np \rightarrow d\pi^\circ$

Pion production leading to a final state deuteron is dominated at intermediate energies by the a_2 amplitude. This is well understood in terms of pion-nucleon rescattering in the P_{33} channel leading to formation of $N\Delta$ in relative s-state. For experiments close to threshold we expect p-wave pion production to be suppressed, offering a better possibility of learning about s-wave pion production. Following recent experiments of $np \rightarrow d\pi^0$ and $\vec{p}p \rightarrow d\pi^+$ near threshold, the relative size of s and p-wave amplitudes has been estimated (see Ref. 4 and D. A. Hutcheon in these proceedings). Within the $NN-\pi NN$ model of Ref. 24 the relative s and p-wave contributions are displayed in Fig. 14. It can be seen that p-wave pion production remains important at least down to $\eta \equiv k_\pi/m_\pi = .2$. On the basis of the simple $\sigma = \alpha\eta + \beta\eta^3$ relation of Eq. 4, Fig. 14 shows that α is a slowly decreasing function of η.

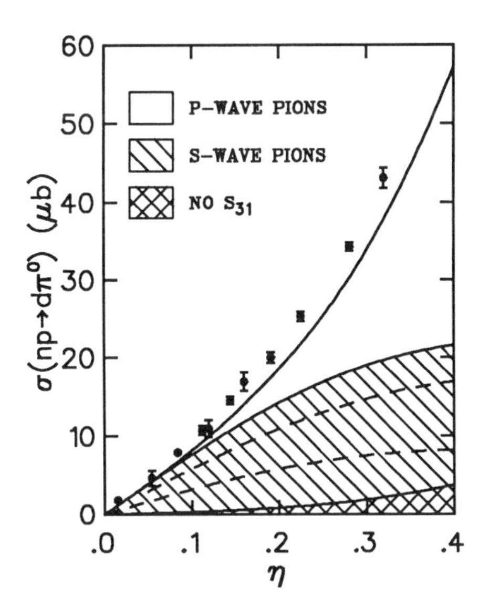

Fig. 14 Partial wave contributions to the total $np \rightarrow d\pi^0$ cross section. Curve close to data results from a full calculation with the $NN-\pi NN$ model of Ref. 24 and has both s and p-wave contributions (amplitudes a_1, a_2, and a_3). All curves in the shaded area refer to s-wave pion production only (amplitude a_1). The lower dashed curve results when πN rescattering is allowed only through S_{31} and N channels; allowing additional scattering through the S_{11} gives the higher dashed curve. Allowing all NN and πN rescattering through any channel *except* the S_{31} channel results in the lowest curve bordering the hashed area.

However the most remarkable feature displayed by Fig. 14 is the collapse of the s-wave cross section when πN rescattering in the S_{31} channel is not allowed. At the same time the S_{31} channel essentially on its own (see lower dashed curve) accounts for only about 2/5 of the cross section. Rescattering in the S_{11} channel is at least as important (but only in combination with S_{31}!). Thus the S_{31} channel acts as a doorway state for s-wave pion production, a result which does not appear to have been noticed previously. The origin of this behavior will be discussed shortly; here we only note that the doorway nature of S_{31} can be partly understood from the dominance of S_{31} rescattering already in the lowest order graph. On the basis of these observations, we can expect low energy $NN \to \pi d$ to play an important role in determining the off-shell behavior of S_{31} and S_{11} interactions.

(b) $pp \to pp\pi^{\circ}$

In contrast to $np \to d\pi^0$, $N\Delta$ formation in relative s-state is forbidden for the reaction $pp \to pp\pi^0$, so one expects greater sensitivity to other reaction mechanisms. With the recent measurement of $pp \to pp\pi^0$ close to threshold (see Ref. 5 and H. O. Meyer in these proceedings), we have an especially good opportunity to study non-delta degrees of freedom in the πNN system. In this case the final two protons are primarily in a $^1S_0$ state and only one amplitude, b_0 of Table I, contributes.

Here we investigate the reaction $pp \to pp\pi^0$ using the same $NN - \pi NN$ model as used above for $np \to d\pi^0$. At this stage, however, we do not do a full calculation of three-body final states. Instead we follow the argument of Gell-Mann and Watson[8] and consider the cross section to be a product of three parts: (i) a matrix element for a $2 \to 2$ process $pp \to \pi^0(pp)_{^1S_0}$ where $(pp)_{^1S_0}$ is some correlated state, (ii) a factor describing the strong final state interaction of the two protons, and (iii) a density of final states factor. It is further assumed that the matrix element of (i) is essentially independent of the internal energy E of the protons and yields a $2 \to 2$ cross section expressible as $\alpha\eta$. An integration over E then leads to the result

$$\sigma'(= \alpha'\eta^2) = c\alpha\eta^2 \qquad (13)$$

for the $pp \to pp\pi^0$ total cross section at threshold. Here c is a factor originating basically from the difference between two and three-body phase space.

Our $NN - \pi NN$ calculation is for the matrix element of (i) above. The correlated pp state is approximated by artificially binding the two protons into a $^1S_0$ deuteron-like state. Apart from this bound state, all other input is exactly the same as for the $np \to d\pi^0$ calculation above. However, because such pp binding is unphysical, we cannot claim any absolute accuracy in the calculation. Nevertheless our results should provide meaningful statements about the relative importance of various rescattering mechanisms. Just as for $np \to d\pi^0$, our calculation can also reveal if the parameter α is itself η dependent. Since the recent IUCF measurement[5] indicates that σ'/η^2 is not a constant close to threshold, knowing the η dependence of the α parameter is especially important.

We find that a change in the binding energy of the pp pair results mainly in a scaling of our cross section. For this reason we treat the proportionality factor c in Eq. (13) as a normalization parameter. In Fig. 15 we effectively compare the parameter α' coming from the IUCF experiment for $pp \to pp\pi^0$ ($\alpha' \approx \sigma'/\eta^2$), and from the $NN-\pi NN$ calculation of $pp \to \pi^0(pp)_{^1S_0}$ ($\alpha' \approx c\sigma/\eta$). A constant factor of 1.8 was used for c. The calculated value of α falls with energy (just as it did for $np \to d\pi^0$); however, the slope around $\eta = 0.2$ is close to zero and is in sharp disagreement with the measured result. Although the inclusion of p-wave pp states would increase the cross section close to $\eta = 1$ (these correspond to the α'' and β' terms in Eq. 4), the discrepancy for $\eta << 1$ might only be overcome through a better treatment of final state interactions as suggested in Ref. 5.

Fig. 15 shows the successive inclusion of multiple scattering through the πN channels N, S_{31}, S_{11}, and through the NN channel $^1P_1$; the full calculation has additional rescattering through the channels $^1S_0$, $^3P_1$, $^1D_2$, $^3D_2$, P_{11}, P_{13}, P_{31}, and P_{33}. The large effect of the $^1P_1$ channel is surprising and calls for further investigation. Fig. 15 also shows the effect of excluding the S_{31} channel from an otherwise full calculation. Just as for $np \to d\pi^0$ the cross section collapses.

We now examine s-wave πN rescattering a little more closely. Since the $\pi^0 p$ scattering amplitude is given by the isoscalar combination $t(\frac{1}{2}) + 2t(\frac{3}{2})$, we know that the physical (i.e. on-shell) $\pi^0 p$ amplitude must be small at low energies (it is exactly zero in the chiral limit). Thus it is sometimes argued that πN rescattering effects for low energy $pp \to pp\pi^0$ are very small.[37] However, this

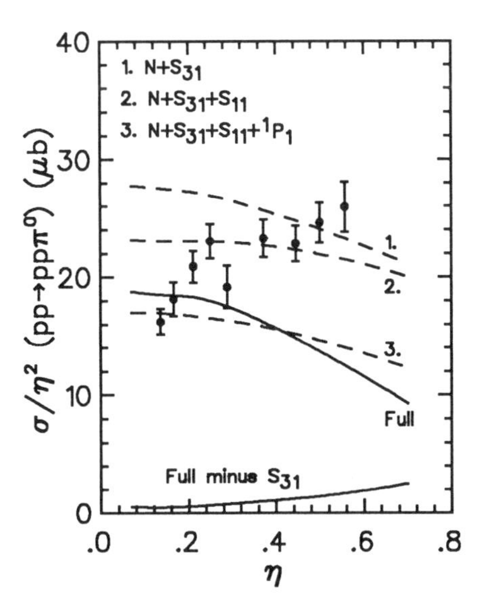

Fig. 15 Partial wave contributions to the quantity σ_{tot}/η^2 for the reaction $pp \to pp\pi^0$. Data of Ref. 5 are compared with a similar quantity coming from an $NN-\pi NN$ calculation where the final two protons are assumed to be bound. The labelling on the dashed curves indicates the 2-body partial wave channels in which multiple scattering is allowed to take place.

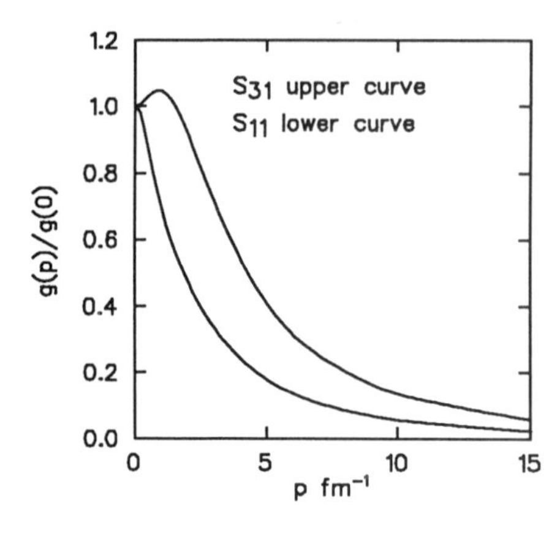

Fig. 16 Comparison of πN S_{11} and S_{31} separable form factors, $g(p)$, used in the present $NN-\pi NN$ calculation. We plot the ratio $g(p)/g(0)$ as a function of πN momentum p in order to facilitate the comparison.

argument ignores the fact that the absorbed pion may be far off-shell, as it is for example in the diagrams of Figs. 3(a) and (c). In our NN–πNN calculation, off-shell effects are included in a way consistent with two- and three-body unitarity, and we find the effects of rescattering to be very important.

Since the S_{11} scattering length is twice as big in magnitude as the one for the S_{31} channel, it may also be surprising that we find S_{31} rescattering to be the prominent one, rather than S_{11}. In order to understand these effects, let us recall that our πN amplitudes are expressed in a fully off-shell separable form as

$$t(p', p; e) = g(p')\tau(e)g(p) \tag{14}$$

where p' and p are the relative momenta, and e is the available kinetic energy in the πN centre of mass system. The quantities $g(p)$ and $\tau(e)$ play the role of form factor and full propagator for a πN quasi-particle. Fits to on-shell data determine the parameters of g. In Fig. 16 we compare the form factors of Thomas[38] for the S_{11} and S_{31} channels which we are using in the $NN-\pi NN$ calculation (here we normalize them to 1 at $p = 0$ for easy comparison). We see that even for a small deviation from $p = 0$ the S_{31} form factor dominates. And it is because of this rapid divergence of S_{11} and S_{31} form factors that the isoscalar combination $t(\frac{1}{2}) + 2t(\frac{3}{2})$ is not small even at threshold when one of the pions is off-shell. Evidently, the higher momentum components of S_{31} also accounts for its dominance in the multiple scattering process.

CONCLUSIONS

It is hoped that in the present discussion of modern theoretical approaches, a curious trend has been noticed. We have generally shown how each succeeding model incorporates new mechanisms over the previous model, so that in principle the phenomenology should become more and more refined. The curious part is that this improvement is not so clear from a comparison of numerical results with experimental data. Rather, the remarkable feature is that the theoretical results are rather similar to each other, essentially independent of the sophistication of the model - in fact they are more similar to each other than to the data! This trend is certainly not morale boosting for those who are trying to gather strength for a yet more sophisticated calculation. There appear to be essentially two possibilities: either we have not yet included *the* mechanism, in which case further heroic calculations probably cannot be avoided; or we are simply doing the whole problem using "the wrong expansion" - including progressively more pions in intermediate state may not be the most convergent way to sum the series. To illustrate the latter case, we emphasize that the idea of both the Argonne-Hannover and $NN-\pi NN$ models was to limit the number of explicit particles to three. The natural extension of these models would then limit the number of particles to four, and so on. Perhaps one needs a different expansion where *some* states with three, four, etc. pions are present at every level of the series.

Another aspect in common between most calculations is their emphasis on "calculating the mechanism". Very little effort has been given to constructing the *input* to such calculations. For example, off-shell effects are often either neglected, or included in a way that is convenient rather than theoretically motivated. Certainly in the case of the Δ resonance we have been successful at getting a qualitative agreement with most data at intermediate energies. Perhaps this experience has encouraged us for too long in looking for other "key" mechanisms.

Our calculations at threshold, however, indicate that the key may lie in a different direction. Within the $NN-\pi NN$ model, we found a remarkable sensitivity of both $np \to d\pi^0$ and $pp \to pp\pi^0$ to the off-shell aspects of s-wave πN amplitudes. Using separable two-body potentials, we find that the S_{31} channel acts as a doorway state for both reactions. This behavior is traced back to the different off-shell behavior of S_{11} and S_{31} amplitudes, see Fig. 16. Of course the crucial question now is, "how realistic are our separable s-wave potentials"? In fact according to the dispersion relation model of Reiner,[39] the πN s-wave interactions are incompatible with a one-term separable approximation. Whether he is right is not the issue here. The point is that such questions must receive more theoretical attention in the future. At low energies the constraints of chiral symmetry are especially important to take into account. [40] Any remaining ambiguities in the off-shell behavior (and there will always be some with strong interactions) need also be tested in other reactions, for example in threshold pion photoproduction.[41] In this sense, understanding low energy pion production in NN collisions and understanding the off-shell behavior of πN amplitudes may be part of the same problem.

REFERENCES

1. M. Betz, B. Blankleider, J.A. Niskanen and A.W. Thomas, "Theories of pion production in nucleon-nucleon collisions", presented at the workshop on Pion Production and Absorption in Nuclei, Bloomington, 1981 - AIP Conference Proceedings Number 79, (American Institute of Physics, New York, 1982).
2. D. B. Lichtenberg, Phys. Rev. **105**, 1084 (1957), and references contained therein.
3. H. Garcilazo and T. Mizutani, "πNN Systems" (World Scientific, Singapore, 1990)
4. D. A. Hutcheon *et al.* Phys. Rev. Lett. **64**, 176 (1990).
5. H. O. Meyer *et al.* Phys. Rev. Lett. **65**, 2846 (1990).
6. A. E. Woodruff, Phys. Rev. **117**, 1113 (1960).
7. D. S. Koltun and A. Reitan, Phys. Rev. **141**, 1413 (1966).
8. M. Gell-Mann and K. M. Watson, Ann. Rev. Nucl. Sci. **4**, 219 (1954).
9. M. Brack, D. O. Riska, and W. Weise, Nucl. Phys. **A287**, 425 (1977); J. Chai and D. O. Riska, Nucl. Phys. **A338**, 349 (1980); O. V. Maxwell, W. Weise, and M. Brack, Nucl. Phys. **A348**, 388 (1980).
10. A. König and P. Kroll, Nucl. Phys. **A356**, 345 (1981).
11. W. Grein, A. König, P. Kroll, M. P. Locher, and A. Švarc, Ann. Phys. (N.Y.) **153**, 301 (1984).
12. M. P. Locher and A. Švarc, J. Phys. G **11**, 183 (1985).
13. B. Blankleider and A. Švarc, Phys. Rev. C **42**, 1623 (1990).
14. A. M. Green and J. Niskanen, Nucl. Phys. **A271**, 503 (1976); A. M. Green, J. Niskanen, and M. E. Sainio, J. Phys. **G 4**, 1055 (1978);
15. J. Niskanen, Nucl. Phys. **A298**, 417 (1978); Phys. Lett. **141B**, 301 (1984).
16. M. Betz and F. Coester, Phys. Rev. C **21**, 2505 (1980); M. Betz and T.-S. H. Lee, Phys. Rev. C **23**, 375 (1981).
17. A. Matsuyama and T.-S. H. Lee, Phys. Rev. C **32**, 516 (1985); *ibid.* 1986; C **34**, 1900 (1986); C **36**, 1459 (1987).
18. H. Pöpping, P. U. Sauer, and Zhang Xi-Zhen, Nucl. Phys. **A474**, 557 (1987); note that the results of this paper have been retracted. More recent calculations give results of similar quality as from other models, P. U. Sauer, private communication.
19. V. S. Varma, Phys. Rev. C **163**, 1682 (1967).
20. I. R. Afnan and A. W. Thomas, Phys. Rev. C **10**, 109 (1974).
21. I. R. Afnan and B. Blankleider, Phys. Rev. C **22**, 1638 (1980).
22. Y. Avishai and T. Mizutani, Nucl. Phys. **A326**, 352 (1979); **A338**, 377 (1980); Phys. Rev. C **27**, 312 (1983).
23. A. W. Thomas and A. Rinat, Phys. Rev. C **20**, 216 (1979).

24. B. Blankleider and I. R. Afnan, Phys. Rev. C **24**, 1572 (1981).
25. I. R. Afnan and R. J. McLeod, Phys. Rev. C **31**, 1821 (1985).
26. N. Giraud, Y. Avishai, C. Fayard, and G. H. Lamot, Phys. Rev. C **19**, 465 (1979); N. Giraud, C. Fayard, and G. H. Lamot, Phys. Rev. C **21**, 1959 (1980); C. Fayard, G. H. Lamot, and T. Mizutani, Phys. Rev. Lett. **45**, 524 (1980); T. Mizutani, C. Fayard, G. H. Lamot, and R. S. Nahabetian, Phys. Lett. **107B**, 177 (1981); Phys. Rev. C **24**, 2633 (1981); T. Mizutani, B. Saghai, C. Fayard, and G. H. Lamot, Phys. Rev. C **35**, 667 (1987); F. Sammarruca and T. Mizutani, Phys. Rev. C **41**, 2286 (1990).
27. T. Mizutani, C. Fayard, G. H. Lamot, and B. Saghai, Phys. Rev. C **40**, 2763 (1989);
28. A. S. Rinat and Y. Starkand, Nucl. Phys. **A397**, 381 (1983); A. S. Rinat and R. S. Bhalerao, Weizmann Institute of Science Report, WIS-82/55 Nov-Ph, 1982.
29. T. Ueda, Prog. Theor. Phys. **76**, 729 (1986); *ibid.* 959.
30. H. Garcilazo, Phys. Rev. Lett. **45**, 780 (1980); **48**, 577 (1982); **53**, 652 (1984); Phys. Rev. C **35**, 1804 (1987); *ibid.* , 1820; C **39**, 942 (1989).
31. W. M. Kloet and R. R. Silbar, Nucl. Phys. **A338**, 281 (1980); Nucl. Phys. **A364**, 346 (1981); J. Dubach, W. M. Kloet, A. Cass, and R. R. Silbar, Phys. Lett. **106B**, 29 (1981); J. Dubach, W. M. Kloet, and R. R. Silbar, J. Phys. G **8**, 475 (1982); Nucl. Phys. **A466**, 573 (1987); J. Dubach, W. M. Kloet, and R. R. Silbar.
32. J. A. Tjon, and E. van Fassen, Phys. Rev. C **34**, 944 (1986).
33. B. K. Jennings and A. S. Rinat, Nucl. Phys. **A485**, 421 (1988); B. K. Jennings, Phys. Lett. B **205**, 187 (1988).
34. N. R. Stevenson *et al.,* Phys. Rev. Lett. **65**, 1987 (1990).
35. B. Blankleider and A. Kvinikhidze, work in progress.
36. I. R. Afnan and B. Blankleider, Phys. Rev. C **32**, 2006 (1985); B. Blankleider, in *Topical Conference on Nuclear Chromodynamics*, edited by J. Qiu and D. Sivers (World Scientific, Singapore, 1988), p. 23.
37. T. Ericson and W. Weise, "Pions and Nuclei" (Oxford University Press, Oxford, 1988).
38. A. W. Thomas, Nucl. Phys. **A258**, 417 (1976).
39. M. J. Reiner, Ann. Phys. (N.Y.) **154**, 24 (1984).
40. J. Gasser, M. E. Sainio, and A. Švarc, Nucl. Phys. **B307**, 779 (1988).
41. S. Nozawa, T.-S. H. Lee, and B. Blankleider, Phys. Rev. C **41**, 213 (1990).

Hadronic η - Production*

W. W. Jacobs
Indiana University Cyclotron Facility and Department of Physics, Bloomington, Indiana 47405 USA, and
Laboratoire National Saturne, 91191 Gif-sur-Yvette Cedex, France

ABSTRACT

Production of the η meson near threshold has been studied with increased interest during the last 6-7 years. The present status of experimental investigation and theoretical understanding of η production by hadronic means in the threshold region is briefly reviewed.

INTRODUCTION

The organizers of this conference on "Particle Production near Threshold" have asked me to discuss hadronic production of the eta meson. Why the η? Following pions, etas are quite logically among the next mesons to study in detail; they are the next heaviest non-strange mesons. Their production cross sections near threshold turn out to be relatively large for nucleon projectiles (see Fig. 1), and η is the second largest inelastic channel (after pions) for π induced reactions. But, while much more is now known about η production than that of many other mesons in this mass range, the stage of development may be compared to that of pion production studies some tens of years ago. On the other hand, the physics of the η is quite different from that of pions, and experimental and theoretical insights are new and potentially interesting for a variety of phenomena. Several recent workshops attest to the current level of interest in the subject.[1-3]

In order to limit the scope of this presentation, I will concentrate on η production in relatively light systems near threshold where the S_{11} N*(1535) resonance appears to dominate both inclusive and exclusive production. A focus will be on what is known about the fundamental, underlying reactions both experimentally and theoretically. Unfortunately, I will not have time to discuss recent interest in η by the heavy ion community,[4] the variety of efforts directed towards "tagging" etas for rare decay studies (but see B. Mayer this workshop, and Ref. 5) the possibility that the η' coupling constants may be related to the fundamental question of the proton spin,[6] that η scattering may be useful in deducing the strangeness content of the nucleon,[7] and many other interesting issues associated with this topic. For the discussion here, the relevant experimental studies come primarily from LAMPF using meson beams and from SATURNE II with beams of nucleons ... at present, few other such laboratories have sufficiently energetic beams (with the exception of BNL and KEK) for these activities.

PHYSICS ISSUES OF THE ETA

η belongs to the same SU(3) pseudoscalar meson nonet as do pions and kaons. Because they share many of the same quantum numbers, η is sometimes thought of as a "heavy π^0," but it differs from π^0 in a variety of ways. Eta is isoscaler and therefore couples only to I = 1/2 baryons (the N* resonances), and is allowed to have strange quark pairs in its wave function. In addition it may have a gluonic component.[8] Because of the heavier quarks, η may be expected to exhibit chiral symmetry violation effects more prominently than the pion. There are two physical η mesons which result from mixing (with angle[9,10] $-20°$) of the pure SU(3) states, $\eta_1 = (u\bar{u} + d\bar{d} + s\bar{s})/\sqrt{3}$ and $\eta_8 = (u\bar{u} + d\bar{d} - 2s\bar{s})/\sqrt{6}$. The lower mass"$\eta$" $= \eta_8 cos(\theta) - \eta_1 sin(\theta)$, at 548.8 MeV is primarily the octet member of the nonet, whereas the larger mass "η'" $= \eta_8 sin(\theta) + \eta_1 cos(\theta)$, at 957.5 MeV is primarily the singlet. Note: unlike the pion, eta is symmetric in its up and down quark components.

Both η mesons have relatively narrow widths (1.19 keV and 208 keV, respectively), but are too unstable to produce "beams" as do pions. η decays primarily (71%) into neutrals, with over half of that going into 2γ. η', on the other hand, decays primarily into η's, ρ's and ω's, and only a few percent of the time into 2γ.[11] The mixing of η and η', as well as the G-parity violating decay, constrain[12] the current quark masses. Isospin mixing of η and π^0 may lead to charge symmetry breaking effects in the nuclear force.[13,14]

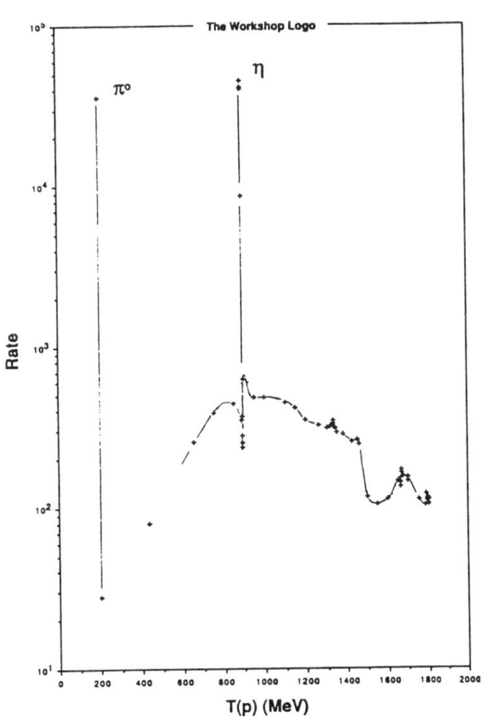

Fig. 1. "Threshold excitation curve"[15] for p + d $\rightarrow$ ^{3}He + X. The large η peak is notable along with that for π^0. The gross structure near $T_p = 1.55$ GeV is not reproduced by recent measurements.[64]

The interaction strength of η with nucleons is expected to be significantly smaller than that for pions. In a quark model picture,[16] an upper bound to the ratio of coupling strengths $(g^2_{\eta NN}/4\pi)/(g^2_{\pi NN}/4\pi)$ is 3/25. Because of its pseudoscalar nature, reduced coupling to nucleons, and larger mass, η is often neglected, or at least its strength poorly determined compared to π^0 in the various OBE models of the nucleon-nucleon interaction. Generally, however, this ratio is found to be slightly larger than the above limit. A compilation by Dumbrajs *et al.*[17] suggests 0.29, which is in accord with the SU(3) relation $G_{\eta NN} = G_{\pi NN}(3-4\alpha)/\sqrt{3}$, with $0.60 < \alpha < 0.66$ constrained by hyperon decay, and an experimentally determined value of 0.26 from baryon exchange studies.[18]

What is the nature of the low energy η-nucleon potential and its coupling to other channels? In an off-shell isobar model treated in a coupled channel formalism, Bhalerao and Liu[19] analysed the $\pi^- p \to n\eta$ data set[20] and found the s-wave ηN interaction to be slightly attractive. The various N* resonances are important to η production in this approach, and relatively strong ηNN^* couplings, deduced from πN phase shifts, are employed. Subsequent analyses using these amplitudes to construct an η-nucleus potential have led to the prediction of bound "η-mesic" states[21] for light nuclei ($A \geq 10$). The strong coupling, and large N* decay branch, to η (with distinct 2γ signature), also make η production studies potentially[22,23] useful as a probe of media effects on N*s, complementing our current understanding of such effects for the Δ. The ηN seems also to couple strongly to two and three π channels, leading to strong "cusp" effects.[24–26] In short, the physics of η production near threshold is intimately tied up with the N*'s resonances and the possibility of strong channel coupling.[25]

η PRODUCTION CLOSE TO THRESHOLD

The $\pi^- + p \to n + \eta$ excitation function peaks near threshold and gives evidence for the strong coupling of η to several N* resonances.[27] Dominant is the coupling to the $S_{11}(1535)$ resonance located only 48 MeV above the ηN threshold, with a decay width of 45-55% into ηN. Slightly higher in energy, the $P_{11}(1710)$ has a branch of 25% into ηN. Several other resonances in this energy region couple to ηN but much less strongly, as displayed in Table 1 ("below threshold," the P_{11} (1440) Roper resonance may also contribute).[11] Near threshold, N* resonances are pictured as doorway states for $\pi N \to \eta N$.

Table 1. Branching ratios for decay of some N^* resonances.[11]

Resonances			πN	ηN	$\pi\pi$N	ΛK	ΣK
N*(1440)	$1/2^+$	P_{11}	50-70%		30-50%		
N*(1520)	$3/2^-$	D_{13}	50-60%	0.1%	40-50%		
N*(1535)	$1/2^-$	S_{11}	35-50%	45-55%	~10%		
N*(1650)	$1/2^-$	S_{11}	55-65%	~1.5%	20-35%	~8%	
N*(1675)	$5/2^-$	D_{15}	35-40%	~1%	60-65%	~0.1%	
N*(1680)	$5/2^+$	F_{15}	55-65%	<1%	35-45%		
N*(1700)	$3/2^-$	D_{13}	5-15%	4%	80-90 %	~0.2%	
N*(1710)	$1/2^+$	P_{11}	10-20%	~25%	<50%	~15%	2-10%
N*(1720)	$3/2^+$	P_{13}	10-20%	~3.5%	<75%	~5%	2-5%
N*(2190)	$7/2^-$	G_{17}	~14%	~3%		~ 0.3%	
N*(2220)	$9/2^+$	H_{19}	~18%	~0.5%		~0.2%	
N*(2250)	$9/2^-$	G_{19}	~10%	~2%		~0.3%	

For nucleon induced η production, vector mesons (e.g., ρ) may be an important channel leading into N*. Thus unlike π^0, the physics of η production near threshold takes place in an environment of strong resonances and may involve vector meson contributions. There is no evidence for any N* coupling to η', and its threshold cross section is significantly smaller than η (as discussed later).

The threshold for the production of eta in the π^-p reaction is $T_\pi = 551$ MeV ($\sqrt{s} = 1488$ MeV). This is just within range of the most energetic pion beams presently available at LAMPF. An excitation function of the cross section for $\pi^-p \to n\eta$ has recently been measured[28] with the new LAMPF BGO-NaI neutrals "η spectrometer".[29] Differential cross sections are plotted as a function of pion beam energy in Fig. 2. The lowest plotted point is roughly 12 MeV above threshold. The solid curve is a fit to the data of the expression $d\sigma/d\omega = aP^*_\eta$, $a = 0.94$ μb/sr(MeV/c)$^{-1}$, where P^*_η is the ^{3}He momentum in the center of mass. This expression is valid only for S-wave production, and along with angular distributions consistent with isotropy, strongly suggest the dominance of the N*(1535) S_{11} resonance at these energies.

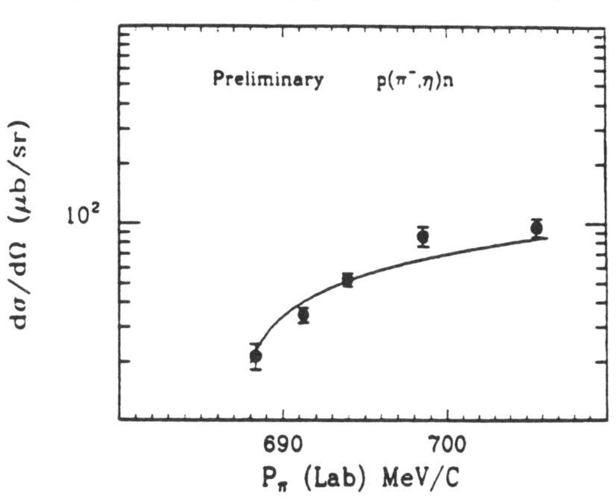

Fig. 2. Preliminary cross section[28] for $\pi^- + p \to \eta + n$ near threshold. The solid curve is a "S-wave" one parameter fit to the data.

Measurements of nucleon-induced η production very close to threshold come from Saclay.[30] Using a tensor polarized deuteron beam of nominal 1.8 GeV bombarding energy, the SPES IV spectrometer was tuned to look at ^{3}He from $d + p \to {}^3He + X$. Except at the lowest energy, 0.2 MeV above threshold, the detected ^{3}He momenta exhibit a double-peak structure, corresponding to forward and backward center-of-mass emission of the η.

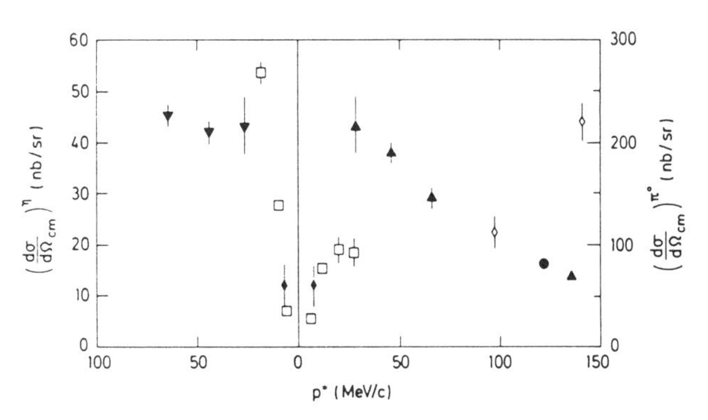

Fig. 3. Center-of-mass cross sections[30] for $d + p \to {}^3He + \eta(\pi^0)$ plotted as solid (open) symbols versus P*, the ^{3}He center-of-mass momentum. Data for forward-going mesons are to the left (η scale), backward emission to the right (compressed π^0 scale).

Cross sections for η corresponding to this forward/backward emission, and results from similar measurements for π^0 production (plotted on a compressed scale), are displayed in Fig. 3. The data points closest to threshold are near the center of the figure, with values of the cross section increasing to the left and right as the energy above threshold increases. There is a distinct qualitative difference between the η and π^0 excitation functions. While the forward/backward cross section ratio is nearly unity for η (up to $P^* \sim 40$ MeV/c), the π^0 data very quickly exhibit an aymmetry. On the other hand, extraction of a spin averaged amplitude for η production near threshold leads to a value comparable (factor 2 less) with that for π^0. The latter is supported by the threshold excitation curve[15] in Fig. 1. These measurements were also made with SPES IV, but in this case tuned for the largest missing mass at each bombarding energy.

NUCLEONIC η PRODUCTION

Very recent data on nucleon-induced η production is of a fairly preliminary nature and I have chosen to discuss these measurements later in a section dealing with "recent topics". Here we concentrate on a familiar "workhorse" of such studies, the $p + d \rightarrow {}^3He + X$ reaction, and theoretical attempts to understand the physics therein. This reaction was first studied over 30 years ago by Abashian, Booth, and Crowe[31] (whence the "ABC" effect, followed by the "DEF" effect[32]) and sporadically several times since.[26] The most recent published systematic data near the threshold region again come from Saclay.[33] They consist of very backward angle ($\sim 180°$) η production excitation functions extending roughly 1.7 GeV above threshold. They are compared to similar excitation functions for "π^0 production" in Fig. 2 of Ref. 33. One can see in this figure that the backward cross sections for η and π^0 production are similar in magnitude and reveal a fair amount of structure (the momentum transfers here are very large, ~ 13 fm^{-1}, for both η and π^0). In the case of π^0, there is a pronounced dip at the position of the opening of the η channel (similar threshold effects have been reported[34] in the t_{20} measurements for $d + p \rightarrow {}^3He + \pi^0$). The data for the η excitation are also shown in Fig. 4a. Indicated in this figure (taken from Ref. 35) are the positions of the N* resonances known to have a large decay width to η. There are cross section maxima near S_{11} and P_{11} (listed in Table I as "1535" and "1710," respectively). There may also be some enhancement at higher excitations (see Ref. 33).

Several attempts have been made to interpret this data theoretically. Two groups[36,37] in particular have extended their calculations for $p + d \rightarrow {}^3He + \pi^0$ to the case of the η. Laget and Lecolley have a microscopic (parameter free) calculation[35] which is shown along with the data in Fig. 4. They consider the basic amplitudes for several relevant diagrams contributing to coherent η production by pd capture. The new ingredient compared to the π^0 calculation is the elementary $\pi N \rightarrow \eta N$ amplitude, whose partial waves are parameterized in terms of the various N* resonances. Because of the larger mass of η relative

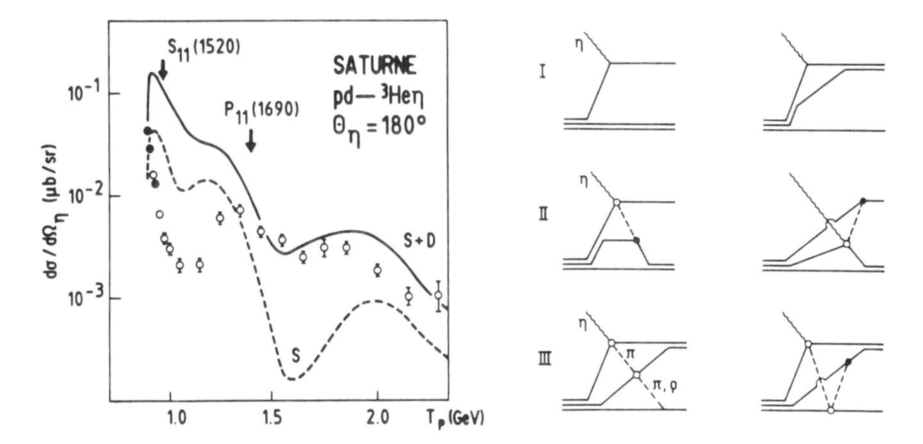

Fig. 4. Cross section data[33] for d + p $\rightarrow{}^3$He + η at 180° vs. bombarding energy are displayed part (a). The curves are the result of calculations by Laget and Lecolley[35] taking into consideration the graphs in (b).

to π^0, the authors find that plane wave and two-body mechanisms, prominent for the π^0 calculation, underestimate the eta cross section by several orders of magnitude. Three body mechanisms, involving meson double scattering processes as depicted in the lower row diagrams in Fig. 4b, appear required in order to supply sufficient cross section. The solid (dashed) curve displayed in the figure takes account of both S and D (S only) parts of the nucleon wave function that interacts with the η. The result, while giving overall the right order of magnitude, misses the data by large factors at energies below 1.5 GeV.

Germond and Wilkin have calculated[38] the p + d $\rightarrow{}^3$He + η process using a p-d cluster model for ^{3}He. The cluster model graph looks similar to the diagrams in row 2 of Fig. 4b, but of course with substantial calculational differences. The two-nucleon sub-process is the pion rescattering graph depicted in Fig. 10. While the "fundamental" two nucleon production amplitudes NN $\rightarrow$ d (d*)π^0 are fairly well known, the equivalent measurements are only now being carried out and analyzed for the case of η (see discussion later). Germond and Wilkin calculate these amplitudes for π- and ρ-induced η production, assuming the ηN amplitudes are dominated by the $S_{11}(1535)$. The ρ contribution was found to dominate, but the η cross section in such a formalism is still underpredicted by about a factor 4 (factor 20 if only π rescattering is included). Particular to the cluster model approach is the lack of any genuine 3-body mechanism which allows the sharing of the momentum transfer among more than two nucleons (and we used to worry about the "single nucleon" mechanism); specification of the high-momentum components of the ^{3}He wavefunction is much more critical and could be the source of some error.

The tensor anaylzing power t_{20} has been measured[30] for the $\vec{d} + p \rightarrow {}^3He + \eta$. It turns out to be relatively flat, negative, and significantly smaller in magnitude than that observed[34,39] for π^0. The π^0 data were understood in the p-d cluster model from the nature of the pion rescattering vertex.[37] Similar arguments would be expected to apply to η, leading to a value of t_{20} at threshold of -1.3 when D state effects are included.[38] The fact that the measurements (extrapolated to threshold, $t_{20} \sim -0.15$) give such a different value suggests the mechanism may not be that of the cluster model, or that the relevant amplitudes and wave functions are more poorly known in the momentum transfer regime for the η. The three-body dominated calculation of Laget and Lecolley[35] does predict a t_{20} around -0.5, not too far from the range of experimental values. The equivalent calculations for π^0, however, where two and three body mechanisms compete are not as successful, predicting strong oscillatory behavior of t_{20}.

η PRODUCTION WITH MESONS

Although data from eta production studies initiated by a variety of mesons exist, we discuss here only those involving pion beams fairly close to production threshold. A systematic program of such studies, ongoing by a group at LAMPF, is the source of nearly all new data in this field. The primary $\pi N \rightarrow N\eta$ reaction very close to threshold was presented in a previous section. We review the more recent exclusive and inclusive data and progress towards a theoretical description.

For pions, ${}^3He$ is the lightest target nucleus for which an exclusive measurement leading to a two-body final state can be made. The reaction $\pi^- + {}^3He \rightarrow t + \eta$ has been studied, both by detecting the outgoing tritons as well as by detecting the 2 γs from η decay. In initial studies[40] with the Large Acceptance Spectrometer (LAS), measurements of the forward going tritons were made, corresponding to very backward η center-of-mass values. For a 680 MeV/c π^- beam, the measured cross sections were found to be on the order of 1 μb/sr.

Shortly thereafter, fairly extensive studies of exclusive $\pi^- + {}^3He \rightarrow t + \eta$ cross sections[41] were initiated at forward angles using the LAMPF "π^0 spectrometer".[42] Indentification of the η's in a spectrum of the invariant mass, reconstructed from the opening angle and energy deposited by the two photons, appears fairly clean. Some of the difficulties, however, of extracting a clear exclusive signal are illustrated by the η excitation energy spectra displayed in Fig. 5. In the top panel, excitation spectra from ${}^3He(\pi^-, \eta)$ are presented at two different bombarding energies, where $E_x = 0$ corresponds to the reaction $\pi^- + {}^3He \rightarrow t + \eta$. Particularly at the higher energy, there are clearly some additional contributions not centered at $E_x = 0$, presumably arising from the t* continuum. An estimation of these effects is provided in the bottom panel, where similar spectra are displayed (taken at the same bombarding energies) for the ${}^3He(\pi^+, \eta)$ reaction which can lead only to 3 proton continuum states. Extracted angular distributions, with error bars including continuum contributions, are displayed in Fig. 6 for bombarding energies 590 - 680 MeV/c. The cross section at the higher energies are forward peaked and quite large, approaching 100 μb/sr. At the lowest bombarding energies (note: 590 MeV/c is below the $\pi^- + p \rightarrow n + \eta$ threshold), the data points are consistent with a flat (isotropic) distribution.

The curves in Fig. 6 are the results of distorted wave impulse approximation (DWIA) calculations using primary $\pi^- + p \rightarrow n + \eta$ amplitudes deduced using two different sets of πN phase shifts. The (π, η) amplitudes are again from Bhalerao and Liu[19] who used an off-shell model in a coupled channel approach to fit the $\pi^- p \rightarrow n\eta$ data set near the threshold energy. The ingredients of the model are that the $\pi^- p \rightarrow n\eta$ reaction proceeds through the N* resonances; the ηnN* couplings are deduced from the πN phase shifts. The resulting DWIA calculations describe the shape of the angular distributions quite well,[28] but are off in magnitude by factors ranging from 1.5 to 3 when the Arndt πN phase shifts are used (factors of 3 to 5 when a compilation from Berkeley is used). Perhaps the reaction has contributing terms that are more "complicated" than described in the DWIA? Are "3-body" mechanisms important as seem to be the case for the $p + d \rightarrow {}^3He + \eta$ discussed earlier?

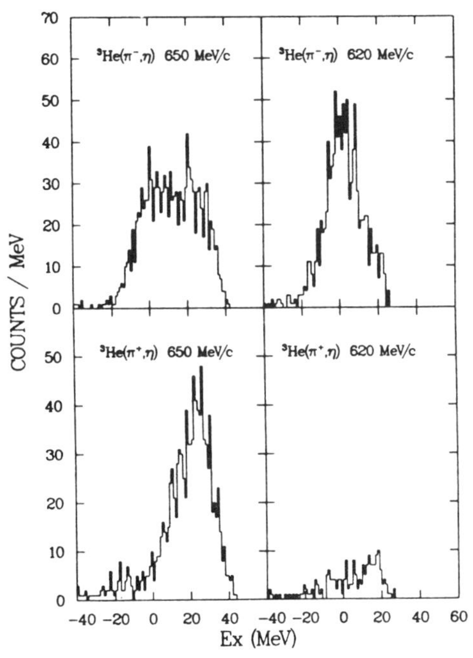

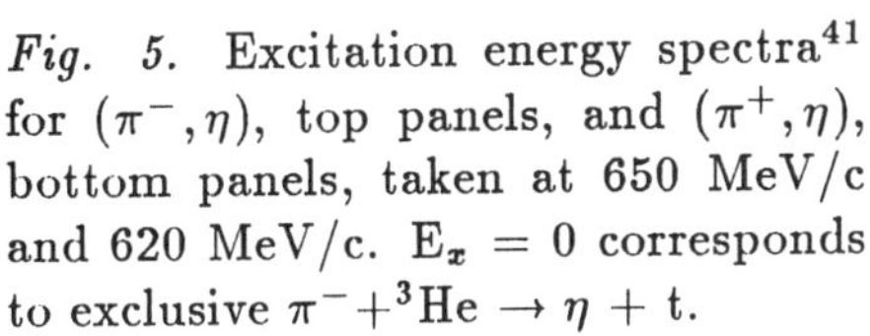

Fig. 5. Excitation energy spectra[41] for (π^-, η), top panels, and (π^+, η), bottom panels, taken at 650 MeV/c and 620 MeV/c. $E_x = 0$ corresponds to exclusive $\pi^- + {}^3He \rightarrow \eta + t$.

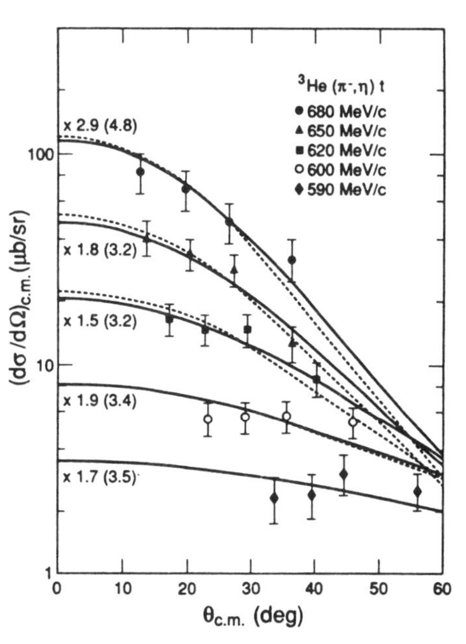

Fig. 6. Cross section angular distribution data[41] for exclusive $\pi^- + {}^3He \rightarrow \eta + t$ transitions at several energies along with DWIA calculations.

The group at LAMPF has also measured a variety of inclusive reactions using the π^0 (and more recently the "η") spectrometer.[28] The simplest and perhaps easiest system to interpret, $\pi d \rightarrow \eta NN$, has been studied and preliminary results for the η kinetic energy spectum from $d(\pi^+,\eta)$ are shown in Fig. 7. The solid curve results from a 3-body phase space calculation normalized to the data. While it describes the general shape of the distribution, some additional interactions may need to be accounted for in order to better describe the low energy region. Data have been taken for the "charge symmetric" inclusive reaction $d(\pi^-,\eta)$. A very preliminary comparison of $d(\pi^\pm,\eta)$ angular distributions appears in Fig. 6 of Ref. 28.

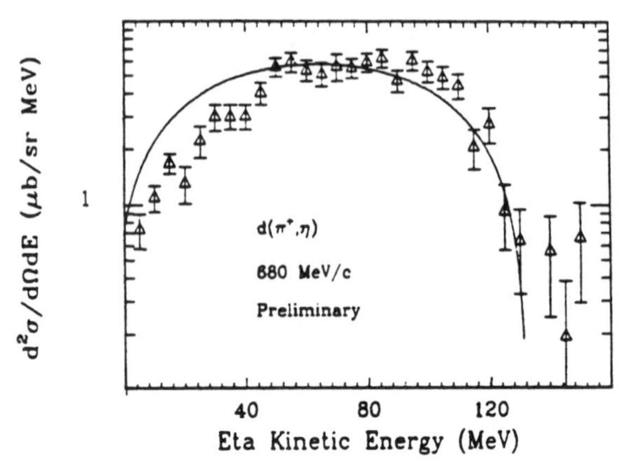

Fig. 7. Preliminary energy spectrum[28] for inclusive $\pi^+ + d \rightarrow \eta + X$ at 680 MeV/c. The curve is a three-body phase-space calculation.

Inclusive measurements for (π^+,η) on a range of target nuclei have been measured for several incident pion bombarding energies. The mass dependence of the differential cross section in the range $0 < (\theta_{lab}) < 30°$ is systematic at each energy, following roughly a power law A^x, with x $\sim$0.47 (as I sight it off Fig. 8 of Ref. 28). Attempts to fit these curves with Glauber type calculations in order to extract a total ηN cross section are probably not reliable at such low energies. Nonetheless, a value of 15 mb is extracted, which is about a factor of two less than the πN total cross section at these energies. This is consistent with what one would expect from a ηNN coupling smaller than πNN.

Shown in Fig. 8 is the η energy spectrum from the $^{12}C(\pi,\eta)$ inclusive reaction, again averaged over forward scattering angles. Calculations with an Inter-Nuclear Cascade (INC) model do a rather poor job of describing this distribution (e.g., see Fig. 7 of Ref.

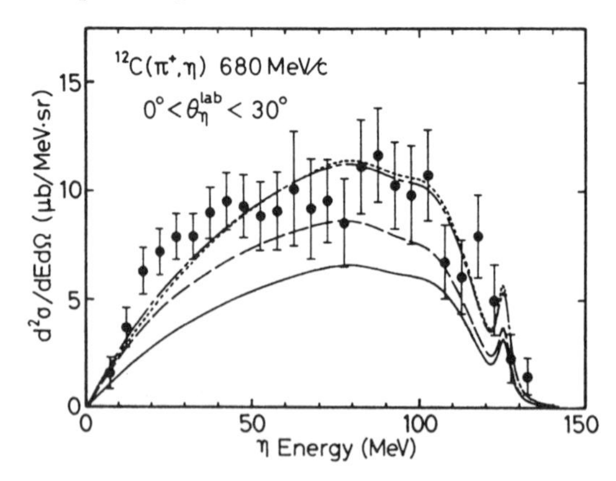

Fig. 8. Inclusive η energy spectrum[28] for $^{12}C(\pi^+,\eta)$ at 680 MeV/c. DWIA calculations[43] are shown, the two lowest use different N* spreading potentials.

28). The curves displayed here in Fig. 8 are the result of very recent DWIA calculations by Kohno and Tanabe[43] employing Green's function methods in order to calculate continuum states. They generate an η-nucleus potential by folding medium-modified scattering amplitudes (obtained following the method of Bhalero and Liu), with nuclear wave functions. The most recent Arndt πN phase shifts were employed in the analysis. Near threshold, the ηN interaction takes place predominantly through the N* resonance $\eta N \rightarrow S_{11}(1535) \rightarrow \eta N$. Additional S_{11} absorptive effects are treated by adding a phenomenological "spreading potential." The solid (dashed) curves in Fig. 8 are the results of calculations with spreading potentials of strength $V_{N^*} = -50 - 50i$ ($V_{N^*} = -50i$). A normalization factor of 1.7 applied to the solid curve leads to the dot-dashed curve, in fair agreement with the data (the dotted curve is a calculation assuming plane waves for the η). As in the exclusive measurements, we see that the DWIA calculation underestimates the data by roughly a factor of two. The shape of the spectrum, however, is not sensitive to the magnitude of the spreading potential.

The above work of Kohno and Tanabe[43] also addresses the question of η-mesic bound states. Measurements at Brookhaven[44] have shown no peak from the $A(\pi^+,p)\eta B$ reaction for ^{16}O at the 3σ confidence level (for a peak width of 9 MeV FWHM, corresponding to 8.7 μb/sr$-$MeV). This is about 1/3 of the cross section predicted by Liu and Haider.[21] The new calculations by Kohno and Tanabe show that detection in singles measurements of these bound, or nearly bound, scattering states will be difficult at best, and that the predicted position, width, and strength of such objects depends upon choices for the real and imaginary parts of the N* spreading potential. In any case, it is clear that coincidence measurements (detecting the decay products of the mesic states) will be required in order to make further progress on this isssue. While such measurements have been attempted at LAMPF,[45] and are being considered at Saturne,[46] it is fair to say that there is no definitive evidence yet (although there have been some "suggestive spectra"[47]). Similar resonant effects may occur in DCX reactions[48] and are being pursued experimentally.[49]

RECENT HADRONIC PRODUCTION AND WORK IN PROGRESS

I suppose that this topic might have been the whole discussion in this presentation. As I have tried to indicate, however, some of the more "recent" history is also quite relevant in this context. For the future, many of the preliminary results from LAMPF I have briefly mentioned are only representative of more extensive measurements that have been made. In particular, the inclusive data may be the next to be published, and should include further studies of energy and target dependences, as well as an investigation of the isotope dependence of π^+ vs. π^- inclusive production. Continued effort on more careful theoretical analyses will be necessary in order to further understand the "primary" $\pi N \rightarrow \eta N$ processes in these production mechanisms. Polarization measurements (some older data exist, but only at higher energies[50]) would be extremely useful in order to pin down the fundamental amplitudes.[8] In the case of nucleonic η production, there is the possibility that new results will come out soon from recent and ongoing experimental programs. I now summarize what some of these efforts are and have been.

We begin with the recent "archeological" work[51] regarding n-p measurements performed at Saclay with SATURNE nearly 15 years ago,[52] originally initiated with the purpose of further investigating the "ABC" anomaly (an enhancement in the $\pi\pi$ mass spectrum close to threshold). Fig. 9 shows the deuteron momentum spectrum from the n + p → d + X reaction taken at a forward angle of 1.5°. At either end of the spectrum, one sees a peaking corresponding to the ABC effect.[31] The contention of the recent work[51] is that the prominent peak in the middle of the spectrum, and absent from spectra taken at larger angles, is mainly due to the n + p → d + η reaction very close to threshold. The explanation is roughly as follows. The nominal neutron beam energy is below threshold, and there is some uncertainty in the angle calibration, yet the harder part of the neutron spectrum (roughly 10%) is expected to be above the 1.976 GeV/c threshold and can produce η's at the most forward angle. A Monte Carlo simulation (for four angles simultaneously) was carried out to describe the η, $\pi^0, 2\pi$, and "3π" contributions to the spectrum. Those contributions are indicated in Fig. 9. In this analysis, the total η production via np → dη is σ_{max} = 110 ± 20 ub, much larger than the analogous π^0 production. This corresponds to an isoscalar threshold amplitude $g_0 = 0.10 \pm 0.02$ fm^2 which is in the range calculated by Germond and Wilkin[53] in a model where the $S_{11}(1535)$ resonance formed by ρ exchange dominates the production process as illustrated Fig. 10. This experimental result, taken under rather marginal conditions, should certainly be carefully remeasured at the earliest opportunity.

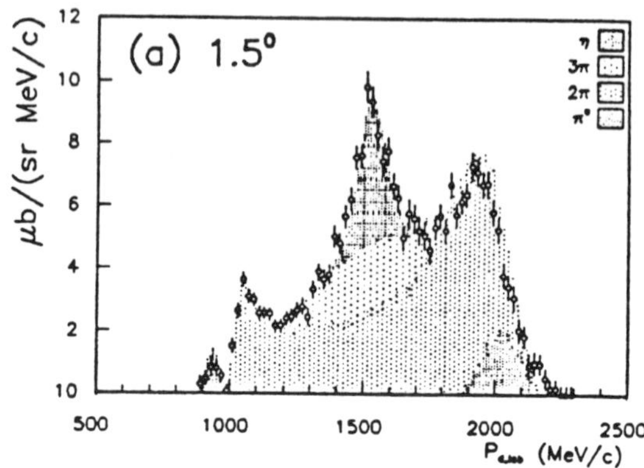

Fig. 9. Differential cross section vs. deuteron momentum for n + p → d + X. Single and multiple pion contributions are indicated by the shaded areas. The peak in the center is attributed[51] largely to X = η.

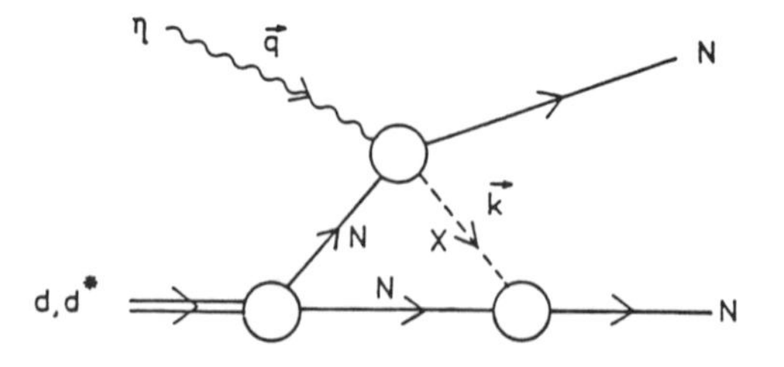

Fig. 10. Diagram of a two nucleon scattering mechanism (X = π, η, ρ) leading to η production.[53]

From the foregoing, it is obvious that it would be extremely interesting to investigate the isospin = 1 channel, pp → ppη, and deduce the corresponding amplitude near threshold (for recent π^0 results, see Ref. 54 as well as H.-O. Meyer, this conference). Such measurements for η have in fact been performed by O. Bing and collaborators [46] by measuring the two emerging protons in the SPES III spectrometer at SATURNE II. Several energies near the threshold region have been investigated, and the data is currently undergoing analysis. A typical two-dimensional plot of the 2 detected proton momenta is shown in Fig. 11 for T_p = 1265 MeV. The ellipse corresponding to the X = η kinematic locus for p + p → p + p + X is clearly visible in the lower left corner, followed by diagonal bands corresponding to 2π and then single pion production near the middle of the figure. The η total cross section is on the order of 1μb. This corresponds roughly (perhaps a little low) to the "most realistic" of the calculations cited above by Germond and Wilkin,[53] where it is expected that isospin factors alone suppress pp → ppη relative to np → dη by about an order of magnitude. The basic NN → nnη process has also been looked at theoretically by Liu and Wellers.[55] They find that the intermediate π and ρ contributions interfere constructively. Experimentally, in addition to looking for η production, O. Bing *et al.* have also run at several higher energies in order to search for η' production near threshold. During my talk I showed a slide hinting at η' production with a cross section about 2 orders of magnitude smaller than the η. Since that time I understand[46] there is now very clear evidence for the η' in pp → ppX at 16.4 MeV above threshold. Presumably this cross section, weak relative to η, results from the lack of any resonant behavior in the η' channel (far above threshold, the η, η' cross sections are of about the same magnitude).[20]

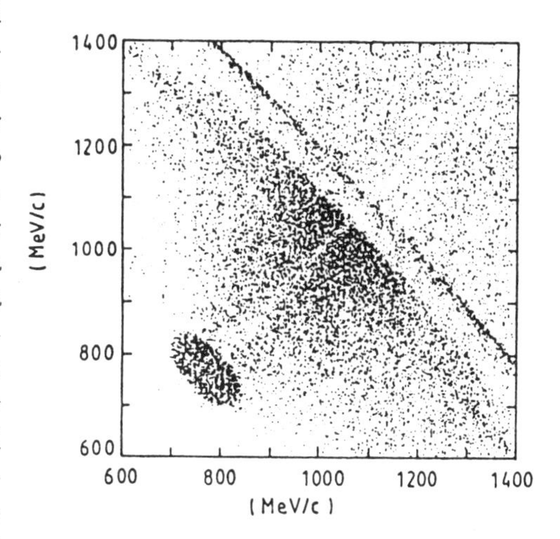

Fig. 11. Momenta for the two detected protons (plotted one vs. the other) from the reaction p + p → p + p + X

Other activities recently carried out with the SPES III are new measurements of the cross section and analysing power angular distributions of $\vec{p}$ + d → ^{3}He + X for a few energies near threshold.[56,57] The aim of the work is to look at these quantities for several of the heavier mesons. The peak from η production is a dominant feature in the missing mass spectrum and the new analysing power information should be among the first results to come out of these studies.

Another source of information about the η-N interaction and the basic NN → NNη process can come from the A(p,η) reaction, and recent theoretical studies for exclusive A(p,p$'\eta$) show that the excited nuclear states should carry the spin-isospin signature of the exchanged meson.[58] While some such exclusive measurements may be possible in the future, inclusive data have already been taken for several energies below the free η production threshold[59] using the PINOT spectrometer at SATURNE II. This "neutrals" spectrometer, developed by a group from Torino, is described in Refs. 60,61. Preliminary results[62] for inclusive η production by protons $T_p \leq 1$ GeV, as function of target mass, are presented in Fig. 12. The straight lines indicate a $A^{2/3}$ power law behavior for all of the points (with the exception of ^{6}Li) on this doubly logarithmic plot. Such a relation suggests a "black disk" behavior of the reaction, although no such explanation is offered. This is a somewhat steeper slope than we encountered in a similar plot for the (π,η) inclusive cross sections in an earlier section.

In short, there are several new results we hope will soon appear, whose preliminary nature I have indicated during this presentation. To those whose work I have failed to mention, I can only offer my sincere apologies. To those who made an extra effort to communicate their most recent results, my sincere thanks.

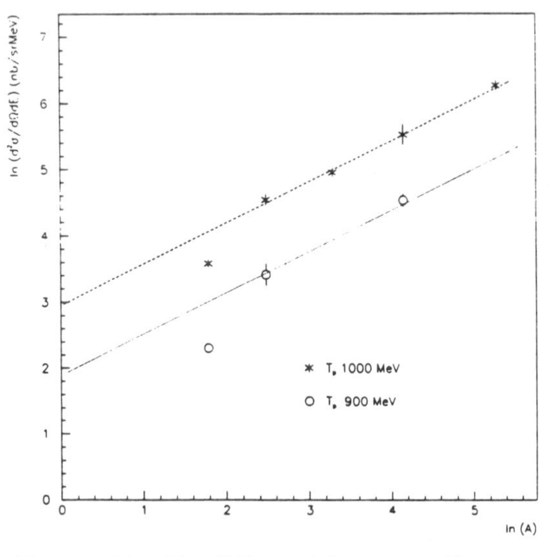

Fig. 12. Doubly differential cross section vs. target mass for inclusive (p,η). Data for two energies are shown.[62] The straight lines indicate a $A^{2/3}$ power law behavior.

SUMMARY REMARKS

The subject of hadronic η production continues to be "hot", although progress is somewhat "deliberate." New published results from the LAMPF effort may soon be forthcoming in terms of the (π,η) inclusive and exclusive measurements we have discussed. In reality, as I understand it, improvements must be made to the energy resolution in the π channel before the benefits of the improved resolution of the "η spectrometer" can be realized, and new exclusive measurements undertaken.

On the nucleonic production side, analysed data on the "fundamental" NN $\rightarrow$ NNη reaction near threshold should soon be available. Similarly, analyzing power data for η (and ω) production in $\vec{p} + d \rightarrow {}^3He + X$ are also being analyzed. Systematic results for (p,η) inclusive reactions on a variety of target nuclei should be forthcoming from the PINOT and perhaps other neutrals spectrometers (e.g., SPES 0) in the not too distant future.

Many fundamental questions remain, however, about the reaction mechanism and how to treat these processes theoretically. The question of strong η coupling to other channels resulting in "cusps" and other phenomenological structures remains to be investigated and understood in detail. Future progress on probing the existence of η mesic states rests upon vastly improved experimental tests. The question as to whether or not the N* resonances, and influence of the media thereupon, can be exploited remains an open question.

As a final note, I should say in the spirit of this conference that several new accelerators will shortly come on line with experiments to address these topics perhaps more succinctly (e.g., CELSIUS, COSY, ...), and thus we can expect an increased emphasis on this physics soon. In the meantime, will someone please (re)measure the np $\rightarrow$ dη cross section discussed in the "archeology" part of the previous section?

* Work supported in part by the US NSF and CEA/IN2P3 Saclay

1. *Workshop on the "Production and Decay of Light Mesons,"* (Ecole Polytechnique, Paris), ed., P. Fleury (World Scientific, Singapore, 1988).
2. *Proc. "Phys. with Light Mesons and 2nd Int'l Wkshp. on π^-N Phys.,"* (Los Alamos, 1987), eds., W.R. Gibbs and B.M.K. Nefkens (LASL, LA-11184-C).
3. *Proceedings Workshop on "Rare Decays of Light Mesons,"* (Gif s/Yvette, 1990), ed., B. Mayer (Editions Frontiers, 1990).
4. A. De Paoli *et al.*, Phys. Lett. **B219**, 194 (1989).
5. Y. Le Bornec, *Proc. of PSI Spring School on "Prospects of Hadronic Phys. at Low Energies,"* (Zuoz, Switzerland, April, 1990), and references therein.
6. A.V. Efremov *et al.*, Phys. Rev Lett, **64**, 1495 (1990).
7. C.B. Dover and P. M. Fishbane, Phys. Rev. Lett. **64**, 3115 (1990).
8. L.C. Liu, in *Proceedings of "3rd Int'l Symposium of Pion-Nucleon and Nucleon-Nucleon Physics,"* (Gachina, April, 1989), p. 299.
9. F.J. Gilman and R. Kauffman, Phys. Rev. **D36**, 2761 (1987).
10. N.A. Roe *et al.*, Phys. Rev. **D41**, 17 (1990).
11. "Review of Particle Properties," Phys. Lett. **B239**, 1 (1990).
12. H. Leutwyler and J. Gasser, Phys. Rep. **87**, 77 (1982).
13. S.A. Coon and B.M. Preedom, Phys. Rev. C **33**, 605 (1986).
14. G.A. Miller, B.M.K. Nefkens, and I. Slaus, Phys. Rep., to be published.
15. F. Plouin, in Ref. 1, p. 114.
16. Brown and Jackson, "The NN Interaction," (North Holland, 1976).
17. O. Dumbrajs *et al.*, Nucl. Phys. **B216**, 277 (1983).
18. H. Becker *et al.*, Nucl. Phys. **B167**, 292 (1980).
19. R.S. Bhalerao and L.C. Liu, Phys. Rev. Lett. **54**, 865 (1985).
20. V. Flaminio *et al.*, CERN-HERA 79-03 (1979).
21. L.C. Liu and Q. Haider, Phys. Rev. C **34**, 1845 (1986).
22. M.G. Huber, Nucl. Phys. **B279**, 249 (1987).
23. M.G. Huber and B.C. Metsch, in Ref. 2, p. 91.
24. R. Bhandari and Y. Chao., Phys. Rev. D **15**, 192 (1977).

25. C. Wilkin in Ref. 1, p. 187, and references therein.
26. F. Plouin, in *Proceedings 17th INS Int'l Symposium on Nucl. Phys. at Intermediate Energy* (Tokyo, 1988), p. 251.
27. R.M. Brown *et al.*, Nucl. Phys. **B153**, 89 (1979).
28. J.C. Peng, *Proc. of the "3rd Int'l Symp. on Pion-Nucleon and Nucleon-Nucleon Physics,"* (Gatchina, 1989), p. 315: *ibid, Proc. of "17th INS Int'l Symp. on Nucl. Phys. at Intermediate Energy,"* (Tokyo, 1988), p. 233.
29. M. Leitch *et al.*, *Proc. "LAMPF Wrkshp. on γ and Neutral Meson Phys.,"* (Los Alamos, 1987), eds., H. Bauer *et al.*, (LA-11177-C), p. 393.
30. J. Berger *et al.*, Phys. Rev. Lett. **61**, 919 (1988).
31. A.Abashian, N.E. Booth, and K.M. Crowe, Phys. Rev. Lett **5**, 258 (1960).
32. J. Banaigs *et al.*, Phys. Lett. **45B**, 394 (1973).
33. P. Berthet *et al.*, Nucl. Phys. **A443**, 589 (1985).
34. C. Kerboul *et al.*, Phys. Lett. **B181**, 28 (1986).
35. J.M. Laget and J.F. Lecolley, Phys. Rev. Lett. **61**, 2069 (1988).
36. J.M. Laget and J.F. Lecolley, Phys. Lett. **B194**, 177 (1987).
37. J.F. Germond and C. Wilkin, J. Phys. **G14**, 181 (1988).
38. J.F. Germond and C. Wilkin, J. Phys. **G15**, 437 (1989).
39. A. Boudard *et al.*, Phys. Lett **B214**, 6 (1988).
40. J.C. Peng *et al.*, Phys. Rev. Lett. **58**, 2027 (1987).
41. J.C. Peng *et al.*, Phys. Rev. Lett., **63**, 2353 (1989).
42. J.C. Peng *et al.*, Nucl. Inst. Meth. **A261**, 462 (1987).
43. M. Kohno and H. Tanabe, Nucl.Phys. **A519**, 755 (1990).
44. R.E. Chrien *et al.*, Phys. Rev. Lett. **25**, 2595 (1988).
45. B.J. Lieb, L.C. Liu, *et al.*, LAMPF P^3-East expt. No. 1022.
46. O. Bing et al., SATURNE Expt. 174; O. Bing, private communication.
47. B.J. Lieb *et al.*, Proc. Int'l Nucl. Phys. Conf., (Sao Paulo, 1989), p. 146.
48. Q. Haider and L.C. Liu, Phys. Rev. C **36**, 1636 (1987).
49. C.L. Morris, C.F. Moore, *et al.*, LAMPF expt. 1140.
50. R.D. Baker *et al.*, Nucl. Phys. **b156**, 93 (1979).
51. F. Plouin, P. Fleury, and C. Wilkin, Phys. Rev. Lett. **65**, 690 (1990).
52. F. Plouin *et al.*, Nucl. Phys. **A302**, 413 (1978).
53. J.F. Germond and C. Wilkin, Nucl. Phys. **A518**, 308 (1990).
54. S. Stanislaus *et al.*, Phys. Rev. C **41**, 1913 (1990).
55. L.C. Liu and F. Wellers, Prog. at LAMPF 1988, LA-11670-PR, P. 103.
56. E. Loireleux, Thesis 1990, Internal Report IPN Orsay, IPNO-T9002.
57. Y. Le Bornec, SATURNE Expt. 186, Y. Le Bornec private communication.
58. L.C. Liu, J.T. Londergan, and G.E. Walker, Phys. Rev. C **40**, 832 (1989).
59. R. Bertini *et al.*, SATURNE Expt. 125, and private communication.
60. E. Chiavassa *et al.*, Nuovo Cimaento **101A**, 805 (1989).
61. G. Caviasso *et al.*, Nuovo Cimaento **103A**, 285 (1990).
62. E. Chiavassa *et al.*, Nucl. Phys. **A519**, 413c (1990).
63. A.M. Bergdolt *et al.*, CRN Strasbourg Rapport D'Activite 1989, p. 83.
64. R. Jahn, R. Siebert, SATURNE Expt. 222; R. Jahn, this conference.

MESON PRODUCTION AND VELOCITY MATCHING

K. Kilian and H. Nann*
Institut für Kernphysik, Forschungszentrum Jülich, D–5170 Jülich, FRG
*Indiana University, Cyclotron Facility, Bloomington, Indiana 74705, USA

Abstract
A sequential mechanism of meson production in proton–nucleus collisions leads to meson production at beam energies below the nucleon–nucleon–meson–threshold. In a cascade–like sequence, a light intermediate particle is created in a first nucleon nucleon collision. In a second step, this intermediate particle produces a heavier outgoing meson by interacting with a second nucleon of the same target nucleus. If besides the light intermediate particle there is a deuteron produced in the first step, this can fuse with the nucleon N from the second step to form a bound final (3N) baryonic system. If the velocities of the baryons produced in the two steps are matched then their relative energy is close to zero. Then the 3N fusion to one nucleus will be enhanced and the meson production cross section for a two body channel should show a bump in the excitation function which has nothing to do with a resonance. Results of Monte–Carlo simulations of this sequential reaction mechanism are shown.

Kinematical considerations

The new storage rings with phasespace cooling like the IUCF cooler in Bloomington, CELSIUS in Uppsala and COSY in Jülich will open possibilities to study meson production close to thresholds. They will provide the high resoultion which is needed here [1]. In elementary proton proton interactions thresholds are interesting since only a few partial waves are contributing so that the complex hadronic process can be more easily interpreted. Meson production on nuclei in p–A (or even in A–A) interactions is more complicated and the above argument for thresholds becomes even more important. Thus in a first step, we will concentrate on the simplest nuclear target, the deuterium.

For two step processes on the two nucleons of the deuterium target we make two simple but important statements concerning the kinematics:

1. There is a reduction of the threshold for particle production already without the necessity of Fermi momentum [2] or cooperative nuclear effects [3].

Take for example η production. Its threshold in p+p interaction lies at 1985.9 MeV/c. If we however first produce a pion in a p+p reaction (step A) and use this pion in a pion–neutron reaction (step B) to make the η, then the threshold is only at 1584.7 MeV/c.

The momentum transfer in both reaction steps is about 900 MeV/c and so high that even in the case when the pion goes only the distance of a few fm inside the deuteron range, we can assume that the reaction partners have still particle properties. Thus normal on–shell particle interaction is still an allowed approximation.

One should expect that two step processes are relatively strong in the "sub-threshold" region (at momenta below the elementary threshold) where quasi-free one step meson production is kinematically suppressed. Examples of threshold momenta for proton induced meson production in one– and two–step processes on a two–nucleon system are given in Table 1. The thresholds for heavy meson production in two–step processes shown in Table 1 lie close to the lowest possible kinematical threshold which is given by the p + (2N) → M + (3N) single–step reaction.

Table I: Threshold momenta in MeV/c for proton induced meson (M) production in a two–step process with intermediate particles (m) and in a single step process. Brackets around the baryonic systems indicate zero relative energy ϵ. This assumption allows to compare the one and two step thresholds. The values are calculated for

step A: $p+n \rightarrow m^o + (n+p)_{\epsilon=0}$ step B: $m^o + p \rightarrow M^o + p$

final heavy meson M	intermed. meson m	two step threshold $N+N \rightarrow m+(2N)$ $m+n \rightarrow M+N$	single step threshold on $(2N)$ target $N+(2N) \rightarrow M+(3N)$	momentum where velocity matching occurs in the $(2N)+N$ exit system
π	γ	676. 9	657. 2	800
	π	776. 5		940
η	γ	1584. 6	1584. 5	1590
	π	1587. 7		1590
	η	1985. 9		2120
ω	γ	2083.	2072.	2120
	π	2080.		2120
	η	2124.		2190
	ω	2669.		2680
ϕ	γ	2621.	2579.	2700
	π	2617.		2700
	η	2579.		2580
	ω	2707.		2780
	ϕ	3405.		

2. The experimentally interesting two body channels with ^{3}He or ^{3}H as final nuclei,

$$p + d \rightarrow (3 \text{ nucleons bound}) + X$$

are strongly enhanced in certain energy and angular regions by the kinematics of two step processes when "velocity matching" between the three nucleons of the three–baryon system occurs.

It is evident that those channels are strongly favoured only when already in the first step a deuteron is formed. In the following we therefore consider as model cases reactions of the type

Step A	$p + p \to \pi^+ + d$	
Step B	$\pi^+ + n \to X + p$	(or $X^+ + n$)
Total	$p + d \to X^0 + {}^3He$	(or $X^+ + {}^3H$)

We can calculate the relative energy ϵ between the deuteron from step A and the proton from step B. ϵ is minimal if the velocities of the two partners match. The invariant mass squared $S(d+p) = (V(d) + V(p))^2$ depends on the four–vectors V of d and p. The excess energy in the d+n exit system, $\epsilon = \sqrt{S} - M(d) - M(p)$ can be used to get an estimate for the fusion probability P into ${}^3He$. The fusion becomes the more probable the smaller the excess energy ϵ is in comparison with the binding energy E_b of the three body system (5.49 MeV for p+d binding in ${}^3He$).

$$P = \exp(-\epsilon/E_b)$$

Exactly the same effects of an enhanced final state interaction in corners of the phase space where relative energies of reaction partners are small, also happen in "recoilless" kinematics. Here a particle is produced at rest in the laboratory system so that it can favourably interact with the residual target nucleus [4].

Examples of velocity matching

This "velocity matching" in sequential meson production is shown in Fig. 1 for the case of the nd $\to$ (3N) + π reaction in which the intermediate particle $m(=\pi)$ is produced at forward angles and then backscattered or undergoes charge exchange in step B. Figures 1a and b show the relative energy ϵ as a function of incident neutron momentum and the probability P for the formation of the final bound (3N) baryonic system, respectively. One sees that the forward going intermediate pions can give rise to small values of ϵ. Folding in the pp $\to$ dπ cross section for step A and restricting the emission of the final pion to very backward angles, the excitation function, depicted in Fig. 1c, is obtained. We predict a cross section enhancement around (1025 ± 100)MeV/c. Data of Rössle et al. [5] for backward produced pions in the nd $\to {}^3He\ \pi^-$ reaction show indeed this behavior (see Fig. 1c). We want to stress that this type of "velocity matching" effect might be misinterpreted as a (3N) resonance in the direct channel. Moreover, also in the p+d $\to$ t+π^+ reaction an enhanced backward emission of pions was observed at T_p = 470 MeV (p_p = 1050 MeV/c) incident energy, which so far no theoretical calculation was able to reproduce (for details see Ref. [6]). The present model of sequential meson production accounts for this enhanced backward emission.

A more realistic description of the two step process has to take into account the full angular– and energy range of reaction partners in both steps. Also one has to consider that the proton and neutron are bound in the deuteron. They have Fermi momenta $\vec{q}(p) = -\vec{q}(n)$ with a continuous distribution and consequently they have off shell masses m^*. The deuteron mass $m(d) = m(p) + m(n) - 2.22$ MeV has to be reproduced by the total p and n energies

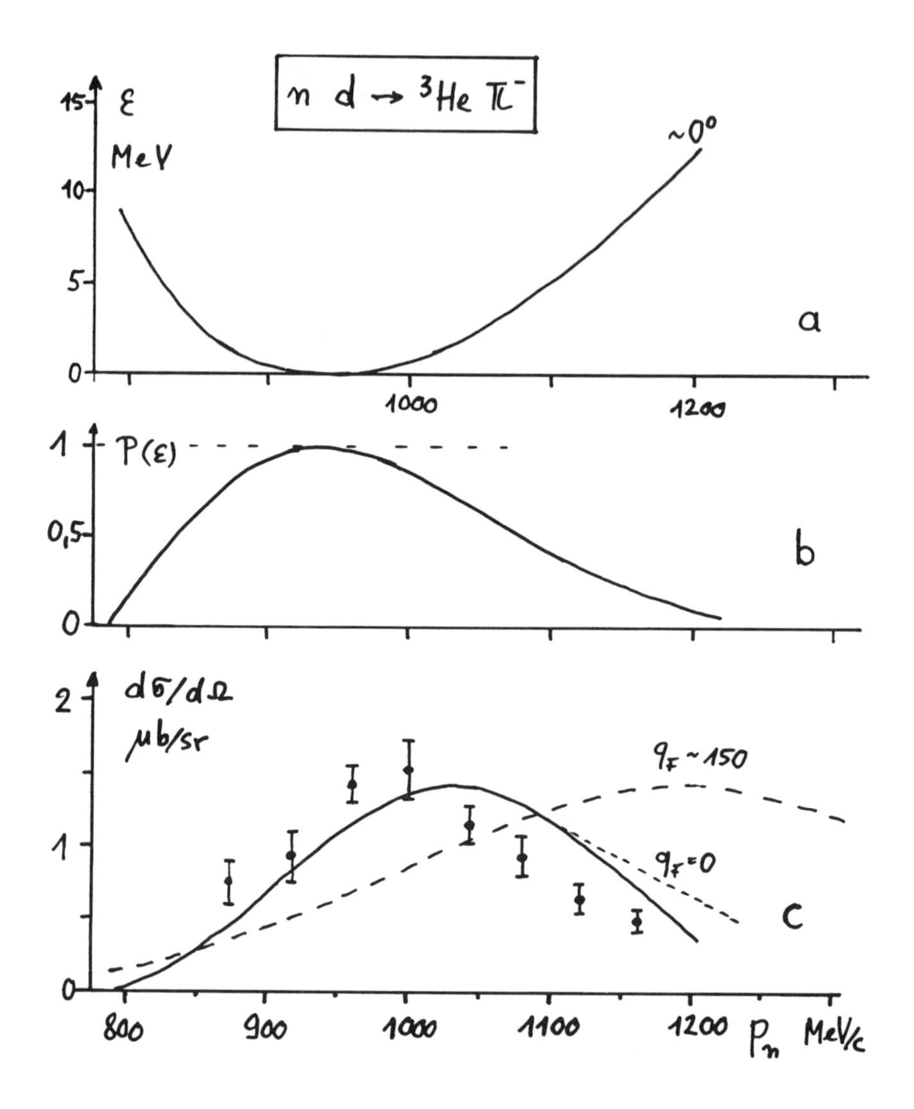

Figure 1: a) Relative energy ϵ of the outgoing p and d in a two step π rescattering process with step A: n+n → π^-d and step B: π^-p→π^-p in quasifree kinematics without Fermi momentum. Forward 0° angles of the outgoing π^- in step A are considered. π^- scattering angles in step B are backward. b) Shows the corresponding fusion probability $P = \exp(-\epsilon/E_b)$ for (p+d)↔^{3}He. c) Comparison of measured n+p→π^- ^{3}He data [5] with a prediction using the fusion probability in Fig. 1b multiplied with the total np→π^-d cross section (solid line) and a prediction with a Monte Carlo Simulation (dashed line) on bound nucleons with Fermi momenta $\langle\vec{q}\rangle$ ~ 150 MeV/c and energy dependent differential cross sections in step A. In step B a constant cross section is used.

$$m(d) = \sqrt{m^*(p)^2 + q^2} + \sqrt{m^*(n)^2 + q^2}\,.$$

If we assume that the off–shell masses of p and n have the same ratio as the on–shell masses, then we can calculate for each q the kinematically relevant fourvectors V of the bound and moving reaction partners in the deuteron target.

$$V(m^*) = (E^*, \vec{q}) = (\sqrt{m^{*2} + q^2}, \vec{q}) = ((m - 1.11\ \text{MeV}), \vec{q})$$

We have done Monte Carlo simulations of several reactions where the result of the first step was used as input for the second step. Fermi momenta with an average of $<|\vec{q}|> = 150$ MeV/c were used. The energy dependent differential cross section for $p+p \to \pi^+ + d$ was included in step A. For step B we assumed an energy independent isotropic (s–wave) differential cross section. The probability for three baryon fusion in the final state was caltulated for each event. The same was done for effects of acceptances in solid angle $\Delta\Omega$ and momentum Δp. The effect of the averaging over all kinematical variables is shown in Fig. 1c for $nd \to \pi^- \, {}^3He$. There is a shift of the maximum towards higher energies. It is worth noting that without Fermi momentum the averaging leads to a prediction very close to the simple 0° approximation in fig. 1c.
A famous example of a reaction with spectacular threshold behaviour is $p^+d \to \eta \, {}^3He$ [7]. A spike appears in the 0° ${}^3He$ production close to the η threshold in $p+d \to {}^3He + \eta$.
This spike has stimulated interest. Two step processes [8] and coupled channel effects [9] have been used to describe properties of this reaction. Its very narrow shape is caused by the narrow momentum acceptance of the used magnetic spectrometer [7]. Independent of this narrowness we find, that there should be a strong enhancement of the total $pd \to {}^3He\,\eta$ two body cross section due to velocity matching just close to threshold. Fig. 2 shows our prediction for a large acceptance measurement of $pd \to {}^3He\,\eta$.

In Fig. 3 we compare a two step simulation and data of the reaction $pd \to X{}^3He$. We simulated the experimental situation described in ref. 7. Here the momentum acceptance of the ${}^3He$ spectrometer sitting at 0° was always tuned such that it accepted ${}^3He$ corresponding to threshold production of the actual maximal possible mass X. As one can see the assumption of a two step process explains the appearance of a wide enhancement of the forward ${}^3He+X$ cross section. The opening of several inelastic channels in step B with rising energy (and rising mass X) should give an increasing total cross section in step B. There should be therefore a slower decrease of the high energy cross section.

Conclusion

The concept of two step interactions with final state fusion of particles which have matched velocities helps to understand experimental results in few body interactions, especially in two body meson production with protons on deuterium close to thresholds.
Another manifestation and a further proof of the two step processes should show up in the three body exit channels

$$p + d \to X^o + p + d \qquad (\text{or } X^+ + n + d).$$

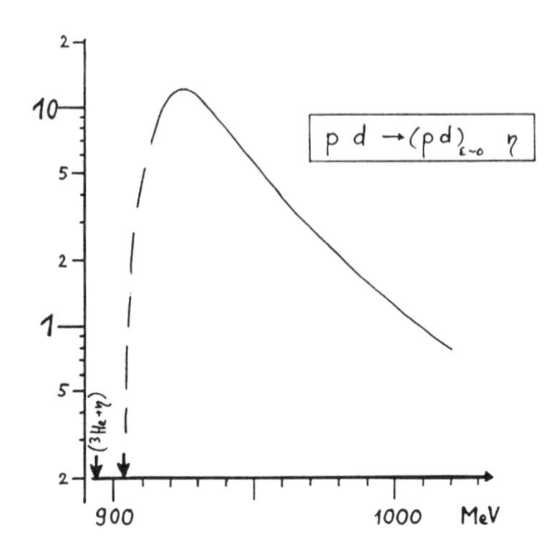

Figure 2: Probability for ^{3}He fusion in a two step process A: $pp \to d\pi^+$ B: $\pi^+ n \to p\eta$ overall: $pd \to (pd)\eta$ (assumptions see text) as function of energy. Note the rather narrow energy band. The arrows indicate the thresholds for pd reactions to ^{3}He η and to pd η. The (off shell) ^{3}He fusion leads to a shift of the calculated excitation function, so that it starts at the ^{3}He η threshold.

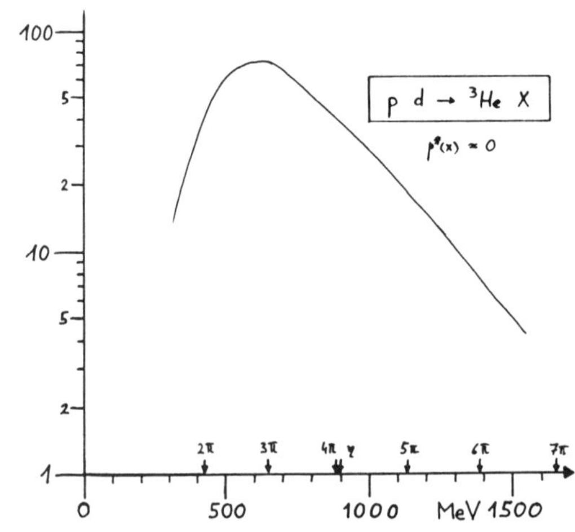

Figure 3: Excitation function for ^{3}He fusion calculated in a two step process (A: $pp \to d\pi^+$ B: $\pi^+ n \to pX$) for the overall reaction $pd \to {}^3HeX$ (see text). The ^{3}He acceptance is tuned such that it always corresponds closely to the maximally possible mass X like in the Saclay experiment [7]. The broad enhancement found in ref. [7] can be qualitatively reproduced by a two step process.

In a Dalitz–plot e.g. of the (squared masses of) subsystems, $M^2(p+d) = (V(p) + V(d))^2$ against $M^2(X+p) = (V(X) + V(p))^2$ there can be very distinct density modulations due to the two step process. They will not extend across the Dalitz plot like a resonance would do. In order to find out the relative importance of two step processes one should therefore study besides the two body channels also the three body reactions ($pd \to X^o pd$, $pd \to X^+ nd$, $pd \to K^+\Lambda^o d$, $pd \to K^o\Sigma^+ d$ etc.) in kinematically complete experiments over the full phasespace.

References

1. K. Kilian et al., Proc. Workshop on Physics at LEAR with Low Energy Cooled Antiprotons, Erice, May 1982, Plenum Press 1984, Ettore Majorana International Science Series Nr. 17, p. 843, edited by U. Gastaldi and R. Klapisch
2. G.F. Bertsch, Phys. Rev. C15 (1977) 713
3. R. Shyam and J. Knoll, Nucl. Phys. A426 (1984) 606
4. K. Kilian, Nuclear Physics with Stored, Cooled Beams, edited by P. Schwandt and O. Meyer, AIP Conf. Proc. Nr. 128 (1985) 319
5. E. Rössle et al., Pion Production and Absorption in Nuclei, edited by R.D. Bent, AIP Conf. Proc. Nr. 79 (1982) 171
6. H.W. Fearing, Phys. Rev. C11 (1975) 1210
7. F. Plouin, Proc. Workshop on Production and Decay of Light Mesons, Paris 1988, p. 114, edited by P. Fleury, World Scientific, Singapore
8. J.M. Laget et al., Phys. Rev. Lett. 61 (1988) 2069
9. C. Wilkin, same proceedings as ref. 7, p. 187

FROM CONTINUUM TO BOUND STATES
-- AND VICE VERSA

G. T. Emery
Bowdoin College, Brunswick, ME 04011

ABSTRACT

A number of physical phenomena related to reaction thresholds are discussed. Some history and some general remarks are included. The principal focus is on Coulomb-bound systems, their formation in nuclear reactions, their structure, and their decay. Among the atom-like systems considered are πA pionic atoms, positronium, pionium ($\pi^+\pi^-$), and multipionic (NπA) atoms.

1. ABOVE AND BELOW THRESHOLD

Consider a reaction in which the particles going out have more rest energy than the incident and target particles. There is then a threshold, a minimum kinetic energy in the entrance channel for the reaction to take place. Several interesting physical phenomena can occur in the vicinity of such a threshold.

Above the threshold, final-state interactions can be important, especially in regions of phase space where the relative velocity of two outgoing particles is such that the interaction between them has a big effect. This can occur because the interaction itself is strong there, because the relative velocity is small, or both. The low-energy nucleon-nucleon interaction is responsible for many interesting FSI effects, including the near-threshold energy dependence of the pp to pp(π^0) cross section studied recently by Meyer and colleagues at IUCF.[1] The Coulomb interaction, energy independent as it is, can have important FSI effects, some of which will be mentioned later.

Final-state interactions can have off-diagonal matrix elements, too, which can show up as **channel coupling**. Consider two outgoing particles, each of which has isospin greater than or equal to one half. Different charge decompositions can contribute to the same T, T_z channel. If isospin is conserved, one expects simple ratios of cross sections, based on isospin Clebsch-Gordan coefficients. For example, in

$$^{12}C + p \begin{array}{l} \nearrow \; ^{13}C + \pi^+ \\ \searrow \; ^{13}N + \pi^0 \end{array}$$

one expects the π^+ cross section to be twice that for π^0, at the same kinematics. The kinematics, however, is never exactly the same. Even if the reaction mechanism conserves isospin, the masses are not the same, leading to different thresholds, and thus at the same bombarding energy different phase-space factors for the different exit channels. These are just the conditions in which T-conserving but non-diagonal operators like $t_1 \cdot t_2$ can effect non-trivial transfers of flux from one exit channel to the other.

Even channels not directly coupled with each other are coupled indirectly by the conservation of flux. Thus, there can be **threshold effects** on observables in other channels when a threshold is crossed. These effects will tend to be more dramatic when either the rate-of-change of the new cross section is large, relative to the total cross section, or in old channels directly coupled to the new channel.

Below threshold, for example just below the threshold for production of a particle, one might expect to see some effect from moderately long lived intermediate states with an "almost real" particle. These would be precursor effects, but not as dramatic as those about which so much thought was expended a few years ago, since what is being pre-cursed here is single-particle, not collective, motion.

Finally, for an attractive Coulomb interaction in the final state there are atomic states just below the threshold, about which we will say a little more later. Some of the general features of these atomic states, and their relation to the Coulomb FSI effects above threshold have been put in context by Ericson.[2] The long range of the electromagnetic interaction and the distance scale of many such systems, intermediate between the nuclear and (electron-) atomic scales, allow one to think of them as **continuum states in the bound spectrum**.

2. FROM THE DEUTERON TO NN SCATTERING

The first work in nuclear physics that connects bound states and continuum states, or at least the first known to me, is the work of Bethe and Peierls on the diplon.[3,4] The story of the naming of the deuteron, including why it was called the diplon for a substantial period in England, has been told by Stuewer[5] and is worth looking at, as are the papers of Bethe and Peierls.

The argument started with the bound state: from the binding energy one knows something about the size of the deuteron, and it is considerably larger than the range of the nuclear interaction, which people at the time knew from an argument of Wigner about the large binding energy of the alpha particle compared with that of the deuteron. The result is that the interaction must be, within its range, very strongly attractive; it must correspond to a very deep well. The well is much deeper than the particle is bound. Then the shape of the wave function inside the well is only very weakly dependent on the energy for which one calculates. If the well is 50, 60, 70 MeV deep, the wave function will not be very different for energy -2.2 MeV from what it is for + an MeV or so. Thus at the edge of the potential the logarithmic derivative will not be very different--that for low energy scattering will be almost the same as for the deuteron. That logarithmic derivative will define the scattering length and thus the low energy cross section. The result is (eq. 11, ref. 4)

$$\frac{d\sigma}{d\Omega} = \frac{1}{\alpha^2 + k^2}$$

where $(\hbar\alpha)^2 = M\epsilon$ and $(\hbar k)^2 = ME$, with M the nucleon mass, ϵ the binding energy of the deuteron and E the CM energy of the scattering system.

Bethe described this episode in a talk at Trieste not so long ago[6], under the subtitle "Solved in the Subway":

> Well our theory was very nice, but it did not agree with experiment! The better the experiments the worse the discrepancy. Finally the solution to this was told to me in 1935 in a subway train in New York by Eugene Wigner. I don't know, I must have been able to hear much better than I do now, which I know is true, and probably Wigner spoke louder than he does now! At any rate, I was able to hear him in the subway, and he said, "Now look here, all that the deuteron tells you is the interaction of neutrons and protons in the triplet state. How do you know how they interact in the singlet state? Probably they interact quite differently."

So, there is a rather early example of connecting the physics on one side of a threshold with that on the other.

3. CONTINUITY OF WAVE FUNCTIONS, ETC.

There is a long history of exploitation of the continuity of the shape of Coulomb wave functions above and below threshold. Near the nucleus the shape of the radial function depends on Z and the orbital angular momentum; in lowest order it is independent of energy, and for quite a distance from the nucleus it depends only weakly on energy. This is as true for Dirac as for Schrödinger radial functions. Beyond the shape, strengths (per unit energy interval) for processes involving the near-nucleus parts of the radial functions are also continuous.

One example of a pair of processes related by a threshold is that of hyperfine structure and internal conversion. If the matrix elements involved in hfs are, for example,

$$\langle\, ns,g \mid H'_{hfs} \mid ns,g \,\rangle$$

where ns gives the atomic and g the nuclear state, those involved in internal conversion are like

$$\langle\, \mathbf{ks},g \mid H'_{hfs} \mid ns,\mathbf{exc} \,\rangle ,$$

where **ks** is now a continuum state. Including off-diagonal hfs makes the connection even more apparent. For magnetic-multipole hfs (and internal conversion) the matrix elements are often dominated by "surface terms", for which the radial integrals are proportional to wave functions evaluated at the nuclear surface, thus making the relative contributions of different atomic states the same in hfs, ML conversion, E0 conversion, and the electron-capture process of beta decay.[7]

Thresholds of systems bound by the strong interaction also show this kind of continuity. We quote one example[8] of a form of title that has been showing up regularly in the last year or so:

Proton mean field in ^{40}Ca between -60 MeV and +200 MeV deduced from a dispersive optical-model analysis.

4. A NOTE ON THE PSEUDOPOTENTIAL

There is an important concept in physics that has been very relevant in relating physical quantities across thresholds, whose history is not very well understood, by me at least. It may be worth looking into. I refer to the idea of the pseudopotential. As far as I know it first appears in the 1954 paper of Deser, Goldberger, Baumann, and Thirring called "Energy Level Displacements in Pi-Mesonic Atoms".[9]

The first part of their argument gives the energy shift of the ns state of the π-A atom in terms of the scattering length in the π-A system:

$$\Delta E_{ns} = - 2\pi (\hbar^2/\mu) \rho_{ns}(0) a_{\pi A} .$$

It is recognized that the scattering length is in general complex, and that the energy shift thus includes the width. The second part of the argument of Deser, *et al.*, is that the π-A scattering length can be expressed as a sum over occupied nuclear states of π-nucleon scattering lengths. Reference 1 of the paper refers to earlier, mostly unpublished work of Fermi and Goldberger.

What followed this famous paper is clearer to me than what preceded it. Six months later there was a paper by Brueckner[10], which pointed out that while the real part of the energy shift is correctly given by the relation above, the width is dominated by the interaction of the negative pion with two nucleons.

Eight pages earlier in the same issue of the Physical Review appeared another famous paper of our field[11], the work of Brueckner's student Leonard Kisslinger, introducing the **k•k'** interaction. Both of these last two papers were bylined from our host institution eighteen miles down the road.

5. SUPPLEMENTARY NOTES ON DIRECT PRODUCTION OF PIONIC ATOMS

Three types of methods have been proposed for the direct production of pionic atoms in nuclear reactions. The pioneer in this area was Tzara, who proposed the **resonant production** of such systems with bremstrahlung photons.[12] Somewhat later I suggested the use of proton beams.[13] One can estimate the cross section, for some cases at least, by extrapolating across the threshold. The 183- and 190-MeV data of Green, *et al.*,[14] for example, on $^{13}C(p,\pi^-)^{14}O_{gs}$, together with the 205-MeV data of Korkmaz, *et al.*,[15] allow a better estimate of the resonant cross section for making the $^{14}O_{gs}$ pionic-atom than was given in ref. 13; the current best estimate for the total production cross section at the peak of the resonance is 300 to 350 nb. In the resonance method one traces out the pionic atom spectrum by scanning the primary accelerator beam.

A second method is **tagged production**, involving a two-body final state in which one body is the pionic atom and the other is a (usually lighter) particle whose spectrum can be measured, thus reflecting the spectrum of pionic atom (and background) states produced. Tagged production by inelastic electron scattering was suggested by Dmitriev[16] and by Koch[17]. Using single-nucleon transfer reactions for tagged production has been explored actively in the last year or so by Yamazaki, Toki, and their collaborators.[18] It is an attractive approach because the relatively large momentum transfer needed in large-j nucleon transfer can be matched against that needed to produce the pion. For more details I refer to the talk of Toki at this meeting.[19] The use of (π^-,π^+) and other reactions has also been explored recently by Nieves and Oset.[20]

Finally there is the possibility of **recoilless production**, as discussed by Kilian,[21] in which there is a magic bombarding momentum at which the pionic atom is left at rest in the laboratory. For details about this intriguing approach I refer to the talk of Machner at this meeting.[22]

The reasons for wanting to produce pionic atoms in these ways have not really changed, and I think are primarily three. (a) Our knowledge of the energy shifts and widths of inner shells, from the ordinary pionic atom methods, are severely limited in Z; for example, the properties of 1s states are not known for Z larger than about 13, of 2p states for Z larger than about 35, etc. The largest measured shifts and widths are in the neighborhood of 25 keV, while predictions based on the best models predict widths of several hundred keV for the 1s states of high-Z pionic atoms.[18] (b) One can make, in nuclear reactions, pionic atoms whose nuclei are considerably more various than the stable and reasonably abundant targets needed for the conventional method. It then becomes possible to think of more extensive studies of isotope and isomer shifts, J-dependence, etc. (c) The region of large-Z inner-shell pionic atoms, where the widths and shifts are large, and where there is an excellent chance of seeing the coupling of atomic and nuclear degrees of freedom, or even "inner" and "outer" states,[23] is so far unexplored, and it looks like a fascinating area.

A resonant production experiment has been approved for the Indiana Cooler, and Meyer and his collaborators have been testing various target and detection techniques. The experiment involves seeing the effect of the ^{14}O 1s pionic atom as a resonance in backward elastic scattering of polarized protons from ^{13}C.[24]

As the widths and shifts get large in pionic atoms one will have to take off-diagonal hyperfine structure into account, which offers some interesting possibilities. One light system that might be interesting and amenable to analysis is the ^{19}Ne pionic atom, where the atomic Lyman series runs from about 235 keV to 325 keV, while there are nuclear excitations of 238 keV (E2) and 275 keV (E1).

One possibility that has not been pursued in recent years is that of the neutral-pion decay of pionic atoms heavier than ^3He. Direct production is required, since there are no stable targets heavier than ^3He with Z greater than N. With the resonant (p,π^-) method, one can start with N = Z + 2 targets to make N = Z - 1 pionic atoms. These can then decay by pion charge exchange to their isobaric analogs in the system with Z = N - 1. A possible example is

$$^{42}Ca + p \rightarrow (^{43}Ti + \pi^-) \rightarrow {}^{43}Sc + \pi^0 \ .$$

The presence of the pionic atom resonance then perturbs the (p,π^0) excitation function. Some preliminary work along these lines has been done.[25]

6. THE EMISSION OF POSITRONIUM

What about other Coulomb-bound systems that might be produced in a nuclear (or particle) reaction or decay? One possible decay mode of excited nuclear levels is positronium emission. I had not realized until reading his memoirs this summer that Andrei Sakharov had published a paper on the effect on pair decay of the Coulomb attraction between e^+ and e^-. The paper is in JETP, reprinted in his Collected Scientific Works, and formed part of his doctoral thesis.[26] There are parts of phase space in which the e^+ and the e^- go off in approximately the same direction with approximately the same velocity, and the probability of the particles doing that is a little bit enhanced by the attractive Coulomb interaction. In the limit they can go off as positronium. I do not know of any situation in which the branching ratio is other than small.

It was almost twenty years ago that Nemenov[27] suggested looking for positronium emission in π^0 decay and estimated the branching ratio. He also considered the (e^+e^-) and $(\mu^+\mu^-)$ branching of η^0. There were follow-up calculations by Nemenov[28] and by Vysotskii[29]. The observation of relativistic positronium, mostly from π^0 decay, was reported by Alekseev, *et al.*,[30] in 1984. More recently the Dubna group has reported a cross section for the interaction of GeV positronium with carbon,[31] and an experimental result for the branching ratio of π^0 into gamma plus positronium.[32] The value found, $(1.84 \pm 0.29) \times 10^{-9}$, agrees with that calculated, 1.6×10^{-9}.

It is interesting that high-energy positronium has been found before low-energy positronium. It had occurred to me that if the 1.8-MeV effect in heavy-ion collisions[33] were connected with an extended neutral particle that perhaps it perhaps it made sense to look also for an extended neutral particle that we already know something about, positronium. Positronium has the disadvantage, in heavy-ion collisions, of being a very extended system, and frangible in electric fields. The Perkins effect[34] is less helpful at low energy than at high. Positronium comes in spin-singlet and spin-triplet forms, and one could think in some circumstances of measuring the singlet-to-triplet ratio, or even the orientation of the triplet spin, to learn something about the transition in which it was produced.

7. THE PRODUCTION AND DECAY OF PIONIUM

Consider the atom whose constituents are π^+ and π^-. Its ground state is Coulomb bound by 1.85 keV and it has a radius of about 400 fm, scaled up and down, respectively, from positronium by the ratio of pion to electron mass. The system can certainly annihilate into two photons, but it was pointed out by Uretsky and Palfrey[35] that it can also undergo a mutual charge exchange into two neutral pions, with a rate determined by the square of the difference of the T=0 and T=2 $\pi\pi$ scattering lengths, $a_0 - a_2$. The $\pi^0\pi^0$ rate is expected to dominate if current knowledge of the scattering lengths is anywhere near correct, and the branching into two photons is then a sensitive measure of that combination of scattering lengths. Properties of these systems, and also πK and KK atoms, have been further considered at Dubna[36-38].

Now, the $\pi\pi$ scattering lengths are supposed to be sensitive indicators of the values of the phenomenological parameters of chiral symmetry breaking.[39] Nann, Vigdor, and others at IUCF have suggested that it may be feasible to produce tagged pionium ($\pi^+\pi^-$ atoms) in a clean way, and to measure the two-gamma branching ratio with enough accuracy to provide significant new information on chiral symmetry breaking.[40] The experiment is tailor-made for a cooled beam. It involves $p + d \rightarrow {}^3He + (\pi^+\pi^-)$; with a proton beam the threshold for the atomic 1s state is 430.447 MeV, while that for an unbound $\pi^+\pi^-$ pair is 3 keV higher. The ^3He goes forward in the laboratory with about 143 MeV of kinetic energy, and with a sufficiently monoenergetic beam one can determine from the ^3He spectrum that atomic states are being produced. One then pictures a follow-up experiment with a large-solid-angle spectrometer for gamma rays, in which one measures the fraction of atoms that decay into two 140-MeV (CM) photons, versus those that decay into four 68-MeV (in two π^0 frames) photons. It is a conceptually beautiful experiment, and it offers the prospect of determining one combination of the $\pi\pi$ scattering lengths in a much cleaner and more precise way than ever could be done in studies of final-state interactions.

8. THE $\pi^-\pi^-$ INTERACTION

How would one determine the T=0 and T=2 $\pi\pi$ scattering lengths separately? Let me finish with a brief description of a sort of blue-sky approach. Consider the Coulomb-bound system that has charge Ze on the nucleus and N negative pions in the atomic 1s state. With fermions it would be silly to consider N greater than 2, but with pions, why not? We do the standard He-atom variational problem. Assume the pionic 1s wave function is given by

$$\phi_i = A \exp(-\zeta r_i/a) ,$$

where ζ is the variational parameter and a the pionic Bohr radius. The kinetic energy of each pion is then given by

$$\langle T_i \rangle = (1/2)\ \zeta^2\ \alpha^2\ mc^2 ,$$

the potential energy due to interaction with the nucleus by

$$\langle V_i \rangle = -\ \zeta Z\ \alpha^2\ mc^2 ,$$

and the Coulomb interaction between each pair of pions by

$$\langle e^2/r_{ij} \rangle = (5/8)\ \zeta\ \alpha^2\ mc^2 .$$

The energy functional is then

$$\langle N|H|N \rangle = N \langle T_i \rangle + N \langle V_i \rangle + (1/2)N(N-1) \langle e^2/r_{ij} \rangle$$

$$= -\ (1/2)\ \alpha^2\ mc^2\ [(1/2)N\zeta^2 - N\zeta Z + (5/16)N(N-1)\zeta]) .$$

It is minimized by making $\zeta = \zeta_{min} = Z - (5/16)(N-1)$, and is then

$$E_{min} = -\ (1/2)\ \alpha^2\ mc^2\ N\ [Z - (5/16)(N-1)]^2 .$$

The binding enery per pion, B(Z,N), is equal to (3.72 keV)$\zeta_{min}{}^2$, and for the particular case Z = 8 is given in Table I.

Now that we have those N negative pions in the 1s state, we can introduce the T = 2 $\pi\pi$ interaction as a perturbation. Using the pseudopotential in the usual way,

$$\Delta E_{ij} = -\ (1/2)\ (\zeta\alpha)^3\ mc^2\ a_2 ,$$

with the scattering length in the traditional units of inverse pion mass. To get the total shift we multiply that by N(N-1)/2. These total energy-level shifts, again for Z = 8, and for the Weinberg[41] value for a_2, are also given in Table I.

Table I. Properties of multipionic atoms with Z = 8 and N negative pions. Column 2 shows the Coulomb binding energy per pion, B(8,N), in keV. Column 3 shows the total energy shift due to the T = 2 $\pi\pi$ interaction, also in keV, in perturbation theory with uncorrelated variational Coulomb wave functions, and using the standard value of the T = 2 $\pi\pi$ scattering length,[41].

N	B(8,N)	Total $\pi\pi$ shift
1	238	0
2	220	0.74
3	202	1.96
4	185	3.44
5	169	5.00
6	154	6.51
7	139	7.85
8	126	8.95

One might worry that the variational calculation has been done with a totally uncorrelated wave function. We can estimate the effect of the Coulomb correlations. The most important correlation[42] is the "r_{ij} term", where for N = 2 the wave function becomes

$$\Psi(\mathbf{r}_1,\mathbf{r}_2) = A \exp[-\zeta(r_1+r_2)/a]\ [1 + \xi r_{12}] \ ,$$

and the value of the new parameter ξ that minimizes the energy is found to be about 1/2. For two particles this correlation reduces ΔE_{ij} by the factor $[1 + (35/8)(\xi/\zeta) + 6(\xi/\zeta)^2]$. For Z = 8 and $\xi = 1/2$ the factor is 0.76.

It is not my expectation that experimental results on the energy levels of multipionic atoms are likely to be available in the near future, but some estimates of their formation probability have been made.[43]

REFERENCES

1. H.O. Meyer, these proceedings; H. O. Meyer, *et al.*, Phys. Rev. Letters 65, 2846 (1990).
2. T.E.O. Ericson, in Nuclear Physics with Stored, Cooled Beams, ed. P. Schwandt and H. O. Meyer (AIP Conf. Proc. No. 128, 1985), p. 172.
3. H. Bethe and R. Peierls, Proc. Roy. Soc. (London) A148, 146 (1935).
4. H.A. Bethe and R. Peierls, Proc. Roy. Soc. (London) A149, 176 1935).
5. Roger H. Steuwer, Am. J. Phys. 54, 206 (1986).
6. H.A. Bethe, in From a Life of Physics (World Scientific, Singapore, 1989), 3; see especially pp. 7-8.
7. G.T. Emery, Ann. Rev. Nucl. Sci. 22, 165 (1972), esp. pp. 174-5.
8. W. Tornow, Z.P. Chen, and J.P. Delaroche, Phys. Rev. C 42, 693 (1990).
9. S. Deser, M.L. Goldberger, K. Baumann, and W. Thirring, Phys. Rev. 96, 774 (1954).
10. K.A. Brueckner, Phys. Rev. 98, 769 (1955).
11. L.S. Kisslinger, Phys. Rev. 98, 761 (1955).
12. C. Tzara, Nucl. Phys. B18, 246 (1970).
13. G.T. Emery, Phys. Lett. 60B, 351 (1976).
14. M.C. Green, *et al.*, IUCF, private communication.
15. E.J. Korkmaz, Ph.D. thesis, Indiana University (1987), and private communication.
16. V.F. Dmitriev, Sov. Phys.-JETP Lett. 14, 81 (1971); Sov. J. Nucl. Phys. 17, 417 (1973).
17. J.H. Koch, Phys. Lett. 59B, 45 (1975).
18. H. Toki and T. Yamazaki, Phys. Lett. 213, 129 (1988); H. Toki, *et al.*, Nucl. Phys. A501, 653 (1989); T. Yamazaki, *et al.*, Nucl. Instrum. Methods A292, 619 (1990), for example.
19. H. Toki, these proceedings.
20. J. Nieves and E. Oset, Phys. Rev. C 42, 690 (1990); preprints.
21. K. Kilian, in Nuclear Physics with Stored, Cooled Beams, ed. P. Schwandt and H.O. Meyer (AIP Conf. Proc. No. 128, 1985), p. 319.
22. H. Machner, these proceedings.
23. J.H. Koch, M.M. Sternheim, and J.F. Walker, Phys. Rev. Lett. 26, 1465 (1971); H. Toki, S. Hirenzaki, and T. Yamazaki, Phys. Lett. B 249, 391 (1990).
24. H.O. Meyer, *et al.*, IUCF Cooler Experiment No. 02.
25. S.M. Aziz, *et al.*, IUCF Sci. & Tech. Report, 1985, pp. 40-42.
26. A.D. Sakharov, Zhur. Eksp. Teoret. Fiz. 18, 631 (1948); Collected Scientific Works (Marcel Dekker, New York & Basel, 1982), p. 255; Memoirs, (Knopf, New York, 1990), pp. 80-81.

27. L.L. Nemenov, Yad. Fiz. 15, 1047 (1972)[Sov. J. Nucl. Phys. 15, 582 (1972)].
28. L.L. Nemenov, Yad. Fiz. 24, 319 (1976)[Sov. J. Nucl. Phys. 24, 166 (1976)].
29. M.I. Vysotskii, Yad Fiz. 29, 845 (1979)[Sov. J. Nucl. Phys. 29, 434 (1979)].
30. G.D. Alekseev, *et al.*, Yad Fiz. 40, 139 (1984)[Sov. J. Nucl. Phys. 40, 87 (1984)].
31. L.G. Afanas'ev, *et al.*, Yad. Fiz. 50, 7 (1989)[Sov. J. Nucl. Phys. 50, 4 (1989)].
32. L.G. Afanasyev, *et al.*, Phys. Lett. B 236, 116 (1990).
33. S.M. Judge, *et al.*, Phys. Rev. Lett. 65, 972 (1990), and references cited there.
34. D.H. Perkins, Phil. Mag. 46, 1146 (1955).
35. J.L. Uretsky and T.R. Palfrey, Jr., Phys. Rev. 121, 1798 (1961).
36. L.L. Nemenov, Yad. Fiz. 41, 980 (1985)[Sov. J. Nucl. Phys. 41, 629 (1985)].
37. G.V. Efimov, M.A. Ivanov, and V.E. Lyubovitskii, Yad Fiz. 44, 460 (1986)[Sov. J. Nucl. Phys. 44, 296 (1986)].
38. A.A. Bel'kov, V.N. Pervushin, and F.G. Tkebuchava, Yad. Fiz. 44, 466 (1986)[Sov. J. Nucl. Phys. 44, 300 (1986)].
39. J.F. Donoghue, C. Ramirez, and G. Valencia, Phys. Rev. D 38, 2195 (1988).
40. H. Nann, S.E. Vigdor, *et al.*, private communication.
41. S. Weinberg, Phys. Rev. Lett. 17, 616 (1966).
42. H.A. Bethe and R. Jackiw, Intermediate Quantum Mechanics, 3rd ed. (Benjamin/Cummings, Menlo Park, CA, 1986), esp. pp. 48-49.
43. S. Hirenzaki, *et al.*, Phys. Lett. B 194, 20 (1987).

THE $pp \rightarrow pn\pi^+$ REACTION NEAR THRESHOLD

W.W. Daehnick, S.A. Dytman, W.K. Brooks, J.G. Hardie, and R.W. Flammang
University of Pittsburgh, Pittsburgh, PA 15260,

J.D. Brown, E. Jacobson, Princeton University, Princeton, NJ 08544

T. Rinckel, IUCF, Indiana University, Bloomington, IN 47405

ABSTRACT

The $pp \rightarrow pn\pi^+$ reaction is one of the basic inelastic nucleon-nucleon reactions, but is has been insufficiently explored near reaction threshold. The motivation for the experiment and the solutions chosen for the experimental difficulties are discussed.

INTRODUCTION

Several basic reactions must be investigated in order to determine the pion-nucleon coupling and the nucleon-nucleon interaction at intermediate energies. Some properties of the NNπ vertex are theoretically accessible through soft pion theorems and PCAC. However, there are unresolved questions about the NNπ coupling strengths now in use[1], and recent results for $\gamma p \rightarrow \pi^0 p$ measurements close to threshold have even cast doubt on the applicability of the soft pion theorem[2]. The range of the pion vertex is poorly constrained, and values used by theorists vary widely. Apart from photo-production, the simplest reactions which are sensitive to the πNN vertex are:

1) NN → NN	2) πN → πN	3) πd → πd
4) NN ↔ πd	5) NN → NNπ	

The first four processes are relatively easy to observe since they lead to two-body final states. Extensive data exist, except at the very lowest energies. However, because of the experimental difficulties, data for reaction (5) are still meagre. Kinematically complete experiments *below* 500 MeV have only recently been performed, primarily at IUCF and TRIUMF, where accelerator or beam improvements permit the observation of ejectiles at small angles.

Some $pp \rightarrow pp\pi^0$ total cross section measurements near threshold have been completed at IUCF and TRIUMF. The $pn \rightarrow pp\pi^-$ reaction is being investigated at TRIUMF with a neutron beam. The remaining variant of type (5) is the $pp \rightarrow pn\pi^+$ reaction which is the subject of experiment CE03 at IUCF. The experiment will measure differential cross sections and analyzing powers at beam energies of 293, 300, and 320 MeV in 1991. This reaction has not been investigated below 392 MeV, which is still 100 MeV above threshold. An understanding of the basic reactions of type (5) is essential for the interpretation of pion creation in heavier nuclei as well as for the nuclear force near the pion creation threshold.

SCIENTIFIC MOTIVATION

In spite of the experimental and theoretical difficulties arising from the three-body final state, this reaction deserves special interest. To begin with, reaction (5) studies a greater range of spin and isospin couplings than the simpler reactions. It samples off-shell characteristics of the vertices in a controlled way. Also, the large negative Q values lead to high momentum transfer, which means the NN wave functions are sampled at short distances. This may uncover system properties which require for their explanation refinements from a quark picture treatment, especially in data for analyzing powers and spin parameters. The particular advantage of measurements near threshold is the restriction of reaction channels, especially the suppression of diagrams involving the Δ(1236) resonance. Near threshold we primarily expect to see contributions from the basic diagrams shown in Fig. 1.

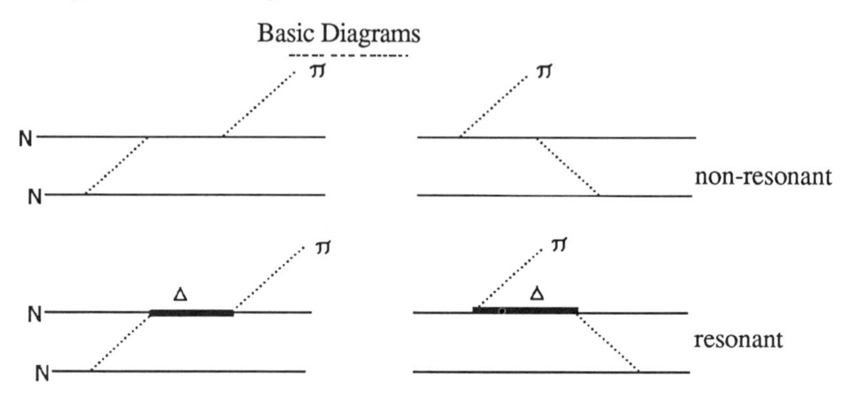

Fig.1. Expected contributions near threshold

For energies above 400 MeV resonant diagrams involving the delta (and eventually higher resonances) are very important and usually dominant. Although this effect simplifies some analyses, it masks the features of the bare NNπ vertex. Calculations using meson theory are more difficult near threshold, since final state interactions become very important.

Lee and Matsuyama[3] have used a coupled channel approach to calculate total cross sections for $pp \rightarrow pp\pi^0$ and $pp \rightarrow pn\pi^+$ for energies down to threshold. Fig. 2 shows their results for two reasonable parameters Λ for the range of the vertex. Previous theoretical descriptions have used values of Λ from 600 to 1200 MeV/c. This uncertainty translates into a 20% effect in the $pp \rightarrow pn\pi$ total cross section. The lowest curve shows a calculation which only considers the Δ(1236) contribution. Near threshold it is reduced by an order of magnitude and suggests that the determination of the non-resonant part is feasible. Cross sections for the three-body final state are sensitive to a range of πN momenta.

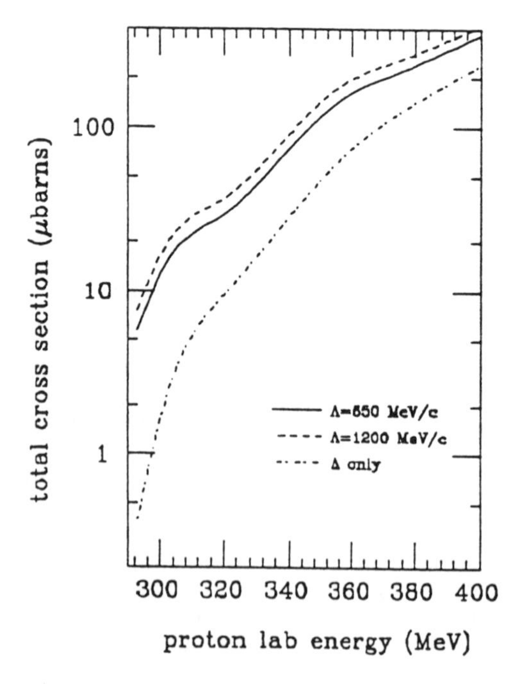

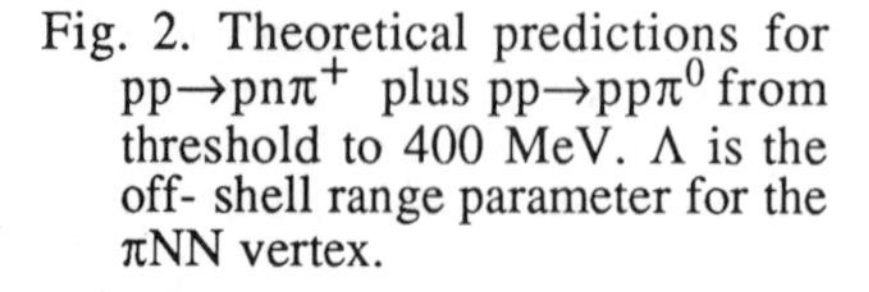
Fig. 2. Theoretical predictions for $pp \rightarrow pn\pi^+$ plus $pp \rightarrow pp\pi^0$ from threshold to 400 MeV. Λ is the off- shell range parameter for the πNN vertex.

A number of measurements are necessary to determine NN inelasticity close to threshold. Data exist for the $pp \rightarrow d\pi^+$ channel, and recently the $np \rightarrow d\pi^0$ cross section was measured. Given the recent IUCF data[4] for $pp \rightarrow pp\pi^0$ and the start of our $pp \rightarrow pn\pi^+$ measurements, it will become attractive and important to refine and expand the theoretical work for this region.

EXPERIMENTAL CHALLENGES AND SOLUTIONS

As is apparent from Fig. 2, the predicted total cross sections for $pp \rightarrow pn\pi^+$ near threshold are of order 5 to 10 μbarn. With the present IUCF gas target and an expected Cooler luminosity of 10^{30} $cm^{-2}sec^{-1}$, the count rate of desired events will be less than 2/sec for a 4π detector geometry. The Cooler is basically a very clean accelerator and makes such measurements possible. Nevertheless, apertures and residual gas scattering in the ring generate a beam halo that can produce high background counting rates near the beam pipe, i.e. at small scattering angles for in-beam experiments. In addition, small-angle cross sections for elastic pp scattering are many orders of magnitude larger than pion production. Therefore, the suppression of background and a large solid angle for the detector were primary design criteria.

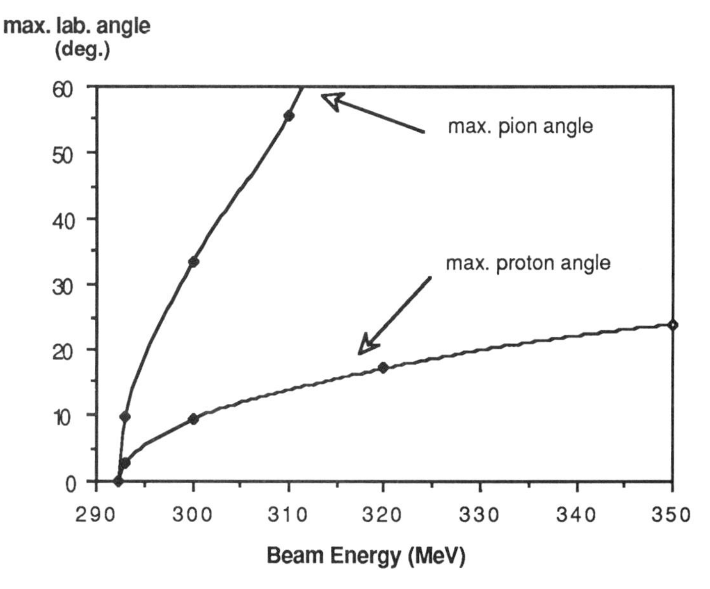

Fig. 3. Kinematic ranges for pions and protons in $pp \rightarrow pn\pi^+$

Neutron and proton ejectiles near threshold are confined to a very narrow cone (see Fig. 3) and have a magnetic rigidity of only 51% of the circulating beam. Hence it is possible to mount the gas target just in front of a large aperture bending magnet and to magnetically separate charged and neutral ejectiles from the circulating beam. We achieve a large effective solid angle for the detectors ($\Omega \geq 2\pi$ to about 3 MeV above threshold) by mounting the hydrogen gas jet target 25 cm in front of the 6° magnet.(see Fig. 4). At threshold, the reaction protons are bent through 12°, twice the angle of the unscattered beam, permitting the intercept of even the smallest proton scattering angles. Similarly, small angle detection of neutrons is accomplished by placing the detectors on the opposite side of the circulating beam.

The apparatus is also capable of detecting pions, although the softest pions tend to decay before reaching the detector. At the higher beam energies, where the pion energies are large enough for reliable detection, pion detection permits additional cross-checks for event reconstruction and for absolute cross sections . Normally, the measurements of θ, φ, and energy for the proton and of θ′, φ′, and time of flight for the neutron determine 6 degrees of freedom for all events, one more than the minimum required to completely specify the three-body final state.

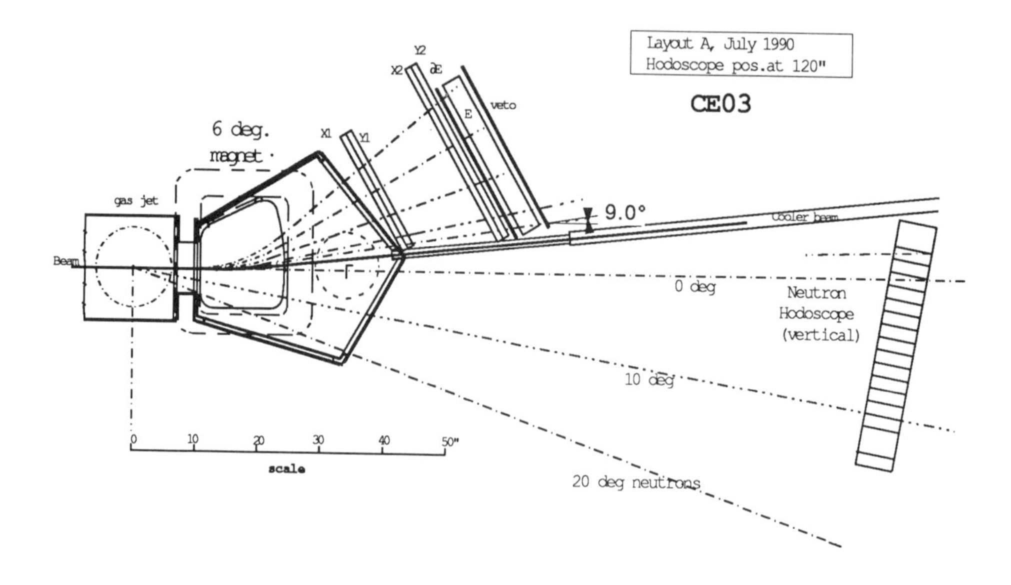

Fig. 4. Top view of the CE03 apparatus. The elements Xi,Yi are multi-wire drift chambers. The other detectors are plastic scintillators.

A special bending magnet with a 12.7 cm usable gap and a suitable vacuum chamber were designed and constructed for the six degree (T-site) bend in the IUCF Cooler Ring. The vertical acceptance angle relative to the gas jet target is ±6.5°, while the horizontal acceptance is ±20°. The exit windows permit ejectile exit angles from the magnet of up to 54° towards the inside of the ring and up to 20° towards the outside. Charged particles exit through a large, Kevlar reinforced window of about 15 mg/cm^2 total thickness. The magnet has been mapped and is being installed.

THE PROTON ARM

The charged particle (proton) arm of the two-arm detector consists of two xy drift chambers (8 wire planes), a fast ΔE scintillator, a set of 7.6 or 15.2 cm thick (E) scintillators to measure the proton energy, and a thin veto scintillator. The two horizontal multi-wire drift chambers have been built and tested. A prototype wire chamber gave a (directly measured) position resolution of 0.24 mm. A similar resolution (0.3 mm) has been deduced for the larger chambers from recent data. Wire spacing is 8 mm in chamber 1 and 10 mm in chamber 2. The readout of the wire chambers utilizes LeCroy 2735DC amplifier/discriminators cards.

The charged particle scintillator assembly for the proton arm uses fast plastic scintillators and has worked satisfactorily from the start. In a "bench" test the E scintillators achieved 2.2% resolution for 95 MeV deuterons. This is better than the resolution expected from ray tracing through the 6 degree bending magnet.

THE NEUTRON ARM

The position-sensitive neutron detector is an array of 14 rectangular scintillator bars, 120 cm long, covering an area of 70 cm x 120 cm. There are four overlapping veto scintillators in front of the hodoscope which will reject charged particles. When a scintillation event occurs in a given bar, the time difference between the arrival of the signals at the phototubes at the ends of the bar gives the (x_1) position of the event along the bar while the position in the perpendicular (x_2) direction is given by noting which bar fired. The neutron energy will be determined from time of flight relative to the proton signal. For the 15 cm depth of the bars and a 7 m flight path, the expected energy resolution is about 4%. The neutron angles will be determined to within an uncertainty of $\pm$ 0.25° when the detector is placed 3 m from the target (its closest anticipated position). In this position the vertical acceptance of the hodoscope is 13° and the horizontal acceptance is 22°, which will cover all angles of interest in the measurements near the pion threshold.

Timing and energy calibrations and stability checks are made continuously with a pulsed (300 ps width) ultraviolet laser whose output is distributed through quartz fiber optics to all scintillators. The optical fibers terminate in small patches of scintillator material which are positioned in the center of each of the detectors. This results in a light pulse originating at a symmetry point and yielding equal, isotropic, and calibrated amounts of light in each photomultiplier.

To date, we have had excellent success with timing calibrations. The light amplitude varies by a few percent over a week, but pulse amplitudes at some PM tubes have varied up to $\pm$20%.

The performance of the hodoscope has been evaluated using radioactive sources, cosmic rays, the pulsed ultraviolet laser, the cyclotron beam and the Cooler beam. The typical value of the position resolution for 90 MeV particles is 2.5 cm. The detection efficiency for 70 MeV neutrons is approximately 20% for our 15 cm thick bars.

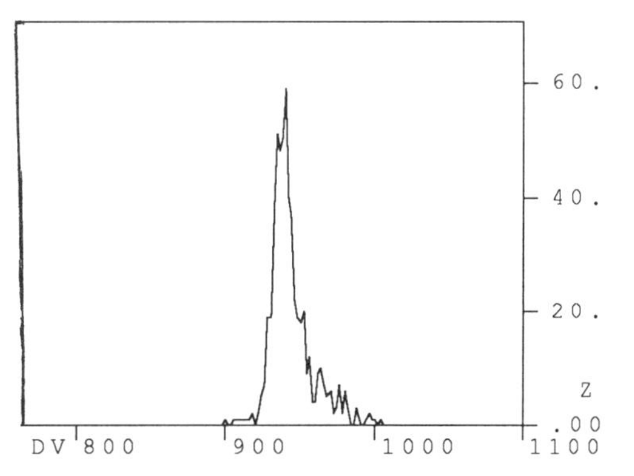

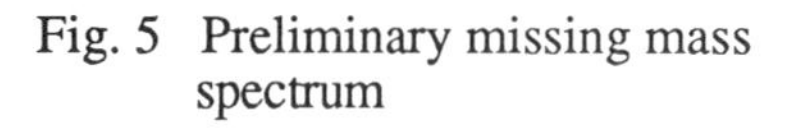
Fig. 5 Preliminary missing mass spectrum

PREPARATION FOR PRODUCTION RUNS

As neither the 6° magnet nor the gas jet target were installed at the time, we have tested the detector at the T-site in two short runs with CH_2 and CD_2 "scraper" targets. Using the CD_2 target, correlated pn pairs were produced. Fig. 5 shows a preliminary missing mass spectrum of the undetected nucleon for pD$\rightarrow$ppn. This result is improving with better E calibrations for the various detectors. More importantly, Monte

Carlo calculations show that the missing mass width strongly depends on the energy of the undetected particle. Widths of about 2 MeV or better are projected for threshold pions. During injection and at the beginning of the acceleration and cooling cycles, the beam was quite large and produced considerable background. However, since the data acquisition gate for our experiment is opened only after the injection and acceleration periods end, we have been able to leave the photomultipliers at full voltage. The wire chambers are more sensitive, hence they are automatically turned down to half the operating voltage by a "dimmer" box until just before the acquisition gate opens. The background seen in our first runs has been low enough not to interfere with the data, but there are two caveats: In the $pp \rightarrow pn\pi$ runs the detectors will be much closer to the beam pipe and the luminosity required for our measurement is a factor of 100 larger.

SUMMARY

At this time, the detector system for the $pp \rightarrow pn\pi^+$ measurements near threshold and the data acquisition software are almost fully completed and tested. A few instrumentation tasks remain. The pulsed UV laser still has to be coupled to the charged particle scintillators, and the unified system permanently mounted and tested. Perhaps the most important remaining task is the development of key on-line checks on the performance of the detector and the beam quality in this relatively new experimental environment.

REFERENCES

1) J.R. Bergervoet et al., Phys. Rev. Lett. **59**, 2255 (1987); A.W. Thomas and K. Holinde, Phys. Rev. Lett. **63**, 2025 (1989).
2) E. Mazzucato et al., Phys. Rev. Lett. **57**, 3144 (1986); R. Beck, contrib. talk at PANIC 1990; S. Nozawa, ibid.
3) T.S.H. Lee and A. Matsuyama, Phys. Rev. C**36**, 1459 (1987) and private communication.
4) H.O. Meyer, IUCF experiment CE01, private communication.

MEASUREMENTS OF PION PRODUCTION IN $\pi^{\pm}p$ INTERACTIONS NEAR THRESHOLD

Omicron Collaboration

G. Kernel, D. Korbar, P. Križan, M. Mikuž, U. Seljak, F. Sever,[1] A. Stanovnik, M. Starič, D. Zavrtanik[2]
J. Stefan Institute and Department of Physics, University of Ljubljana, YU-61111 Ljubljana, Yugoslavia

C.W.E. van Eijk, R.W. Hollander, W. Lourens[3]
Delft University of Technology, Delft, The Netherlands

E.G. Michaelis[4]
CERN, CH-1211 Geneva 23, Switzerland

N.W. Tanner
Nuclear Physics laboratory, Oxford University, Oxford OX1 3RH, UK

S.A. Clark,[5]
Rutherford and Appleton Laboratory, Chilton, Didcot OX11 0QX, UK

J.V. Jovanovich
Department of Physics, University of Manitoba, Winnipeg, Manitoba R3T 2N2, Canada

J.D. Davies, J. Lowe, S.M. Playfer[6]
Department of Physics, University of Birmingham B15 2TT, UK

ABSTRACT

Results of full-kinematics measurements of $\pi^-p \to \pi^-\pi^+n$, $\pi^-p \to \pi^-\pi^0p$ and $\pi^+p \to \pi^+\pi^+n$ in the threshold region are compared with current-algebra calculations. Discrepancies can be largely attributed to isobar production, but even the threshold behaviour is not adequately described by existing calculations.

[1]Present address: DPHN CEN-Saclay, F-91191 Gif-sur-Yvette, France
[2]Present address: CERN EP, CH-1211 Geneva 23, Switzerland
[3]Present address: State University Utrecht, Utrecht, The Netherlands
[4]Deceased
[5]Present address: DEC, RDL 2B Queens House, Forbury road, Reading RG1 3JH, UK
[6]Present address: Institut für Mittelenergiephysik der ETH Zürich, CH-5232 Villigen PSI, Switzerland

INTRODUCTION

The low energy $\pi\pi$ scattering is in many respects one of the most interesting hadronic processes. Its simplicity and connection to chiral symmetry makes it a sensitive tool of low-energy QCD. As a means to study the pion-pion scattering the production of pions, $\pi p \rightarrow \pi\pi N$, close to the reaction threshold seems to be particularly suitable[1]. Historically, current algebra and PCAC made a number of firm predictions about the behaviour of the pion-nucleon and pion-pion interactions at low energy. Explicitly the amplitudes for $\pi p \rightarrow \pi\pi N$ at threshold were prescribed in terms of a single parameter ξ which is related to the structure of the chiral symmetry breaking term in the Lagrangian[2]. Weinberg's Lagrangian leads to a value of $\xi = 0$, whereas models put forward by Schwinger[1] attribute to ξ the values +1, +2, and -2. In QCD the chiral symmetry breaking occurs via the quark mass term ($\xi = 0$). A systematic non-perturbative QCD based description of low energy processes, has been developed as an expansion in powers of external momenta and of quark masses[3]. While pion-pion scattering has been examined in the framework of this theory and substantial corrections to the current algebra predictions were established, no calculation was performed to find the corresponding prediction for $\pi p \rightarrow \pi\pi N$ at threshold.

There exist a multiplicity of measurements of the cross-section of $\pi^- p \rightarrow \pi^-\pi^+ n$ and a few measurements of the other channels (see Ref.[4] for a review). The threshold behaviour of the amplitudes has been rendered uncertain by the lack of adequate statistics and complete kinematical information. To supplement the existing data we have performed good statistics, full kinematics measurements of the reactions $\pi p \rightarrow \pi\pi N$ using π^- and π^+ beams at several energies near threshold. The experiments were carried out with the Omicron spectrometer at the CERN SC. The apparatus and its performance were described in detail in Ref.[5]. Results of measurements of the reactions $\pi^- p \rightarrow \pi^-\pi^+ n$, $\pi^- p \rightarrow \pi^-\pi^0 p$ and $\pi^+ p \rightarrow \pi^+\pi^+ n$, were reported elsewhere[6]. In this contribution we compare our cross-sections with current-algebra calculations and give an account for discrepancies.

RESULTS

From the many attempts to calculate cross-sections for $\pi p \rightarrow \pi\pi N$ reactions[2,7] within the framework of current algebra, the most elaborate is the one of Arndt et al[8]. In their work, integrated cross-sections for all the $\pi p \rightarrow \pi\pi N$ reaction channels are calculated by summing diagrams up to the tree-level. The cases of $\xi = 0$ and $\xi = -2$ are evaluated. Comparison of our integrated cross-sections with their calculations is shown in Fig. 1.

The measured values for $\pi^- p \rightarrow \pi^-\pi^+ n$ are well above the $\xi = 0$ line which would result in a negative value of ξ. In our differential distributions we observed, however, the influence of a broad s-wave $\pi\pi$ resonance with a mass lying outside the invariant-mass region covered by our experiment. The fit of a relativistic

Breit-Wigner resonance was very unstable and gave values around 600 MeV for the mass. The resulting width was comparable to the mass. On the other hand, we see no significant effects of Δ and ρ production in our data. Thus, we concluded that a broad s-wave resonance in the $\pi^-\pi^+$ system enhances the $\pi^-p \to \pi^-\pi^+n$ cross-section even very close to threshold. Its influence makes this reaction unsuitable for simple studies neglecting isobar production.

The agreement of the measured $\pi^-p \to \pi^-\pi^0p$ cross-section with the $\xi = 0$ tree-level calculation is remarkable. All the data points lie within 1σ from the prediction while the cross-section spans more than two orders of magnitude. No signs of isobar production were seen in the differential distributions, neither in the $\pi\pi$ nor in the πp systems, although the small number of events (up to 430) makes these statements somewhat weak. The absence of the structure in the $\pi^-\pi^0$ invariant-mass distributions fixes the $\pi\pi$ resonance isospin to $I = 0$.

The behaviour of the $\pi^+p \to \pi^+\pi^+n$ cross-section is twofold. Up to 360 MeV/c it lies above the $\xi = 0$ line, indicating a positive value for ξ. At higher momenta it flattens off and gets even below the $\xi = -2$ prediction. This kink in the cross-section we belive to be connected with the influence of the Δ^+ production in the final state, of which we see evidence in our differential distributions at momenta above 360 MeV/c. No other structure except for the reflection of the Δ^+ resonance is observed in the $\pi^+\pi^+$ system, confirming the isospin assignment to the $\pi\pi$ resonance.

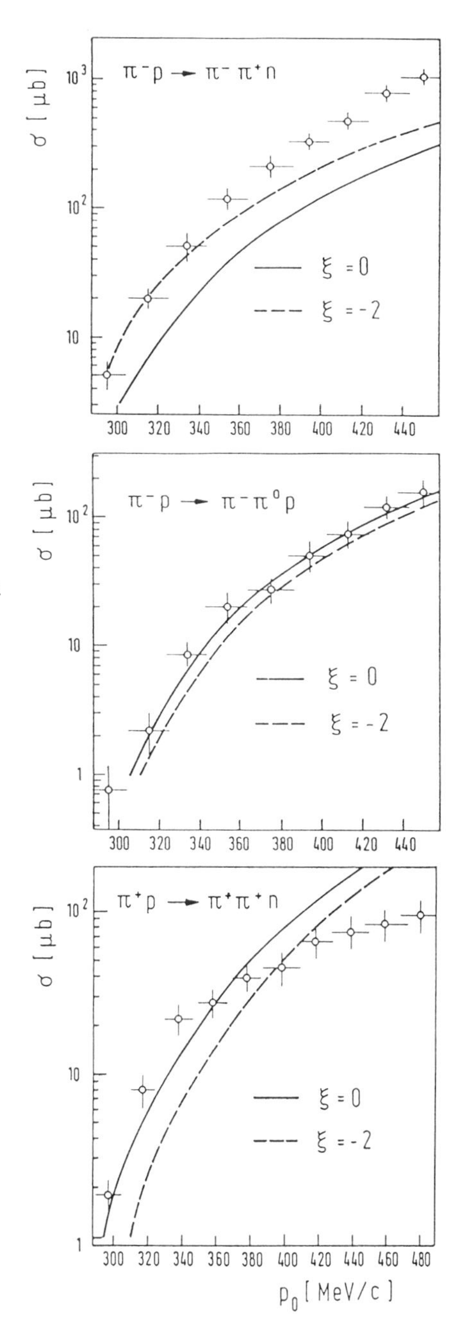

Figure 1: Measured integrated cross-sections in comparison with calculations of Arndt et al[8].

DISCUSSION

From the comparison made we conclude that the $\pi\pi$ system at low energies cannot be described by simple current-algebra calculations, the $\pi p \to \pi\pi N$ reaction bearing additional complications in terms of the πN isobars. A resonance-like behaviour is observed in the $\pi^-\pi^+$ system. We were unable to define precisely the mass and width of the resonance but the spin isospin and parity values of $I = 0, J = 0^+$ were determined. The absence of $\pi\pi$ resonances with $I = 1$ and $I = 2$ yields better agreement of $\pi^- p \to \pi^-\pi^+ n$ and $\pi^+ p \to \pi^+\pi^+ n$ cross-sections in the region where the influence of πN isobars is not significant. Even there, current algebra predictions differ for both reactions considered. As non-perturbative QCD calculations give substantial corrections to current-algebra $\pi\pi$ scattering-lengths, they might also clarify the threshold behaviours of the $\pi p \to \pi\pi N$ reaction cross-section. We hope that our data present enough motivation to theoreticians to perform the necessary calculations.

In order to clarify the behaviour of the $\pi\pi$ system in the $I = 0$ state a full kinematics measurement of $\pi^- p \to \pi^0\pi^0 n$ reaction in the threshold region would be highly appreciated. In addition, we have started to extract the fourth charged channel, $\pi^+ p \to \pi^+\pi^0 p$, for which we can quote a preliminary integrated cross-section of $(189 \pm 8 \pm 28)\mu b$ at $p_0 = 418\text{MeV/c}$.

REFERENCES

1. S. Weinberg, Phys. Rev. Lett. **17**, 616 (1966); Phys. Rev. **166**, 1568 (1968), J. Schwinger, Phys. Lett. **24B**, 473 (1967).

2. M.G. Olsson, L. Turner, Phys. Rev. Lett. **20**, 1127 (1968); Phys. Rev. **181**, 2141 (1969).

3. J. Gaser, H. Leutwyler, Phys. Lett. **B125**, 321 (1983); Phys. Lett. **B125**, 325 (1983); Ann. of Phys. **158**, 142 (1984); Nucl. Phys. **B250**, 465 (1985); Nucl. Phys. **B250**, 517 (1985); Nucl. Phys. **B250**, 539 (1985).

4. D.M. Manley, Phys. Rev. **D30**, 536 (1984).

5. G. Kernel et al., Nucl. Instr. and Meth. **214**, 273 (1983); Nucl. Instr. and Meth. **A244**, 376 (1986).

6. G. Kernel et al., Phys. Lett. **B216**, 244 (1989); Phys. Lett. **B225**, 198 (1989); Subm. to Zeit. Phys. (1990).

7. L.N. Chang, Phys. Rev. **162**, 1497 (1967),
W.F. Long, J.S. Kovacs, Phys. Rev. **D1**, 1333 (1970).

8. R.A. Arndt et al., Phys. Rev. **D20**, 651 (1979).

MEASUREMENT OF THE $^{12}C(p,\pi^0)^{13}N_{g.s.}$ REACTION USING RECOIL DETECTION

J. Homolka, W. Schott, W. Wagner, W. Wilhelm
Physik Department, Technische Universität München

R.D. Bent, M. Fatyga, R.E. Pollock
Indiana University Cyclotron Facility, Bloomington, IN 47405

M. Saber, R.E. Segel
Northwestern University, Evanston, IL 60201

P. Kienle
Gesellschaft für Schwerionenforschung, Darmstadt, Germany

K.E. Rehm
Argonne National Laboratory, Argonne, IL 60439

ABSTRACT

Cross sections for the reaction $^{12}C(p,\pi^0)^{13}N_{g.s.}$ have been measured at 153.5, 166.1, 186.0 and 204.0 MeV bombarding energy by detection of the ^{13}N recoil ions. The shapes of the differential cross section angular distributions agree well with those for the $^{12}C(p,\pi^+)^{13}C_{g.s.}$ reaction obtained by pion detection. The ratio of the (p,π^+) to (p,π^0) total cross section is close to two, as expected from isospin invariance, at 153.5, 186.0 and 204.0 MeV, but deviates significantly from this at 166.1 MeV bombarding energy. The (p,π^+) and (p,π^0) cross sections at the lower energies are both larger than expected from an extrapolation of the higher energy data.

INTRODUCTION

Proton–nucleus pionic capture is a large momentum transfer process ($q \geq$ 450 MeV) even close to threshold ($T_p \approx 140$ MeV). The reaction is therefore sensitive to nuclear dynamics at short distances. Part of the motivation for studying the highly inelastic (p,π) process is the possibility that the participating nucleons may overlap substantially during the production process and that quark degreees of freedom might therefore have to be treated explicitly for a full understanding of the process.

EXPERIMENT

The $^{12}C(p,\pi^+)$ and (p,π^0) reactions were studied using the recoil–ion detection setup at the IUCF, which employs the QQSP magnetic spectrograph at 0° to the beam axis and a heavy–ion detector in the focal plane. The method and earlier results obtained for the $(^3He,\pi^+)$ reaction have been described previously.[1–3] With the recoil method, it is possible to measure differential cross sections for both the $^{12}C(p,\pi^+)^{13}C_{g.s.}$ and $^{12}C(p,\pi^0)^{13}N_{g.s.}$ reactions simultaneously using the same setup.

RESULTS

In Fig. 1, the differential cross sections for the $^{12}C(p,\pi^0)^{13}N_{g.s.}$ reaction obtained using the recoil detection method are shown. The dashed curves are angular distributions calculated from pion data for the $^{12}C(p,\pi^+)^{13}C_{g.s.}$ reaction at 170, 183 and 190 MeV [5], multiplied by a factor of 0.5 according to isospin invariance. Coulomb effects were taken into account by fitting energy independent R-matrix elements to the measured $d\sigma/d\Omega$ and A_y values and describing the energy dependence by phase space and Coulomb barrier penetration factors.

The shapes of the (p,π^0) angular distributions measured by the recoil method are in good agreement with those obtained for (p,π^+) by pion detection, but the absolute values of the recoil cross sections at 154 and 166 MeV are significantly larger than those derived from the (p,π^+) data. To treat this effect quantitatively, the (p,π^+) angular distributions were expanded in terms of Legendre polynomials, and these polynomial expansions were then normalized to the (p,π^0) angular distribution data shown in Fig. 1 by varying the first term in the expansion, which is the total cross section.

Fig. 2 shows the resulting total cross sections for the $^{12}C(p,\pi^0)^{13}N_{g.s.}$ reaction obtained by recoil detection, multiplied by a factor of 2, and the $^{12}C(p,\pi^+)^{13}C_{g.s.}$ reaction obtained by pion detection, plotted vs the pion momentum in units of $m_\pi c$ ($\eta = p_\pi/m_\pi c$).

Phase space and Coulomb barrier penetration calculations fitted to the (p,π^+) data at η=0.55, 0.71 and 0.78 are shown by the solid and dashed lines for the (p,π^+) and (p,π^0) reactions, respectively. At small η values (0.33-0.38), the measured cross sections are larger than expected from an extrapolation of the higher energy data by phase space and Coulomb barrier calculations. The ratio of the (p,π^+) to (p,π^0) total cross sections is close to two, as expected from isospin invariance, except at $\eta = 0.55$, where there is a discrepancy of the order of two that does not appear to be explained by Coulomb effects.

DISCUSSION

The deviation of the energy dependence of the (p,π^+) and (p,π^0) total cross sections at low energies from that predicted by phase space and Coulomb barrier penetration effects alone implies that the assumption of constant R–matrix elements is wrong. More data are needed to confirm the $\eta = 0.55$ data and determine in more detail the shapes of the (p,π^+) and (p,π^0) excitation functions in this energy range. Of particular importance would be a determination of the width of any observed resonance-like structure.

REFERENCES

1. J. Homolka et al., Nucl. Instr. & Meth. in Phys. Res. **A260**, 418 (1987)
2. W. Schott et al., Phys. Rev. **C 34**, 1406(1986)
3. J. Homolka et al., Phys. Rev. **C 38**, 2686(1988)
4. F. Soga et al., Phys. Rev. **C 24**, 570 (1981)
5. M.C. Green, Internal Report No. 83-3a, Ph.D. Thesis, Indiana University, August 1983
6. M.P. Locher, M.E. Sainio, and A. Svarc, in Advances in Nuclear Physics, vol. 17, ed. J.W. Negele and E. Vogt (Plenum Press, New York, 1986) p. 47
7. G.T. Emery, Phys. Lett. vol. 60B (1976) 351
8. D.V. Bugg, Comments Nucl. Part. Phys. vol. 12 (1984) 287
9. R.R. Silbar, ibid, p. 177

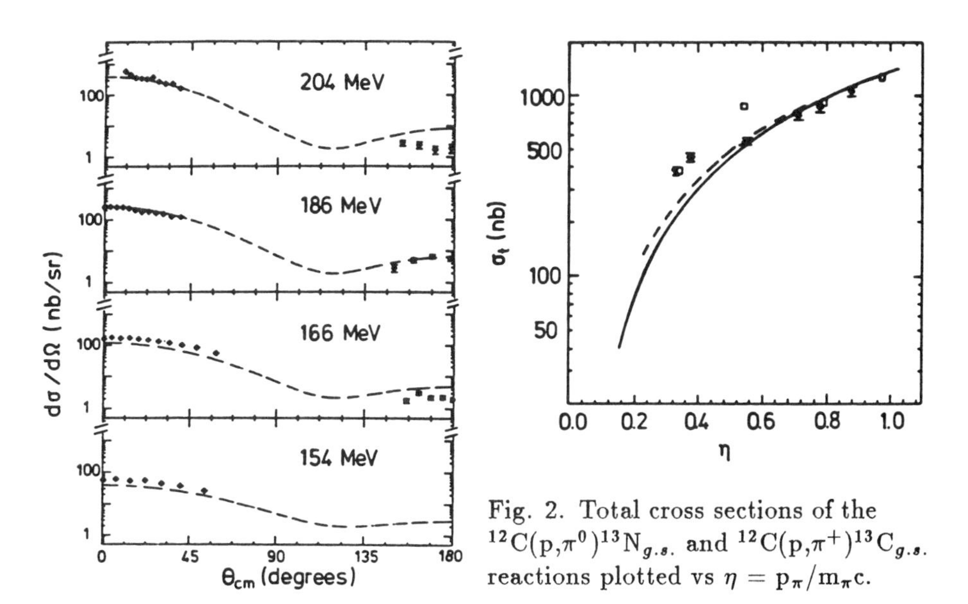

Fig. 1. Differential cross sections of the $^{12}C(p,\pi^0)^{13}N_{g.s.}$ reaction at 154, 166, 186 and 204 MeV.

Fig. 2. Total cross sections of the $^{12}C(p,\pi^0)^{13}N_{g.s.}$ and $^{12}C(p,\pi^+)^{13}C_{g.s.}$ reactions plotted vs $\eta = p_\pi/m_\pi c$.

SEARCH FOR EXOTIC STATES T=2, Q=+3 AND T=3, Q=+4 IN THE 1 GEV PROTON-NUCLEUS COLLISIONS

L.M.Andronenko(a), M.N.Andronenko(a), A.A.Kotov(a), W.Neubert(b), D.M.Seliverstov(a), I.I.Strakovsky(a,c), L.A.Vaishnene(a), V.I.Yazura(a)

a - Leningrad Nuclear Physics Institute, Gatchina, Leningrad 188350 USSR
b - Central Institute for Nuclear Research, Rossendorf, P.B. 19, Dresden 8051 DDR
c - TRIUMF 4004 Wesbrook Mall, Vancouver B.C. V6T 2A3 Canada

The search for unusual states (that could be, for example, purely gluonic, quark - gluonic or multiquark) has become an important problem in hadron spectroscopy. However, it is a difficult task, because theory, as a rule, can hardly say anything definite about their expected properties. There are many experimental attempts in search of long-lived dibaryonic states. But, unfortunately, the question of the existence of such exotic states remains open up to now. In the case of T=2 with $M < 2M_N + m_\pi$ theory can give some information that allows one to apply results of the experiments and obtain strict limits on the production of these states[1]. This conclusion could be expanded to the T=3 case[2].

The main aim of this talk is to present a preliminary result of the search for exotic states T=2, Q=+3 (as 2Li) and T=3,Q=+4 (as 3Be) in pA interactions. The experiment was performed with a 1 GeV proton beam from the Gatchina synchrocyclotron. The integrated beam charge was ~1 mC. We used Ag, Ni and Au as targets. Two twin axial ionization chambers of 320 mm active length and 70 mm diameter provided identification of fragments[3]. The chambers were placed at 30 and 126 degrees with respect to the beam axis. Each branch consisted of two parallel plate avalanche counters to TOF measurements. The analysis was allowed the determination of the maximum of Bragg curve, range and energy (in the interval E = 1 - 16 MeV) of fragments traversing both chambers. The figure shows on-line information on the range vs energy of fragments from Ag. This method allowed to identify of fragments up to Z = 17 with $\Delta Z/Z<0.025$, $\Delta R/R <0.025$ and $\Delta E/E<0.010$.

The experimental limit for the 2Li production

cross section, summing up three targets Ag, Ni and Au, ~ 5 nb/sr. This value depends on assumption about the lifetime τ . In our case τ > 20 ns. This conclusion has received some support from recent experiments (for details see Ref.1). On the other hand, the experimental limit for the 3Be production cross section is ~ 10 nb/sr.

We plan more detailed measurements using the method under discussion to obtain experimental limits on the level of exotic state production cross section ~ 1 nb/sr for τ < 5 ns in the energy interval of fragments E = 1 - 30 MeV.

REFERENCES

1. Ya.I.Azimov and I.I.Strakovsky, Sov. J. Nucl. Phys. 51, 614 (1990); Preprint TRIUMF, TRI-PP-89-87,1989.
2. A.B.Kaidalov, private communication.
3. A.A.Kotov et al. Exp. Tech. Phys. 36, 513 (1988); A.A.Kotov, private communication.

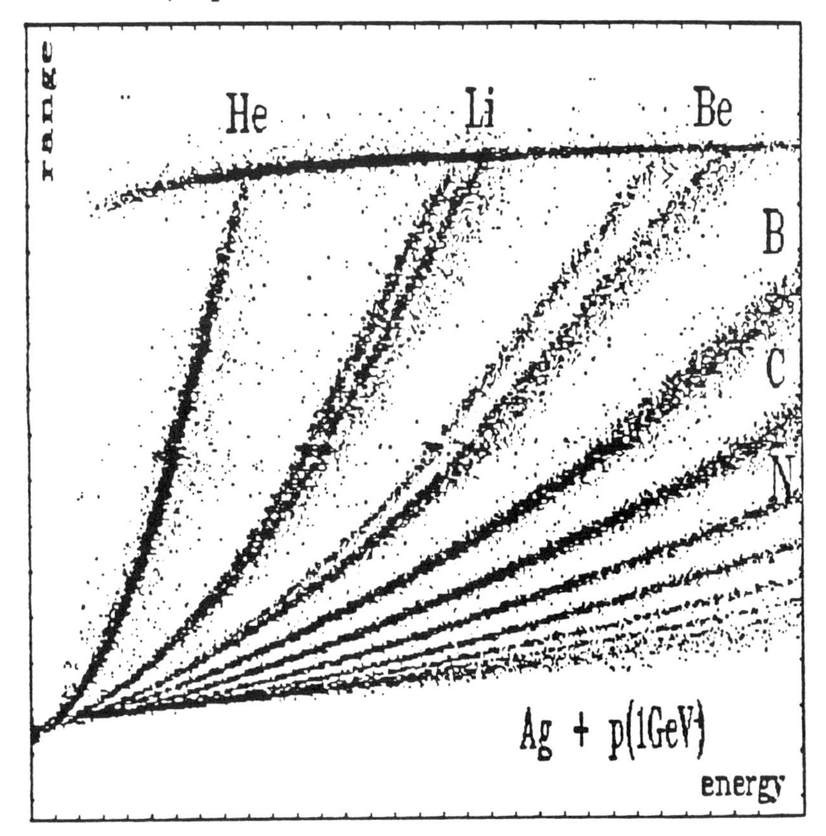

DIBARYONS AND THRESHOLDS (REVIEW)

Igor Strakovsky
Leningrad Nuclear Physics Institute, Gatchina, Leningrad 188350 USSR and
TRIUMF, 4004 Wesbrook Mall, Vancouver, B.C., V6T 2A3 Canada

In this talk we limit our discussion of the modern experimental situation to the search for a nonstrange system with baryon number B=2 in the I=1 channel at not-very-high energies $\sqrt{s}$=1.88-3.02 GeV. In the framework of phenomenological approach we used partial-wave analysis (PWA) technique as a method of reconstruction of amplitudes of coupled channels. It is of importance not only that results of separate PWAs for different reactions agree with each other, but it is even more important that PWA results obtained independently for coupled channels are also in agreement. As an illustration of the self-consistency of results of independent PWA fits to different reactions it is convenient to consider the total cross sections in pure pp spin states. $\Delta\sigma$ for the reaction pp--> π pN obtained by a subtraction $\Delta\sigma_{pp}^{in}$ [1] - $\Delta\sigma_{pp\to\pi d}$ [2] are in good agreement with recent ANL measurements[3].

The association of dibaryons with thresholds is discussed in general. Let us emphasize that the Breit-Wigner poles, obtained in different phenomenological analyses, are in fact the simplest parameterization for the NΔ cut contribution. Thus the nature of the singularities discussed here remains unknown.

The current status of dibaryon resonances is reviewed. The most important observation is that several partial amplitudes require resonances with approximately equal masses (M~2.17 GeV) and widths (Γ~100 MeV). Such solutions for $^1D_2$, $^3F_3$ and may be P NN partial waves have been obtained independently by different authors using different methods and different data sets[2,4,5]. As a result, all the noncontradictory NN, π d and π^+d-->pp PWAs yield the branching ratio for $^1D_2$ and P waves in the π d channel $\Gamma_{\pi d}/\Gamma_{tot}$~0.3 ($\Gamma_{pp}/\Gamma_{tot}$ ~0.1) while for the $^3F_3$ state respectively ~0.04 (~0.17).

In the framework of the general analysis on the

base of unitarity conservation we would like to discuss some raw and vague conclusions that have been made based as a rule on the high-quality experimental data[6,7]. More details analysis of the situation is given in papers[8,9].

Since detailed information already exists on different reactions at energies close to the delta isobar region ($\sqrt{s}$~2.17 GeV), we shall concentrate on the energy range $\sqrt{s}$ >2.3 GeV where only a few data exist. Our interest focuses on the nontrivial points in the energy behavior of quantities in both elastic and inelastic channels near 2.4, 2.7 and 2.9 GeV in $\sqrt{s}$.

The situation in the longlived-dibaryon sector will be discussed with special interest in the energy range up to the threshold of the pion production $\sqrt{s}$~2.015 GeV. This is just the case when theory can give some information that allows one to apply results of the experiments and obtain strict limitations on the T production[10].

Open questions and future directions of the experimental activities are discussed.

REFERENCES

1. R.A.Arndt et al. Phys. Rev. D28, 97 (1983).
2. I.I.Strakovsky et al.Sov.J.Nucl.Phys. 40, 273 (1984).
3. A.B.Wicklund et al. Phys. Rev. D35, 2670 (1987).
4. R.A.Arndt et al. Phys. Rev. D35, 128 (1987).
5. N.Hiroshige et al. Prog. Theor.Phys. 72, 1287 (1984).
6. A.Saha et al. Phys. Rev. Lett. 51, 759 (1983); D.B.Barlow et al. Phys. Rev. C37, 1977 (1988).
7. R.L.Shypit et al. Phys. Rev. C40, 2203 (1989).
8. I.I.Strakovsky, Proc. 3rd Intern. Symposium on πN and NN Physics. Gatchina, 1989. v.2, p.158.
9. M.G.Ryskin and I.I.Strakovsky, Phys. Rev. Lett. 61, 2384 (1988).
10. Ya.I.Azimov and I.I.Strakovsky, Sov. J. Nucl. Phys. 51, 614 (1990); Preprint TRIUMF, TRI-PP-89-87, 1989.

SESSION D

Meson–Meson and Meson–Baryon Interactions

STRUCTURE AND PRODUCTION OF DEEPLY BOUND PIONIC ATOMS

HIROSHI TOKI
Department of Physics, Tokyo Metropolitan University
Setagaya, Tokyo 158, Japan
and
RIKEN, Hirosawa,Wako, Saitama 351-01, Japan

ABSTRACT

We study the structure and production of deeply bound pionic atoms in heavy nuclei, which are found quasi-stable due to the repulsive pion-nucleus optical potential and the attractive Coulomb potential. The bound pion forms a pionic halo just outside of nucleus. We discuss then the use of pion transfer reactions such as (n,p) and (d,^{2}He) to form these states. In addition, we study other processes such as (n,d) and (d,^{3}He) and the use of the inverse kinematics for formation of deeply bound pionic atoms.

INTRODUCTION

In the standard method, pionic x-rays are measured to study the pionic atoms. The x-ray emission cascade, however, ceases at the "last" orbit, where the width is around 1keV, before reaching the bottom of the pionic atom due to pion absorption. Hence, deeper states than the "last" orbit, which are called deeply bound pionic atoms, are never observed; i.e. the 1s states are not observed above Z>13 with Z being the nuclear charge number. A question would be raised then whether or not deeply bound pionic atoms exist at all. If yes, then what is the structure of these states and how can we observe them? It is very interesting to study deeply bound pionic atoms, which should carry informations of nuclear structure and pion-nucleus interaction. Hence, we shall discuss the structure and production of deeply bound pionic states.

STRUCTURE OF DEEPLY BOUND PIONIC ATOMS

We answer the above questions by solving the Klein-Gordon equation with a proper pion-nucleus optical potential,

$$[-\Delta + m^2 + 2mV_{opt}]\Psi = [\omega - V_c]^2\Psi$$

Here, m denotes the reduced mass and ω the eigenvalue. The Coulomb interaction V_c includes the effects of the finite charge distribution and the vacuum polarization. We assume that the pion-nucleus optical potential V_{opt} obtained from existing pionic atom data, such as the one obtained by Seki-Masutani[1], are valid also for deeply bound pionic atoms.

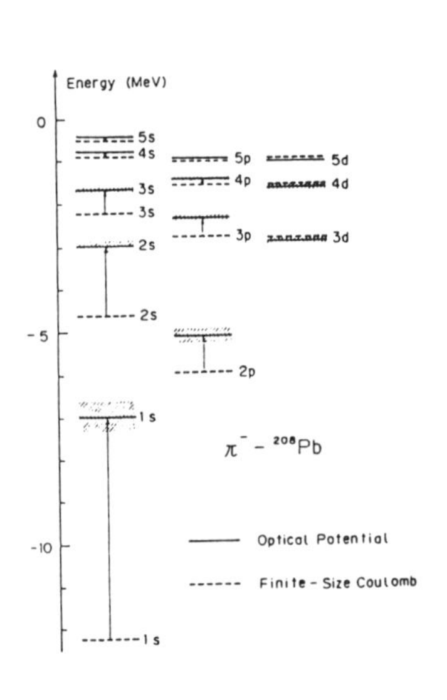

Fig. 1
Energy level of π-^{208}Pb, in which the width for each state is indicated by the hatched area.

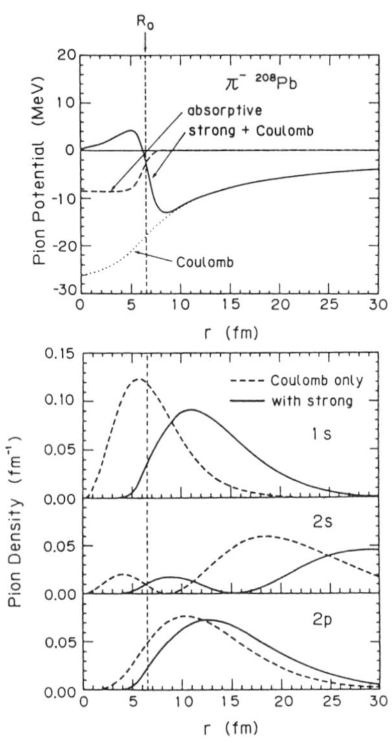

Fig. 2
Pion-nucleus potential for π-^{208}Pb and the density distributions for 1s, 2s and 2p states calculated with and without the optical potential.

We show in Fig.1 calculated results for deeply bound pionic atom levels of ^{208}Pb ,which are obtained by Toki et al.[2,3] The width for the

1s state is found about 0.6MeV, which is much smaller than the level spacing and hence it is quasi-stable. The reason for this finding is seen in the optical potential, which is repulsive and forms a pocket together with the Coulomb attraction just outside of the nucleus as shown in Fig.2. The pion forms a halo outside of the nucleus. This is an interesting result, since the properties of deeply bound pionic atoms are extremely sensitive to the nuclear surface, in particular the neutron skin, the information of which is hard to get otherwise.

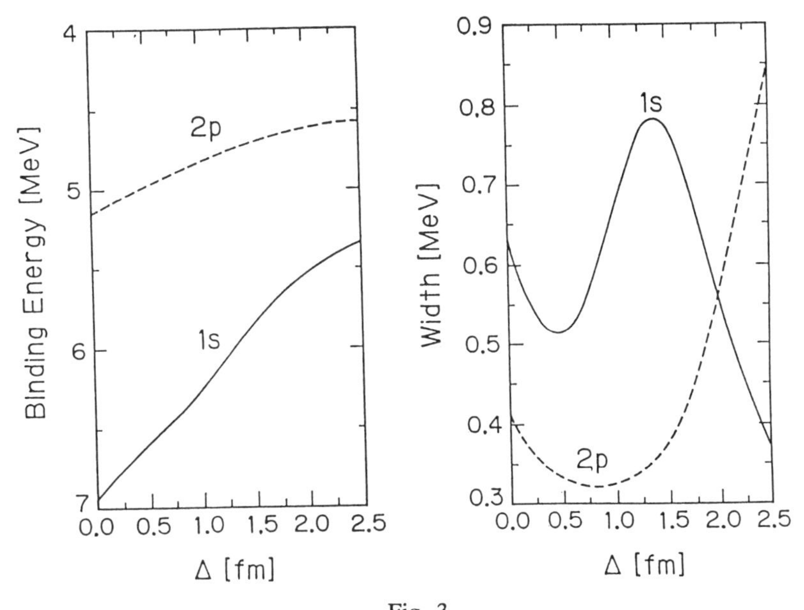

Fig. 3
The real and the imaginary parts of pionic atom energies for 1s and 2p states of π-^{208}Pb as function of the neutron skin.

This fact motivates us to look into the properties of deeply bound pionic atoms as function of the neutron skin Δ, which is defined as the difference between the neutron and the proton radii; $\Delta = R_n - R_p$.[4] Shown in Fig.3 are the real and the imaginary parts of the energies of 1s and 2p states in ^{208}Pb as a function of Δ. The real part increases monotonically with Δ, while the imaginary part decreases first and then increases with Δ. For 1s, it decreases again. This behavior is related with the property of the pion-nucleus optical potential which changes characteristically due to the strong isovector potential with Δ as shown in Fig.4. It would be interesting to observe such a change experimentally, which may be possible by the use of the inverse kinematics for production of pionic atoms on unstable nuclei, which will be mentioned in the next section.

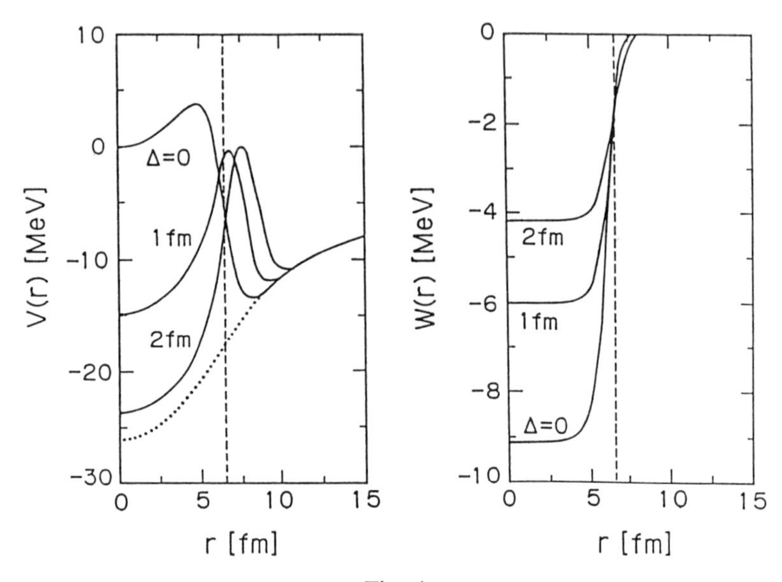

Fig. 4
The pion nucleus potential for π-^{208}Pb as function of the radial coordinate for various neutron skin parameter Δ.

PRODUCTION OF DEEPLY BOUND PIONIC ATOMS

Since the standard method does not populate deeply bound pionic atoms, we have to investigate possible methods in order to produce these states. Toki et al. have proposed the use of the pion transfer reactions such as (n,p) and (d,^{2}He) at intermediate energy.[2,3] Following the suggestion, experiments on ^{208}Pb have been performed for (n,p) at TRIUMF[5] and (d,^{2}He) at SATURNE[6]. Clear signals for pionic states have not been identified and the upper bound of the (n,p) cross sections are found to be 30μb/sr at T_n = 420MeV.[5] The distorted wave calculations provide small cross sections, which are consistent with the experimental finding.[5] The reason for the small (n,p) cross sections is found to be the large angular-momentum mismatch inherent in the pion transfer reactions.[5] The momentum transfer is about 200MeV/c, which would correspond to the angular momentum transfer of qR=L~7 when the (n,p) reaction takes place at the nuclear surface. For small L, the reaction takes place in the deep interior and hence the dominant contribution is removed by the distortion on the in- and out- going nucleon. Hence, pion transfer reactions as (n,p) and (d,^{2}He) may not be suited for production of deeply bound pionic atoms, unless the experimental energy resolution is largely improved or the spectra are taken in coincidence with signals of pionic atoms.

We have to look for other methods. The conditions to fulfill for suitable methods are; 1. the cross section has to be large, 2. the S/N ratio has to be large.

Nieves and Oset have studied electromagnetic processes as (γ,π^+) and (e,e').[7,8] They found that the cross section for (γ,π^+) for the largest case was ~70nb/sr/MeV and S/N was about 20%. On the other hand, for (e,e'), S/N was found too large, though the cross sections were not small. Hence, it was concluded that the use of (γ,π^+) is prefered than (e,e'). They emphasize the coherence of the (γ,π^+) process, in which all the nucleons in the target can participate for production of pionic atoms, which makes this reaction more suited than others.

We stick to use the strong interaction probes and look into the (n,d) reactions leading to pionic atoms with a neutron hole.[9] In these reactions, the picked up neutron can take care of the large momentum transfer, which is the reason to make the (n,p) reaction cross sections very small. We estimate the cross sections using the effective number approach, which is known to work extremely well.[10] The effective numbers are calculated by using the pionic atom and neutron hole wave functions with the distortion effects in terms of the eikonal approximation. The elementary cross sections for $n + n \rightarrow d + \pi^-$ are taken from the charge conjugate reaction $p + p \rightarrow d + \pi^+$ date. Due to the kinematical reason, the elementary cross sections in the laboratory system are very large; 5~20mb/sr at the forward angles and they show the resonant behavior caused by delta isobar.

We study therefore the incident energy of T_n=600MeV, which corresponds to the peak energy of the elementary cross section. The cross sections are found reasonablly large, when the neutron hole is made in $i_{13/2}$ orbit. This is because the large momentum transfer at this energy; q~200MeV/c, matches the large angular momentum of the neutron orbit at the nuclear surface where the reaction takes place. The reaction cross sections for this case are found, therefore, reasonably large even after the distortion effect is taken into account, in the range of 50μb/sr for deeply bound pionic atom formation as shown in Fig. 5. Including all the contributions from possible neutron hole states as p and f orbits, we find the (n,d) energy spectrum as shown in Fig. 6. By comparing the two figures (Figs. 5 and 6), we see that the spectrum is dominated by the formation of pionic atoms with the $i_{13/2}$ neutron hole.

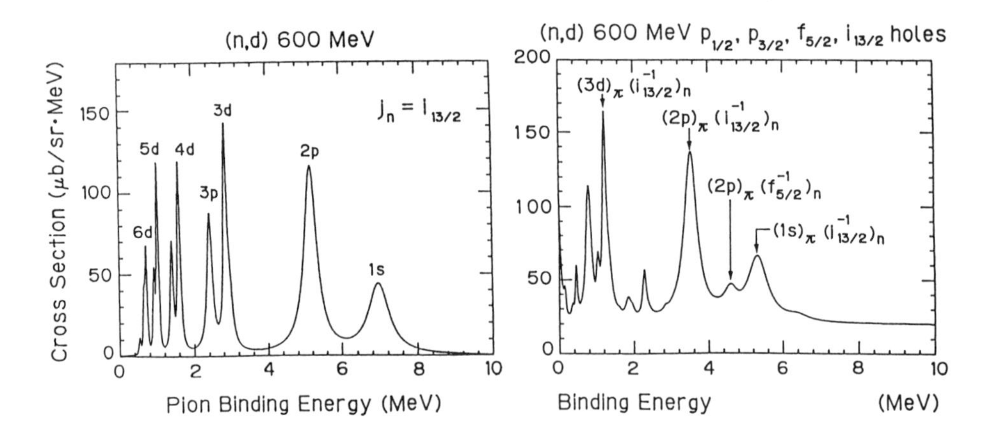

Fig. 5
(n,d) reaction cross sections leading to various pionic atom states with an $i_{13/2}$ neutron hole in ^{208}Pb at zero degree as function of pion binding energy at the incident energy T=600MeV.

Fig. 6
(n,d) reaction cross sections leading to various pionic atom states with several neutron hole states in ^{208}Pb as function of pion binding energy at the incident energy T=600MeV.

Another interesting case studied for (n,d) reactions is the incident energy at T_n=350MeV, where the momentum transfer is close to zero; the recoilless kinematics. In this case, the cross sections of the 'substitutional' states are large; i.e. pionic p-states with various principal quantum number with neutron 3p-states as shown in Fig. 7. If we want to use this kinematics in order to excite the pionic s-states, we should choose nuclei with a neutron s-state being close to the Fermi surface as those with the neutron number N=82. The energy spectrum at T_n=350MeV with all the neutron hole contributions is shown in Fig. 8.

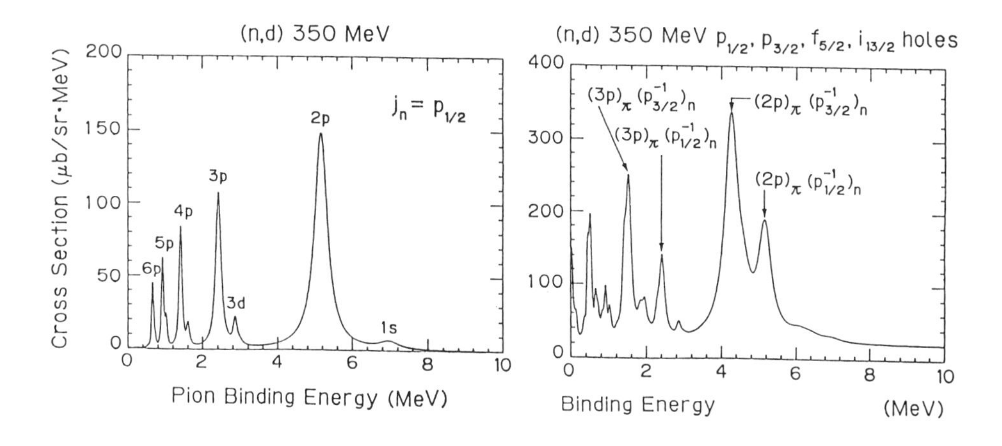

Fig. 7
(n,d) reaction cross sections leading to various pionic atom stattes with an $p_{1/2}$ neutron hole in ^{208}Pb at zero degree as function of pion binding energy at the incident energy T=350MeV.

Fig. 8
(n,d) reaction cross sections leading to various pionic atom states with several neutron hole states in ^{208}Pb as function of pion binding energy at the incident energy T=350MeV.

The next question for (n,d) reactions is about the back ground cross sections in the range of E_x=140MeV excitation energies. If the background cross sections at the excitation energy of pionic atoms; E_x~140MeV, are found small, the (n,d) reactions should be suited for excitation of pionic atoms. Since it is difficult to estimate the back ground theoretically, we looked for experimental data. There exist only spectra of (p,d) reactions at intermediate incident energies with the excitation energies up to E_x~30MeV, where the back ground cross sections are about 1~10μb/sr/MeV.[11] A recent survey experiment at

TRIUMF for (n,d) reactions at E_n=400MeV in the region of the pion threshold energy demonstrates that the quasi-pion production peak is clearly seen above the threshold energy after a flat back ground just below it.[12] This fact indicates that the pion production process is a dominant one around this excitation energy. Theoretical calculations for pionic atom production and quasi-free pion production in the same framework are found of similar size. Hence, (n,d) reactions seem to be suited for production of deeply bound pionic atoms.

Although suited theoretically, (n,d) reactions need secondary beams and therefore it is difficult to do high resolution experiments. Hence, we have looked into (d,^{3}He) reactions also, which are similar to (n,d) reactions. The elementary cross sections are by an order of magnitude smaller than (n,d) reactions due to the form factor effect in a complex projectile system. The effective numbers are found similar. Hence, the cross sections for pionic atom formation would be by an order of magnitude smaller than (n,d) reactions. However, the beam intensity of deuteron is by far larger than that of neutron and it seems (d,^{3}He) reactions should be extremely good, in particular, for good resolution experiments.

It may be interesting to use (p,pp) reactions for pionic atom formation, which need to measure two protons in coincidence. Since the elementary cross sections p+n -> p+p+ π^- are smaller by an order or two than those of n+n -> n+p+π^-, the cross sections for pionic atom formation are expected to be smaller than the case of (n,np) or (n,d) reactions. However, a high luminosity and high resolution facility as the Cooler Ring should make these reactions feasible.

We have studied also the use of the inverse kinematics, in which energetic heavy ion beam is bombarded on target of light element and recoiling light element after pion transfer reactions is observed.[13] This method enables us to measure spectra of pion transfer reactions with good resolution, when used with a good quality beam such as the one at ESR/SIS at GSI. We have studied then the inverse kinematics for (d,^{3}He) reactions for pionic atom formation, which are possible also with good resolution[14]. In addition, the use of the inverse kinematics makes it possible to produce pionic atoms on unstable nuclei.

ACKNOWLEGEMENT

The author is grateful to T. Yamazaki, S. Hirenzaki and R. Hayano for fruitful collaborations on the subject presented here. He appreciates also illuminating discussions with E. Oset and J. Nieves.

This is an invited paper presented at the IUCF Topical Conference "Particle Production Near Threshold" held at Nashville, Indiana in Sept. 30 - Oct. 3, 1990.

REFERENCE

1. R. Seki and K. Masutani, Phys. Rev. C27 (1983) 2799
2. H. Toki and T. Yamazaki, Phys. Lett. B213 (1988) 129
3. H. Toki, S. Hirenzaki, T. Yamazaki and R. Hayano, Nucl. Phys. A501 (1989) 653
4. H. Toki, S. Hirenzaki and T. Yamazaki, to appear in Phys. Lett. (1990)
5. M. Iwasaki et al., submitted to Pys. Rev.C (1990)
6. T. Yamazaki et.al., SATURNE proposal (1989)
7. J. Nieves and E. Oset, Valencia preprint (1990)
8. J. Nieves and E. Oset, Valencia preprint (1990)
9. H. Toki, S. Hirenzaki and T. Yamazaki, INS preprint (1990)
10. C. Dover, L. Ludecking and G. Walker, Phys. Rev. C22 (1980) 2073
11. G.R. Smith et.al., Phys. Rev. C30 (1984) 593
12. K.P. Jackson and A. Miller, private communication (1990)
13. T. Yamazaki, H. Toki, R. Hayano and P. Kienle, Nucl. Instr. and Meth. A292 (1990) 619
14. T. Yamazaki, R. Hayano and H. Toki, INS preprint (1990)

EXPERIMENTAL INVESTIGATION OF LOW-LYING STATES OF PIONIC ATOMS*

W. B. Amian, P. Cloth, A. Djaloeis, D. Filges, D. Gotta, K. Kilian, H. Machner, H. P. Morsch, D. Protic, G. Riepe, E. Roderburg, P. von Rossen, P. Turek, K. H. Watzlawik

Institut für Kernphysik, Forschungszentrum Jülich, Germany

L. Jarczyk, J. Smyrski, A. Stralkowski

Institut of Physics, Jagellonian University, Krakow, Poland

A. Budzanowski, H. Dabrowski, I. Skwirczynska

Institut of Nuclear Physics, Krakow, Poland

H. Plendl

Florida State University, Tallahassee, USA

J. Konijn

NIKEF K, Amsterdam, Netherlands

ABSTRACT

We propose to study pionic atoms in low-lying states. The pions will be produced with the help of recoil free kinematics at small energies in the laboratory. A dedicated detector will be applied allowing the measurement of the width as well as the energy shift of these states.

---

*Talk presented by H. Machner

1. Introduction

The study of the meson nucleon interaction in nuclei is of great interest, since modifications of the free interaction due to the nuclear medium are expected. Much more precise informations from bound states can be obtained than from continuum scattering states. The conventional technique of studying x-rays following π^--capture fails for low-lying states in heavy nuclei because when the hadronic interaction becomes strong this interaction leads to annihilation. Therefore, the intensity of the electromagnetic interaction is then very weak and unmeasurable. Only recently the 1s state could be measured with this technique for nuclei up to ^{28}Si with the help of an elaborate set-up including anti-compton shielded detectors[1]. The measured widths have large error bars compared to measurements for lighter nuclei. An alternative method of studying pionic atoms is the direct production of the pion in such low-lying states. For a further discussion we refer to the contribution delivered by Toki to this conference.

The probability to capture a π^- into an atomic state P can be calculated from the overlap integral of the wave functions to be

$$P \propto \exp(-E_\pi/B) \qquad (1.1).$$

It seems therefore natural to produce the pions with as small momenta as possible according to what was proposed by Kilian[2].

2. Recoilless Kinematics

The central idea of the experiment is to study the elementary reaction

$$p + n \rightarrow (2p + \epsilon) + \pi^- \qquad (2.1)$$

in nuclei with the residual system acting as a spectator. This is just what is well known as quasi-free reaction. Different is the simultaneous pion production. The two protons - sometimes referred to as ^2He - carry the whole beam momentum with them. The beam momentum at which the kinematics allows zero momentum

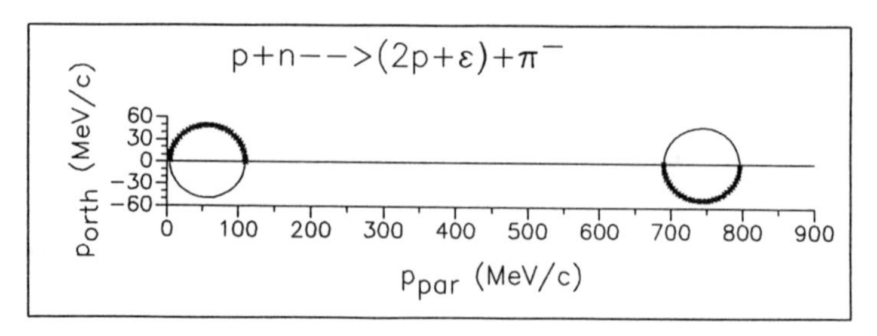

Fig. 1: Kinematical loci for the reaction $p+n \to {}^2He+\pi^-$ at a momentum of 800 MeV/c. To the maximal longitudinal ^2He momentum corresponds a π momentum of zero. To the largest orthogonal momentum up for ^2He corresponds the largest momentum down for the π.

for the pion is for the present reaction 829 MeV/c and is easily available at COSY. This kinematical situation is sketched

in Fig. 1. From this figure we take the information that recoilless π production is possible and that the ^2He particles are emitted into an angle smaller than 87 mr.

It is well known, that there is an anticorrelation between the two protons with a most probable relative momentum of $\Delta p \approx 20MeV/c$. This leads to an increased angular range of $\Theta(^2He) \leq 137mr$.

3. Detector

A detector for the investigation of the above mentioned reaction has to meet the following requirements. It should have a sufficiently large angular acceptance. To reconstruct the missing mass of the pionic system four momentum vectors of the two protons have to be measured. To achieve a sufficient missing mass resolution the direction of each proton should be measured with an accuracy of 2 mr and the energy resolution should be in the order of $\Delta E/E \approx 100KeV$. These requirements will be obtained by a combination if the existing magnetic spectrometer BIG KARL[3] and the GERMANIUM WALL presently under development. The primary beam will be separated from the reaction products in the first magnetic dipole of the magnetic spectrometer being in its 0° position. The primary beam will leave the spectrometer through a hole in the yoke of the first dipole and will then be dumped. Protons will be transported to the focal plane of the spectrograph if they had entered its acceptance of 110 mr horizontally and 56 mr vertically. Protons not meeting these constraints will be detected in the GERMANIUM WALL which covers a range of 400 mr in both directions. The direction of the protons detected in the magnetic system will be either by an induction drift chamber situated in front of the first magnetic quadrupole or by measuring the track in the focal plane by a stack of drift chambers. The latter method will be possible by upgrading the existing instrument with a third magnetic quadrupole which will allow ray tracing.

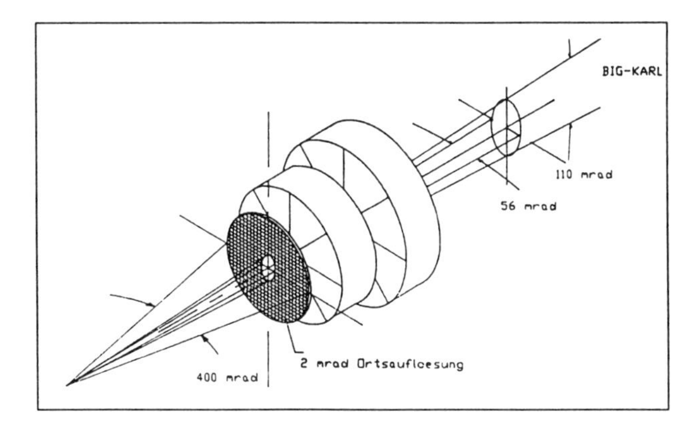

Fig. 2: Shown is the GERMANIUM WALL. It consist of three germanium detectors each having an elliptical hole corresponding to the acceptance of the magnet spectrograph BIG KARL. The first detector has strips on its front side and orthogonal to these strips on its back side as well. The strips will form a grid with 200×200 *pixels. The following two Ge-detectors allow for the total energy measurement of protons up to 100 MeV.*

The GERMANIUM WALL is shown in Fig. 2. It consists of three detectors made from high purity germanium (HPGe). They all have an elliptical hole which corresponds in its size to the solid angle acceptance of the magnet spectrograph. The first one with a thickness of 1 mm has 200 strips on the front side in x-direction and 200 strips on the back side in y-direction. This allows a position measurement with the accuracy of 2 mr. The following two detectors have thicknesses of 10 mm each. This set up allows the total energy measurement of protons up to 100 MeV. The energy resolution of such a detector system using standard electronics is sufficient for the present purpose. A description of the fabrication and tests of strip detectors made from germanium can be found in Ref.[4].

In order to reduce the background from nuclear reactions with multi body final states, which is expected to be some hundred times more intense than the reaction of interest, a HADRON BALL will be used. This is a small 4π detector consisting of 32 phoswitch detectors. With this detector the anticorrelation behavior of particles following pion absorption will be detected which is not expected to exist in the background reaction.

4. Monte Carlo Calculations

In order to study the resolution obtainable in the missing mass we have performed Monte Carlo calculations. Input were the above discussed spatial and energy resolution. It was found that the target thickness

has the dominant influence on the possible resolution. In Fig. 3 are results shown for ^{40}Ca targets of 40 and $1 mg/cm^2$. Also shown are the positions expected for the atomic 2s,1p state as obtained from the Klein Gordon equation for a point like charge and the atomic 1s state built up upon a nuclear excitation of 2 MeV, which is typical for the nuclei of interest. Obviously it will be possible to resolve these states in the case of the thin target while it will be impossible in the case of the thick target.

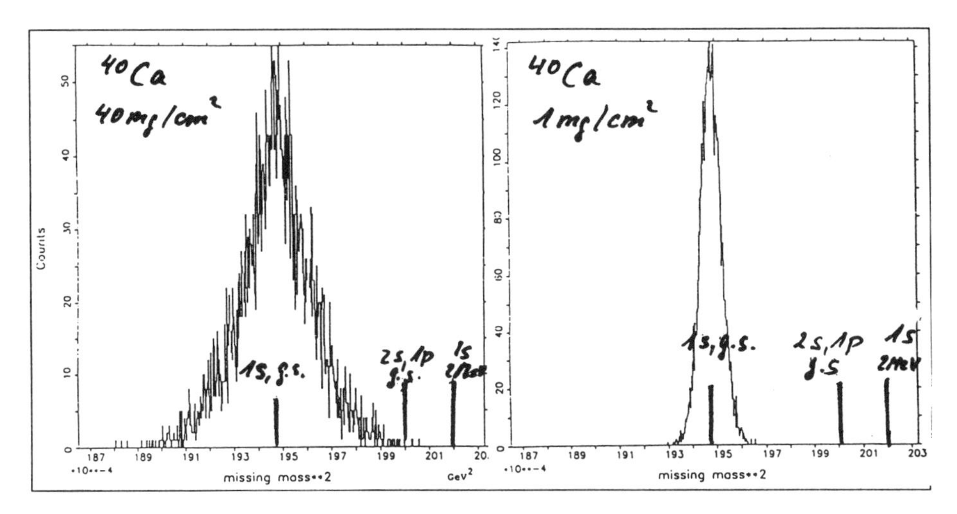

Fig. 3: Monte Carlo calculations for the missing mass resolution of the setup discussed above and for different target thicknesses.

REFERENCES

1. A. Taal et al.,*Nucl. Phys.* **A 511**, 573 (1989)

2. K. Kilian, COSY Note 61 (1986)

3. S. Martin et al., *Nucl. Instr. Methods in Physics Research* **A 214**, 281 (1983)

4. A. Hamacher et al., *Nucl. Instr. Methods in Physics Research* **A 295**, 128 (1990)

SHORT RANGE STRUCTURE OF HADRON AND NUCLEAR WAVE FUNCTIONS AT HIGH X*

Paul Hoyer and Stanley J. Brodsky**
Stanford Linear Accelerator Center
Stanford University, Stanford, California 94309

ABSTRACT

We discuss the short-range structure of hadronic and nuclear wave functions expected in QCD. In addition to the "extrinsic" contributions associated with radiation from single partons, there is an "intrinsic" hardness of the high-mass fluctuations of the wave function due to the spatial overlap of two or more partons.

We argue that intrinsically-hard partons, having large mass and/or large transverse momentum, will dominate in the region of large Feynman x_F. Their rescattering in nuclear targets is expected to be larger than for extrinsically-hard partons, leading to a suppressed production cross section for hadrons scattering on heavy nuclei. Experimental evidence for this exists for open charm, J/ψ, and Υ production at large x_F.

The effects of intrinsic hardness may be particularly striking in nuclear wave functions, where the overlap of partons belonging to different nucleons can give rise to cumulative ($x > 1$) phenomena. The data on backward cumulative particle production from nuclei supports the existence of an intrinsically-hard component in nuclear wave functions. Partons at large x_F may also be associated with the enhanced subthreshold production of particles observed in hadron-nucleus and nucleus-nucleus collisions.

We discuss the evidence for anomalies in the large angle $pp \to pp$ cross section near the charm threshold. Arguments are presented that charmonium states may bind to nuclei through the QCD Van der Waals force. This would lead to a striking signal in charm production near threshold.

INTRODUCTION

At high energies, most scattering processes only involve states that were formed long before the collision takes place. Consider the Fock expansion of a meson in QCD,

$$|h\rangle = |q\overline{q}\rangle + |q\overline{q}G\rangle + \ldots + |q\overline{q}Q\overline{Q}\rangle + \ldots \tag{1}$$

where $q(Q)$ refers to a light (heavy) quark and G to a gluon. The individual Fock components in (1) have "lifetimes" Δt (before mixing with other components)

* Work supported by the Department of Energy contract DE–AC03–76SF00515.
** Work supported by the Academy of Finland. Permanent address: Department of High Energy Physics, University of Helsinki, Finland

which can be estimated from the uncertainty relation $\Delta E \Delta t \sim 1$. At large hadron energies E the energy difference becomes small,

$$\Delta E \approx \frac{1}{2E} \left(m^2 - \sum_i \frac{m_i^2 + p_{Ti}^2}{x_i}\right) \tag{2}$$

where $x_i = (E_i + p_{Li})/(E + p_L)$ is the fractional (light-cone) momentum carried by parton i. Fock components for which $1/\Delta E$ is larger than the interaction time have thus formed before the scattering and can be regarded as independent constituents of the incoming wave function. At high energies only collisions with momentum transfers commensurate with the center of mass energy, such as deep inelastic lepton scattering ($Q^2 \sim 2m\nu$) and jet production with $p_T \sim \mathcal{O}(E_{cm})$ produce states with lifetimes as short as the scattering time.

The above arguments show that a typical scattering process is essentially determined by the mixture of incoming Fock states, *i.e.*, by the wave functions of the scattering particles. This is true even for collisions with very heavy quarks or with particles having very large p_T in the final state, provided only that the momentum transferred in the collision is small compared to E_{cm}. The cross sections for such collisions are thus determined by the probability of finding the corresponding Fock states in the beam or target particle wave functions; *cf.* Eq. (1). An example of this is provided by the Bethe-Heitler process of e^+e^- pair production in QED. A high energy photon can materialize in the Coulomb field of a nucleus into an e^+e^- pair through the exchange of a very soft photon. The creation of the massive e^+e^- pair occurs long before the collision and is associated with the wave function of the photon. The collision process itself is soft and does not significantly change the momentum distribution of the pair. Similarly, heavy quark production in hadron collisions is at high energies ($E_{cm} \gg m_Q$) governed by the hard (far off energy-shell) components of the hadronic wave functions.

The Structure of Intrinsically Hard States

The leading extrinsic contribution to heavy quarks in a hadronic wave function is one gluon splitting into a heavy quark pair, $G \to Q\overline{Q}$ (Fig. 1a). We call this contribution extrinsic since it is independent of the hadron wave function, except for its gluon content. The extrinsic heavy quarks are, in a sense, "constituents of the gluon". The extrinsic heavy quark wave function has the form

$$\Psi^{extrinsic}(q\overline{q}Q\overline{Q}) = \Gamma_G \; T_H(G \to Q\overline{Q}) \; \frac{1}{E\Delta E} \tag{3}$$

The square of the gluon amplitude Γ_G gives the ordinary gluon structure function of the hadron. The gluon splitting amplitude T_H is of order $\sqrt{\alpha_s(m_Q^2 + p_{TQ}^2)}$, and ΔE is the energy difference (2). The integral of the extrinsic probability $|\Psi^{extrinsic}|^2$ over p_{TQ}^2 for $p_{TQ} \lesssim \mathcal{O}(m_Q)$ brings a factor of m_Q^2. Hence we see that the probability of finding extrinsic heavy quarks (or large p_T) in a hadronic wave function is actually independent of the quark mass (or p_T). This is related to the quadratic divergence of the quark loop in Fig. 1b. The production cross section of the $Q\overline{Q}$ pair is still damped by a factor $1/m_Q^2$, this being the approximate transverse area of the pair.

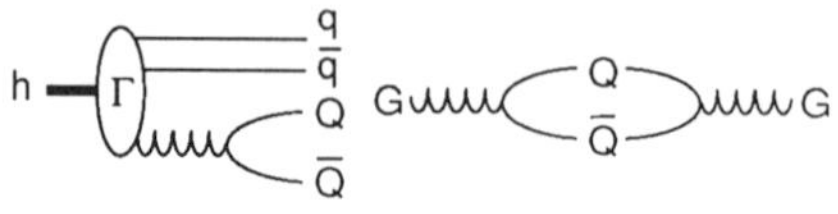

Figure 1. (a) Gluon splitting gives rise to extrinsic heavy quarks in a hadron wave function. The pointlike coupling to the gluon implies that all quark masses and all transverse momenta are generated with equal probability. (b) In the squared amplitude, this is seen as a quadratic divergence of the quark loop.

Intrinsic heavy quark Fock states[1] arise from the spatial overlap of light partons. Typical diagrams are shown in Fig. 2. The transverse distance between the participating light partons must be $\lesssim \mathcal{O}(1/m_Q)$ for them to be able to produce the heavy quarks. The wave function of the intrinsic Fock state has the general structure

$$\Psi^{intrinsic}(q\overline{q}Q\overline{Q}) = \Gamma_{ij}\ T_H(ij \to Q\overline{Q})\ \frac{1}{E\Delta E} \tag{4}$$

Here Γ_{ij} is the two-parton wave function, which has a dimension given by the inverse hadron radius. $T_H(ij \to Q\overline{Q})$ is the amplitude for two (or, more generally, several) light partons i, j to create the heavy quarks, and ΔE is the energy difference (2) between the heavy quark Fock state and the hadron. A sum over different processes, and over the momenta of the light partons, is implied in (4). In renormalizable theories such as QCD, the amplitude T_H is dimensionless. Hence, up to logarithms, the probability $|\Psi^{intrinsic}|^2$ for intrinsic heavy quarks is of $\mathcal{O}(1/m_Q^2)$ (after the p_T^2 integration). This is smaller by $1/m_Q^2$ as compared to the probability (3) for extrinsic heavy quarks,[1,2] as is true of higher twist. The relative suppression is due to the requirement that the two light partons be at a distance $\lesssim 1/m_Q$ of each other in the intrinsic contribution.

In contrast to the extrinsic contribution (3), which depends only on the inclusive single gluon distribution, an evaluation of the intrinsic Fock state (4) requires a knowledge of multiparton distributions amplitudes. In particular, we need also the distribution in transverse distance between the partons. Our relative ignorance of the multiparton amplitudes Γ_{ij} for hadrons[3] makes it difficult to reliably calculate the magnitude of the intrinsic heavy quark probability. We can, however, estimate[1] the distribution of intrinsic quarks from the size of the energy

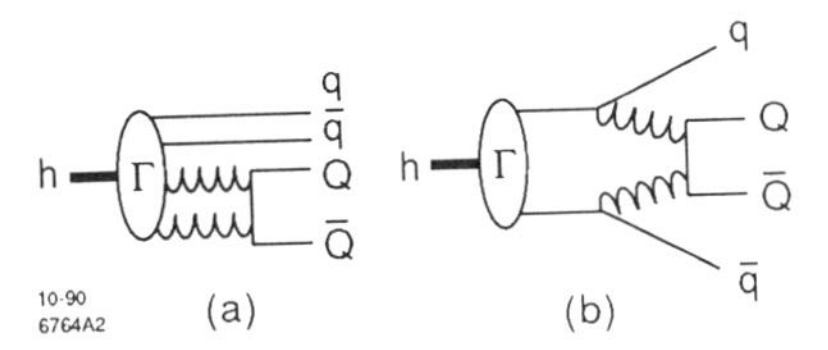

Figure 2. Intrinsic heavy quark contributions to a hadronic wave function, generated by (a) gluon fusion and (b) light quark scattering. The large mass of the produced quark implies that the participating light partons must be at a small transverse separation.

denominator ΔE, as given by (2). It is clear that those Fock states which minimize ΔE, and hence have the longest lifetimes, also have the largest probabilities. In fact, taking

$$|\Psi^{intrinsic}|^2 \sim 1/(\Delta E)^2 \ , \tag{5}$$

one finds that the maximum is reached for

$$x_i = \frac{\sqrt{m_i^2 + p_{Ti}^2}}{\sum_i \sqrt{m_i^2 + p_{Ti}^2}} \ , \tag{6}$$

implying equal (longitudinal) velocities for all partons. The rule (5) has been found to successfully describe the hadronization of heavy quarks.[5,6]

Using the probability (5), we see from (6) that partons with the largest mass or transverse momentum carry most of the longitudinal momentum. This has long been one of the hallmarks of intrinsic charm. We also note that the intrinsic heavy quark states have a larger transverse size than the extrinsic ones, although both tend to be small, of $\mathcal{O}(1/m_Q^2)$. The extrinsic heavy quarks are produced by a single (pointlike) gluon (Fig. 1), whereas the intrinsic mechanism is more peripheral (Fig. 2). This means that rescattering and absorption effects for intrinsic states produced on heavy nuclei will be relatively more significant, compared to that for extrinsic states. In addition to the heavy quarks Q, such rescattering may affect the light partons involved in the intrinsic state (*e.g.*, the quarks q in Fig. 2(b). These light quarks tend to be separated by a larger transverse distance than the heavy quarks, further enhancing the rescattering.

Consider now the formation of intrinsic heavy quark states in nuclear wave functions. At high energies, partons from different nucleons can overlap, provided only that their transverse separation is small. Thus the partons which create intrinsic heavy quarks in Fig. 2 can come from two nucleons which are separated by a longitudinal distance in the nucleus. Now it is reasonable to assume that partons belonging to different nucleons are uncorrelated, *i.e.*, that the two-parton amplitude Γ_{ij} in Eq. (4) is proportional to the product $\Gamma_i\Gamma_j$ of single parton amplitudes. Hence the amount of intrinsic charm in nuclei may be more reliably calculated than for hadrons. The probability for intrinsic charm will increase with the nuclear path length as $A^{1/3}$. Moreover, the total longitudinal momentum of the intrinsic quark pair, being supplied by two different nucleons, can be larger than in a single hadron, and can in fact exceed the total momentum carried by one nucleon.

All that we have said above concerning heavy quark Fock states applies equally to states with light partons carrying large transverse momentum. Extrinsic and intrinsic mechanisms for generating large p_T in hadronic wave functions are shown in Fig. 3. Using Eq. (5) as a guideline for the probability of intrinsic hardness, we see in fact that the parton mass and p_T appear in an equivalent way. We again expect that the intrinsic mechanism will be dominant at large x_F, and in particular in the cumulative ($x_F > 1$) region of nuclear wave functions.

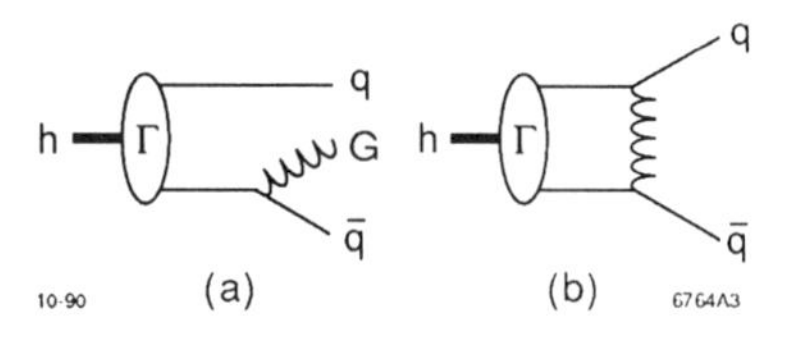

Figure 3. (a) An extrinsic contribution to large transverse momentum partons in a hadron and (b) an intrinsic contribution.

The possibility of parton fusion has been considered previously in the context of the evolution of parton distributions with momentum transfer (Q^2).[7,8] At very large Q^2 and small x, the number of gluons can become large enough to force them to overlap and coalesce. Our emphasis here is different. We are interested in rare phenomena at large x, where processes involving two or more gluons and valence quarks can give dominant effects, even though the likelihood for such fluctuations is small. The colliding partons in Figs. 1–3 are to be thought of (in a first approximation) as nearly on-shell, and having small p_T. Only the part of the processes in Fig. 3 leading to large p_T partons is to be considered as a new contribution to the wave function. In particular, the fusion of two partons into one (*e.g.*, $qG \to q$), which cannot give large p_T, is a part of the non-perturbative wave functions Γ, and hence does not contribute to intrinsic hardness.

Charm Production

The concept of intrinsic charm was originally inspired by experiments[9] showing unexpectedly abundant charm production at large $x_F = 2p_{charm}/E_{cm}$. When extrapolated to small x_F, the data suggested total charm cross sections in the millibarn range, far beyond the predictions (20 – 50 μb) of the standard QCD gluon fusion process (*cf.* Fig. 4(a)). Later data with good acceptance at low x_F showed that the total charm cross section actually is compatible with the gluon fusion process.[10] Nevertheless, more evidence was also obtained showing that charm production at large x_F, albeit a small fraction of the total cross section, still is larger than expected.[11] The large x_F data also shows correlations (leading particle effects) with the quantum numbers of the beam hadron that are incompatible with gluon fusion.[12]

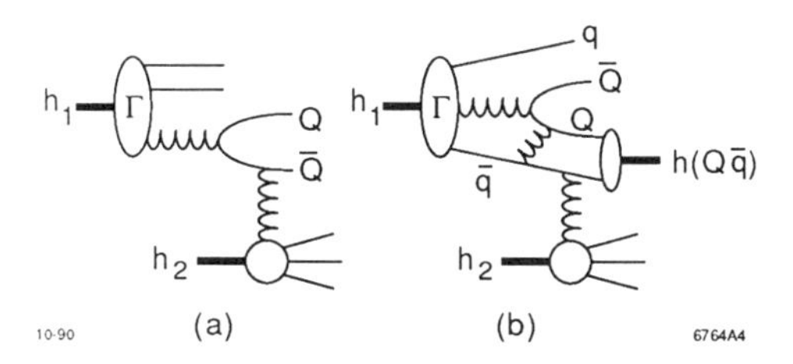

Figure 4. (a) The gluon-gluon fusion process in QCD. At high energies, the extrinsic $Q\overline{Q}$ pair preforms in the incoming wave function and is put on mass-shell by a soft gluon from the target. (b) An example of intrinsic heavy quark production. The heavy quark can get additional momentum from a light valence quark, and the produced hadrons at large x_F may get quantum numbers that are correlated to those of the valence quark (leading particle effect). The scattering can be from one of the light partons involved in the intrinsic state.

The intrinsic charm production mechanism (Fig. 4(b)) is expected to be smaller than the extrinsic one, due to the $1/m_Q^2$ suppression from the requirement of spatial overlap of initial light partons. However, at sufficiently large x_F the intrinsic mechanism will dominate, because the momentum of several incoming partons can be transferred to the heavy quarks. Our present, improved understanding of intrinsic charm, as outlined above, will allow a more quantitative theoretical discussion of these phenomena than was possible heretofore. Such an analysis will also become increasingly meaningful as the data on hadroproduced charm at large x_F improves.

Experiments on charm production from nuclear targets have shown an anomalous dependence on the nuclear number A. If the open charm (D, Λ_c) cross section is parametrized as

$$\frac{d\sigma}{dx_F} \propto A^{\alpha(x_F)} \tag{7}$$

then $\alpha(x_F \sim 0.2) \sim 0.7 \ldots 0.9$ is obtained.[10,13] For heavy nuclei ($A \approx 200$) this means a factor of $2 \ldots 3$ suppression in the cross section, compared to the leading QCD expectation ($\alpha = 1$). In this respect, the charm production data is quite different from that of massive μ-pair production, for which α is found to be very close to 1.[14]

For J/ψ production, the data on the x_F-dependence of α is particularly detailed,[15,16] showing a remarkable decrease from $\alpha = 1$ near $x_F = 0$ to $\alpha =$

0.7...0.8 at large x_F. The data at different beam energies agree, implying that Feynman scaling is valid. It is possible to show that the nuclear suppression is not due solely to the shadowing of the nuclear target structure function.[17] The effects of the target structure function can be eliminated by forming cross section ratios at a given value of the fractional momentum (x_2) of the target parton. This does not eliminate the target effects seen in the data, however, implying that the suppression does not factorize into a product of beam and target structure functions, as expected in leading twist. The target dependence thus must be due to a higher twist effect, *i.e.*, one that is of $\mathcal{O}(1/m_Q^2)$, compared to the leading (factorizable) QCD process. This is supported by preliminary data on Υ production,[16] which shows a significant but weaker nuclear suppression than for J/ψ production.

At high energies, the $c\bar{c}$ quarks do not have time to separate significantly inside the nucleus. Thus the J/ψ forms only after the charm quarks have left the nuclear environment, and the suppression cannot be related to the size of the J/ψ wave function.[18] This is also supported by preliminary data showing that the nuclear suppression for the $\Psi(2S)$ and the J/ψ is the same.[16] The $c\bar{c}$ state itself has a finite size, of $\mathcal{O}(1/m_Q^2)$, and could lose some momentum due to rescattering. Due to the rapid decrease of the cross section at large x_F, the trend of this effect is to make α decrease with x_F as observed. However, it is difficult to explain the magnitude of the x_F-dependence of α without assuming the loss of a large fraction of the momentum of the $c\bar{c}$ system.

A natural explanation of the increase of the nuclear suppression in J/ψ production with x_F is provided by the existence of two production mechanisms, the extrinsic and intrinsic ones.[19] As discussed above, intrinsic charm production is damped by a factor $1/m_c^2$, but can still dominate the small gluon fusion cross section at large x_F. Since the intrinsic heavy quark state tends to have a larger transverse size than the extrinsic one, it will suffer more rescattering in the nucleus. The x_F-dependence of α can then be understood as reflecting the increasing importance of intrinsic Fock states at large x_F.

In conclusion, the present experimental evidence for the existence of intrinsic charm is suggestive. However, the theoretical and experimental situation must improve for definite conclusions to be made. More quantitative studies of the intrinsic charm wave function, using multiparton distributions, coupled with better data on open charm at large x_F, should improve the situation in the near future.

The Intrinsic Hardness of Nuclear Wave Functions

We noted in Section 1 that intrinsic hardness should be enhanced in nuclear wave functions, due to the increased probability for spatial overlap of light partons from different nucleons. All of the data on charm production discussed above was obtained with beams of ordinary hadrons, and the experimental acceptance generally limited the observations to the forward ($x_F > 0$) hemisphere. This data thus reflects the importance of charm in the wave functions of the beam particles. An important exception to this is the EMC measurement of the charm structure function of the Fe nucleus.[20] An enhancement over the extrinsic photon-gluon contribution was observed at large x_F, but the limited statistics prevented a firm conclusion.

Several features of scattering on nuclear targets show that the nucleus cannot always be treated as a collection of ordinary nucleons. Measurements of deep inelastic lepton scattering have revealed[21,22,23] deviations of the nuclear structure functions from those of free nucleons, both at very small and at intermediate values of x (the "EMC Effect"). There are also indications[24] that the quark distributions in nuclei extend beyond $x = 1$. Unusual states of the nucleus could be involved as well in the production of large p_T particles in hadron–nucleus collisions, where the yield is known to increase faster than the nuclear number A (The "Cronin Effect").[25,26]

The most direct evidence for an enhancement of the nuclear structure function at large x comes from the so-called "Cumulative Effect".[27] Cumulative particles are defined as hadrons produced in the fragmentation region of a nucleus which have $x_F > 1$, *i.e.*, they carry more momentum than the individual nucleons (apart from Fermi motion effects). In practice, experiments are mostly done by scattering a variety of particles (leptons, hadrons and nuclei) on stationary nuclei, and observing hadrons that are moving backward in the laboratory. A simple kinematical exercise shows that at sufficiently high beam energies, the energy E_h and longitudinal momentum p_h^L of a hadron h produced on a free stationary nucleon must satisfy

$$x \equiv \frac{E_h - p_h^L}{m_N} \leq 1 \tag{8}$$

where m_N is the nucleon mass and $p_h^L < 0$ in the backward direction. The variable x defined by (8) is the usual (light-cone) fractional momentum, which is equivalent to the Feynman momentum fraction x_F of h in the CM system. This equivalence is strictly true for infinite beam momentum; a number of alternative definitions of x have been used in order to take finite energy effects into account. The difference between the various definitions will not be important for our qualitative discussion below.

Cumulative particle production has been seen in many experiments using a variety of beam particles and energies, up to values of $x = 4$ or so. To a first approximation, Feynman scaling (*i.e.*, independence of beam energy) sets in already at quite low energies, $p_{beam} \sim 2\ GeV$ (Fig. 5(a)). The shape of the cumulative hadron distribution is insensitive to the type of beam particle used. These features suggest that the cumulative particle distribution reflects properties of the nuclear wave function.

The laboratory momenta of the cumulative particles range well beyond 1 GeV, making a description in terms of ordinary Fermi motion unlikely. If a nucleon basis is used in the wave function it would be necessary, in this energy range, to include in an essential way also N^* and Y^* excitations.[29] In fact, many

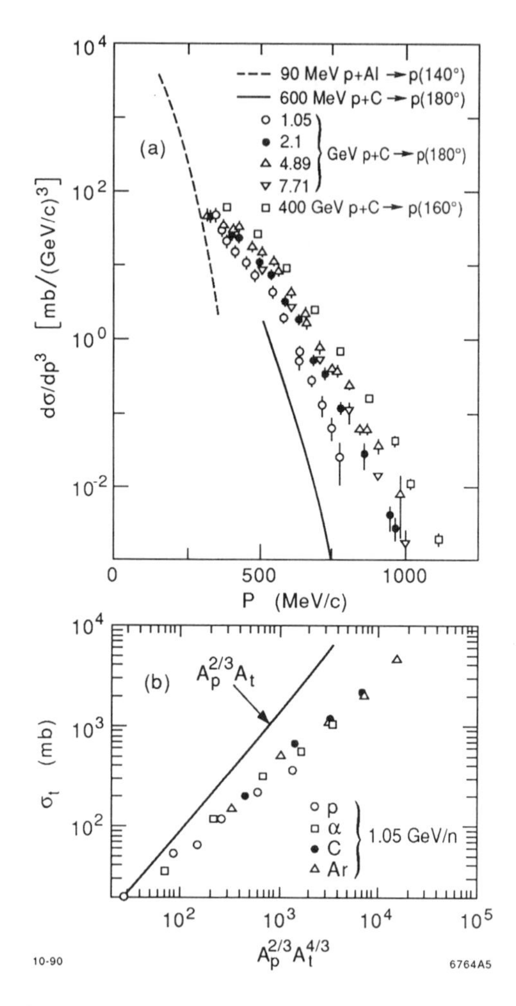

Figure 5. (a) Laboratory momentum distributions of cumulative protons produced by protons scattering on carbon and aluminum nuclei. In an analogy to the Rutherford experiment, the backscattering of 1 *GeV* protons from a beam of 2 *GeV* protons suggests encounters with small structures within the nucleus. (b) Dependence on the atomic number of the target (A_t) and projectile (A_p) for cumulative protons in the target fragmentation region. The data were fitted to a gaussian momentum distribution with a total rate parametrized by σ_t, which scales when plotted as a function of $A_p^{2/3}A_t^{4/3}$. Data and further references in Ref. 28.

arguments[27] point to the cumulative phenomena being linked to short-distance features of the nuclear wave function. The momenta of several nucleons in a nucleus have to be combined in order to produce the cumulative particles observed at the highest values of x. This presumably requires a close spatial correlation between the nucleons. Such short-distance effects in the nuclear wave function are best described in terms of quark and gluon degrees of freedom.[30]

The dependence of the cumulative particle distribution on the atomic number of the target nucleus is at least as fast as A^1, and is compatible with $A^{4/3}$ for cumulative protons at lower energies[28,31] (Fig. 5(b)). An A-dependence this strong suggests that the production of the cumulative partons is a volume effect, with little absorption of the outgoing quanta. An $A^{4/3}$ dependence is what one would naively expect for intrinsic hardness, given that the small size of the hard cluster implies a suppression of rescattering in the nucleus, and taking into account the factor $A^{1/3}$ enhancement from the transverse overlap (of two partons) along the nuclear diameter. For nuclear projectiles, the dependence on the atomic number of the projectile is compatible[28,31] with $A_{proj}^{2/3}$. For $A_{proj} < A_{targ}$, this is also in accord with naive expectations, since the projectile presumably can put intrinsically hard clusters on their mass shell throughout a region of transverse space proportional to the area of the projectile.

Direct evidence that the cumulative phenomenon is associated with small transverse size is provided by the p_T-distribution of the produced hadrons.[32,33] The average p_T^2 grows rapidly with x, and reaches 2 GeV^2 for pions at $x = 3$ (Fig. 6(a)). This is expected for the intrinsic configurations (4), since ΔE depends on p_T^2/x (see Eq. (2)). Note that although the individual partons in an intrinsically-hard cluster (*cf.* Fig. 3) have large transverse momenta, the total transverse momentum of the cluster is small. Hence in a case such as J/ψ production, where both intrinsic quarks are incorporated in the same final hadron, much of the large p_T cancels out. On the other hand, when an intrinsic quark combines with a low p_T spectator the final hadron will carry large p_T. The experimental result that cumulative protons tend to have smaller p_T and larger cross section at a given x may be due to more intrinsic partons getting incorporated in the protons than in the pions.[30]

A remarkable feature of the cumulative x-distributions is that their shape is quite similar for all observed particles: protons, positive and negative pions and kaons. Thus, *e.g.*, the ratio between the K^- and π^- yields[34] shown in Fig. 6(b) is constant over the measured range $1.5 \leq x \leq 2.5$. This differs from the fragmentation of single nucleons,[35] for which this ratio decreases as $x \to 1$. The magnitude of the K^+ yield is[34] also much higher than would be naively expected. The heaviest nuclear targets produce roughly equal numbers of K^+ and π^+ mesons at $x \geq 1.5$.

For intrinsically-hard quarks we noted in Section 1 that the x-distribution should be similar for all quarks in a given range of p_T or quark mass, according to Eqs.(5) and (6). At the x-values considered here, the typical p_T-values are larger than, or at least comparable to, the strange quark mass (*cf.* Fig. 6(a)). Hence the π and K mesons produced by intrinsic u, d and s quarks are expected to have similar x-distributions, as observed. The K^+ mesons can get their momenta

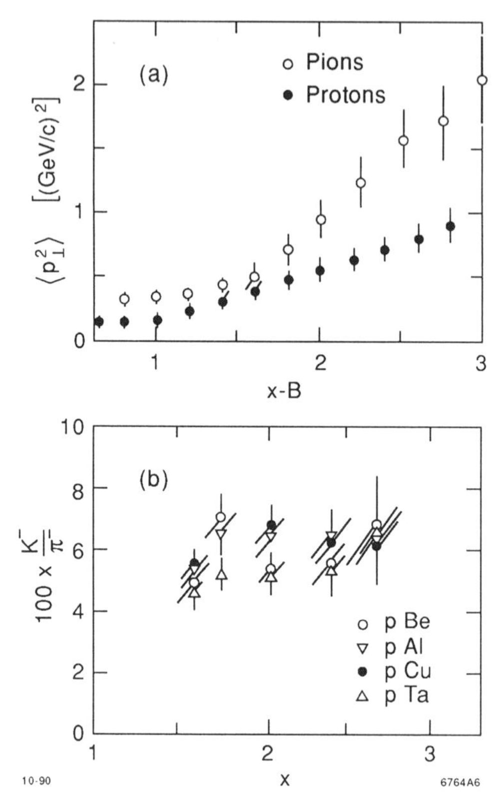

Figure 6. (a) The mean square transverse momentum of cumulative pions (o) and protons (•) produced by 10 GeV protons on Ta and Pb. The scale of the x-axis is offset by $B = 1$ for the protons ($B = 0$ for the pions). Data from Ref. 32. (b) The ratio of cumulative K^- to π^- production on several nuclei as a function of x. Data from Ref. 34.

from intrinsic u valence quarks. Since the creation of an $s\bar{s}$ pair is not suppressed at the relatively large p_T-scale involved, we can understand the equality of the K^+ and π^+ meson rates. The production of a K^- meson at large x, on the other hand, requires an energetic $\bar{u}$ or s sea quark. In this case momentum must be transferred from the valence quarks and gluons according to Fig. 2. Hence it is not surprising that the rate of K^-mesons is suppressed by about a factor 20 in the cumulative region, as seen in Fig. 6(b).

Our interpretation of the cumulative phenomena in terms of an enhancement in the nuclear structure function for $x > 1$ is compatible with some earlier suggestions.[26,27,36,37] Models of multiquark bags have been used to provide a unified explanation of the EMC, Cronin, and Cumulative Effects. An analysis of the EMC Effect in fact suggested the existence of a small admixture in nuclear wave functions of "collective" sea quarks, which are as energetic as the valence quarks.[38] The multiquark bag models do not, however, predict the probability for bag formation, nor the x-distributions of the quarks in the bag. The properties of the intrinsically-hard component of nuclear wave functions, on the other hand, can be calculated from perturbative QCD in terms of the known quark and gluon distribution functions of nucleons. An immediate consequence is that the multiquark correlations must have a small transverse range, implying an increase of the average p_T at large x, as observed in the data (Fig. 6(a)).

Other puzzles involving fast nuclear fragments, which also may be related to intrinsic hardness, include the production of particles from nuclei below threshold for collisions on free nucleons. For example, subthreshold production of antiprotons has been observed both in $p + Cu$ and $Si + Si$ collisions.[39] While the $\overline{p}$ rate was thought to be understood for the $p + Cu$ data, based on the high cumulative momenta being interpreted as due to Fermi motion, it turned out that the corresponding calculation underestimated the rate for $Si + Si$ collisions by three orders of magnitude. In our view, the cumulative momenta should be discussed at the parton level. The rate for $\overline{p}$ production may then proceed much more favorably through, *e.g.*, the $gg \to p\overline{p}$ reaction, whose threshold is just $2m_p$ in the center-of-mass.

EFFECT OF HEAVY QUARK THRESHOLDS ON ELASTIC pp SCATTERING

One of the most unusual ways of identifying the effect of heavy quark thresholds is to study *elastic* scattering of hadrons at large angles. Through unitarity, even a threshold cross section of only 1 μb for the production of open charm in pp collisions will have a profound influence on the $pp \to pp$ scattering at $\sqrt{s} \sim 5\ GeV$, because of its very small cross section at 90°. The production of charm at threshold implies that there is a contribution with massive, slow-moving constituents to the pp elastic amplitude which can modify the ordinary PQCD predictions, including dimensional counting scaling laws, helicity dependence, angular dependence, and especially the "color transparency" of quasi-elastic pp scattering in a nuclear target.

It is possible to use a nucleus as a "color filter"[40,41] to separate and identify the threshold and perturbative contributions to the scattering amplitude. If the interactions of an incident hadron are controlled by gluon exchange, then the nucleus will be transparent to those fluctuations of the incident hadron wavefunction which have small transverse size. Such Fock components have a small color dipole moment and thus will interact weakly in the nucleus; conversely, Fock components with slow-moving massive quarks cannot remain compact. They will interact strongly and be absorbed during their passage through the nucleus. In fact, large momentum transfer quasi–exclusive reactions[4] are controlled in perturbative QCD by small color–singlet valence–quark Fock components of transverse size $b_\perp \sim 1/Q$; initial–state and final–state corrections to these hard reactions are suppressed. Thus, at large momentum transfer and energies, quasi–elastic exclusive reactions are predicted to occur uniformly in the nuclear volume, unaffected by initial or final state multiple–scattering or absorption of the interacting hadrons. This remarkable phenomenon is called "color transparency."[18]

Thus perturbative QCD predicts that the quasi–elastic scattering cross section will be additive in proton number in a nuclear target.[42] There are two conditions which set the kinematic scale where PQCD color transparency should be evident. First, the hard scattering subprocess must occur at a sufficiently large momentum transfer so that only small transverse size wavefunction components $\psi(x_i, b_\perp \sim 1/Q)$ with small color dipole moments dominate the reaction. Second, the state must remain small during its transit through the nucleus. The expansion distance is controlled by the time in which the small Fock component mixes with other Fock components. By Lorentz invariance, the time scale $\tau = 2E_{\overline{p}}/\Delta\mathcal{M}^2$ grows linearly with the energy of the hadron in the nuclear rest frame, where $\Delta\mathcal{M}^2$ is the difference of invariant mass squared of the Fock components. Estimates for the expansion time are given in Refs. 41, 43, and 44.

The only existing test of color transparency is the measurement of quasi–elastic large angle *pp* scattering in nuclei at Brookhaven.[45] The transparency ratio is observed to increase as the momentum transfer increases, in agreement with the color transparency prediction. However, in contradiction to perturbative QCD expectations, the data suggests, surprisingly, that normal Glauber absorption seems to recur at the highest energies of the experiment $p_{\rm lab} \sim 12\ GeV/c$. Even more striking is that this is the same energy at which the spin correlation A_{NN} is observed to rise dramatically:[46] the cross section for protons scattering with their spins parallel and normal to the scattering plane is found to be four times as big as the cross section for anti-parallel scattering, which is again in strong contradiction to PQCD expectations.

In Ref. 47 it was noted that the breakdown of color transparency and the onset of strong spin–spin correlations can both be explained by the fact that the charm threshold occurs in *pp* collisions at $\sqrt{s} \sim 5\ GeV$ or $p_{\rm lab} \sim 12\ GeV/c$. At this energy the charm quarks are produced at rest in the center of mass, and all of the eight quarks have zero relative velocity. The eight-quark cluster thus moves through the nuclear volume with just the center-of-mass velocity. Even though the initial cluster size is small (since all valence quarks had to be at short transverse distances to exchange their momenta), the multi-quark nature and slow speed of the cluster implies that it will expand rapidly and be strongly absorbed in the nucleus. This Fock component will then not contribute to the large-angle quasi-elastic *pp* scattering in the nucleus: It will be filtered out.

The charm threshold effect will be strongly coupled to the *pp* $J = L = S = 1$ partial wave.[47] (The orbital angular momentum of the *pp* state must be odd since the charm and anti–charm quarks have opposite parity.) This partial wave predicts maximal spin correlation in A_{NN}. Hence, if this threshold contribution to the $pp \to pp$ amplitude dominates the valence quark QCD amplitude, one can understand both the large spin correlation and the breakdown of color transparency at energies close to charm threshold. Thus the nucleus acts as a filter, absorbing the threshold contribution to elastic *pp* scattering, while allowing the hard scattering perturbative QCD processes to occur additively throughout the nuclear volume.[41] One also observes a strong enhancement of A_{NN} at the threshold for strange particle production, which is again consistent with the dominance of the $J = L = S = 1$ partial wave helicity amplitude. The large size of A_{NN} observed at both the charm and strange thresholds is striking evidence of a strong effect on elastic amplitudes due to threshold production of fermion–antifermion pairs.

Nuclear Bound Quarkonium

According to the above, a slow-moving heavy quark system produced near threshold may be expected to experience strong final state interactions in the nucleus. This has interesting implications for the production of charmonium at threshold in a nuclear target. In this case it is possible that the attractive QCD van der Waals potential due to multi-gluon exchange could actually bind the η_c to light nuclei. Consider the reaction $\overline{p}\alpha \rightarrow {}^3H(c\overline{c})$ where the charmonium state is produced nearly at rest. (See Fig. 7.) At the threshold for charm production, the incident particles will be nearly stopped (in the center of mass frame) and will fuse into a compound nucleus because of the strong attractive nuclear force. The charmonium state will be attracted to the nucleus by the QCD gluonic van der Waals force. One thus expects strong final state interactions near threshold. In fact, it is argued in Ref. 48 that the $c\overline{c}$ system will bind to the ${}^3H$ nucleus. It is thus possible that a new type of exotic nuclear bound state will be formed: charmonium bound to nuclear matter. Such a state should be observable at a distinct $\overline{p}\alpha$ center of mass energy, spread by the width of the charmonium state, and it will decay to unique signatures such as $\overline{p}\alpha \rightarrow {}^3H\gamma\gamma$. The binding energy in the nucleus gives a measure of the charmonium's interactions with ordinary hadrons and nuclei; its hadronic decays will measure hadron–nucleus interactions and test color transparency starting from a unique initial state condition.

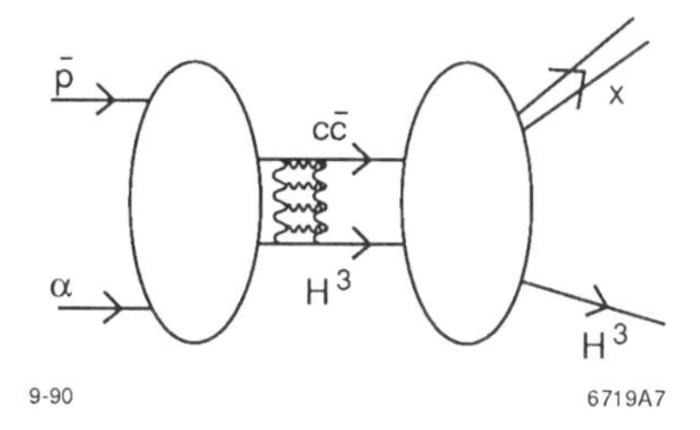

Figure 7. Formation of the $(c\overline{c}) - {}^3H$ bound state in the process $\overline{p}\alpha \rightarrow {}^3HX$.

In QCD, the nuclear forces are identified with the residual strong color interactions due to quark interchange and multiple–gluon exchange. Because of the identity of the quark constituents of nucleons, a short–range repulsive component is also present (Pauli–blocking). From this perspective, the study of heavy quarkonium interactions in nuclear matter is particularly interesting: due to the distinct flavors of the quarks involved in the quarkonium–nucleon interaction there is no quark exchange to first order in elastic processes, and thus no one–meson–exchange potential from which to build a standard nuclear potential. For the same reason, there is no Pauli–blocking and consequently no short–range nuclear repulsion. The nuclear interaction in this case is purely gluonic and thus of a different nature from the usual nuclear forces.

The production of nuclear–bound quarkonium would be the first realization of hadronic nuclei with exotic components bound by a purely gluonic potential. Furthermore, the charmonium–nucleon interaction would provide the dynamical basis for understanding the spin–spin correlation anomaly in high energy pp elastic scattering.[47] In this case, the interaction is not strong enough to produce a bound state, but it can provide a strong enough enhancement at the heavy–quark threshold characteristic of an almost–bound system.[49]

ACKNOWLEDGEMENTS

Some of the material presented above is based on collaborations with G. de Teramond and I. Schmidt. PH wishes to thank the SLAC theory group for its warm hospitality during his sabbatical and the organizers of this meeting for their invitation and for an interesting conference.

REFERENCES

1. S. J. Brodsky, P. Hoyer, C. Peterson, and N. Sakai, Phys. Lett. 93B, 451 (1980);
   S. J. Brodsky, C. Peterson, and N. Sakai, Phys. Rev. D23, 2745 (1981).
2. S. J. Brodsky, H. E. Haber, and J. F. Gunion, in *Anti–pp Options for the Supercollider,* Division of Particles and Fields Workshop, Chicago, IL, 1984, edited by J. E. Pilcher and A. R. White (SSC–ANL Report No. 84/01/13, Argonne, IL, 1984), p. 100;
   S. J. Brodsky, J. C. Collins, S. D. Ellis, J. F. Gunion, and A. H. Mueller, published in Snowmass Summer Study 1984, p. 227.
3. Some constraints are known for the meson distribution amplitude $\phi_M(x, Q) = \Gamma_{q\overline{q}}$ for valence $q\overline{q}$ components of the wave function. See, *e.g.*, Ref. 4.
4. S. J. Brodsky and G. P. Lepage, in *Quantum Chromodynamics*, edited by A. H. Mueller, (World Scientific, 1990.)
5. C. Peterson, D. Schlatter, I. Schmitt and P. M. Zerwas, Phys. Rev. D27, 105 (1983).
6. R. J. Cashmore, *Proc. Int. Symp. on Prod. and Decay of Heavy Flavors,* Stanford 1987 (E. Bloom and A. Fridman, Eds.), p.118.
7. L. V. Gribov, E. M. Levin and M. G. Ryskin, Nucl. Phys. B188, 555 (1981) and Phys. Rep. 100, 1 (1983);
   A. H. Mueller and J. Qiu, Nucl. Phys. B268, 427 (1986).
8. F. E. Close, J. Qiu and R. G. Roberts, Phys. Rev. D40, 2820 (1989).
9. A. Kernan and G. VanDalen, Phys. Rep. 106, 297 (1984), and references therein.
10. S. P. K. Tavernier, Rep. Prog. Phys. 50, 1439 (1987);
    U. Gasparini, *Proc. XXIV Int. Conf. on High Energy Physics*, (R. Kotthaus and J. H. Kühn, Eds., Springer 1989), p. 971.

11. S. F. Biagi, *et al.*, Z. Phys. C28, 175 (1985);
P. Chauvat, *et al.*, Phys. Lett. 199B, 304 (1987);
P. Coteus, *et al.*, Phys. Rev. Lett. 59, 1530 (1987);
C. Shipbaugh, *et al.*, Phys. Rev. Lett. 60, 2117 (1988);
M. Aguilar-Benitez, *et al.*, Z.Phys. C40, 321 (1988).

12. M. Aguilar-Benitez, *et al.*, Phys. Lett. 161B, 400 (1985) and Z. Phys. C31, 491 (1986);
S. Barlag, *et al.*, Z. Phys. C39, 451 (1988) and CERN–PPE/90–145 (1990).

13. M. MacDermott and S. Reucroft, Phys. Lett. 184B, 108 (1987);
H. Cobbaert, *et al.*, Phys. Lett. 191B, 456 (1987), *ibid.*, 206B, 546 (1988) and Z. Phys. C36, 577 (1987);
M. E. Duffy, *et al.*, Phys. Rev. Lett. 55, 1816 (1985);
M. I. Adamovich, *et al.*, CERN–EP/89–123 (1989).

14. K. J. Anderson, *et al.*, Phys. Rev. Lett. 42, 944 (1979);
A. S. Ito, *et al.*, Phys. Rev. D23, 604 (1981);
J. Badier, *et al.*, Phys. Lett. 104B, 335 (1981);
P. Bordalo, *et al.*, Phys. Lett. 193B, 368 (1987).

15. Yu. M. Antipov, *et al.*, Phys. Lett. 76B, 235 (1978);
M. J. Corden, *et al.*, Phys. Lett. 110B, 415 (1982);
J. Badier, *et al.*, Z. Phys. C20, 101 (1983);
S. Katsanevas, *et al.*, Phys. Rev. Lett. 60, 2121 (1988).

16. D. M. Alde, *et al.*, Phys. Rev. Lett. 64, 2479 (1990) and Los Alamos preprint LA–UR–90–2331 (1990);
C. S. Mishra, *et al.*, Contribution to the XXVth Rencontres de Moriond, Les Arcs (1990), Fermilab–Conf–90/100–E (May 1990).

17. P. Hoyer, M. Vänttinen and U. Sukhatme, Phys. Lett. 246B, 217 (1990).

18. A. H. Mueller, *Proc. XVII Recontre de Moriond* (1982);
S. J. Brodsky, *Proc. XIII International Symposium on Multiparticle Dynamics*, Volendam (1982);
S. J. Brodsky and A. H. Mueller, Phys. Lett. 206B, 685 (1988), and references therein.

19. S. J. Brodsky and P. Hoyer, Phys. Rev. Lett. 63, 1566 (1989).

20. J. J. Aubert, *et al.*, Nucl. Phys. B213, 31 (1983).
See also E. Hoffmann and R. Moore, Z. Phys. C20, 71 (1983).

21. J. J. Aubert, *et al.*, Phys. Lett. 123B,275 (1983);
J. Ashman *et al.*, Phys. Lett. 202B, 603 (1988);
M. Arneodo *et al.*, Phys. Lett. 211B, 493 (1988).

22. R. G. Arnold *et al.*, Phys. Rev. Lett. 52, 727 (1984).

23. G. Bari, *et al.*, Phys. Lett. 163B,282 (1985).

24. I. Savin, *Proc. XXII Conf. on High Energy Physics,* Leipzig 1984 (A. Meyer and E. Wieczorek, Eds.) Vol II, p. 251;
W. P. Schütz, *et al.*, Phys. Rev. Lett. 38, 259 (1977).

25. J. W, Cronin, *et al.*, Phys. Rev. D11, 3105 (1975);
    C. Bromberg, *et al.*, Phys. Rev. Lett. 42, 1202 (1979).
26. A. V. Efremov, V. T. Kim and G. I. Lysakov, Sov. J. Nucl. Phys. 44, 151 (1986);
    S. Gupta and R. M. Godbole, Phys. Rev. D33, 3453 (1986), Z. Phys. C31, 475 (1986) and Phys. Lett. 228B, 129 (1989).
27. For experimental and theoretical reviews of the cumulative effect, see:
    V. S. Stavinskii, Sov. J. Part. Nucl. 10, 373 (1979);
    V. B. Gavrilov and G. A. Leksin, preprint ITEP 128–89 (1989);
    A. V. Efremov, Sov. J. Part. Nucl. Phys. 13, 254 (1982);
    L. L. Frankfurt and M. I. Strikman, Phys. Rep. 76, 215 (1981) and Phys. Rep. 160, 235 (1988).
28. J. V. Geagea, *et al.*, Phys. Rev. Lett. 45, 1993 (1980).
29. For data on N^* production, see B. S. Yuldashev, *et al.*, WISC-EX-90-310 (1990), and references therein.
30. The fact that the number of cumulative nucleons is much larger than the number of cumulative pions (see Ref. 27) does, however, imply that the recombination of several quarks must be taken into account when using the parton basis to describe cumulative nucleons.
31. All the available data cannot, however, be fitted with a simple A^{α} power law dependence. See also:
    Yu. D. Bayukov, *et al.*, Phys. Rev. C20, 764 (1979) and Sov. J. Nucl. Phys. 42, 238 (1985);
    S. Frankel, *et al.*, Phys. Rev. C20, 2257 (1979);
    N. A. Nikiforov, *et al.*, Phys. Rev. C22, 700 (1980);
    M. Kh. Anikina, *et al.*, Sov. J. Nucl. Phys. 40, 311 (1984);
    G. R. Gulkanyan, *et al.*, Sov. J. Nucl. Phys. 50, 259 (1989).
32. S. V. Boyarinov, *et al.*, Sov. J. Nucl. Phys. 46, 871 (1987).
33. A. I. Anoshin, *et al.*, Sov. J. Nucl. Phys. 36, 400 (1982);
    A. M. Baldin, *et al.*, Sov. J. Nucl. Phys. 39, 766 (1984);
34. S. V. Boyarinov, *et al.*, Sov. J. Nucl. Phys. 50, 996 (1989).
35. P. Capiluppi, *et al.*, Nucl. Phys. B79, 189 (1974).
36. A. M. Baldin, Nucl. Phys. A434, 695 (1985).
37. A. V. Efremov, Sov. J. Nucl. Phys. 24, 633 (1976);
    G. Berlad, A. Dar and G. Eilam, Phys. Lett. 93B, 86 (1980);
    C. E. Carlson, K. E. Lassila and U. P. Sukhatme, preprint WM-90-115 (1990).
38. A. V. Efremov, Phys. Lett. 174B, 219 (1986);
    A. V. Efremov, A. B. Kaidalov, V. T. Kim, G. I. Lykasov and N. V Slavin, Sov. J. Nucl. Phys. 47, 868 (1988).
39. J. B. Carroll, *et al.*, Phys. Rev. Lett. 62, 1829 (1989);
    A. Shor, *et al.*, Phys. Rev. Lett. 63, 2192 (1989), and references therein.

40. G. Bertsch, S. J. Brodsky, A. S. Goldhaber, and J. Gunion, Phys. Rev. Lett. 47, 297 (1981).
41. J. P. Ralston and B. Pire, Phys. Rev. Lett. 61, 1823 (1988), University of Kansas preprint 90–0548 (1990).
42. By definition, quasi–elastic processes are nearly coplanar, integrated over the Fermi motion of the protons in the nucleus. Such processes are nearly exclusive in the sense that no extra hadrons are allowed in the final state.
43. B. K. Jennings and G. A. Miller, Phys. Lett. B236, 209 (1990). and University of Washington preprint 40427-20-N90 (1990).
44. G. R. Farrar, H. Liu, L. L. Frankfurt, M. I. Strikman, Phys. Rev. Lett. 61, 686 (1988).
45. A. S. Carroll, *et al.*, Phys. Rev. Lett. 61, 1698 (1988).
46. G. R. Court, *et al.*, Phys. Rev. Lett. 57, 507 (1986).
47. S. J. Brodsky and G. de Teramond, Phys. Rev. Lett. 60, 1924 (1988).
48. S. J. Brodsky, G. de Teramond, and I. Schmidt, Phys. Rev. Lett. 64, 1011 (1990).
49. The signal for the production of almost–bound nucleon (or nuclear) charmonium systems near threshold is the isotropic production of the recoil nucleon (or nucleus) at large invariant mass $M_X \simeq M_{\eta_c}$.

LOW ENERGY PSEUDOSCALAR-PSEUDOSCALAR INTERACTIONS IN THE NONRELATIVISTIC QUARK MODEL

John Weinstein

Department of Physics, University of Tennessee, Knoxville, TN, 37996, USA

ABSTRACT

We give a brief introduction to the nonrelativistic quark model and then present results from a coupled-channel quark model analysis of light-quark S-wave pseudoscalar-pseudoscalar scattering and production processes. Annihilation mixing to s-channel $q\bar{q}$ scalar resonances and quark exchange interactions are found to be important. Contributions from the exchange of t-channel mesons are not required to obtain agreement with the data. Intermeson forces are important in all channels and, in particular, lead to isospin 0 and 1 $K\overline{K}$ bound states analogous to the deuteron; we identify these with the $S^*(975)$ and $\delta(980)$ scalar resonances.

1. INTRODUCTION

The $qq\bar{q}\bar{q}$ system has been studied in many different models over the past few decades.[1] In this talk we use the nonrelativistic quark model, which has been very successful in the meson[2,3] and baryon[4] sectors, and discuss the $qq\bar{q}\bar{q}$ predictions in this model.[5–8] We find most ground states to be pairs of pseudoscalar mesons which interact with each other through both quark exchange and, in allowed channels, $q\bar{q}$ annihilation and creation over ranges of several fermi. We find that these intermeson interactions are important in all pseudoscalar-pseudoscalar [PP] S-wave systems and in $I = 0$ and 1† they are strong enough to form $K\overline{K}$ bound states.[9] We call these states $K\overline{K}$ molecules and identify them with the $S^*(975)$ and $\delta(980)$ scalar resonances,* an assignment which clarifies many outstanding issues. Good agreement with experimental data is obtained without including effects due to t-channel meson exchange.

The $I = 0$ channel has also recently been studied by several other groups: one used a $\pi\pi$ - $K\overline{K}$ coupled-channel model based on a separable potential

† We denote isospin by I rather than T.

* We retain the notation S^* and δ for these two states as the new nomenclamature is intended for $q\bar{q}$ states.

formalism,[10] a second considered a model in which there are both t- and s-channel meson exchanges,[11] and a third carried out a direct K-matrix analysis of the scattering data.[12] These studies all support the conclusion that the forces in this channel lead to the formation of a weakly bound $I = 0$ $K\overline{K}$ state of the type predicted in refs. 5 and 6.

In addition to predicting $K\overline{K}$ bound states the results of our quark model studies may help solve many of the outstanding issues in low-energy hadron dynamics such as the controversial spin-parity assignments of many resonances, the interactions of broad overlapping states coupled through continuum channels, the non-appearance of certain resonances or final states naively expected in particular reactions, and the search for glueballs and hybrids which will require that we understand the hadronic backgrounds involving these coupled-channel interactions.

The model we use is a coupled-channel matrix Schrödinger equation with the individual sectors labelled by isospin and strangeness. The quark-exchange and annihilation potentials appearing in these equations are extracted from a nonrelativistic quark model analysis of the $qq\bar{q}\bar{q}$ wavefunction[5,6] and from a simple $^3P_0$ SU(3) vertex model[3] respectively. It is encouraging that such a simple formulation leads to detailed solutions which beautifully describe much of the complicated physics seen in these scalar channels.

In section 2 below we introduce the the nonrelativistic quark model, in section 3 we describe our treatment of $qq\bar{q}\bar{q}$ systems within this model, in section 4 we discuss the results and compare them with experiment, in section 5 we review the $K\overline{K}$ molecule interpretation of the S^* and δ, in section 6 we introduce the production model and compare some results with experiment, and in section 7 we conclude.

2. THE NONRELATIVISTIC QUARK MODEL

This section is a review of the most basic features of the nonrelativistic quark model.[2–6] It is a "QCD inspired model" and attempts to include the essential charateristics of the QCD Lagrangian while remaining soluble. Although it is often criticised for its faults it has the virtue of successfully predicting hundreds of meson[2] and baryon[4] masses, lifetimes and branching ratios[3] based on about a dozen parameters. These successes suggests that we might profitably change our response towards its shortcomings from insisting that it cannot work to understanding why it does work.

The model is based on a Hamiltionian formalism with one-body and two-body terms:

$$H^{\mathrm{NRQM}} = \sum_i m_i + \sum_i \frac{p_i^2}{2\,m_i} + \sum_{i>j} H_{ij} \tag{1}$$

where the m_i are the constituent quark masses, the $\frac{p_i^2}{2\,m_i}$ their nonrelativistic kinetic energies, and the two-body H_{ij} terms contain a linear confining piece as found by lattice gauge theory and a short-range "QED" type interaction which models the color fields at short range in analogy to the photon field:

$$H_{ij} = b r_{ij} \vec{\mathrm{F}}_{\mathrm{i}} \cdot \vec{\mathrm{F}}_{\mathrm{j}} + \frac{\alpha_{\mathrm{QCD}}}{\alpha_{\mathrm{QED}}} H_{ij}^{\mathrm{QED}} \vec{\mathrm{F}}_{\mathrm{i}} \cdot \vec{\mathrm{F}}_{\mathrm{j}}. \tag{2}$$

The $\vec{\mathrm{F}}_{\mathrm{i}}$ are related to the 8 3×3 SU(3) lambda matricies by

$$\mathrm{F}_i^a = \begin{cases} +\frac{1}{2}\lambda_i^a & i = \text{quark} \\ -\frac{1}{2}\lambda_i^{*a} & i = \text{antiquark} \end{cases} \tag{3}$$

and they describe the strength of the $3 \times 3 \times 8$ possible quark-quark-gluon color couplings.

The QCD states are product wave functions with color, flavor, spin, and space dependence; their matrix elements with the Hamiltonian yield the hadron masses and wave functions.

In order to apply this model to the $qq\bar{q}\bar{q}$ sector we first calculate the $q\bar{q}$ spectra for the light quarks (u, d, and s) and fit the Hamiltonian parameters. We then apply this Hamiltonian to properly antisymmeterized variational $qq\bar{q}\bar{q}$ trial wave functions which have the freedom to fall apart into $(q_1\bar{q}_3)(q_2\bar{q}_4)$ or $(q_1\bar{q}_4)(q_2\bar{q}_3)$ pseudoscalar-pseudoscalar systems.

3. THE COUPLED CHANNEL MODEL

If one considers the ground or metastable states of the scalar $q_1q_2\bar{q}_3\bar{q}_4$ system it is clear that there are at least three possible qualitatively different configurations: one is "baryonium" in which all four particles are in a single hadron, a second is a system of two lightly bound mesons, and a third is two free pseudoscalar mesons.

Although baryonium states have been predicted in several models[13] there is no clear evidence establishing their existence. Fall-apart modes of these states have also been discussed extensively, particularly in the P-matrix formalism.[14]

It is clear that two mesons can bind only if there are attractive interactions between them. A possible short-range mechanism is one gluon exchange which converts two color singlet mesons into two color octet mesons in an overall color singlet state. The octet-octet state is, however, a linear combination of meson pairs in each of the two singlet-singlet color combinations $(q_1\bar{q}_3)_1(q_2\bar{q}_4)_1$ and $(q_1\bar{q}_4)_1(q_2\bar{q}_3)_1$. Thus we see that one gluon exchange and quark exchange are related and lead to $P_1P_2 \leftrightarrow P_1'P_2'$ transitions. In principle the quark-exchange processes will also transform pseudoscalar pairs into $J = 0$ vector pairs. In our quark model analysis, however, the vector-vector components of the ground-state $qq\bar{q}\bar{q}$ wave function are very small and we have omitted these channels.

Quark exchange may, for example, transform an initial $(n\bar{s})(s\bar{n})$ $K\overline{K}^{I=0}$ system into an $(n\bar{n})(s\bar{s})$ $\eta\eta$, $\eta\eta'$ or $\eta'\eta'$ pair, where n represents an up or down quark. Furthermore, since the physical η and η' flavor states are $\approx (\eta_{n\bar{n}} - \eta_{s\bar{s}})/\sqrt{2}$ and $(\eta_{n\bar{n}}+\eta_{s\bar{s}})/\sqrt{2}$ respectively (where $\eta_{n\bar{n}} \equiv (u\bar{u}+d\bar{d})/\sqrt{2}$ and $\eta_{s\bar{s}} \equiv s\bar{s}$), two sequential exchanges can lead to transitions such as $\pi\pi \to \eta\eta \to K\overline{K}$ in which all $n\bar{n} \leftrightarrow s\bar{s}$ processes occur inside the maximally mixed η and η' wave functions. This reaction chain can take place in the central interaction region even below the thresholds for $\eta\eta$ or $K\overline{K}$ production. Diagonal potentials such as $\pi\pi \leftrightarrow \pi\pi$ are also generated by the quark-exchange mechanism in all PP channels except for $K\overline{K} \leftrightarrow K\overline{K}$.

From our variationally determined nonrelativistic quark model ground-state $qq\bar{q}\bar{q}$ wave functions we have extracted strong, short-range, gaussian-like intermeson quark-exchange potentials which connect different PP channels and are dominantly produced by the $\vec{S}_i \cdot \vec{S}_j$ hyperfine interaction.[5,6] For numerical tractability we replace these potentials with square wells which give similar low energy phase shifts and have a common range $a = 0.8$ fm (see ref. 6 for details). Given the uncertainties inherent in this analysis we believe that this is a reasonable first approximation.

It is widely believed that the variational technique is better suited to finding bound-state energies than to determining the underlying wave functions or potentials. With the aid of accurate numerical studies we have found that one can determine to a good approximation the true wave function if the variational wave function is given the freedom to respond to very small changes in the energy. We have demonstrated in a toy model that simple input potentials of one or two Gaussians lead to variational wave functions which accurately reproduce the input potentials when the Schrödinger is solved for $V[\psi(r)]$.[15]

Transitions between PP pairs can also go through $q\bar{q}$ annihilation to a $^3P_0$ $q\bar{q}$ scalar meson $[S]$ followed by $q\bar{q}$ creation. The s-channel mesons involved are the $I=1$ $a_0(1300)$, $I=\frac{1}{2}$ $K_0^*(1410)$, $I=0$ $n\bar{n}$ $f_0(1300)$, and $I=0$ $s\bar{s}$ $f_0'(1500)$. Each PPS vertex is modelled by a square well potential of range $a = 0.8$ fm and depth set by the product of the channel dependent flavor Clebsch-Gordon coefficient, a kinematic factor, and an SU(3) constant which is taken from the decay analysis of Kokoski and Isgur[3] and gives their $f_0(1300) \to \pi\pi$ width in a two channel approximation.

Another set of potentials are needed to determine the $q\bar{q}$ scalar meson wave functions. We model these with P-wave $(\mu r^2)^{-1}$ cores added to square well potentials of range $a = 0.8$ fm and depths chosen to fix the $^3P_0$ masses at their appropriate values.[6,7] Quark confinement is imposed in these channels by a 10 GeV potential in the exterior region $r > a$.

From this discussion it is clear that, unless all of the off-diagonal potentials are negligible, one cannot consider PP systems in isolation from the other accessible $P'P'$ and scalar systems; a coupled-channel approach is required.[16] To model this at low energies we use the radial coupled-channel Schrödinger equation:

$$(\mathbf{K} + \mathbf{M} + \mathbf{V})\,\mathbf{u}(r) = \mathrm{E}\,\mathbf{u}(r) \tag{4}$$

with the potential matrix $\mathbf{V}$ determined as described above and given symbolically as

$$\mathbf{V} = \left(\begin{array}{c|c} \quad & \quad \\ \hline \quad & \quad \end{array} \right).$$

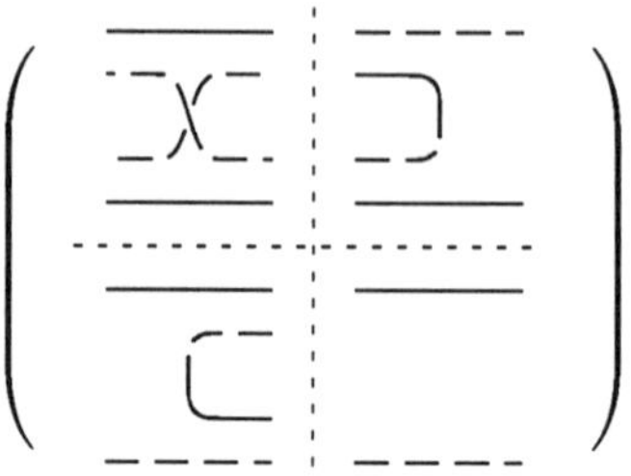

The elements of the kinetic energy matrix are $(\mathbf{K})_{ij} = -\delta_{ij}\,\frac{1}{2\mu_i}\,\frac{\partial^2}{\partial r^2}$, $\mathbf{M}$ is a diagonal matrix whose entries are the sum of the meson masses in the PP channels or the sum of the constituent quark masses in the scalar $q\bar{q}$ channels, and E is the nonrelativistic invariant energy.

The radial wave functions $\mathbf{u}(r)$ in eq. (4) are 1, 4, and 7-component vectors whose entries represent all of the interacting states; they take the form

$$\mathbf{u}^{I=2}(r) = (\mathrm{u}_{\pi\pi^{I=2}}),$$

$$\mathbf{u}^{I=\frac{3}{2}}(r) = (\mathrm{u}_{K\pi^{I=\frac{3}{2}}}),$$

$$\mathbf{u}^{I=1}_{strangeness=\pm 2}(r) = (\mathrm{u}_{KK}),$$

$$\mathbf{u}^{I=0}(r) = \begin{pmatrix} \mathrm{u}_{\pi\pi^{I=0}} \\ \mathrm{u}_{K\overline{K}^{I=0}} \\ \mathrm{u}_{\eta\eta} \\ \mathrm{u}_{\eta\eta'} \\ \mathrm{u}_{\eta'\eta'} \\ \mathrm{u}_{f_0(1300)} \\ \mathrm{u}_{f_0'(1500)} \end{pmatrix},$$

$$\mathbf{u}^{I=\frac{1}{2}}(r) = \begin{pmatrix} \mathrm{u}_{K\pi^{I=\frac{1}{2}}} \\ \mathrm{u}_{K\eta} \\ \mathrm{u}_{K\eta'} \\ \mathrm{u}_{K_0^*(1410)} \end{pmatrix}, \quad \text{or} \quad \mathbf{u}^{I=1}(r) = \begin{pmatrix} \mathrm{u}_{\eta\pi} \\ \mathrm{u}_{K\overline{K}^{I=1}} \\ \mathrm{u}_{\eta'\pi} \\ \mathrm{u}_{a_0(1300)} \end{pmatrix},$$

according to the channel of interest.

With all operators in eq. (4) specified we integrate numerically to find the wave functions $\mathbf{u}(r)$ at each scattering energy E. Knowing the wave functions enables us to determine each PP scattering S-matrix.

4. SCATTERING SOLUTIONS

Experimental phase shift data has been published for the single-channel $\pi\pi^{I=2}$ system.[17] We find that the phase shift generated with the extracted $\pi\pi^{I=2}$ quark-exchange potential underestimates this data. The nonrelativistic quark model is known, however, to underestimate meson radii by a factor of about 2. As the ranges of the quark-exchange potentials and annihilation potentials are largely determined by the meson radii we have allowed ourselves the freedom to increase the (common) range of all potentials by a factor of 2 and scale down the strength of the modified quark-exchange potentials by a second common factor which is determined to be 0.5 (with an estimated uncertainty of about ± 0.1) from the threshold $\pi\pi^{I=2}$ phase shift data. This procedure leads to the square well ranges of $a = 0.8$ fm quoted above; the $qq\bar{q}\bar{q}$ derived potentials had ranges of about $a = 0.4$ fm. The resulting $\pi\pi^{I=2}$ phase shift is shown in fig. 1 along with the data and the phase shift given by a relativistic dispersion relation.

While this rescaling is perhaps the most ad-hoc part of our analysis we note that the repulsive $\pi\pi^{I=2}$ potential is a direct and nontrivial consequence of the $qq\bar{q}\bar{q}$ analysis. We consider the signs and the relative strengths of all of the extracted potentials to be predictions of the model and have simply rescaled all of them by the same factors. Thus the rescaling represents one new parameter in the model; the range expansion is chosen to be 2 and the strength factor is fitted.

This proceedure has been criticised as a serious defect of this calculation; however, the solutions of the coupled-channel Schrödinger equation beautifully describe many diverse phenomena, as we shall see below, and indicate that there is much to be learned from the model. The rescaling enters, moreover, when the results of one calculation, the study of the $qq\bar{q}\bar{q}$ ground-state wave function, is used in another calculation, the coupled-channel Schrödinger equation.

The $I = \frac{3}{2}$ channel has a repulsive $K\pi$ interaction and leads, with no adjustments of the parameters, to the phase shift shown in fig. 2,[18] in fine agreement with the data. The other single-channel sectors are the exotic

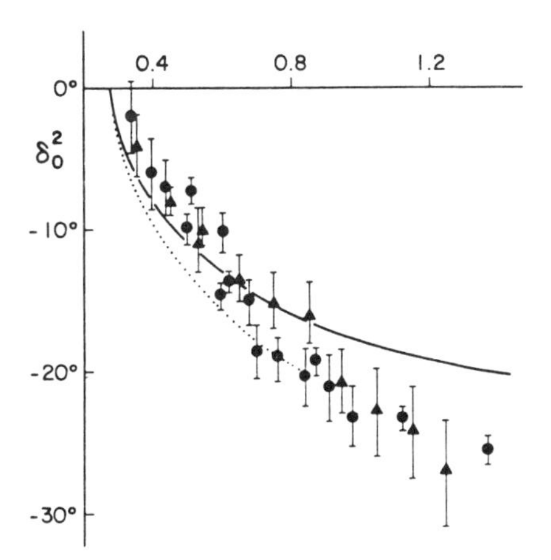

Fig.1 The experimental $I=2$ $\pi\pi$ phase shift[16] compared to theory; the dotted curve corresponds to using a relativistic dispersion relation.

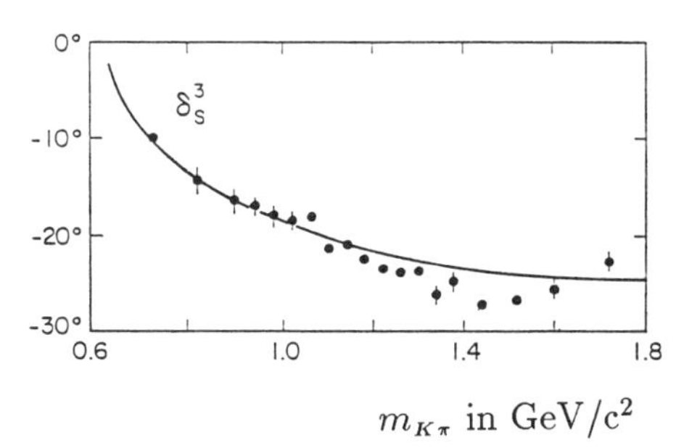

Fig.2 The experimental $I = \frac{3}{2}$ $K\pi$ phase shift [17] compared to theory.

$I = 1$ strangeness=± 2 KK and $\overline{KK}$ channels whose predicted phase shift[8] is shown in fig. 3. There is no data, as yet, for this process. The $I = 0$ $(K^+K^0 - K^0K^+)/\sqrt{2}$ and $\pi\pi^{I=1}$ systems are not discussed here because they are antisymmetric in flavor and cannot be S-wave.

A standard description for PP scattering through an s-channel scalar is the nonrelativistic Breit-Wigner formulation which gives the phase shift in terms of the PP kinetic energy and the mass and width of the scalar. In fig. 4 we plot the Breit-Wigner phase shift for $K\pi$ scattering through an s-channel resonance with $m = 1.420$ GeV/c^2 and $\Gamma = 0.380$ GeV/c^2. Also plotted is the phase shift generated by a restricted 2-channel solution of eq. (4) in which the $K\pi$ system is coupled to a K_0^* of mass 1.470 GeV/c^2 via the annihilation potential $V_{K\pi\leftrightarrow K_0^*} = 0.406\,\Theta(0.8\,\mathrm{fm} - r)$ GeV and with the quark-exchange potential set equal to zero. The close agreement between the two descriptions demonstrates their underlying similarities. Note, however, the difference in the s-channel resonance mass between the two approaches.

In figs. 5 and 6 we compare the $I = \frac{1}{2}$ $K\pi$ elastic phase shift and Argand diagram from the model[7] and from experiment.[18,19] These indicate a strongly attractive $K\pi$ potential for which there is direct evidence in the form of low mass $K\pi$ enhancements.[20] The quality of the agreement shown in these figures seems satisfactory in all but perhaps one respect: although there are indications of significant inelasticity at $K\eta'$ threshold in the data of Estabrooks *et al.*[18] as predicted by the model, the more recent accurate data of Aston *et al.*[19] remains near the unitarity circle over the whole kinematic range considered here.

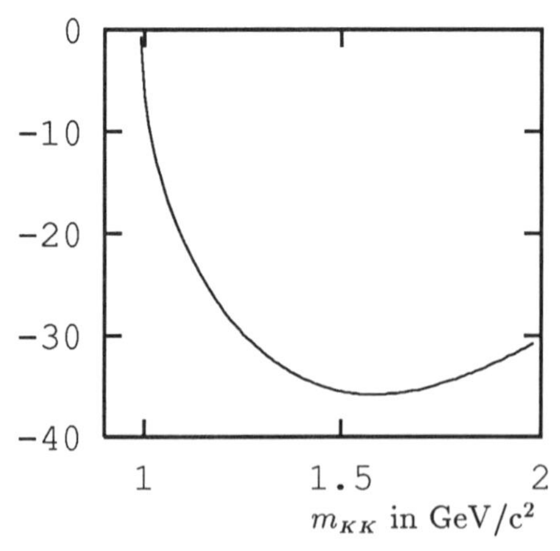

Fig.3 The predicted phase shift for strangeness=± 2, I=1 KK.[5]

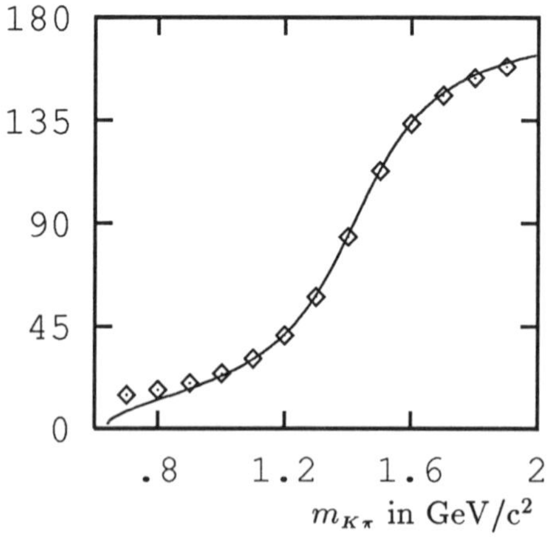

Fig.4 The Breit-Wigner phase shift ($\diamond$) and the 2-channel phase shift from eq. (4) (lines) with no quark-exchange potentials.

The coupled-channel interactions have, as in the 2-channel problem above, significantly reduced the apparent K_0^* mass from the input single-channel bare value of 1470 MeV/c^2 to agree with the experimental value of 1412 $\pm$ 4 $\pm$ 5 MeV/c^2.[21] A similar change occurs in the width: in the "narrow resonance approximation" the K_0^* would have a width of about 500 MeV/c^2.[7] This is reduced by the coupled-channel interactions to agree with the experimental value of 294 $\pm$ 10 $\pm$ 21 MeV/c^2. (Such effects appear to be particularly strong in this channel as the K_0^* mass is near the $K\eta'$ threshold.) This shows the danger inherent in relating the bare meson masses and widths from quark model predictions to the experimental values and suggests that a coupled-channel "interface" is required for a more realistic comparison.

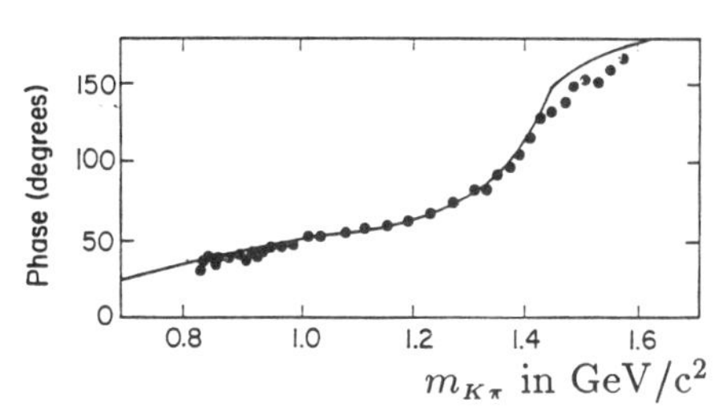

Fig.5 The $I = \frac{1}{2}$ $K\pi$ phase shift (in degrees) from Aston *et al.*[18] (•) and from the model (line).

Figs. 7 show the separate effects of the two main contributions to the $I = \frac{1}{2}$ phase shift. In fig. 7(a) the quark-exchange potentials have been set to zero, thereby isolating the contributions of the $^3P_0$ resonances to the

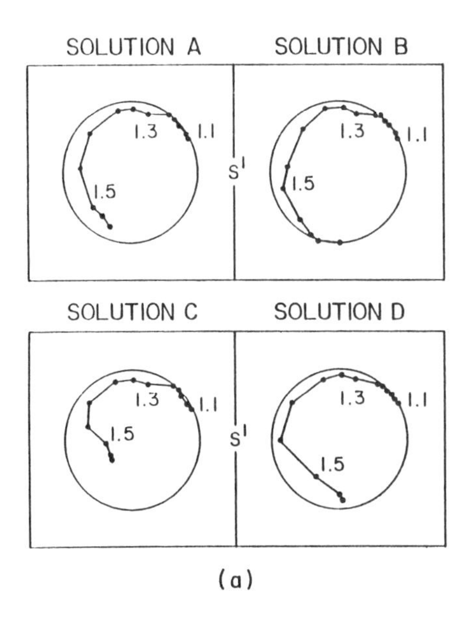

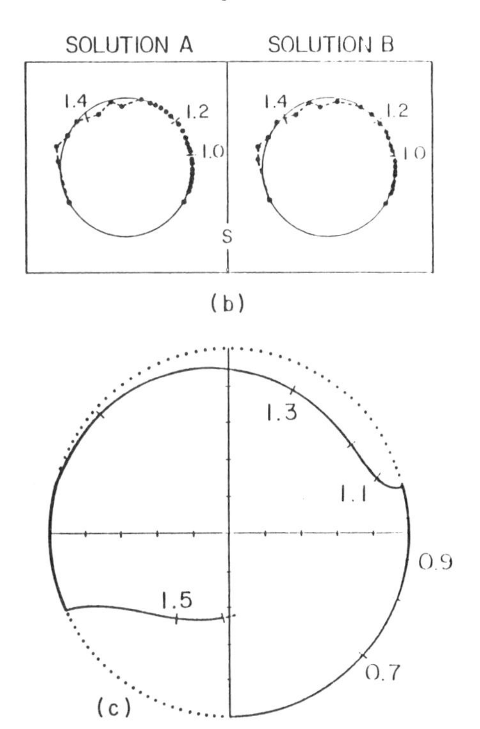

Fig.6 (a) The four experimental $I = \frac{1}{2}$ $K\pi$ Argand diagrams from Estabrooks *et al.*;[17] (b) the two solutions of Aston *et al.*;[18] (c) and the solution of the model.

phase shift and showing that the quark-exchange potentials are an essential ingredient of the physics. Fig. 7(b) shows the phase shift due to the quark-exchange potentials alone. We see that the very substantial $K\pi$ attraction from these potentials is "half the physics" at low $K\pi$ masses, but it is usually lumped into an "effective range approximation" polynomial background amplitude which places the spotlight on resonance physics. Forces analogous to these "background" forces would also be expected to play an important role in the understanding of nuclear forces.

We now consider solutions to the 4-channel $I = 1$ system.[5,6] The elastic $\eta\pi$ phase shift found by solving eq. (4) is shown in fig. 8(a). As data on elastic $\eta\pi$ scattering is not available we compare it with the phase shift of an elastic Breit-Wigner $\eta\pi$ resonance of mass 983 MeV/c^2 and width 57 MeV/c^2, the parameters of the $\delta(980)$. From this comparison we see that the coupled-channel interactions produce a phase shift which closely follows the Breit-Wigner form below $K\overline{K}$ threshold but deviates from it above threshold; consistent with the observed properties of the δ. The $^3P_0$ $q\bar{q}$ $a_0(1300)$ does not stand out in the phase shift diagram but can be associated with the "loop" in the complex $\eta\pi \to \eta\pi$ transition matrix between about 1200 and 1400 MeV/c^2 (fig. 8(b)). This is consistent with the known difficulty in observing the a_0.[22] We identify the rapid fall of $\delta_{K\overline{K}}$ at $K\overline{K}$ threshold (fig. 8(a)) with an effective $K\overline{K}$ binding potential and the $K\overline{K}$ bound state with the $\delta(980)$. The rapid change in the $\eta\pi$ elastic phase shift around 980 MeV/c^2 is due to $\eta\pi$ coupling to the $K\overline{K}$ molecule. It is well known that threshold effects can make broad resonances appear lighter and narrower; thus one might argue that the structure at 980 MeV/c^2 is actually a highly modified $a_0(1300)$.[23] In the $K\overline{K}$ molecule picture the δ is an extra non-$q\bar{q}$ $I = 1$ scalar state produced by coupled-channel interactions and the $a_0(1300)$ exists as a separate state. The confirmation of the $a_0(1300)$ is therefore a crucial test of our picture.

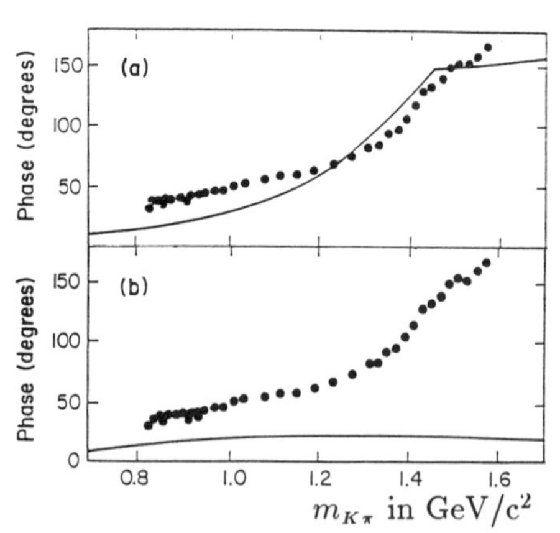

Fig.7 (a) The phase shift resulting from only $K\pi/K\eta/K\eta' \to K_0^*$ annihilation couplings. (b) The phase shift due to the quark-exchange potentials only.

Detailed information on $\pi\pi^{I=0}$ elastic scattering exists[24] and we can compare our 7-channel $\pi\pi$ phase shift directly with the data (fig. 9(a)). The major features are reproduced by the model below 1400 MeV/c^2. Above this energy the model will be less reliable; relativistic dynamics, $\pi\pi'$ and related channels, and $J = 0$ vector-vector channels are some of the factors missing from our analysis. The narrow S^* effect seen around 980 MeV/c^2

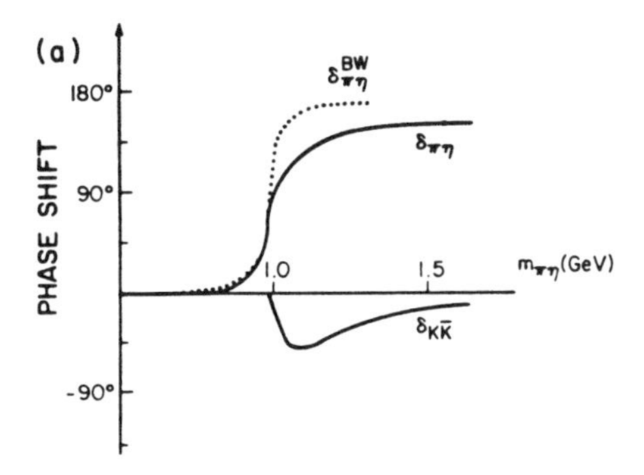

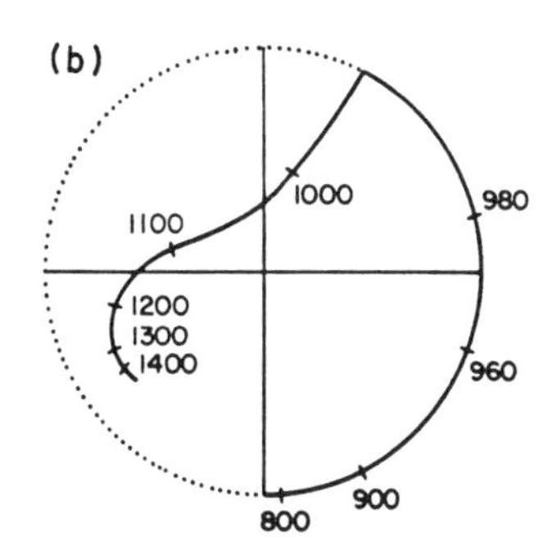

Fig.8 (a) The elastic $\eta\pi$ phase shift from a 4-channel solution to eq. (4) (solid curve) is compared to the Breit-Wigner phase shift for $\eta\pi$ scattering through a resonance with the mass and width of the $\delta(980)$ (dotted curve). Also shown in (a) is the elastic $K\overline{K}$ phase shift obtained from eq. (4). The drop in $\delta_{K\overline{K}}$ at $K\overline{K}$ threshold is due to the binding $I=1$ $K\overline{K}$ potential. In (b) the elastic $\eta\pi$ transition matrix element is shown.

is due to the $\pi\pi$ system coupling to a $K\overline{K}$ bound state that is induced by the coupled-channel interactions. We identify this state with the $S^*(975)$ scalar meson. In figs. 9(b) and (b′) we compare the experimental $I = 0$ $\pi\pi \rightarrow \pi\pi$ transition matrix with that of the coupled-channel model, noting that the experimental errors in the phase (for example, at $K\overline{K}$ threshold) are not shown in the corresponding Argand curve. Although there are minor differences in detail between the experiment and the model they are in very good qualitative agreement. The loop in the Argand diagram between $\approx$ 1200 and 1400 MeV/c^2 is due to the $f_0(1300)$ and a small loop just above 1500 MeV/c^2 (not shown because of the questionable validity of the model above $\approx$ 1400 MeV/c^2) is due to the $f_0'(1500)$.

The narrow S^* is also seen as a falling elastic $K\overline{K}$ phase shift, indicative of a binding $\overline{K\overline{K}}^{I=0}$ potential (figs. 10(a) and (b) compare experiment and theory).

In fig. 11 we show the wave functions found by solving eq. (4) at $m_{\pi\pi} = 0.975$ GeV/c^2; the dominance of the $K\overline{K}$ component is clear. From the wave functions we can find the probability for the interacting system to be in each of its possible components in the central region, given unit amplitude for the incoming $\pi\pi$ wave function. In fig. 12 we plot these probabilities as function of $m_{\pi\pi}$. The $K\overline{K}$ component of the S^* is very mass dependent and, just below $K\overline{K}$ threshold the system is essentially all $K\overline{K}$. Above $K\overline{K}$ threshold one can compare $K\overline{K}$ to $\pi\pi$ final state probabilities.[15] The f_0 and f_0' amplitudes also show interesting behavior, the amplitude for the f_0 is zero at $K\overline{K}$ threshold while the f_0' has a cusped peak at this energy. The $\eta\eta$, $\eta\eta'$ and $\eta'\eta'$ probabilities are small and we have omitted them from fig. 12 for clarity.

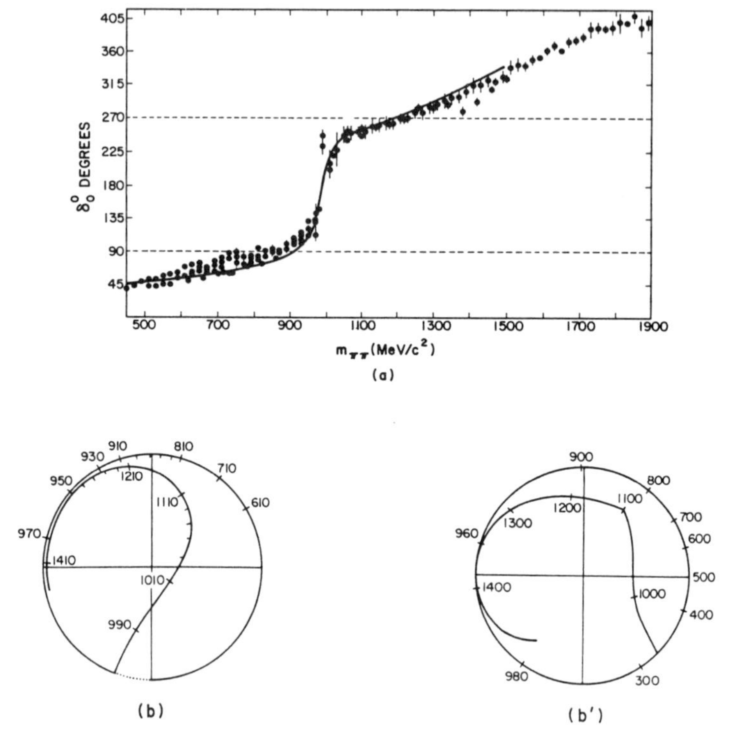

Fig.9 (a) The elastic $I=0$ $\pi\pi$ phase shift from the coupled channel equation is shown with experimental data.[11] The major features of the data below ≈ 1400 MeV are reproduced. We also compare the (b) experimental[11] and (b′) model $\pi\pi \rightarrow \pi\pi$ transition matrix element.

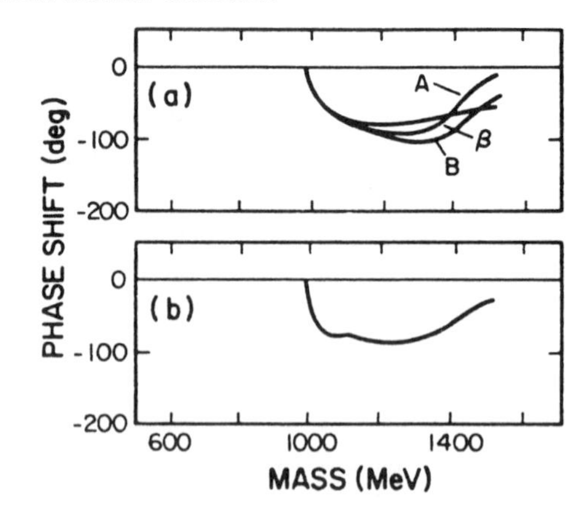

Fig.10 The elastic $K\overline{K}$ phase shift from the experiment (b) and the model (c) are similar. The drop in $\delta_{K\overline{K}}$ at $K\overline{K}$ threshold is due to the $I=0$ binding $K\overline{K}$ potential.

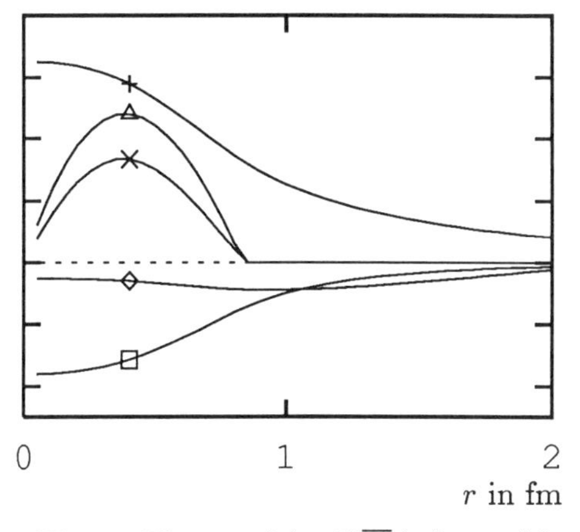

Fig.11 The $\pi\pi\,(\diamond)$, $K\overline{K}\,(+)$, $\eta\eta\,(\square)$, $f_0\,(\times)$ and $f_0'\,(\triangle)$ wave functions (= u(r)/r) at $m_{\pi\pi} = 0.975$ GeV/c^2. Note the large $K\overline{K}$ component. The $\eta\eta'$ and $\eta'\eta'$ components are very small and are omitted for clarity.

.5 1.5

$m_{\pi\pi}$ in GeV/c^2

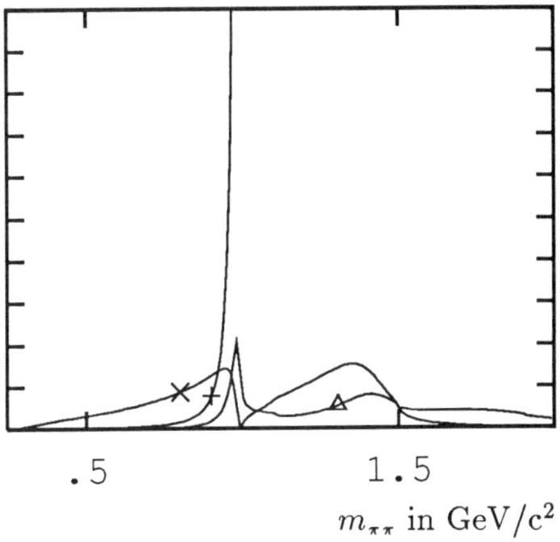

Fig.12 Probabilities for the $K\overline{K}\,(+)$, $f_0\,(\times)$ and $f_0'\,(\triangle)$ components in the central region given unit flux for $\pi\pi$. The $\eta\eta$, $\eta\eta'$ and $\eta'\eta'$ components are small and are omitted for clarity.

5. THE S^* AND δ AS $K\overline{K}$ MOLECULES

The assignment of the S^* and δ to the $q\bar{q}$ scalar nonet is problematical[1] and in this section we support our claim that the $K\overline{K}$ molecule interpretation of the S^* and δ solves many of these problems.

The narrow experimental widths of the S^* and δ, 34$\pm$6 and 57$\pm$11 MeV/c^2 respectively,[25] disagree with the expectations of the naive $^3P_0$ $q\bar{q}$ model.[2,3] A comparison of the $\pi\pi$ and $\eta\pi$ phase shifts immediately below the $K\overline{K}$ threshold with Breit-Wigner phase shifts indicates that the coupled-channel widths of the S^* and δ are $\approx$ 38 and $\approx$ 60 MeV/c^2 respectively, consistent with the observed widths.[26]

The (14) candidate states for the $^3P_0$ nonet are the $\delta(980)$, $a_0(1300)$, $S^*(975)$, $f_0(1300)$, $f_0'(1500)$, $G(1590)$ and $K_0^*(1410)$ so this sector is overpopulated. The S^* and δ are nearly degenerate in mass which suggests that the S^* has almost no $s\bar{s}$ content. They sit just below $K\overline{K}$ threshold, as would be expected for weakly bound $K\overline{K}$ molecules, and are about 300 MeV/c^2 too light relative to the other P-wave mesons to fit into the expected pattern of weak spin-orbit splittings (which predicts that the $^3P_0$, $^3P_1$, and $^3P_2$ mesons with the same flavor wave functions should differ in mass by less than $\approx$ 100 MeV/c^2). If the $a_0(1300)$ is confirmed and the S^* and δ are $K\overline{K}$ molecules then the $^3P_0$ $n\bar{n}$ states will agree with quark model expectations in both mass and number. The very small $\vec{L}\cdot\vec{S}$ splittings in these reassigned ^3P_J nonets will be consistent with those observed in the baryon sector.[4]

The SU(3) prediction for $\Gamma(S^* \to \pi\pi)/\Gamma(\delta \to \eta\pi)$ is about 4 for $(u\bar{u} \pm d\bar{d})/\sqrt{2}$ content; in the coupled-channel $K\overline{K}$ molecule picture we find ≈ 0.6 in agreement with the observed value of 0.6 ± 0.2.

Another SU(3) prediction is that $\Gamma(S^* \to K\overline{K})/\,\Gamma(S^* \to \pi\pi) \approx p_K/(3\,p_\pi)$, with $p_K(p_\pi)$ the kaon(pion) momentum, for $S^* = (u\bar{u}+d\bar{d})/\sqrt{2}$. The data does not support this expectation; above $K\overline{K}$ threshold the S^* decays predominantly to $K\overline{K}$, as expected for the high mass tail of a $K\overline{K}$ bound state. Similar remarks apply to the ratio $\Gamma(\delta \to \eta\pi)/\Gamma(\delta \to K\overline{K})$.

The two photon widths of the S^* and δ^0 have been calculated,[27] assuming both $q\bar{q}$ and $K\overline{K}$ assignments, with the results

$$\Gamma(S^* \to \gamma\gamma) = \begin{cases} 4.5 & \text{keV for } (u\bar{u} + d\bar{d})/\sqrt{2} \\ 0.33 & \text{keV for } s\bar{s} \\ 0.6 & \text{keV for } K\overline{K} \end{cases}$$

and

$$\Gamma(\delta^0 \to \gamma\gamma) = \begin{cases} 1.5 & \text{keV for } (u\bar{u} - d\bar{d})/\sqrt{2} \\ 0.6 & \text{keV for } K\overline{K}. \end{cases}$$

Experimental measurements are

$$\Gamma(S^* \to \gamma\gamma)\; BR(\pi\pi) = 0.19 \pm 0.05 \pm .12 \text{ keV} \qquad \text{(a)}$$

$$\Gamma(\delta^0 \to \gamma\gamma)\; BR(\eta\pi) = 0.19 \pm 0.07\ {}^{+0.10}_{-0.07} \text{ keV} \qquad \text{(b)}$$

$$= 0.29 \pm 0.05 \pm 0.14 \text{ keV} \qquad \text{(c)}$$

for (a) Mark II, (b) Crystal Ball and (c) Jade,[28] consistent with the $K\overline{K}$ but not the $(u\bar{u}+d\bar{d})/\sqrt{2}$ content. The $s\bar{s}$ assignment for the S^* favored by this measure is inconsistent with the near degeneracy of the S^* and δ masses as mentioned above. Although the absolute widths for the $K\overline{K}$ molecule interpretation are larger than experiment by a factor of about 3 for both the S^* and δ^0, the ratio of the $\gamma\gamma$ widths is predicted to be 1: the S^* and δ are $(K^+K^- \pm K^0\overline{K}^0)/\sqrt{2}$ and the $\gamma\gamma$ widths depend only on the K^+K^- content. The ratio is independent of the details of the wave functions in both these models and is therefore a crucial test:

$$\frac{\Gamma(S^* \to \gamma\gamma)}{\Gamma(\delta^0 \to \gamma\gamma)} = \begin{cases} \frac{25}{9} & \text{for } (u\bar{u} \pm d\bar{d})/\sqrt{2} \\ 1 & \text{for } K\overline{K}\ . \end{cases}$$

Since $BR(S^* \to \eta\pi)$ and $BR(\delta \to \eta\pi)$ are both ≈ 0.8[25] they approximately cancel in the experimental ratio. Thus the data clearly favors the $K\overline{K}$ picture, although a definitive measurement would be of great interest.

6. PRODUCTION MODEL AND SOLUTIONS

To discuss the effect of the coupled-channel interactions on pseudoscalar pair production we consider the decay $J/\psi \to \omega\pi\pi$, which has a strong, wide enhancement at ≈ 500 MeV/c^2 in the $\pi\pi$ invariant mass distribution and no S^* signal (fig. 13(a)).[30] The ω is a narrow $(u\bar{u} + d\bar{d})/\sqrt{2}$ resonance and one expects to find a $(u\bar{u} + d\bar{d})/\sqrt{2}$ pair recoiling from it. Following SU(2) pair creation in the recoiling $q\bar{q}$ pair a $\pi\pi$ system can form and undergo coupled-channel interactions as it propagates from the origin to the asymptotic region. Although this is not $\pi\pi$ scattering, the $\pi\pi$ pair is formed inside the potentials, it is clear that it must be related to the scattering dynamics.

We use an inhomogeneous coupled-channel equation based on eq. (4) to solve this "production problem":[30,31]

$$(\mathbf{K} + \mathbf{M} + \mathbf{V})\, \mathbf{u_s}(r) = \mathrm{E}\, \mathbf{u_s}(r) + \mathrm{s}\, \delta(r) \tag{5}$$

The solutions $\mathbf{u_s}(r)$ have the same form as the solutions $\mathbf{u}(r)$ of eq. (4). The "source" vector s is chosen to be nonzero in one entry only and models the "creation" of a particular channel of $\mathbf{u_s}$ by giving that channel a nonzero radial wave function at the origin. A linear combination of $\mathbf{u_s}(r)$ and the homogeneous solutions $\mathbf{u}(r)$ of eq. (4) can be constructed to yield a solution $\mathbf{u}_+(r)$ which has only outgoing waves. The amplitude to find a given system in the asymptotic region, after a given channel is created at the origin, is determined from the outgoing solution $\mathbf{u}_+(r)$ and is interpreted as the probability amplitude for the process. Phase space times the modulus of the probability amplitude squared is proportional to the rate for the process. We can generate the probability amplitude, and hence the rate, as a function of the invariant mass of the detected system, resulting in a spectrum which should be proportional to the experimental invariant mass distribution. If the energy rises above the threshold for two or more open channels then the relative rate from a common initial channel into each asymptotic channel is fixed by the outgoing solution $\mathbf{u}_+(r)$.

To illustrate this procedure we consider the decays $J/\psi \to \omega\pi\pi$ and $J/\psi \to \omega K\overline{K}$. We assume the production of a $\pi\pi$ pair at the origin, although we anticipate that the $f_0(1300)$ and the other PP channels will also contribute and that the details of the production spectra will change when these contributions are added.[31] The unknown relative normalization of the experimental and model spectra can be determined by fitting the model rate to the experimental rate at any arbitrary energy. For $J/\psi \to \omega\pi\pi$ we constrain the low mass enhancement in the model to have the same area as the experimental spectrum. This normalization is then used to predict the absolute $K\overline{K}$ spectrum in $J/\psi \to \omega K\overline{K}$. Thus, recoiling against the ω, we predict the shape of the $\pi\pi$ spectrum and both the shape and normalization of the $K\overline{K}$ spectrum.

Some previously unresolved problems with these spectra are the $\pi\pi$ threshold effect in $\omega\pi\pi$,[32,33] the absence of an S^* signal in $\omega\pi\pi$, and the $K\overline{K}$ threshold effect in $\omega K\overline{K}$. As seen in figs. 13(a) and (b), the model qualitatively fits the data and suggests that these decays may be better understood if the coupled-channel interactions are taken into account.

In the related decays $J/\psi \to \phi\pi\pi$ and $J/\psi \to \phi K\overline{K}$ we assume the creation of a $K\overline{K}$ pair at the origin and find the amplitude for $\pi\pi$ or $K\overline{K}$ to appear in the asymptotic region. We again expect that the $f_0'(1300)$ and the other PP channels will contribute.[31] The unknown overall normalization is determined by fitting the $K\overline{K}$ threshold effect ($m_{K\overline{K}} < 1.2$ GeV/c^2) from the model to the experimental spectrum (fig. 13(c)). This normalization is then used to predict the $\pi\pi$ spectrum (fig. 13(d)). Thus, recoiling against the ϕ, we predict the shape of the $K\overline{K}$ spectrum and both the shape and normalization of the $\pi\pi$ spectrum.

Some apparent inconsistencies with these spectra are the lack of a $\pi\pi$ threshold enhancement in $\phi\pi\pi$[32] (given the enhancement seen in $\omega\pi\pi$ and elsewhere), the strong S^* signal in $\phi\pi\pi$ (which appeared to violate the OZI rule), and the $K\overline{K}$ threshold effect in $\phi K\overline{K}$. As seen in figs. 13(c) and (d) the model qualitatively resolves these issues, demonstrating the importance of the coupled-channel interactions and stressing the need to include them in data analysis.

Coupled channel interactions also appear to play a major role in the $\Delta I = \frac{1}{2}$ rule in the weak decay $K \to \pi\pi$.[34] The $\pi\pi$ system can be made in either $I = 0$ or 2, corresponding to $\Delta I = \frac{1}{2}$ or $\Delta I = \frac{3}{2}$. Experimentally one finds the ratio of amplitudes

$$\left[\frac{\Gamma(K_s^0 \to \pi^0\pi^0)}{\Gamma(K^+ \to \pi^0\pi^+)}\right]^{\frac{1}{2}} \approx 25$$

which is far from the naive standard model prediction of

$$\left[\frac{\Gamma_{\Delta I=\frac{1}{2}}}{\Gamma_{\Delta I=\frac{3}{2}}}\right]^{\frac{1}{2}} = \sqrt{2}\ .$$

A comparison of these ratios reveals that $\pi\pi^{I=2}$ production is suppressed relative to the $\pi\pi^{I=0}$ state. From the $\pi\pi$ phase shifts at the kaon mass one infers that $I = 0$ π pairs have an attractive interaction (fig. 9(a)) and $I = 2$ π pairs have a repulsive interaction (fig. 1), leading respectively to the enhancement and the suppression of the wave function at the origin. Our production solutions for $\pi\pi$ creation at the kaon mass predict an additional factor of ≈ 4 in the expected amplitude ratios. When included with other perturbative QCD corrections the long standing problem of the $\Delta I = \frac{1}{2}$ rule is largely resolved.

The spin-parity and line shape of the $\eta/\iota(1440)$ have been discussed extensively and it is now proposed that there may be 3 resonances in this mass region.[35] In earlier papers[36] we examined this decay in the approximation that the $K\pi$ potential was neligible. Although we now know that this potential is strong the calculation shows that final state interactions[37] can lead to dramatic effects in such processes. In figs. 14 we show effective mass distributions for $\iota \to K\overline{K}\pi$ in 4 scenarios, phase space, phase space with $K\overline{K}$ interactions, K^* isobar decay, and K^* isobar decay in which $K\overline{K}$ interactions increase the width of the K^*. In the decays with these interactions the effective mass distributions create signals in the $\delta(980)$ region and could lead to a misleading interpretation of the data. The clear message is that the intermeson potentials cannot safely be ignored.

In fig. 15 we show how final state interactions might affect the trajectories of particles as they leave the interaction region so that they no longer reveal the spin structure of the vertices to a standard spin-parity analysis; the rescattering effects must first be taken into account.

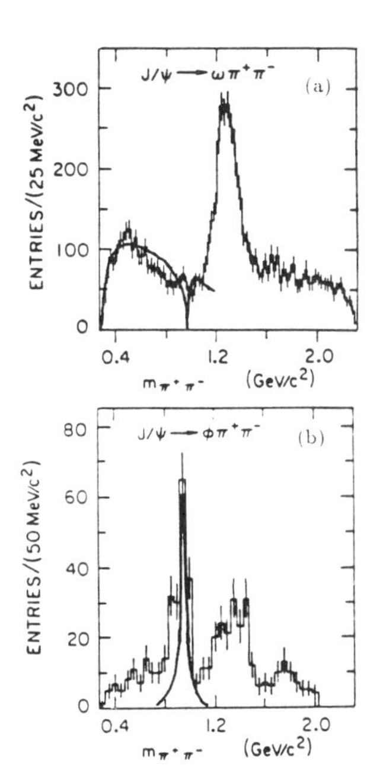

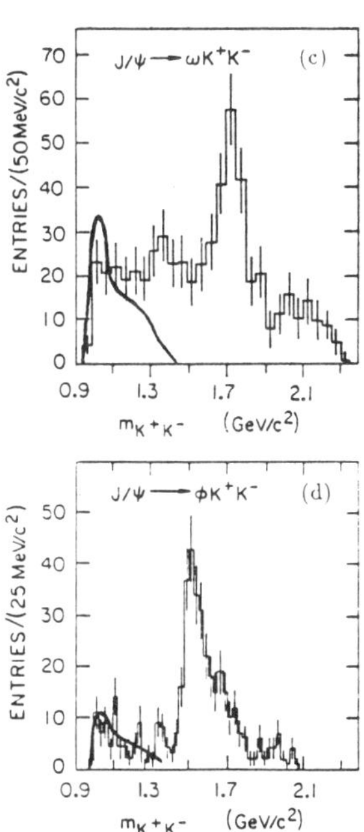

Fig.13 Some puzzles in J/ψ decay may be resolved by coupled channel interactions: the predicted spectra are the solid curves which show • the $\pi\pi$ threshold peak in $\omega\pi\pi$ (a) but not in $\phi\pi\pi$ (b) • no S^* peak in $\omega\pi\pi$ (a) but a strong S^* peak in $\phi\pi\pi$ (b) • a threshold enhancement in both $\omega K\overline{K}$ (c) and $\phi K\overline{K}$ (d). The shape of these spectra are predicted by the model, and the normalizations of (b) and (c) are determined by (d) and (a) respectively. These are preliminary spectra; not all production mechanisms are included.

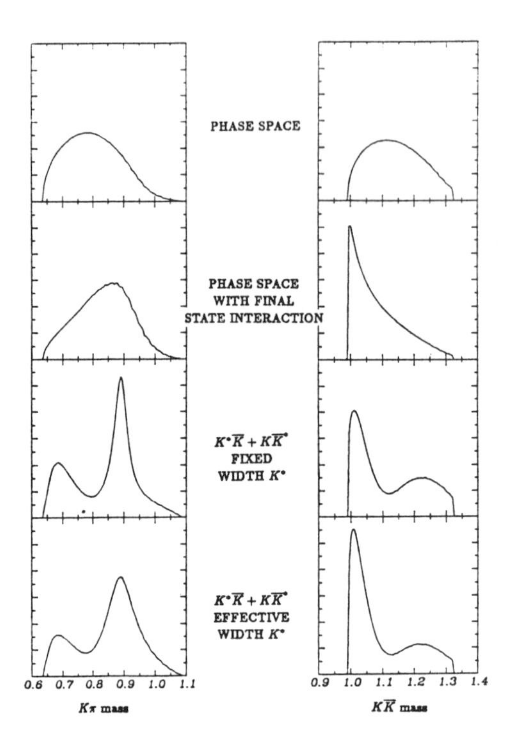

Fig.14 Effective mass distributions for $\iota \to K\overline{K}\pi$ in 4 decay scenarios.

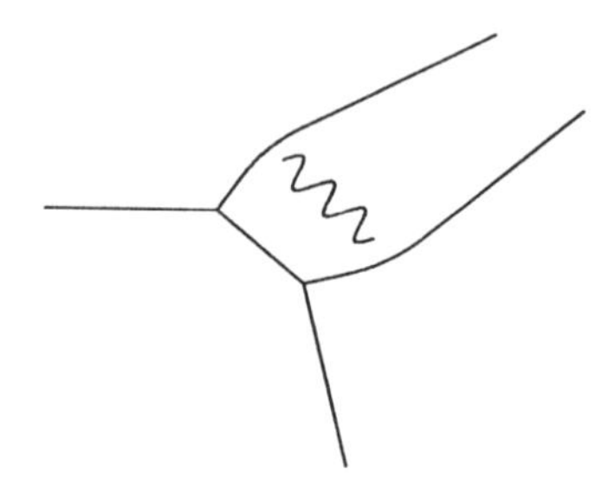

Fig.15 A cartoon showing that coupled channel interactions (represented by the squiggle) may divert the final state particles from their spin-parity dependent decay vertex distributions.

7. CONCLUSIONS

The proton and neutron are known to bind as qqq clusters so it is not surprising that mesons should also have analogous strong "nuclear" forces and form bound states of $q\bar{q}$ clusters.

To investigate the $qq\bar{q}\bar{q}$ systems we have used the nonrelativistic quark model and the variational principle to determine the ground-state wave functions and from these we have extracted S-wave $P_1P_2 \leftrightarrow P_1'P_2'$ potentials which depend on quark exchange. A model for the $P_1P_2 \leftrightarrow S$ couplings provides additional interactions which also lead to meson-meson potentials. We find that t-channel meson exchange effects are not required to explain the data. The coupled-channel Schrödinger equation is used to study the low-energy behaviour of these meson-meson interactions.

In $I = 2$ the quark-exchange potentials predict the correct sign of the phase shift and we use this data to rescale these potentials. The resulting $I = \frac{3}{2}$ $K\pi$ phase shifts are found to agree with the data and we predict the unmeasured strangeness ± 2 $I = 1$ KK phase shift.

In the $I = \frac{1}{2}$ $K\pi$ system there are no PP bound states and we find that the predictions are in good agreement with the data. Both the exchange interaction (which in isolation leads to "background" phase shifts) and annihilation effects are important and we find significant shifts in the masses and widths predicted by the naive quark model due to the coupled-channel interactions.

The $I = 0$ and 1 solutions lead to the prediction that two mesonic analogues of the deuteron exist; $K\overline{K}$ systems form bound states which we identify with the S^* and δ scalar resonances. With this assignment many of the properties of the S^* and δ such as their masses, full widths, two photon widths, and branching fractions both below and above $K\overline{K}$ threshold, are clarified. The systematics of the $^3P_0$ $q\bar{q}$ sector and mass comparisons with the $^3P_1$ and $^3P_2$ $q\bar{q}$ sectors are also improved by the $K\overline{K}$ molecule assignments.

In addition to creating the S^* the $I = 0$ coupled-channel interactions reproduce the $\pi\pi$ elastic scattering phase shift data from threshold to ≈ 1400 MeV/c^2 . In $I = 1$ these interactions lead to predictions about the elastic $\eta\pi$ phase shifts.

The same model can be applied to production processes where problems in certain $J/\psi \rightarrow VPP$ channels may be resolved. A long standing puzzle of the $\Delta I = \frac{1}{2}$ rule in the nonleptonic weak decay $K \rightarrow \pi\pi$ is largely explained.

The wide-ranging successes of this model in the scalar sector are most encouraging. At present we are continuing our study of the dynamics of coupled-channels in pseudoscalar pair production in all isospin channels. We plan to apply it to other sectors such as vector-vector in $J = 0$, 1, and 2[38] (There are unexplained threshold effects seen in $\gamma\gamma \rightarrow V_1 V_2^{J=2}$.), the vector-pseudoscalar systems (Is the $E(1420)$ a KK^* threshold effect?), and meson-baryon systems (Is the $\Lambda(1405)$ a $\overline{K}N$ molecule and are $\gamma N \rightarrow N\pi$[39] and $N\pi \rightarrow N\eta$[40] ammenable to this analysis?). An analysis of the proposed electrically bound $\pi^+\pi^-$ system[41] is also possible. Charmed $qq\bar{q}\bar{q}$ studies and a three-body coupled-channel equation relevant to the $K\overline{K}\pi$ and $\eta\pi\pi$ decays of the E, ι, and D mesons are two less straightforward but interesting extensions of this research.

Many of the present problems in 1-2 GeV/c^2 spectroscopy may be artifacts of our neglect of coupled-channel hadron-hadron interactions; the correct classification and identification of the $q\bar{q}$ states, glueballs, and hybrids in this mass region may prove impossible until we have understood these processes.

ACKNOWLEDGEMENTS

I am pleased to thank Ted Barnes, Frank Close, and Nathan Isgur for their many invaluable contributions.

This research was partially supported by the U.S. Department of Energy, Division of High Energy Physics, under Contract DE-AC05-840R21400, and the UTK/ORNL Science Alliance.

REFERENCES

1. For a history of the $qq\bar{q}\bar{q}$ systems see refs. 5-7, 9-12, and 14, and references therein. For a review of the history of scalar mesons see L. Montanet, Rep. Prog. Phys. **46**, 337 (1983) and references therein.
2. S. Godfrey and N. Isgur, Phys. Rev. **D32**, 189 (1985).
3. R. Kokoski and N. Isgur, Phys. Rev. **D35**, 907 (1987).
4. S. Capstick and N. Isgur, Phys. Rev. **D34**, 2809 (1986).
5. J. Weinstein and N. Isgur, Phys. Rev. Lett. **48**, 659 (1982); Phys. Rev. **D27**, 588 (1983).
6. J. Weinstein and N. Isgur, Phys. Rev. **D41**, 2236 (1990).
7. J. Weinstein and N. Isgur, UTK-90-01, to appear in Phys. Rev. **D**, and references therein.
8. J. Weinstein, UTK-90-07, to be submitted to Phys. Rev. **D**.
9. At least two experimental groups have previously suggested that the S^* or δ were $K\overline{K}$ bound states: A. Astier *et al.*, Phys. Lett. **25B**, 294 (1967); A.B. Wicklund *et al.*, Phys. Rev. Lett. **45**, 1469 (1980); A.B. Wicklund, *Proc. of the 16th Rencontre de Moriond*, Les Arcs, France, 1981 *ed.* J. Trân Thanh Vân, (Editions Frontieres, Dreux, France, 1981), p. 339.
10. F. Cannata, J.P. Dedonder, L. Leśniak, Z. Phys. **A 334**, 457 (1989).
11. An s-channel plus t-channel meson exchange model has been applied to $\pi\pi$ and $K\pi$ scattering, see J. Speth, these proceedings; D. Lohse, J.W. Durso, K. Holinde, J. Speth, Phys. Lett. **234B**, 235 (1990).
12. K.L. Au, D. Morgan, and M.R. Pennington, Phys. Lett. **167B**, 229 (1986); Phys. Rev. **D35**, 1633 (1987); M.R. Pennington, to appear in *Proc. of the Reinfels Workshop, Hadron 90.*
13. See, for example, Chan Hong Mo and H. Høgaasen, Phys. Lett. **72B**, 121 (1977); Nucl. Phys. **B136**, 401 (1978); Phys. Lett. **72B**, 400 (1977); **76B**, 634 (1978); R.L. Jaffe, Phys. Rev. **D15**, 267, 281 (1977).
14. R.L. Jaffe and F.E. Low, Phys. Rev. **D19**, 2105 (1979); R.L. Jaffe and M.P. Shatz, CALT-68-775; R.P. Bickerstaff, Philos. Trans. R. Soc. London **A309**, 611 (1983); Phys. Rev. **D27**, 1178 (1983).
15. J. Weinstein, in preparation.
16. There are many coupled channel studies, see for example N. Törnqvist; Phys. Rev. Lett. **49**, 624 (1982) and references therein; S.M. Flatté Phys. Lett. **63B**, 224 (1976); K.M. Watson, Phys. Rev. **88**, 1163 (1952); R.N. Cahn and P.V. Landshoff, Nucl. Phys. **B266**, 451 (1986); W. Lockman,

Proc. of the IIIrd Int. Conf. on Hadron Spectroscopy, Ajaccio *ed.* F. Binon, J.-M Frère, J.-P. Peigneux; L. Lesniak, *ibid.* For a study of these systems in a multichannel Schrödinger equation without the $0^-0^- \leftrightarrow 0^-0^-$ potentials see E. van Beveren, to appear in *Proc. of the Reinfels Workshop, Hadron 90.*

17. The data of Fig. 5 are a combination of the compilation of F. Wagner in *Proc. of the XVII Int. Conf. on High Energy Physics*, London, 1974 (Rutherford Lab, Chilton, Didcot, England, 1974), p. II-27 and the data of J. Prukop *et al.*, Phys. Rev. **D10**, 2055 (1974). See also D. Cohen *et al.*, Phys. Rev. **D7**, 661 (1973); W. Hoogland *et al.*, Nucl. Phys. **B126**, 109 (1977); W. Hoogland *et al.*, Nucl. Phys. **B69**, 109 (1974).
18. P. Estabrooks *et al.*, Nucl. Phys. **B133**, 490 (1978).
19. D. Aston *et al.*, Nucl. Phys. **B296**, 493 (1988).
20. F. Feld-Dahme, Ludwig-Maximilians-Universität, Ph.D. thesis, Munich (1987).
21. Reanalysis of the K_0^* data of ref.18 leads to a K_0^* mass of $1412 \pm 4 \pm 5$ MeV/c^2, Aston *et al.*, *Proc. IIIrd Conf. on Part. and Nucl. Phy.*, Rockport, Maine, (1988), *ed.* G.M. Bunce.
22. M. Boutemeur, *Proc. of the IIIrd Int. Conf. on Hadron Spectroscopy*, Ajaccio, *ed.* F. Binon, J.-M. Frère, J.-P. Peigneux.
23. See N. Törnqvist and S.M. Flatté in ref. 16.
24. G. Grayer *et al.*, Nucl. Phys. **B75**, 189 (1974). For other measurements see A.A. Belkov *et al.*, JETP Lett. **29**, 597 (1979); L. Rosselet *et al.*, Phys. Rev. **D15**, 574 (1977); V. Srinivasan *et al.*, Phys. Rev. **D12**, 681 (1975); N.M. Cason *et al.*, Phys. Rev. **D28**, 1586 (1983).
25. Particle Data Group, Phys. Lett. **239B** (1990).
26. Recent measurements suggest that the S^* width may be ≈ 60 MeV/c^2, A. Palano, to appear in *Proc. of the Reinfels Workshop, Hadron 90.*
27. T. Barnes, Phys. Lett. **165B**, 434 (1985, and *Proc. of the VIIth Int. Workshop on Photon-Photon Collisions*, Paris (1986), ed. A. Courau and P. Kessler (World Scientific, Singapore, 1986), p. 25; K. Dooley, in preparation.
28. See, for example, G. Gidal *Proc. VIII Int. Workshop on $\gamma\gamma$ Collisions*, Shoresh, Israel, 1988. Recent suggestions indicate that these widths have been underestimated, D. Morgan, and M.R. Pennington, RAL-90-030.
29. See, for example, L. Kopke and N. Wermes, Phys. Rep. **174**, 68 (1989).
30. J. Weinstein, *Proc. of the τ-charm Factory Workshop*, SLAC, (1989), *Proc. of the IIIrd Int. Conf. on Hadron Spectroscopy*, Ajaccio *ed.* F. Binon, J.-M Frère, J.-P. Peigneux.
31. J. Weinstein and N. Isgur, in preparation.
32. H.G. Dosch and D. Gromes, Z. Phys. **C34**, 554 (1987).

33. G. Kernel *et al.*, Omicron Collaboration, CERN PPE/90- , Sept 1990, submitted to Nucl. Phys. **A;** D. Zavartanik, these proceedings.
34. N. Isgur, K. Maltman, J. Weinstein, T. Barnes, Phys. Rev. Lett. **64**, 161 (1990).
35. J. Dowd, to appear in *Proc. of the Reinfels Workshop, Hadron 90.*
36. M. Frank, N. Isgur, P.J. O'Donnell, and J. Weinstein, Phys. Lett. **158B**, 443 (1985); Phys. Rev. **D32**, 2971 (1985).
37. K.M. Watson, in ref. 16.
38. A study of the $I = 2,\ J = 2\ \rho\rho$ system is nearly complete, G. Grondin, E. Swanson, J. Weinstein, and T. Barnes, in preparation.
39. D. Drechsel, these proceedings.
40. W.W. Jacobs, these proceedings.
41. G.T. Emery, these proceedings.

THRESHOLD EFFECTS IN MESON–MESON AND MESON–BARYON SCATTERING

J. Speth
Institut für Kernphysik (Theorie)
KFA Jülich, D–5170 Jülich, Germany

ABSTRACT

We investigate within a generalized meson–exchange model meson–meson and meson–baryon scattering. We concentrate our investigations on the behavior of such systems when a new particle channel opens. Specifically we demonstrate these effects at the $K\bar{K}$–threshold in $\pi\pi$–scattering and the $\bar{K}N$-threshold in $\pi\Sigma$–scattering.

1. INTRODUCTION

It is generally accepted that quantum chromodynamics (QCD) is the fundamental theory of strong interaction. Therefore, in principle, the hadron-hadron interaction is determined by the quark–gluon dynamics. Unfortunately, little is known about QCD solutions in the low–energy (non–perturbative) regime, where most of the nuclear and medium energy physics experiments take place. In this region, however, there are indications [1] that most of the dynamics can be understood in terms of color–neutral objects, *i.e.* nucleons, mesons and isobars. Therefore in the low energy region (low compared to the perturbative QCD–regime) the strong interaction can probably be described to a very good approximation within the framework of meson–exchange interactions. The basic ingredients of such models are meson–baryon–baryon and meson–meson–meson vertices which represent an effective description of the very complicated, and mathematically yet intractible, multi–quark and gluon exchanges. These vertices contain coupling constants and form factors which parameterize the finite size of the hadrons. It is obvious that the meson exchange model is limited in its applicability, and that one expects corrections due to the underlying quark structure. The interesting question, however, is where do we see those "quark effects" in nuclear and medium energy physics? The range of its validity is clearly connected with the size of the confining region of the quarks. As the radii of hadrons are given not only by the size of the confining region of the quarks, but also by the extension of the surrounding meson cloud, the range of applicability of the meson–exchange models may be much larger than one expects from oversimplified estimates based on the hadron size.

In the past few years the Bonn NN–potential [2], which is based on suitably chosen meson–nucleon–nucleon and meson–nucleon–delta vertices, has been generalized to include the exchange of strange mesons [3,4], enabling the calculation of $N\Lambda$– and $N\Sigma$–potentials as well as the treatment of the KN– and $\bar{K}N$–systems [5]. More recently also meson–meson scattering has been investigated within the same model [6]. In this contribution we investigate specifically the $\pi\pi$– and $\bar{K}N$–system, where it has been shown that a coupled—channel approach is necessary because there exists a strong coupling to various channels. Our main interest here is the behavior of these systems at the thresholds, when a new particle channel opens. Such questions may be investigated experimentally with the new cooler facilities because these effects happen in the close vicinity of the threshold. In the first part of my talk, I shall discuss $\pi\pi$– and Kπ–scattering which are closely connected to each other, but which differ strongly as far as the coupling to other channels is concerned. In the second part I will show that such an anomalous threshold behavior exists also in baryon–meson systems.

2. MESON–MESON SCATTERING

MESON–MESON INTERACTION

The meson–exchange model which successfully describes the baryon–baryon interaction at low and medium energies can easily be extended to meson—meson scattering. All the formulas given e.g. in [2] are also valid for meson—meson scattering if we replace the corresponding baryon quantum numbers by meson quantum numbers and remember the rules for identical bosons where they have to be applied. The final formulas can be found in ref. [6]. This extended meson–exchange interaction model allows us to:

(i) investigate the strong interaction between two pseudoscalar meson in the frame–work of the meson–exchange model;

(ii) analyse the $\pi\pi$ and Kπ scattering data in a coupled–channel approach using this meson–exchange interaction;

(iii) explain the resonance f_0 (975) in the $\pi\pi$ system in a natural way as a quasi–bound state in the coupled K$\overline{\text{K}}$ channel;

(iv) estimate the masses of the "genuine" scalar mesons (with and without strangeness) which belong to the scalar meson nonet, and which can be interpreted as one quark–one antiquark systems, to be around 1400 MeV and 1600 MeV, respectively.

We restrict our investigations to the energy range below 1.5 GeV because we neglect possible coupling to N$\overline{\text{N}}$ or two vector–meson channels. Since the K$\overline{\text{K}}$ channel will turn out to be of extreme importance in the case of $\pi\pi$ scattering, we will treat these channels in a coupled–channel framework.

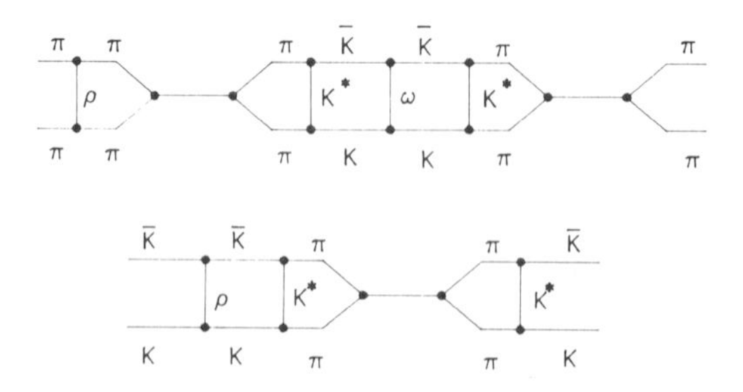

Fig. 2.1: Coupled–channel approach for $\pi\pi$–scattering.

The corresponding coupled channels are shown in Fig. 2.1. Of course, below the $K\overline{K}$ threshold this coupling occurs only virtually in the $\pi\pi$ system, whereas $K\overline{K}$ scattering, because of real transitions to the $\pi\pi$ system, is inelastic for all energies. The quantum numbers of the s–channel mesons are given by the quantum numbers of the considered $\pi\pi$–channel.

PION–PION SCATTERING

We restrict our discussion on results of pion–pion scattering to the scalar isoscalar channel $I^G(I^{PC})=0^+(0^{++})$ because only this channel is connected with strangeness. The experimental $\pi\pi$ phase shift in this channel (δ_0^0) shows a 180° jump in a narrow energy range around E ≃ 980 MeV. This resonance ($f_0(975)$) has originally been interpreted as a member of the scalar $q\bar{q}$ nonet. Jaffe [6] has pointed out that this interpretation leads to difficulties and suggested that f_0 (975) and a_0 (980) might be 4 quark states (2 quark–2 antiquark), whereas the genuine $q\bar{q}$ scalar mesons should be several hundred MeV higher in energy. We will show that in our model the spectrum below 1 GeV is essentially given as a correlated two–pion and two–kaon system.

In a first approach we solved the T–matrix equation in the $\pi\pi$ channel only. The effects of other mesons are included as box diagrams in the quasipotential V(z). Since the coupling to the $\eta\eta$–channel is very weak, we consider only the coupling to the $K\overline{K}$ pair, which is shown in Fig. 2.2. The effect of this box diagram on the

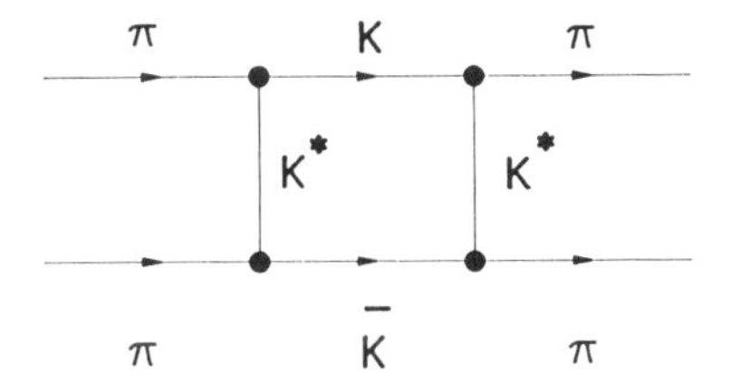

Fig. 2.2: $K\overline{K}$–box–diagram

δ_0^0 phase shift is small, as demonstrated in Fig. 2.3. The dashed–dotted line is the result if one considers ρ– and ϵ–exchange (in the t–channel) only. The dashed line is the result, where the open $K\overline{K}$ channel is included, but no explicit resonance behavior is included as is seen in the data. In order to explain the experimental phase shift one has to consider explicitly a scalar meson (pole graph) in the s–channel, the coupling constant and mass of which are appropriately chosen to reproduce the experimental data.

The t–channel interaction between $K\overline{K}$ pairs for I=0 is rather strong and attractive because all contributions add coherently. Therefore we expect also a larger effect from the $K\overline{K}$ channel if we take this interaction into account. The effects due to the channel coupling on the phase shift and inelasticity are indeed very strong, as demonstrated in Fig. 2.3 and Fig. 2.4. First of all we see

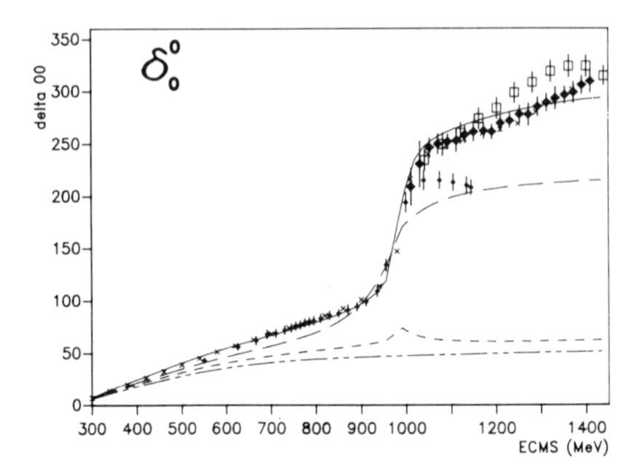

Fig. 2.3: The predictions of four different models for I=0, J=0 $\pi\pi$ phase shifts (see explanation in the text).

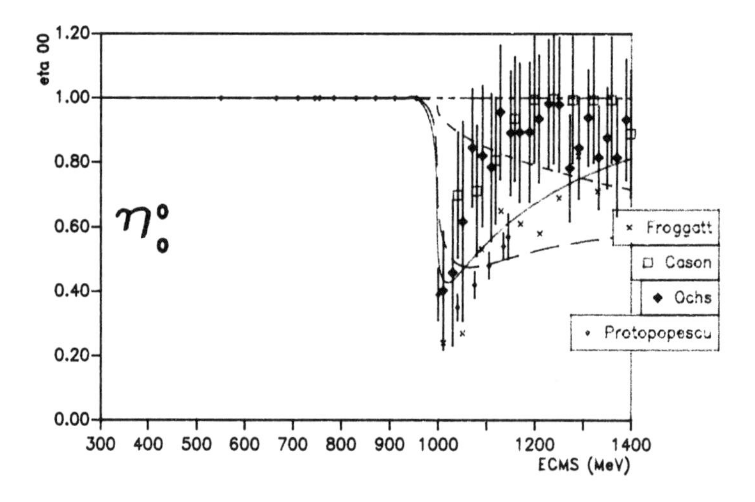

Fig. 2.4: Predictions of the same four different models as in Fig. 3.3 for the elasticity parameter in the I=0, J=0 $\pi\pi$ channel.

in Fig. 2.3 a large jump in the phase shift which comes from a quasi–bound $K\overline{K}$ pair. Because of the attractive interaction in the t–channel, this state is shifted by about 5 MeV below threshold, which gives rise to a strong resonance–like behavior in the phase shift. From Fig. 2.4 we notice that this increase of the phase shift has only little influence on the elasticity η_0^0. It is important to mention that so far only the interaction in the t–channel has been taken into account. We refer to this as Model I. We do not need to consider a genuine scalar resonance around [1 GeV] in order to reproduce the experimental phase shift from the $\pi\pi$ threshold up to 1 GeV. Beyond 1 GeV, however, the theoretical results deviate qualitatively from the more recent experimental phase–shift analyses. As we explain the phase jump at 980 MeV as a quasi–bound $K\overline{K}$ pair, it is obvious that the further increase of the phase beyond 1 GeV should be connected with the genuine scalar meson which is a member of the scalar SU(3) nonet. (Actually one expects two scalar mesons–one a member of the octet and the other a singlet.) Such scalar mesons will give rise to pole terms in the s–channel. The particle data group lists a scalar meson f_0 (1400) (the previous $\epsilon(1300)$) at 1400 MeV with a width of 150–400 MeV that decays mainly into two pions. We consider in the following calculation a scalar meson (which we refer to as ϵ–meson), the bare mass and the coupling constant of which we take as free parameters that are adjusted to fit the experimental phase–shift beyond 1 GeV. The full lines in Figs. 2.3 and 2.4 indicate the theoretical results in this approach. The mass of the ϵ–meson extracted from this analysis using an Argand plot is approximately 1400 MeV and the width is broad – about 600 MeV.

KAON–PION SCATTERING

Our model can also be applied in a straightforward way to the $K\pi$ system. Besides the strangeness as a new degree of freedom, kaons have isospin I=1/2 and no definite G–parity. Therefore the isospin of the $K\pi$ system is half–integer and the various channels are characterized by the quantum numbers $I(J^P)$. The $K\pi$ system is of special interest because, at least in principle, the $K\pi$ interaction is fully determined from the previous investigations and the SU(3) symmetry relations. In the calculations we report, the meson–exchange interaction in the t–channel is indeed the same as the one used in the $\pi\pi$–system, whereas the pole contributions in the s–channel have to be adjusted separately. Actually our calculation gives information about the genuine $I(J^P)=1/2(0^+)$ meson with strangeness, usually denoted as κ. Because of the isospin 1/2 of the kaon, the scalar meson κ (the analog of the scalar ϵ–meson discussed in the previous section) appears in the I=1/2, $J^P = 0^+$ channel of the $K\pi$ system. High–statistics $K\pi$ data are become available [7], from which very reliable $\delta_0^{1/2}$ phase–shifts in the "scalar" channel have been obtained. In Fig. 2.5 the results of our calculation are compared with the experimental data. In the present case Model I again denotes the approach in which only the meson–exchange interaction in the t–channel was included, with no channel coupling. The only possible open channel below 1.5 GeV is the $K\eta$ system, which can be neglected beause of the weak coupling of the η to the other mesons. The theoretical results calculated within Model I are shown as a dashed line. It is clear from that figure that the t–channel interaction is by far the most important part of the $K\pi$ interaction below 1 GeV. We want to point out again that there are no free parameters in Model I; it is exactly the same t–channel interaction as used in $\pi\pi$–scattering.

In $K\pi$ scattering we also have the ambiguity of the coupling of the scalar meson to the two pseudoscalar mesons. In the actual calculation we used a cut off Λ=3 GeV and a bare mass of m^o= 1600 MeV. The corresponding physical mass of scalar meson K is m_κ= 1430 MeV with a width of Γ= 290 MeV. The scalar channel of the $K\pi$ system is actually simpler than in the $\pi\pi$ system, because here we expect only one genuine $(q\bar{q})$ scalar meson. The good agreement between the experimental and theoretical $\delta_0^{1/2}$ phase shift

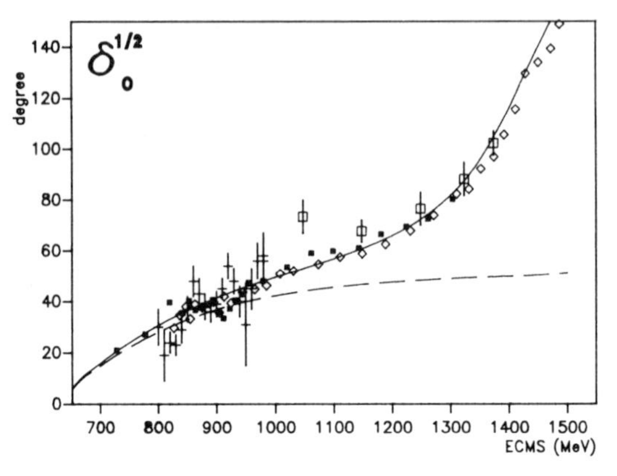

Fig. 2.5: The I=1/2, J=0 $K\pi$ phase shift with (full line) and without (dashed line) the κ–meson contribution.

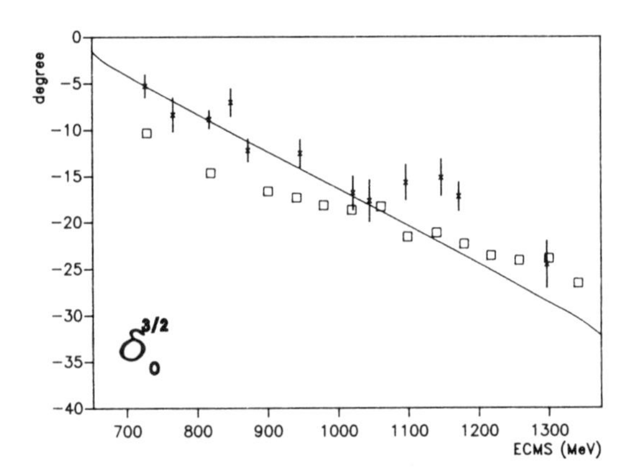

Fig. 2.6: The I=3/2, J=0 $K\pi$ phase shift.

strongly supports our model. We should, however, also bear in mind that channel coupling may become important around 1.5–1.6 GeV where the $\rho K*$ and $\omega K*$ thresholds lie. Such investigations are in progress.

A further excellent test for the meson–exchange interaction in the t–channel is the $\delta_0^{3/2}$–phase shift. Like the δ_0^2–phase shift in $\pi\pi$–scattering, all of the effects come from the t–channel interaction. The good agreement between theory and experiment shown in Fig. 2.6 indicate that our model of the interactions works well, especially since all of the relevant parameters have been fixed by the fit in the $\pi\pi$ sector.

3. $\bar{K}$N SCATTERING AND THE Λ(1405) RESONANCE

The $\bar{K}$N system represents a special challenge since, already at the $\bar{K}$N threshold, it is a system of three coupled channels: $\bar{K}$N, πΛ, πΣ.

In previous studies, the low–energy $\bar{K}$N data (below 300 MeV/c kaon lab momentum) can be described satisfactorily, partly because in this energy region the $\bar{K}$N cross section is not dominated by resonances, but decreases monotonically with increasing scattering energy. However, due to the poor quality of existing $\bar{K}$N data (measured cross sections have large error bars, and polarization data do not exist at all in this energy region), the parameters cannot be determined uniquely. These ambiguities have serious consequences for corresponding investigations in nuclei; i.e., they make a reliable interpretation of such data almost impossible. Also, precisely due to these uncertainties in the free $\bar{K}$N interaction, it is even today not completely clear whether the empirical Λ(1405) resonance is a quasibound $\bar{K}$N state or is essentially a genuine three–quark resonance.

Therefore, real progress in theoretical understanding requires a more unique determination of $\bar{K}$N interaction models. Of course more and better data would be extremely helpful in this respect. However, in the low–energy region conventional bubble chamber experiments have already reached their limit. There are, however, proposals for new conceptions which have so far not been realized.

Thus, in view of the current poor experimental situation, one has to look for another way to fix the $\bar{K}$N interaction uniquely. Indeed, instead of concentrating on the $\bar{K}$N system only, a promising and theoretically appealing alternative is to aim at a combined description of many different hadronic reactions, based on the same underlying picture and using the same calculational scheme as consistently as possible. This requirement rules out phenomenological models from the beginning, poses additional constraints on the model parameters, and leads ultimately to a unified description of various hadronic processes. For several years, this program has been pursued by the Jülich group. Starting from the Bonn meson–exchange NN interaction [2], an analogous and consistent scheme has been used to describe the hyperon–nucleon (ΛN, ΣN) [4], nucleon–antinucleon [8], and, finally, the KN interaction [5]. (A corresponding investigation of the πN system is under way.)

Now we turn our attention to the $\bar{K}$N interaction. Indeed, in the meson exchange picture this interaction is strongly constrained by the KN interaction derived before, since a considerable part of both interactions is related by G–parity transformation. With these especially strong constraints, it is an interesting question whether the resulting $\bar{K}$N interaction can still describe the existing low–energy data. Precisely due to these constraints, we will be able to address the important question about the nature of the Λ(1405).

We have treated the $\bar{K}$N interaction as a system of three coupled channels ($\bar{K}$N, πΛ, πΣ) and have, mainly to assure consistency with the KN model, taken into account the effect of further channels (K*N, K*Δ, KΔ) by adding the corresponding "box–diagrams". The various contributions to the quasi–potential V(z) are shown in Fig. 3.1.

	$\bar{K}N$	$\Lambda\pi$	$\Sigma\pi$
$\bar{K}N$	N $\bar{K}$; σ,ω,ρ; N $\bar{K}$	Λ π π Λ; K*, N; N $\bar{K}$ N $\bar{K}$	Σ π π Σ; K*, N; N $\bar{K}$ N $\bar{K}$
$\Lambda\pi$		Λ π; σ; Λ π	Σ π; ρ; Σ π
$\Sigma\pi$			Σ π; σ,ρ; Σ π

Fig. 3.1. The coupled–channel approach to $\bar{K}N$ scattering.

In passing we mention that one has to consider also "pole graphs". Fortunately, these contributions are far away from the $\bar{K}N$ threshold and do therefore not influence our conclusion about the Λ(1405). In Fig. 3.2 we compare our theoretical results with experimental values. The agreement of the theory especially with the new data is quite good.

Let us now turn to the $\bar{K}N$ threshold and the problem of the Λ(1405) resonance. The strange resonance Λ(1405) has isospin 0 and $J^P = \frac{1}{2}^-$ and is found in s–wave Σπ scattering. There is still debate in the literature whether this state has to be interpreted either as a genuine 3–quark resonance or as a quasibound $\bar{K}N$ state, which would correspond to a 4–quark–1–antiquark state in a constituent quark–model. Our model definitely favours the latter interpretation since the diagonal $\bar{K}N$ interaction is strongly attractive due to the combined effect of σ–, ω– and ρ–exchange, which all add coherently in the

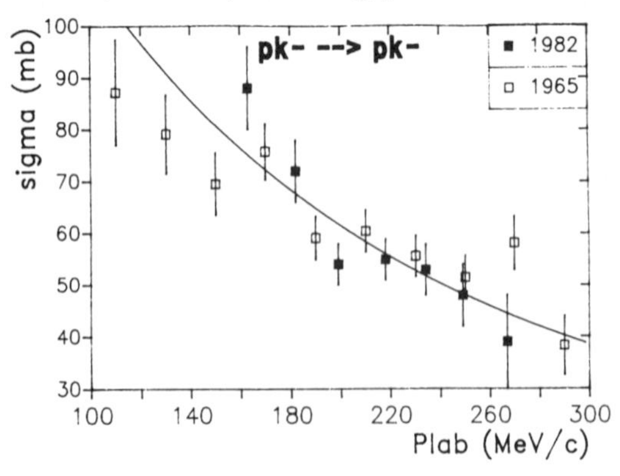

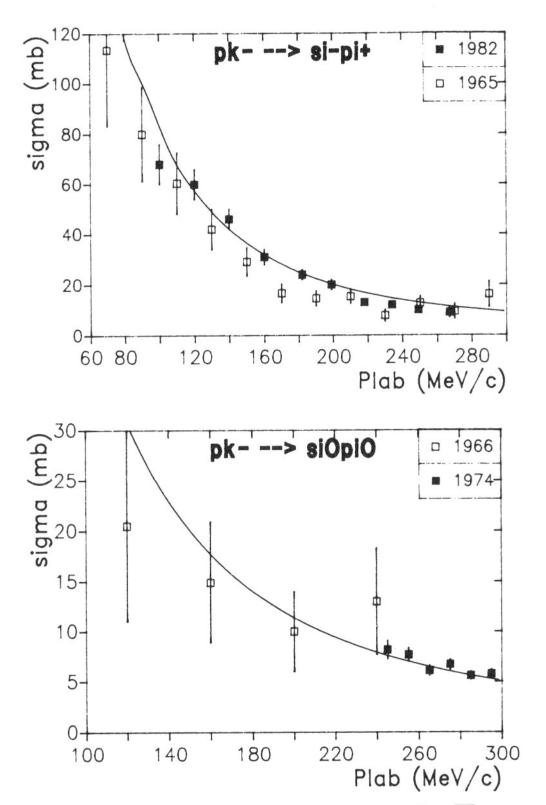

Fig. 3.2. Various total cross–sections for $\overline{\text{K}}$N–reactions.

relevant channel. Indeed, our model uniquely predicts a resonance in the diagonal $\pi\Sigma$ amplitude, even at the right place, i.e., without readjusting parameters. This is demonstrated in fig. 3.3, which shows the corresponding mass spectrum, obtained automatically in our coupled–channel calculation. If we omit the diagonal $\overline{\text{K}}$N interaction but keep the transition potentials ($\pi\Sigma\rightarrow\overline{\text{K}}$N) the resonance behaviour disappears almost completely, which shows that the resonance is indeed a quasibound $\overline{\text{K}}$N state.

We thus conclude that, in the framework of our model, no explicit three–quark resonance has to be included in order to describe the empirical resonance at 1405 MeV. On the contrary, a reasonable reproduction of the scattering data above the $\overline{\text{K}}$N threshold unavoidably implies a resonant behaviour of the S_{01} partial wave below threshold, at the right position. This point has also been investigated in ref. [9], where both a separable potential and a vector–meson exchange model have been used, leading in both cases to the quasi–bound $\overline{\text{K}}$N state.

To add to this already resonant S_{01} wave an additional pole graph at about 1.4 GeV is not reasonable since it would lead to two resonances just

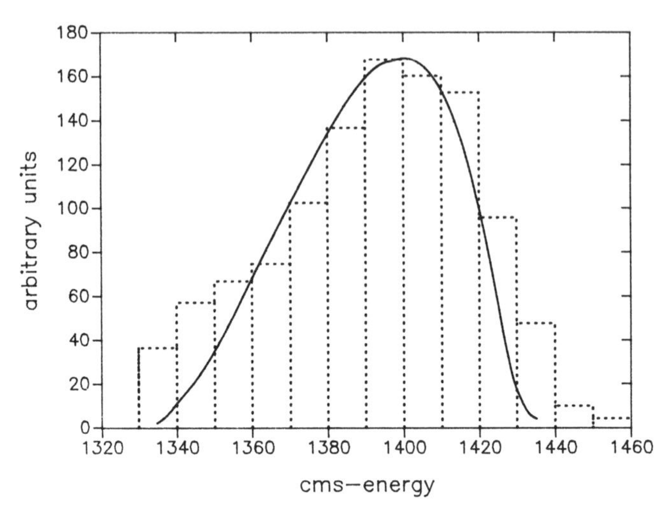

Fig. 3.3. The $\Lambda(1405)$ mass spectrum . We compare the prediction of model I in the diagonal $\pi\Sigma$ channel, i.e., the quantity $|T_{\Sigma\pi}|^2 q_{c.m.}$, with experimental data.

below the $\overline{K}N$ threshold, which is not supported by experiment. Thus the genuine $\frac{1}{2}^-$ quark–model state should appear at higher energies. Possible candidates are the $\Lambda(1670)$ and the rather broad and uncertain $\Lambda(1800)$. Indeed, such an interpretation is much more in line with quark–model calculations. For example, in calculations with the MIT bag, a $\frac{1}{2}^-$ state appears which lies about 100 MeV higher than the corresponding $\frac{3}{2}^-$ state connected with the empirical $\Lambda(1520)$.

SUMMARY AND CONCLUSIONS

In order to establish reliably the validity of the meson–exchange picture for low (and possibly medium) energy hadronic systems, corresponding interactions have to be described in a consistent meson theoretic scheme. Only in this way will one have the chance to isolate possible discrepancies and thus be able to trace them back to genuine quark–gluon effects.

According to the general program of the Jülich group, we have presented meson–exchange models for the $\overline{K}N$–interaction as well as for the $\pi\pi$– and $K\pi$–interaction. One of the basic questions of these models concerns the interaction at short distances, where we expect deviations from the meson–exchange picture due to the underlying quark structure.

In this contribution, we concentrated on effects which appear in the vicinity of thresholds. Such effects can be investigated with the new cooler synchrotrons. Therefore we may hope that these interesting problems can be solved within the next few years.

I thank K.Holinde, J. Durso, and V. Brown for valuable discussions.

REFERENCES

1) E. Witten, Nucl. Phys. B160 (1979) 57.
2) R. Machleidt, K. Holinde and C. Elster, Physics Reports 149 (1987) 1.
3) R. Büttgen, K. Holinde, B. Holzenkamp and J. Speth, Nucl. Phys. A450 (1986) 403.
4) B. Holzenkamp, K. Holinde and J. Speth, Nucl. Phys. A500 (1989) 485.
5) R. Büttgen, K. Holinde and J. Speth, Phys. Lett. B163 (1985) 305.
R. Büttgen, K. Holinde, A. Müller–Groeling, J. Speth and P. Wyborny, Nucl. Phys. A506 (1990) 586.
6) D. Lohse, J.W. Durso, K. Holinde and J. Speth, Phys. Lett. B234 (1990) 235; Nucl. Phys. A516 (1990) 513.
7) D. Aston et al., Nucl. Phys. B296 (1988) 493.
8) Th. Hippchen, K. Holinde and W. Plessas, Phys. Rev. C39 (1989) 761.
9) P.B. Siegel and W. Weise, Phys. Rev. C38 (1988) 2221.

THE WIDTH OF BOUND ETA IN NUCLEI

H.C. Chiang*, E. Oset** and L.C. Liu***

*Institute of High Energy Physics, Academia Sinica, Beijing, China
**Departamento de Física Teórica and IFIC, Universidad de Valencia, Spain
***T-2 Los Alamos National Lab., Los Alamos New Mexico 87545, U.S.A.

We have used a model for the $\eta N \rightarrow \eta N$ interaction based on the dominance of the N* (1535) pole [1] and from it we have derived the η nucleus optical potential and investigated the bound states of η in nuclei and their decay width. The work follows closely the delta-hole model used to study pion nucleus interaction around resonance [2].

The key ingredient in the approach is the N* selfenergy in the medium which we study in detail. The N* decay channels into $N\pi$, $N\eta$ and $N\pi\pi$ are properly evaluated and the medium modifications due to the interaction of the π and η with the nucleus are accordin ly considered. As a consequence one is including systematically the one body and two body mechanisms for η absorption in nuclei. The real part of the N* selfenergy can not be evaluated reliably and is left as a free parameter. The results for the binding energies and widths depend upon the assumed value of this N* real potential. If one assumes $V_{N^*} = 50\ \rho/\rho_o$ [MeV], repulsive. Together with $V_N = -50\ \rho/\rho_o$ [MeV] for the nucleons, the N* pole contribution is drastically reduced and one obtains η nuclear states lossely bound and with narrow widths. However in such case, since the N* pole contribution is so much decreased, other terms in the ηN interaction would be more important and the results are not very reliable.

Assuming $V_{N^*} = V_N$ one obtains more binding of the η and also larger widths. In table I we show the results obtained for several nuclei.

TABLE I

Nucleus	n, l,	B[MeV]	Γ[MeV]
^{12}C	1 0	8.1	18.9
^{40}C	1 0	20.9	27.1
	1 1	4.5	19.2
^{208}Pb	1 0	31.0	29.7
	1 1	23.8	28.0
	2 0	12.1	24.5
	2 1	1.4	19.8

As we can see, the widths are considerably larger than the separation between the levels. In this case one does not obtain any narrow structure for the η nucleon states. If one assumes $V_{N^*} = 0$ both the binding and the widths decrease but the widths are still larger than the separation between levels. Only if are assumes $V_{N^*} \simeq 50\ \rho/\rho_0$ [MeV] one obtains some states with narrow widths. However, as indicated before, this latter case is the weakest one theoretically.

In summary we tentatively conclude that it is unlikely that only narrow structures corresponding to bound eta states in nuclei will be observed experimentally. Conversely, the observation of any such narrow states would put severe constraints on the potential of a N* inside the nucleus.

REFERENCES

[1] R.S. Bhalerao and L.C.Liu, Phys. Rev. Lett. 54(1985)865.
[2] E.Oset, H.Toki and W.Weise, Phys. Reports 83(1982)281.

PRODUCTION OF DEEPLY BOUND PIONIC STATES IN HEAVY NUCLEI

J. Nieves and E. Oset

Departamento de Física Teòrica and IFIC, Universidad de Valencia and CSIC. E-46100 Burjassot (Valencia) Spain.

We have investigated several reactions in order to produce deeply bound pionic states so far unobserved. These are states like the 1s or 2p in the Pb region for which standard potentials provide widths which are smaller than the separation between the levels. This is particularly true of a new potential which solves the anomalies in the pionic atoms[1].

We have investigated the (n, p), (γ, π^+), (π^-, π^+), (e, e') reactions, all of them producing an extra π^- bound in the nucleus. The signal for the production of the pionic atom would appear as a peak in the cross section $d\sigma/d\Omega\, dE$ over the background from the corresponding inclusive processes. We have calculated the cross sections for the excitation of the atomic states as well as the inclusive background. For the (n, p) reaction a model is taken for the elementary process which reproduces the $nN \rightarrow pN\pi^-$ reaction. This differs from the model of Toki and Yamazaki[2] where the neutron is assumed to produce a π^- and a proton and the π^- (off shell) is trapped by the nuclear plus Coulomb field of the nucleus. When proper distortion of the p, n wave functions is taken into account we obtain values for the cross sections at the peaks of the pionic atoms which are about 5% of the background for the 2p state of ^{208}Pb and much smaller for the 1s states. These numbers are much lower than the predictions of ref.[2] even corrected by the distortion[3]. The cross sections at the peak which we obtain for the 2p state at energies $T_n \simeq 1$ GeV are of the order of 0.02 mb/sr/MeV over a background of about 0.5 mb/sr/MeV. This would explain why the experiment at TRIUMF has failed to see any clear signal[4]. We propose that two neutrons back to back, sharing the energy of the pion are detected in coincidence, since they would be an indication of the absorption of the pion. However the contribution of the background to this window would be extremely small.

The (γ, π^+) reaction offers the best ratio of signal to background at $E_\gamma = 450$ MeV. The cross sections are however small, of the order of 0.024 μb/sr/MeV for the 1s state and 0.071 μb/sr/MeV for the 2p state of ^{208}Pb. They represent 7% and 20% respectively of the background from inclusive (γ, π^+).

The other two reactions (π^-, π^+) and (e, e') have the problem that, because there is a change of charge in the nucleus they only allow the valence nucleons to particpate and one looses the coherence factor A^2 of the other two reactions. One obtains cross sections of the order of 10^{-3} times the background, which makes them unsuited for the experimental search.

REFERENCES

1)J. Nieves, E. Oset and C. García-Recio, submitted to Phys. Rev.
2) H. Toki and T. Yamazaki, Phys. Lett. B213 (1988) 129.
3) T. Yamazaki, invited talk et Pasadena 1989, INS preprint, Tokyo.
4) M. Iwasaki et al. to be published.

An ηNN QUASI-BOUND STATE

T. Ueda
Faculty of Engineering Science , Osaka University, Toyonaka, Osaka, Japan, 560

ABSTRACT

The question whether an ηNN quasi-bound state exists is theoretically investigated. One solves the three-body equation for the πNN and ηNN coupled systems. The coupling occurs through the S_{11} resonance which connects with both the πN and ηN channels. One finds a pole structure near ηNN threshold ($\sqrt{s} = 2430$ MeV) in the $I = 0, J^P = 1^- \ \eta d$ scattering amplitude.

INTRODUCTION

So far the possibility of the existence of πNN bound states is argued by several authors. However since the most important interaction for binding the system is the $\pi N - P_{33}$ resonance one, this P-wave nature prevents the system from being bound by the large centrifigal repulsion. Recent experimental investigation gives negative results to the existence.

In turn the ηNN system has different property from the πNN system. In this system the most important attractive interactions are the $\eta N - S_{11}$ and $NN - {}^3S_1$. Namely both interactions are of the S-wave nature giving no centrifugal repulsion and provide much more possibility for the bound ηNN state. Since the S_{11} resonance couples both to the πN and ηN channels, the ηNN system couples neccessarily to the πNN system. The ηNN bound state should decay strongly into the πNN system.

The method treating the coupled plural three-body system has been developed by the present author and others. The method has been applied to $\pi NN - \rho NN$[1], $NNN - NN\Delta - \pi dN$[2] and other systems in some approximation. In the present $\eta NN - \pi NN$ case we solve the equation exactly within the limitation of the input interactions being several important ones.

FORMULATION and CALCULATION

The three-body equation is written in the operator form as follows.

$$X = Z + Z\tau X, \tag{1}$$

where Z is the particle rearrangement term and τ the propagation term of the system with a spectator particle and an interacting pair. Eq.(1) is the integral equation for the amplitude X in the momentum space. The Z and τ have the standard form of the three-body theory[2]. They are expressed by the form-factor $g_\alpha(p)$ of the input separable potential:

$$V_{\alpha\beta}(p, p') = \lambda g_\alpha(p) g_\beta(p'), \tag{2}$$

where p and p' are the final and initial relative momenta respectively. The suffices α, β and γ specify both the particle and angular-momentum channels.

In Table 1 the particle channels and the locations of the non-vanishing matrix elements for Z and τ are indicated.

The input two-body interactions involve the $\pi N - P_{11}, \pi N - P_{33}, NN - {}^3S_1$ and $\pi N - \eta N$ S_{11} potentials. The first three potentials take the one channel form ($\alpha = \beta$) in eq.(2), while the last one does the two channel one for the πN and ηN channels. All the potentials are made by fitting to the phase shift analysis results.

One takes into account the angular-momentum channels in Table 2 and calculates the ηd and πd scattering amplitudes with $I = 0, J^P = 1^-$ and $I = 1, J^P = 1^-$, respectively. One notes that the P_{33} two-body potential does not work in the former case. Furthermore the $NN - {}^1S_0$ potential does not work in either case.

Table 1 The particle-channel and the locations of nonvanishing Z and τ matrix elements. The interacting pairs in the particle channel are parenthesized.

Channel	$(N_2\pi)N_1$	$(N_1\pi)N_2$	$(N_1N_2)\pi$	$(N_2\eta)N_1$	$(N_1\eta)N_2$	$(N_1N_2)\eta$
$(N_2\pi)N_1$	τ	Z	Z	τ		
$(N_1\pi)N_2$	Z	τ	Z		τ	
$(N_1N_2)\pi$	Z	Z	τ			
$(N_2\eta)N_1$	τ			τ	Z	Z
$(N_3\eta)N_2$		τ		Z	τ	Z
$(N_1N_2)\eta$				Z	Z	τ

Table 2 The angular-momentum channels for (A) the $I = 0, J^P = 1^-$ and (B) the $I = 1, J^P = 1^-$ states. The first row indicates the quantum numbers of the interacting pair. The "$S3$" and "$L3$" represent the total spin and the orbital angular-momentum, respectively, possessed by the interacting pair and the spectator.

(A)	P_{11}	S_{11}	S_{11}	${}^3S_1 - {}^3D_1$		(B)	P_{11}	S_{11}	S_{11}	P_{33}	${}^3S_1 - {}^3D_1$	
$S3$	0	1	1	1	1	$S3$	0	1	1	2	1	1
$L3$	1	0	2	0	2	$L3$	1	0	2	1	0	2

RESULT, DISCUSSION AND CONCLUSION

The ηd scattering amplitude of the $I = 0, J^P = 1^-$ state has very surprizing feature around the ηNN threshold energy region as is shown in figs.1 (A) and 2. The phase shift starts from 180° at the ηd threshold: $E_\eta = -2.223$ MeV, where E_η represents the total energy subtracted by the ηNN threshold energy 2430 MeV. This suggests the existence of a bound state. If the πNN channel would not exist, the Levinson theorem assures this statement. To make sure, one investigates whether any pole exists near the threshold in the complex energy plain. One finds a pole at $E_\eta = 1.27 - i0.90$ MeV in the first Riemann sheet of the complex energy plain. One may interpret the pole as follows. If the πNN channel is gradually switched off, the pole moves in the negative direction along the real energy axis with some small and negative imaginary part and locates

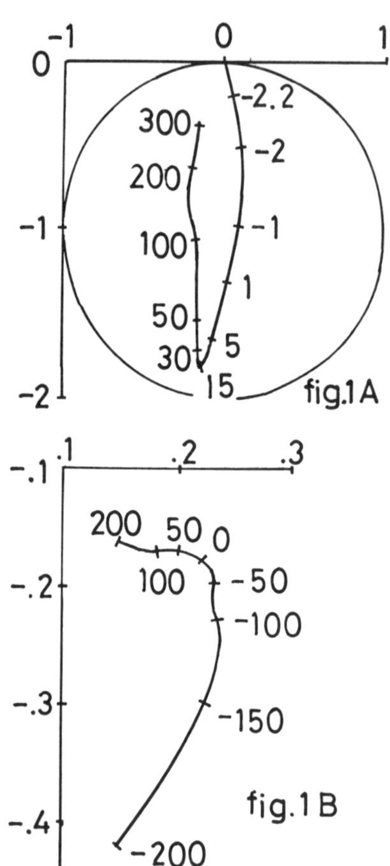

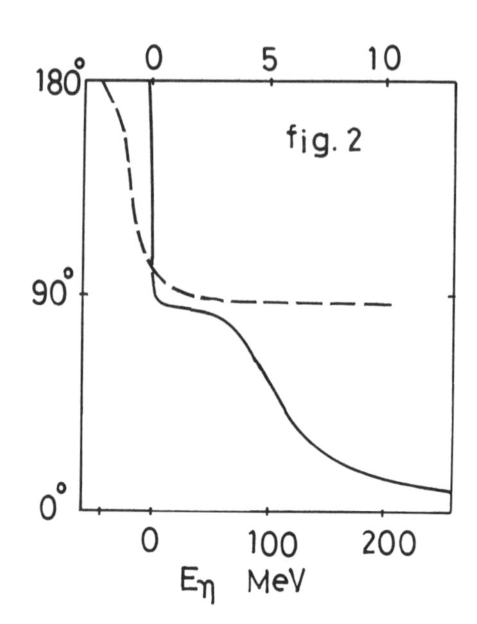

Fig.1 The Argand plots of (A) the ηd and (B) the πd scattering amplitudes. The associated numbers indicate E_η in units of MeV.

Fig.2 The phase shift of ηd scattering vs. E_η. The solid and the broken curves should be read by the down and upper abscissa respectively.

finally on the axis with no πNN channel as a stable bound state. One remarks here that the most important origin for the attraction of the ηNN system arises from the mechanism of the nucleon rearrangement with the $\eta N - S_{11}$ and $NN - {}^3S_1$ interactions in the initial and final two-body channels.

On the other hand one calculates the πd scattering amplitude for the $I = 1, J^P = 1^-$ state. As is shown in fig.1(B) the η threshold effect is less remarkable on it than the ηd amplitude case. However one observes apparantly anormalous behavior of the amplitude in this case too.

In conclusion one finds the $I = 0, J^P = 1^-$ quasi-bound state near the ηd threshold ($\sqrt{s}$=2430 MeV) in the calculation of the coupled $\eta NN - \pi NN$ system. The bound state can be investigated experimentally through the reactions: $\gamma d \to \eta d(E_\gamma = 633$ MeV), $np \to \eta d(E_p = 1260$ MeV), $\pi d \to (\eta d)\pi(E_\pi = 590$ MeV), etc. where the energies parenthesized indicate the incident kinetic energy in the laboratory system. Let me emphasize that it is worthwhile to explore this quasi-bound state experimentally.

REFERRENCES

1 T. Ueda, Phys. Lett. **B141** 157 (1984).
2 T. Ueda, Nucl. Phys. **A505** 610 (1989).

NARROW BOUND STATES IN LIGHT AND MEDIUM Σ HYPERNUCLEI

P. Fernández de Córdoba, E. Oset
Departamento de Física Teórica and IFIC, Universidad de Valencia–CSIC
46100 Burjassot (Valencia) Spain.
and
L.L. Salcedo
Departamento de Física Moderna, Facultad de Ciencias, Universidad de Granada,
18001 Granada, Spain.

We have evaluated the Σ nucleus optical potential with a minimum of theoretical input, relying particularly on phenomenology. With this potential we have looked for the bound states in several nuclei, with the result that in light and medium nuclei several states have widths smaller than the separation between neighbouring states.

The starting point is a fit to the Σ^- atom data with a potential of the type

$$\begin{aligned} V_{opt}(\vec{r}) &= -U(\vec{r}) - iW(\vec{r}) \\ U(\vec{r}) &= U\,\rho(\vec{r})/\rho_o \\ W(\vec{r}) &= W\rho(\vec{r})/\rho_o \end{aligned} \tag{1}$$

we have taken proton densities from experiment and corrected them for the proton finite size. The results of our best fit give $U=(31\pm4)$ MeV, $W=(15\pm2)$ MeV, which agree with those of ref.[1], $U=28\pm3$ MeV, $W=15\pm2$ MeV. The number for W agrees remarkably well with the results of the low density theorem which states

$$W(\vec{r}) = \frac{1}{2} < \sigma(\Sigma^- p \to \Lambda n)\, v_{rel} > \rho_p(\vec{r}) = \frac{1}{4} < \sigma\, v_{rel} > \rho(\vec{r}) \tag{2}$$

Since we have $<\sigma v_{rel}> = 17.3$ mb from experiment, in the kinematical regime where we more, eq. (2) provides $W=14.5$ MeV, versus the value 15 ± 2 MeV from the fit to the atoms.

One could take the potential (1) and evaluate the energies and widths of the bound states. However this would ignore the important density dependence in $W(\vec{r})$ as ρ increases. Indeed one knows from the analysis of the nucleon–nucleus optical potential at the same energies as in the present problem that, as $\rho \to \rho_o$, $W(\vec{r})$ does not grow dinearly with ρ but presents some saturation[2]. We have done the same study here and found that $W(\vec{r})$ for the hypernuclear case also saturates as a function of ρ. One reason for it is the Pauli blocking, but the most important is the effect due to the polarization of the nucleus by the spin–isospin interaction responsible for the

$\Sigma N \rightarrow \Lambda N$ transition[3]. However the real part remains remarkably close to the linear function in ρ of eq. (1). The results of our calculation give a function $W(\vec{r})$

$$W(\vec{r}) = 15\frac{1}{5.2}\ \mathrm{arctg}\ (5.2\ \rho(\vec{r})/\rho_o)[\mathrm{MeV}] \qquad (3)$$

which at low densities agrees with the results from Σ atoms (see eq. (1)), and at high results saturates value densities, $\rho \simeq \rho_o$, saturates to a constant value. With this potential and U = 31 MeV, like in the case of Σ atoms, we have carried out calculations for Σ^+, Σ^- and Σ^o bound states. We observe that, while with the linear potential in ρ of the Σ atoms the widths of the states are of the order of 20−30 MeV for most of the states, exceeding the separation between the levels, the widths with the saturating potential are around 7 MeV or smaller and much narrower than the separation between the levels. We see in table 1 some results for different hypernuclei with the saturating potential

	Σ^- (units in MeV)				Σ^o				Σ^+			
	2p		1s		2p		1s		2p		1s	
	B	Γ	B	Γ	B	Γ	B	Γ	B	Γ	B	Γ
^{12}C	1.7	4.1	14.9	6.9	___		11.2	6.8	___		7.5	6.6
^{16}O	5.0	5.4	17.0	7.0	1.2	5.0	12.5	6.9	___		8.0	6.8
^{24}Mg	10.8	6.5	22.1	7.4	5.3	6.3	15.9	7.3	___		9.7	7.2

Although some states, like the 2p in ^{12}C for Σ^-, would not be observable (the half width exceeds the binding), other states should be perfectly identifiyable. For some lighter nuclei the 1s state should also show up with narrow widths and larger binding than the half width. Obviously the nuclear excitations superposed to the Σ states could blur a clean identification in some cases.

We have also investigated the recently identified ${}^4_\Sigma He$ state[4]. For isospin reasons, W = 20 MeV for the Σ^+ pnn system, 15 MeV for the Σ^o npp system and 10 MeV for the Σ^-pnn system. The real part is also different in these cases but more difficult to control theoretically. We can find a ${}^4_\Sigma He$ bound state (Σ^+pnn or Σ^onpp) with a width of about 5 MeV and ~ 3 MeV binding, as shown by experiment, provided the real part is $U \simeq 45-50$ MeV. However with the linear potential in ρ the widths are ~ 16 MeV. On the other hand, for the Σ^-pnn system we find B < 100 KeV provided that U < 24 MeV, in which case there would be no visible peak in the experiment as is the case in ref.[4].

REFERENCES

1) C.J. Batty et al. Phys. Lett. 74B (1978) 27.
2) S. Fantoni, B.L. Friman and V.R. Pandharipande, Nucl. Phys. A399 (1983) 51.
3) E. Oset, L.L. Salcedo and R. Brockmann, Phys. Reports in print.
4) R.S. Hayano et al., Phys. Lett. B231 (1989) 355.

SESSION E

Future Experimental Developments

A 0° FACILITY IN THE COSY RING

W. Borgs, M. Büscher, D. Gotta, H.R. Koch, W. Oelert, H. Ohm,
O.W.B. Schult, H. Seyfarth, K. Sistemich
Institut für Kernphysik, Forschungszentrum Jülich,
D–5170 Jülich, F.R. Germany

J. Ernst, F. Hinterberger
Institut für Strahlen– und Kernphysik, Universität Bonn,
D–5300 Bonn, F.R. Germany

V. Koptev
Leningrad Institute of Nuclear Studies, USSR Academy of Sciences,
Gatchina, USSR

S.V. Dshemuchadze, V.I. Komarov, M.G. Sapozhnikov, B.Zh. Zalyhanov,
N.I. Zhuravlev
Joint Institute of Nuclear Research, Dubna, USSR

H. Müller
Zentralinstitut für Kernforschung, Rossendorf, Dresden,
German Democratic Republic

P. Birien
EP Division, CERN, CH–1211 Geneva, Switzerland

ABSTRACT

A 0° facility is being developed for the study of meson production in proton–nucleus collisions and other investigations at intermediate bombarding energies at an internal target position (TP2) of COSY. A set of three laminated magnets will provide a closed orbit bump for the proton beam and deflection as well as momentum analysis for ejectiles with $p_0/30 \doteq p \doteq p_0/2$ (p_0 = projectile momentum). This facility will enable the spectroscopy of rare ejectiles like K^+ mesons with a momentum resolution of $\doteq 1\%$. It also allows studies of ejectiles emitted under 180° like protons from deuteron breakup.

INTRODUCTION

The production of mesons through the bombardment of nuclei with protons at intermediate energies is a process which provides valuable insight into elementary reaction mechanisms. Thus, the production[1] of K^+ mesons at projectile energies below the nucleon–nucleon threshold shows the influence of high Fermi momenta or of cluster formation. For a detailed understanding of these processes it is indispensable to measure the momenta and angular distributions of the produced mesons and to compare these data with theoretical predictions. In the case of the sub–threshold K^+ production the momentum distribution at $\sim 0°$ is expected to shed light[1,2] on the reaction mechanism. To understand a reaction like the deuteron breakup through proton bombardment it is important to investigate in detail the spectrum of backward emitted protons.

The proton beam of COSY offers a unique possibility for this type of studies at the internal target position TP2[3]. The beam has a high momentum resolution and its energy can be varied easily. High luminosity can be achieved with thin targets. This enables the study of reactions with low cross sections and, nevertheless, avoids contaminant two–step processes in different target nuclei as well as the deterioration of the momentum resolution through energy loss of the projectiles and ejectiles inside the target. Magnetic separation is needed for the momentum analysis, e.g. of K^+ mesons which are produced orders of magnitude less frequently than the background of protons and pions. The flight path of the K–mesons from the target to the detectors must be short to avoid large decay losses. For this purpose a 0^0 facility is under development.

LAYOUT OF THE 0^0 FACILITY

The aim of the "0^0 facility" is to separate ejectiles which are emitted under forward or backward direction with respect to the projectile beam and to analyse their momenta. It will consist of three magnets, see Fig. 1. The central magnet (D2) acts as the separator and analyser for the ejectiles. It focusses the particles with identical momenta horizontally. The two smaller magnets (D1 and D3) provide a closed orbit bump of the projectile beam. The magnetic lengths of D1 and D3 are half of that of D2, so that the system is achromatic.

The magnets have to be ramped with the COSY ring. Thus, they have to be laminated. The ejectile momenta of interest for the different studies range from $\sim p_0/30$ to $p_0/2$ (p_0 = projectile momentum). The requirements on the 0^0 facility can be fulfilled most economically if D2 is identical to the magnets of the COSY ring. However, it is better to use an independent power supply so that the magnetic rigidity in the facility can be chosen optimally. In this way a high degree of flexibility concerning the ejectile momenta is achieved. An inversion of the field direction allows the analysis of negatively charged particles under 0^0 or of positively charged ejectiles under 180^0.

It will be possible to move D2 (horizontally) perpendicular to the beam direction, so that the proper closed beam bump for each experiment can be used. For baking out the vacuum chamber D2 can be removed. Thus full use can be made of the 9 cm gap of the COSY type magnet. The flight path of the produced particles is minimal: the losses of K^+ mesons due to their decay between the target and the detectors will be $\doteq$ 70%.

DETECTORS

Since the studies with the 0^0 facility should allow the determination of the momentum distributions even of rare ejectiles like K^+ mesons, the detectors have to enable particle identification and the measurement of the momenta in the presence of a large background. The detector setup is outlined in Fig. 2. At the focal plane the vacuum chamber has a window of about 30 μm stainless steel. Immediately behind this window a first multiwire chamber determines the position of the particles and – together with a second one – measures the angle of the particles. This allows to trace back the path to the target and to eliminate scattered particles.

Time–of–flight and ΔE–E measurements are needed to select the particles of interest from background. In K^+ studies, the different ranges of protons, kaons and pions with identical momenta can be exploited for particle identification. The protons can be absorbed in scintillators and other absorbing material. A Čerenkov counter behind these absorbers can be used to stop the K^+ mesons and identify the monoenergetic muons which is emitted in their decay. A veto counter behind this Čerenkov detector suppresses the pions. The fact that the μ^+ radiation is delayed with respect to the wire chamber pulses helps to select the K^+ mesons.

It is planned to use detectors which are 40 cm high and 100 cm long in order to optimize the accepted solid angle, resolution and costs. The total solid angle of D2 and the detector arrangement will amount to 6 msr. The expected average momentum resolution is $\Delta p/p \sim 5 \cdot 10^{-3}$. It will be mainly limited by the scattering of the particles in the steel window of the vacuum chamber, but also the width of the target and the position and angle resolution of the detector setup may contribute. For particles with momenta of about 400 MeV/c like the backward emitted protons in the deuteron breakup, a momentum resolution of $2 \cdot 10^{-3}$ should be possible.

EVENT RATES

Strip, fiber, jet, cluster–jet or frozen–pellet targets can be used. They will be moved into the beam or switched on after acceleration of the protons in COSY. Thicknesses of about 10^{14} atoms/cm^2 can be reached with cluster–jet targets[4]. Fiber and ribbon targets with thicknesses of $\sim 10^{18} - 10^{20}$ atoms/cm^2 have been tested[5,6] as internal targets. With gaseous or pellet targets, e.g. of H atoms, the cooled COSY beam can be used. Luminosities of more than $10^{31}/\text{cm}^2\cdot\text{s}$ are possible. For a 100 μg/cm^2 ribbon target of carbon a lumonosity of the order of $10^{34}/\text{cm}^2\cdot\text{s}$ is expected. Here the "hot" beam of COSY would be used. The width of the target is then smaller than the beam and it is optimized for good momentum resolution. The target will be placed between D1 and D2 or between D2 and D3 for studies under 0^0 and 180^0, respectively.

The overwhelming fraction of the particles reaching the detector will be scattered protons; pions will be about a factor of 10 less frequent. The count rate that the detector setup will be able to handle is estimated to amount to 10^6/s which corresponds to 1 event per revolution of the COSY beam. If the number of background particles and K^+ mesons is assumed to be proportional to σ_{total} (^{12}C + p) and σ_{total} (K^+) then 3 detected K^+ mesons/min are expected. These rates will provide sufficiently clean momentum spectra within less than 1 day. A similar measuring time is expected for the study of protons under 180^0 from the deuteron breakup.

It is planned to perform also studies on the pion and eta production in proton–nucleus collision, e.g. in order to check whether the intermediate Δ excitation occurs in the projectiles or in the target nucleus[7] and to test the role[8] of N^*, respectively. Because of the expected high intensities this type of reactions provide good testing conditions for the 0^0 facility and the detector arrangements.

REFERENCES

1. V.P. Koptev, S.M. Mikirtych'yants et al., JETP 67, 2177 (1988)
2. J. Zofka, Ceskeslovocka Akademie Vued, CSSR, private communication
3. Cooler Synchrotron COSY Jülich, User Guide, May 1990 (K. Kilian et al. eds.)
4. C. Ekström, Proc. Conf. Electronuclear Physics with Internal Targets, Standorf 1989 (World Scientific, Singapore 1990), p. 171 (G. Arnold ed.)

5. C. Ekström, CERN Accelerator School, Uppsala 1989, Report CERN 90–04, p. 184 (S. Turner ed.)
6. H.R. Koch, G. Riepe et al., Nucl. Instr. Meth. A271, 375 (1988)
7. V.V. Abaev, A.B. Gridnev et al., Akad. Nouk SSSR No. 569 (1980)
8. L.–C. Liu, J.T. Londergan et al., Phys. Rev. C40, 832 (1989)

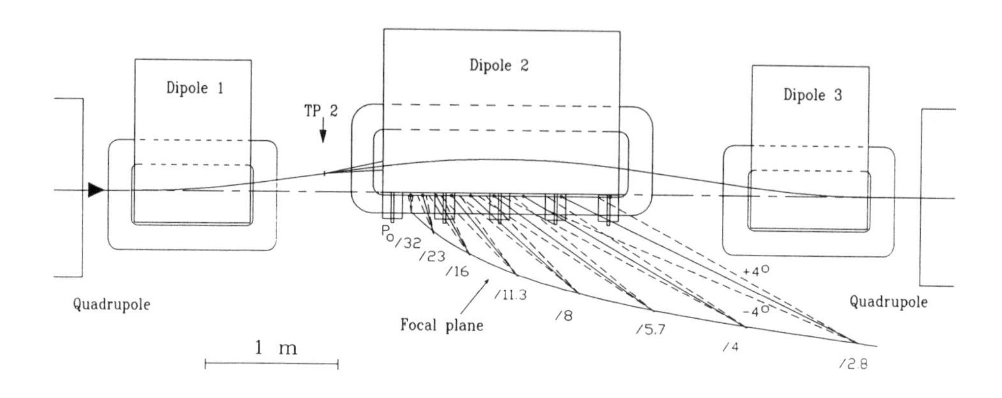

Fig. 1: Sketch of the proposed 0° facility between the quadrupoles in the straight section of COSY. The closed orbit bump corresponds to a deflection radius of 7 m.

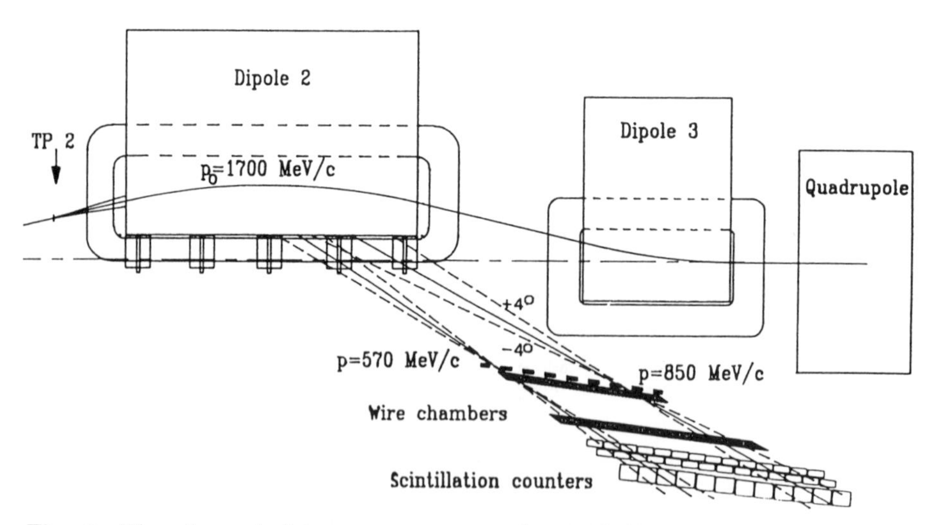

Fig. 2: The planned detector system at the 0° facility. The focal plane is indicated as a dashed line. The first and second wire chamber yield the momentum and angle information, respectively. The scintillation and Čerenkov counters serve for particle identification. Nota bene: a deflection radius of 4 m is assumed so that high momenta can be studied.

CHICANE/K300 SPECTROMETER AT THE IUCF COOLER

D.W. Miller and G.P.A. Berg
Indiana University Cyclotron Facility, Bloomington, IN 47405

ABSTRACT

A new three-magnet Chicane combined with an existing K300 Spectrometer are planned for installation in the S-straight section of the IUCF Cooler ring. Some classes of physics experiments accessible to this system and a description of the experimental layout are given.

INTRODUCTION

In order to detect with good resolution lower-rigidity reaction products and recoil nuclei produced in nuclear reactions in the IUCF Cooler Ring, and to allow for the tagging of neutrons for the study of secondary reactions, a new spectrometer system is planned. It will consist of a three-magnet Chicane combined with a modified existing K300 Spectrometer inserted in the S-straight section of the Cooler ring. The plan was developed by an IUCF user working group[1], and approved by the IUCF Long Range Planning Committee. Design work and planning for experiments are continuing.

The Chicane/K300 system is designed for a maximum reaction particle rigidity of 1.9 Tm at 0° lab, or slightly over one-half of the maximum Cooler beam rigidity of 3.6 Tm. Examples of possible physics experiments accessible to this proposed system include good-resolution charge-exchange studies in the (n,p) ["β^+"] direction, elastic scattering of polarized neutrons, spin correlation studies, symmetry breaking tests, and γ, π, $\pi\pi$ and pπ production processes. These are discussed briefly below.

Charge-exchange reactions in the (p,n) ["β^-"] direction up to 200 MeV have been studied very successfully at IUCF since the mid-1970s. The Cooler ring, incorporating synchrotron acceleration, extends the proton energy available for these (p,n) studies up to 500 MeV. However, with the addition of the planned Chicane/K300 spectrometer system, two complementary reactions in the β^+ direction become feasible up to 145 MeV/nucleon (determined by the maximum deuteron energy in the ring). Both the (n,p) reaction [using tagged neutrons], and the (d,2p) reaction [detecting the two coincident protons in the $^1S_0$ state ("^{2}He")], should be possible with some five times better resolution than obtained in previous charge-exchange studies. These two reactions in the β^+ direction single out transitions from isospin T $\rightarrow$ T+1 states, and spin changes of $\Delta S = 0,1$ for (n,p) and $\Delta S = 1$ for (d,^{2}He). This selectivity, plus the good resolution available, should allow a variety of measurements of importance to calculations regarding double-beta decay and astrophysics, and to studies of ground-state correlations, spectroscopy of high spin and high isospin states, isovector giant resonances, etc. Measurement of tensor analyzing powers in the ($\vec{d}$,^{2}He) reaction should also provide unique signatures of the spins of the final states in these investigations.

An additional application of the tagged polarized neutrons which will be available from this Chicane/K300 system is the study of neutron elastic scattering. The data available on heavier elements at medium energies are rather sparse, due to the difficulty in separating inelastic groups from elastic scattering. However, there has been recent renewed interest in these data, because new Dirac and density-dependent calculations predict large differences in σ and A_y between proton and neutron elastic scattering. The aim will be to obtain good elastic neutron scattering data for comparison with these calculations, and for use in (p,n) analyses.

Several possible investigations are being considered using hydrogen isotopes both for the Cooler beam and for the internal targets, with the Chicane/K300 system for detection of He reaction products down to 0°. Spin correlation studies of $\vec{p} + \vec{d} \rightarrow {}^3\text{He} + \gamma$ should be strongly sensitive to various pieces of the physics in the reaction as well as to the ^{3}He wave function. Symmetry breaking tests are also proposed very near threshold in which the reaction He product is detected with good resolution in a small forward cone, corresponding to 4π in the center of mass frame. The p + d $\rightarrow$ ^{3}He + $(\pi^+\pi^-)_{1S}$ reaction is of interest as a possible method of seeking information on chiral-symmetry breaking parameters from the $(\pi^+\pi^-)$ atom. A search for charge-symmetry breaking attributable to π^0 - η^0 mixing is also being considered using the d + d $\rightarrow$ ^{4}He + π^0 reaction near threshold.

The Chicane/K300 system will also be well adapted to measurements of recoil nuclei from various reactions, similar to the experiments to be carried out in the near future using the 6° magnet in the T-section of the Cooler ring. In these experiments, simultaneous observations of the recoils in the same apparatus from competing processes, e.g. $(p,\pi^\pm)$ and (p,π^0), can be carried out. This results in a large solid angle compression from the center-of-mass to the lab system. Another plan is to study recoils from the (p,γ) reaction, including analyzing-power measurements, which provide equivalent information regarding the (γ,p) reaction, but at a much higher rate.

Multiparticle reactions will also be convenient with the Chicane/K300 system. For example, studies like $(p,p\pi^+)$ would detect the high rigidity particle in the K300, while the lower rigidity positive particle(s) could be detected with separate detectors placed inside the ring next to the large second magnet of the chicane.

EXPERIMENTAL ARRANGEMENT AND SPECIFICATIONS

Most of the physics experiments discussed above rely on detection of the outgoing particles of interest down to 0° relative to the Cooler beam. Because the strength of the 6° magnet in the IUCF Cooler T section is not adequate to provide enough separation between the beam and outgoing protons to accommodate a 0° spectrometer, the proposed Chicane (see Fig. 1) is to be installed in the Cooler S region. After completion of a second Siberian Snake experiment (CE15) in the Cooler S region, this straight Cooler section will become available for the installation of the Chicane magnets. In the S region two magnets CM-1 and CM-3 are needed to bend the beam out of the straight section and back into the ring. The Chicane magnet CM-2 serves as the particle separation magnet. All three Chicane magnets have laminated yokes and pole pieces because they will be ramped with the beam energy in the same way as all other ring magnets.

The Chicane arrangement shown in Fig. 1 is for the maximum beam bending angle of 24° in CM-2. This provides sufficient separation of beam deuterons up to 290 MeV and ^{2}He protons of slightly over half the beam momentum to allow magnetic analysis of the protons in the subsequent spectrometer system with large solid angle and medium resolution. Calculations show that a Chicane bending angle up to 24° is compatible with Cooler operation if certain provisions are made. These include vertically focusing entrance and exit edges of CM-1 and CM-3 with an angle of 12° (to keep the transition energy constant) and adjustments of a series of quadrupoles in the main ring.

Momentum analysis of the reaction products, produced in an internal ring target ahead of CM-2, is accomplished in the magnet system CM-2, the quadrupole Q (with large horizontal and vertical opening) and the dipole D. The existing K300 dipole at IUCF will be used as dipole D. Because of the separation magnet CM-2, the first element of the spectrometer system is a dispersive element. This is an unusual arrangement which requires a particularly large horizontal quadrupole opening. For the same reason the horizontal acceptance angle is reduced as compared to the original K300 system. Low momentum recoils from reactions are accessible at backward angles on the left side of CM-2 because of its semi-circular field boundary. Fig. 1 shows, e.g., the path of recoil particles with magnetic rigidities $B\rho$ of 0.2 Tm.

Characteristic properties of the preliminary layout of the Cooler spectrometer are shown in Table 1. The solid angle is somewhat smaller than desired for the study of the (d,^{2}He) reaction, since this reaction has a characteristic low diproton detection efficiency, but it is a reasonable compromise. The alternative would be the construction of a much more expensive system with larger vertical gap, requiring a new dipole instead of utilizing the existing K300 dipole.

An important feature of the proposed Chicane is the flexibility in mechanical motion. The beam bending angle of the Chicane can be varied. This is accomplished by a translation of the system CM-2/Q/D perpendicular to the unperturbed beam direction in the straight section. The quadrupole-dipole can be rotated around Center 1 and Center 2 shown in Fig. 1. This allows analysis of particles with different scattering angles and magnetic rigidities. The translational motion allows analysis of particles with lower momentum at more forward spectrometer angles. This reduces the range of the necessary spectrometer rotation around Center 1. The Cooler vacuum with typical pressures of 10^{-9} Torr and the spectrometer vacuum in the range of 10^{-6} Torr have to be separated, e.g. by a thin foil. It should be noted that the target is indicated in the above figures by its position only. Space requirements vary widely with the type of target, e.g. gas jet or carbon fiber.

In summary, the planned Chicane/K300 spectrometer system should greatly expand the physics opportunities on the IUCF Cooler ring from present experiments (which emphasize high-resolution variable beam energy in the ingoing channel and few nucleon systems, using modest resolution detectors), to experiments utilizing good resolution detection of outgoing reaction particles from simple or complex systems. This single basic setup should be able to cover a wide range of interesting physics investigations, from application of good resolution techniques to studies like charge exchange, to much more exotic investigations like symmetry-breaking effects. It should also be well suited to a user facility like IUCF, providing opportunities for a variety of one-PhD-thesis size experiments.

REFERENCES

1. G.P.A. Berg, D.W. Miller, B.D. Anderson, J.W. Watson, D. Frekers, C.A. Gagliardi, K.H. Hicks, E.R. Sugarbaker, S.W. Wissink, (with consultants R.D. Bent, L.C. Bland, D. DuPlantis, C.D. Goodman, R.E. Pollock, P. Schwandt, and K. Solberg), Report of the IUCF Cooler Magnetic Spectrometer Working Group, 2 December 1989, Indiana University Cyclotron Facility Internal Report 487 (unpublished).

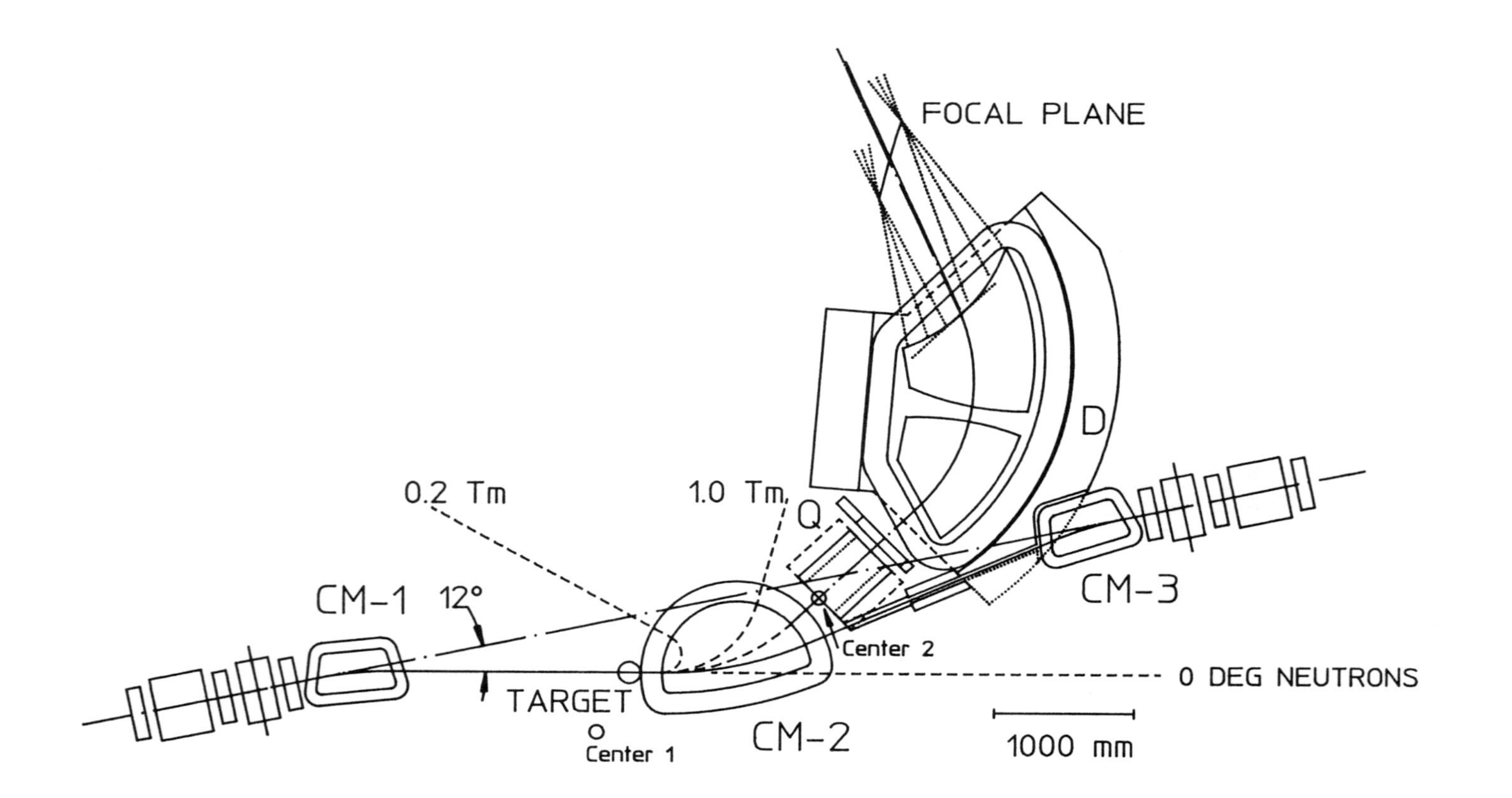

Fig. 1 Magnet spectrometer in Cooler S region. The spectrometer is shown in its most forward position where it can analyze 1.9 Tm protons at 0 degree.

Table 1
Characteristic Data of COOLER, CHICANE and K300 DQD Spectrometer

Cooler:	Projectile	Energy (MeV)
	p, $\vec{p}$	30 - 480
	d, $\vec{d}$	20 - 290
	Emittance,cooled $\epsilon_x = \epsilon_y \sim 0.05\pi$ mm · mrad	
	Momentum resolution p/Δp(fwhm) ~ 50,000 (45 MeV p)	
Target:		
	Spot size, cooled	$2x_o = 1$ mm
		$2y_o = 1$ mm
	Dispersion:	$B_{16} = 4.56$ cm/%
Chicane system	**CM1-CM2-CM3**	
CM1, CM3:	Deflection angle	0° – 12°
	Max. field	$B_{max} = 1.44$ T
	Max. beam rigidity	$(B\rho)_{max} = 3.6$ Tm
CM2:	Beam:	
	Deflection angle	0° – 24°
	Max. field	$B_{max} = 1.44$ T
	Max. beam rigidity	$(B\rho)_{max} = 3.6$ Tm
	Reaction products:	
	Angular range	0° – 40°
	Magn. rigidity range	$B\rho = 0.2$ Tm – 1.9 Tm
Spectrometer:		
	Angular range	0° – 25°
	Mean radius	$\rho_o = 1.3$ m
	Max. particle rigidity	$(B\rho)_{max} = 1.9$ Tm(0°), 2.55 Tm(> 12°)
	Max. proton energy	160 MeV(0°), 270 MeV(> 12°)
	Solid angle	$d\Omega \leq 7.5$ msr, (elliptical)
	Acceptance angle	$d\Theta = \pm 60$ mrad, $d\Phi = \pm 40$ mrad
	Length of focal plane	60 cm
	Momentum range p_{max}/p_{min}	1.20
	Tilt angle of focal plane	39° - 42°
	Horizontal magnification	0.30
	Vertical magnification	~ 33 (max. in center of focal plane)
	Momentum dispersion	$S_{16} = 3.1$ cm/%
	Ratio	$D/M_x = 7300$ mm
	Resolving power	$p/\delta p = 7300$ (for 1 mm object)
	Flight length, central ray	534 cm

NEAR THRESHOLD TWO MESON PRODUCTION IN HADRONIC FUSION REACTIONS*

Rainer Jahn
Insitut für Stahlen- und Kernphysik
University of Bonn, FRG

Abstract

An approved and funded exclusive COSY experiment is presented, which focusses on near threshold two meson production via the reactions $p+d \rightarrow^3 He+\pi^+\pi^-$ and $p+d \rightarrow^3 He+K^+K^-$. It takes advantage of the high quality of the cooled external COSY beam and the existing spectrometer BIG KARL. The setup consists of a vertex wall and a scintillator cylinder and endcap covering a 4π solid angle. The large efficiency and high resolution of this detection method will yield precision data on the low energy (T < 50 MeV) meson-meson interaction and probe into questions like the ABC-effect and $K\overline{K}$ molecule. The detector further allows a measurement of possible radiative ϕ (1020) decay, which will direcly probe the strange quark content of the $f_0(975)$. Existing inclusive data as well as first results of a very recent 'semi-exclusive' experiment performed at SATURNE will also be presented.

Introduction

In the quark model, mesons are the most elementary constituents of hadronic matter since they simply consist of quark-antiquark pairs. Experimental information about the interaction of mesons among each other is thus of fundamental interest. There is, however, a lack of high precision data on low energy meson- meson interactions and on near threshold production. This region is of special interest for the following reasons :

→ The low energy pion-pion interaction seems to have strongly attractive but nonresonant components. Enhancements in the production of low invariant mass pion pairs have been reported in a variety of experiments including the classic "ABC" effect[1], which has so far only been measured inclusively.

$\rightarrow$ A search can be made for a $K\overline{K}$ bound state via a measurement of the K^+K^- angular distribution from threshold across the $\phi(1020)$ peak. By using the ϕ as a reference phase, one should be able to determine whether the s-wave $K\overline{K}$ phase near threshold is rising (as expected for the high mass tail of a Breit-Wigner resonance) or falling (as appropriate to a channel with a bound state just below threshold).

$\rightarrow$ In an analysis using a K-matrix coupled channel formalism including the $\pi\overline{\pi}$ and $K\overline{K}$ channels Au et al.[2] claim there are three resonance poles in the 1 GeV mass region. They conclude that the $f_0(975)$ (former S_0) consists of two close and narrow states $S_1(991)$ and $S_2(988)$. They suggest that the $S_2(988)$ is a natural candidate for the regular $s\bar{s}$ ground state. Direct confirmation that the $S_2(988)$ is built of strange constituents would be the observation of the radiative decay $\phi \rightarrow \gamma S_2$. This process should provide γ-rays of some 30 MeV with a spread of only a few MeV from the ϕ width. The authors further speculate that the $S_1(991)$ with a width of about 42 MeV is a candidate for the lightest glueball, which should couple equally to the pion and kaon pairs. Therefore, a precision measurement of the $K\overline{K}/\pi\overline{\pi}$ branching ratio will be of interest.

The exclusive COSY - experiment

The detection method, which will be utilized, is depicted schematically in Fig.1. It takes advantage of the low emittance ($\leq 1\ \pi$ mm mrad) of the cooled external COSY-beam and the existing high resolution spectrometer BIG KARL. The vertices of the two mesons are measured by a detector wall, and the ^3He-nuclei are identified in the magnetic spectrometer. Kinematic completeness is obtained in this way. The small beam spot ($\leq$ 1mm) available at COSY enables a precise vertex reconstruction by simply measuring the planar coordinates of the mesons passing through the detector wall. The following observables will be obtained: $\vec{p}_{3He}$, T_{3He}, $\vec{p}_{m1}$, $\vec{p}_{m2}$. In addition, energy conservation yields the sum of the kinetic energies of the two mesons, their masses can be evaluated and the mesons identified. In a three body final state this sum can only be $2m_\pi$ or $2m_K$.

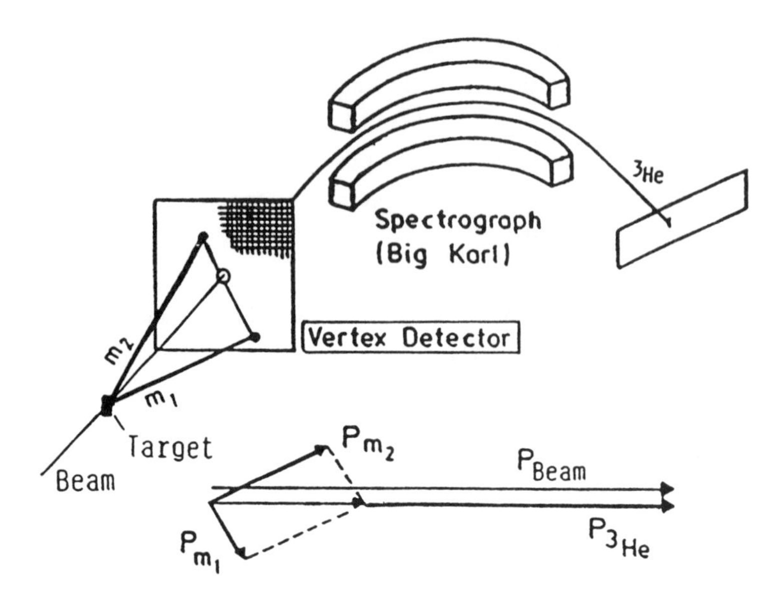

Fig.1 : Schematic sketch of the detection method

Therefore, a search for possible radiative ϕ-decay will become feasible via the reaction $pd \rightarrow^3 He + (\phi \rightarrow KK\gamma)$. The γ -rays can be identified as missing total kinetic energy. Near reaction thresholds the mesons will be emitted into a cone in the forward direction of the laboratory system. The vertex detector will have an opening angle of ± 45 deg. and thus accept all mesons up to 50 MeV c.m. energy above the reaction thresholds. However, a measurement of the $K^+K^-/\pi^+\pi^-$ branching ratio in the 1 GeV missing mass region requires the detection of pions with a high relative energy, which are emitted almost back to back in the laboratory system. Therefore an arrangement of scintillators ("pion can") will be placed around the 3 mm liquid deuterium target, which covers the region not seen by the vertex detector. It consists of a cylinder, subdivided radially into 18 stripes, and an endcap of the same granularity. The vertex detector consists of four layers of scintillating fibers. Two layers (20x20 cm^2, 1mm ⊘ fibers) will be positioned at 10cm distance from the target, the other two layers (40x40 cm^2, 2mm ⊘ fibers) at 20 cm from the target. The scintillating fibers will be read out by novel 64-fold photomultipliers. First tests with a small fiber arrangement and the multiple multipliers have been performed and are not discouraging.

The ^{3}He nuclei can uniquely be identified by their time of flight along the 15 m distance between the target and the focal plane detector of BIG KARL. The on-line trigger will be a double hit in the vertex detector in coincidence with an event in the focal plane of the appropriate ^{3}He time of flight. With the 3mm LD_2-target and an averaged cooled beam intensity of 10^9 protons/s, the luminosity is $L=1.5 \cdot 10^{31}$ cm^2s^{-1}. Thus a cross section of 10 nb will yield $1.3 \cdot 10^4$ events per day. The energy resolution of the obtained missing mass spectra is expected to be better than 1 MeV/c^2.

Recent Results

Very recently an experiment[3] was performed at SATURNE at the SPES IV spectrometer which focussed on near threshold meson production via the reaction $p + d \rightarrow^3 He + X$. The incident beam energy and the ^{3}He momentum were matched in such a way, that the mesons were produced at rest in the c.m. system. The run consisted of two parts. The first was an extension and partly a remeasurement of the inclusive 'threshold excitation function'[4]. The reaction was measured at more than 70 different beam energies ranging from 1.4 GeV to 2.0 GeV. A preliminary analysis shows clear enhancements of the cross section near the $K\overline{K}$- and the ϕ - thresholds. The second part of the run was a first step towards exclusivity. Coincidences between a fourfold segmented circular scintillator, positioned around the beam axis near the target, and the ^{3}He in the focal plane of SPES IV were measured. With the 4 cm LD_2target beam intensities up to 10^{10}protons/pulse could be used without overloading the counters in the scattering chamber. Coincidences have been observed and data analysis is underway.

References

1. A. Abashian et al., Phys. Rev. 132, 2296 (1963)
2. K.L. Au et al., Phys. Rev. D35, 1663 (1987)
3. SATURNE exp.222, Collaboration IPN Orsay, LN Saturne, ISKP Bonn
4. F. Plouin, Proc. Workshop on Production and Decay of Light Mesons, Paris, March 1988, World Scientific (1988), p. 114

* Work supported by Bundesministerium für Forschung und Technologie

EXCITATION FUNCTION MEASUREMENTS WITH INTERNAL TARGETS[1]

J. Bisplinghoff and F. Hinterberger
Institut f. Strahlen- und Kernphysik, Universität Bonn, D 5300 Bonn, Germany

Abstract

An experiment is described which is presently being prepared to be done at internal target station TP2 of COSY. It focusses on (but is not confined to) spin averaged as well as spin dependent observables in elastic p-p scattering with an eye on dibaryonic resonances. The detector concept, status of construction and Monte Carlo simulations are described, feasibility is discussed, and accuracy/sensitivity estimates are given.

Excitation function measurements with internal targets at cooler rings are well suited to provide data accurate enough for phase shift analysis and thus rigorous comparison to theoretical predictions. Furthermore, such measurements are a very sensitive method to prove or disprove the existence (and quantum numbers) of short lived intermediate excitation modes. Currently, the EDDA[2] collaboration is setting up an experiment along these lines to be done at COSY, the cooler synchroton now under construction at KFA Jülich. The experiment is designed to study the excitation functions

$$p + p\,(\rightarrow B_2\,?) \rightarrow p + p \quad \text{and} \quad p + p\,(\rightarrow B_2\,?) \rightarrow d + \pi^+$$

in the multipass technique so as to minimize the effect of random instabilities. With internal targets, the required fast beam energy variation is easily achieved by the computer control of COSY or by measuring during acceleration. Both spin-averaged and spin-dependent cross sections are to be measured in small energy steps at internal target station TP2, and with kinetic energies ranging from 250 MeV to 2.5 GeV.

The motivation for the proposed experiment is threefold. It rests with the search for intermediate excitation modes, the sensitivity of the method chosen, and the need for excitation functions permitting phase shift analysis.

[1] supported by BMFT

[2] J. Bisplinghoff, J. Ernst, D. Eversheim, F. Hinterberger, R. Jahn, U. Lahr, C. Lippert, T. Mayer-Kuckuk, E. Millich, F. Mosel, H. Scheid, F. Schwandt, D. Theis, W. Wiedmann (Institut f. Strahlen- u. Kernphysik, Universität Bonn); B. Metsch, H.R. Petry (Institut f. Theoretische Kernphysik, Universität Bonn); B. v. Przewoski (IUCF Bloomington); M. Gasthuber, H. Rohdjess, W. Scobel, L. Sprute (Universität Hamburg); W. Amian, P. Cloth, P. Dragovitsch, V. Drüke, D. Filges, D. Prasuhn, P. v. Rossen (KFA Jülich); H. Paetz gen. Schieck (Universität Köln)

Theoretically, intermediate excitation modes such as dibaryonic resonances or "states" have been predicted in the framework of both meson exchange and 6 quark bag models by several authors [1]. For instance, a rather elaborate calculation has been performed by Lomon [2], predicting a singlet s-wave resonance at a p-p invariant mass of 2.8 GeV/c^2 with a total width of less than 50 MeV. Moreover, the predictions in the 6 quark bag picture and the meson exchange approach are at variance with one another [3], so that the underlying question addressed is that of confinement, and therefore rather fundamental.

Experimentally, the situation is characterized by considerable frustration and must be considered an open question. Narrow signals which might be interpreted as evidence for dibaryon states have been observed in a number of reactions [1,4-8]. None of them is, however, considered as establishing the existence of a dibaryonic resonance as yet.

Excitation function measurements are a good probe into these questions, as they are highly sensitive to resonances even in the case of a weak elasticity. This is because the effect of a weak resonance amplitude is strongly increased by interference with the nonresonant scattering amplitude. We estimate that a narrow dibaryonic resonance with an elasticity larger than about 1% should be detectable in the experiment under discussion. This estimate is based on scores of excitation functions computed from phases [10] that have been arbitrarily modified to reflect narrow resonances. Strong resonance excursions with a characteristic dependence on angle and resonance quantum numbers show up even in the spin-averaged cross sections $d\sigma/d\Omega$. Figs. 1 and 2 give some examples.

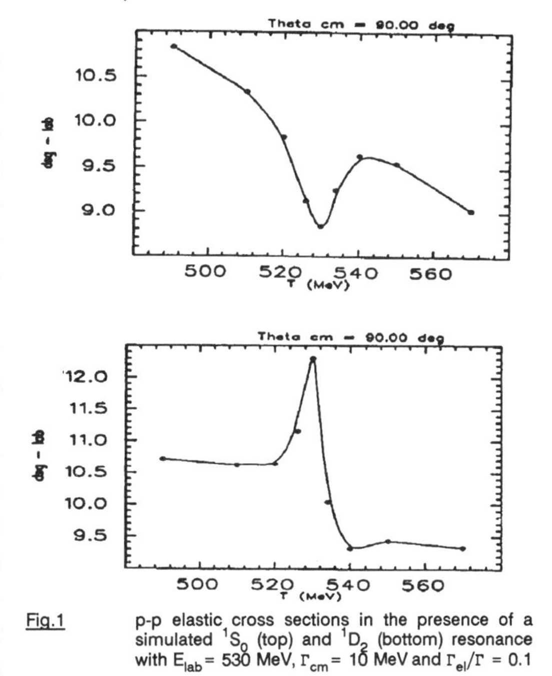

Fig.1 p-p elastic cross sections in the presence of a simulated $^1S_0$ (top) and $^1D_2$ (bottom) resonance with E_{lab} = 530 MeV, Γ_{cm} = 10 MeV and Γ_{el}/Γ = 0.1

Quite generally, the p + p excitation function is not well investigated experimentally at energies above 1.2 GeV. There are but two [8,9] excitation function studies where cross sections or polarization observables were measured systematically in small energy steps with high relative accuracy. The

EDDA experiment seeks to improve on this situation, so that phases as a function of bombarding energy and spin quantum numbers can be accurately extracted.

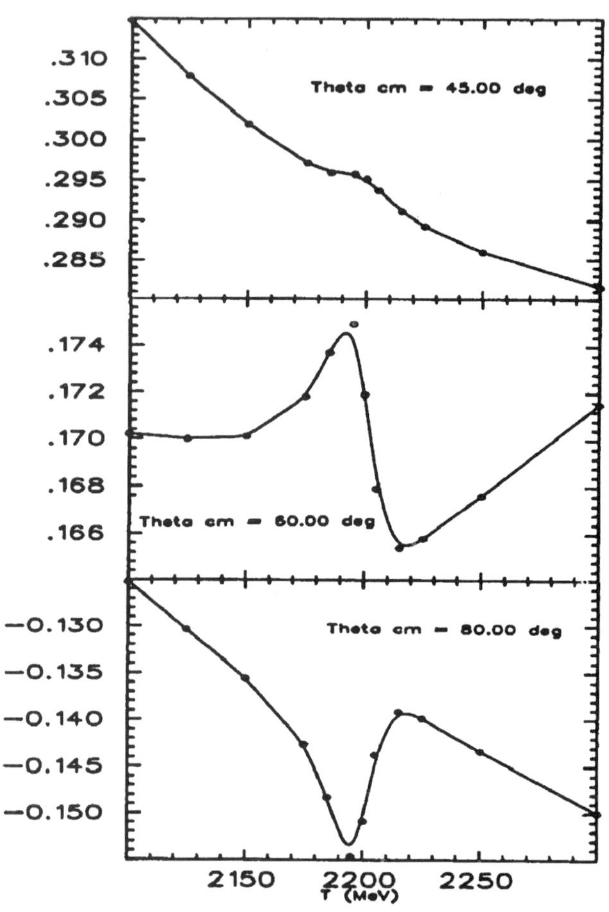

Fig.2 p-p elastic analyzing power in the presence of a simulated $^1S_0$ resonance with E_{lab} = 2.2 GeV, Γ_{cm} = 10 MeV, Γ_{el}/Γ = 0.1

The experiment is - for practical reasons - to be divided into two phases:

In the first phase, spin-averaged cross sections $d\sigma/d\Omega$ are to be measured. For these measurements, a hydro-carbon plastic fiber target can be used without unsuitably adverse effects [11] on the stability of the internal beam both during acceleration above some 200 MeV and while recirculating at the "flat top". Polypropylene fibers[12] of 14 μm diameter have been carbon coated to achieve a conductivity required for secondary electron emission to serve as a luminosity monitor (among other schemes). Sample targets are due to undergo testing in a low energy proton beam shortly. Carbon fiber targets will be required for background subtraction. The luminosity is estimated to be some 10^{30} $cm^{-2}s^{-1}$ at the lowest internal beam intensities compatible with beam control.

In the second phase, spin observables are to be measured, using a polarized COSY beam and/or a polarized atomic beam target. Once both are available, four spin observables can be measured, the analyzing power A and the polarization correlations A_{NN}, A_{SS} and A_{LS}. Only A_{LL} remains basically inaccessable. Luminosities are estimated to be about 10^{29} $cm^{-2}s^{-1}$ at the highest internal beam intensities compatible with the space charge limit.

The detector concept is tailored to the exit channels under study (p+p and $d+\pi^+$) and schematically depicted in Fig. 3: The beam tube at internal target station TP2 is to be surrounded by 32 plastic scintillator bars which run parallel to the beam axis. This inner cylinder-shaped detector layer is surrounded by an outer cylinder consisting of 32 plastic scintillator rings (actually 64 half-rings), each corresponding to a $\Delta\theta_{cm} \approx 5°$ width. Obviously, coincident signals occuring in opposing scintillator bars may be interpreted as indicating coplanarity with the beam and thus a two body exit channel. This can be used as a "two body trigger". Moreover, a hit in some half-ring scintillator can be used to have a (reprogrammable) fast logic "predict" which half-ring should have fired in case of either kind of wanted event. This provides a second, "kinematic trigger", which is , however, valid only for a limited energy range. Extensive Monte Carlo simulations were done on the trigger efficiency, using the FOWL and GEANT codes. They have shown that reprogramming the kinematic trigger every 0.5 GeV or so makes for a high combined trigger efficiency, accepting some 95% of the wanted events and suppressing unwanted ones to below some 1-3% leakage.

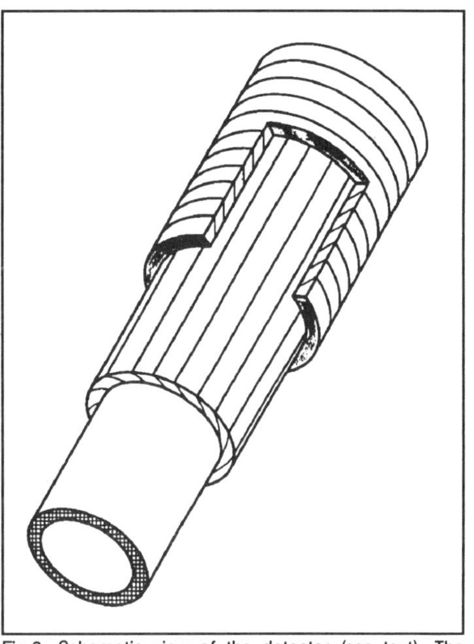

Fig.3 Schematic view of the detector (see text). The target is located at the up stream end of the cylindrical scintillator array.

The discrimination of p+p or $\pi^+ + d$ exit channel events against the three body background on a trigger basis alone is good enough to

- sort the elastic and $\pi^+ + d$ triggers into $\Delta\theta_{cm} \approx 5°$ wide scaler bins,
- read the scalers out some every 10 ms, corresponding to some 2 MeV energy gain during acceleration
- and thus obtain a rough multi pass excitation function on line.

This on line data is rough in the sense that it is based on the spatial resolution provided by the trigger logic and the scintillator granularity alone.

The scintillaters are, however, to provide better spatial resolution than their granularity indicates. They will not be given rectangular or quasi-rectangular cross sections. Instead, they are to overlap one another in what might be called a position sensitive light splitting scheme. Under this scheme, each particle originating from the target will produce light pulses in two neighbouring bars and in two neighbouring rings. The ratios of the amounts of light produced in neighbouring scintillator elements can then be used to determine the path, at which the particle traverses the detector, much more accurately (factor 5) than would be possible on the basis of the detector granularity alone. The idea is evident from Fig. 4 and has been successfully tested at DESY.To what degree position sensitive light splitting will be used has yet to be decided, along with the exact detector geometry. Its inherent need for digitization of amplitudes restricts it to application to every 10th or 100th event obtained at the high count rates associated with a fiber target. This restriction does not apply to the measurement of spin observables with the polarized atomic beam target. The increased precision offered by the light splitting scheme will certainly help the detailed off-line analysis, however, and be instrumental in establishing a dibaryonic resonance, should any promising structure show up in the excitation functions.

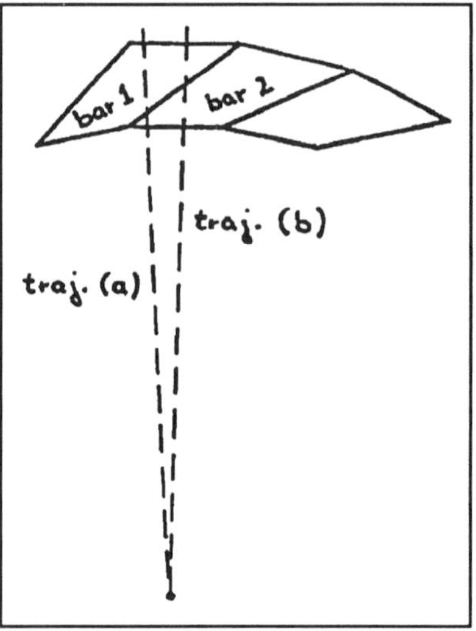

Fig.4 Position sensitive light splitting in neighbouring elements of a scintillator array

Detector construction will begin early next year, and the experiment should be ready to take at least test data by summer of 1992.

References

1) M.P. Locher, M.E. Sainio and A.Svarc, Adv. Nucl. Phys. 17, 47 (1986) and references therein
2) E.L. Lomon, J. de Physique C2, 329 (1985)
3) M.G. Huber, B.C. Metsch and H.G. Hopf, Lecture Notes in Physics 243, 499 (1985)
H. Hofestädt, S. Merk and H.R. Petry, Z. Physik A326, 391 (1987)
4) B. Tatischeff et al., Phys. Rev. C36, 1995 (1987)
5) L. Santi et al., Phys. Rev. C38, 2466 (1988)
6) B. Bock et al., Nucl. Phys. A459, 573 (1986)
7) R. Bertini et al., Phys. Lett. 162B, 77 (1985)
F. Lehar, private communication
8) H. Shimizu et al., submitted to Phys.Rev.Lett
9) M. Garcon, Nucl. Phys. A 445, 669 (1985)
10) R.A. Arndt et al., Phys.Rev. D28, 97 (1983), Computer Code "SAID"
11) F. Hinterberger and D. Prasuhn, Nucl. Instr. Meth. A279, 413 (1989)
12) courtesy Farbwerke Hoechst AG, Frankfurt/Main, Germany

RECOIL ION DETECTION SYSTEM FOR THE IUCF COOLER

J. D. Brown and E. R. Jacobsen
Princeton University

R. E. Segel
Northwestern University

G. Hardie
Western Michigan University

R. D. Bent, G. P. A. Berg, H. Nann, R. E. Pollock
Indiana University Cyclotron Facility

K. E. Rehm
Argonne National Laboratory

J. Homolka
Technische Universität München

INTRODUCTION

The detection of neutral particles or exotic ejectiles in reactions initiated by medium energy probes poses several experimental problems. Efficiency of detection is often quite small, as is the angular range that can be covered by a fixed detector. In addition, there are difficulties associated with the decay of a light reaction product or the coincident detection of two light products in a three body final state. Some of these problems can be surmounted by the detection of the heaviest of the reaction products. For example, in the case of proton induced neutral pion production on carbon, one has the choice of detecting either the two photons from the π^0 decay or the residual nitrogen nucleus. It is clear that accurate energy and position measurements can be more readily made for the latter. Further advantages result from simple kinematic considerations. In high momentum transfer reactions, such as (p, γ) and (p, π), the recoil ions are emitted in a small forward cone, and hence a small detector array in the laboratory covers a large center-of-mass angular range. This kinematic focussing results in high detection efficiency. In many cases, the magnetic rigidity of the recoil ions is much smaller than that of the projectile, thereby facilitating the magnetic separation of the recoil ions from the beam. Often, it is possible to detect nearly all charge states of the heavy ejectile. Finally, if the momentum acceptance of the detection system is large enough, one can study several processes simultaneously. For example, in the case of pion production, with a judicious choice of the detection system one can detect the ejectiles from positive, negative and neutral pion production simultaneously. A recoil-ion detection system is currently under construction and will be installed in the IUCF Cooler ring in 1991, in preparation for the experimental program outlined below.

DETECTOR SYSTEM

One of the main advantages of detecting recoils is the forward kinematic focussing. In order to separate the recoils from the large beam flux, it is necessary to use magnetic deflection. Therefore, the initial element in the system will be a large magnet placed just downstream of a target location in the Cooler ring. This magnet will be adjusted to deflect the beam in the Cooler through 6 degrees. Recoils of lower energy and rigidity will be deflected much further. A schematic of this magnet together with the detection system is shown in Figure 1.

In order to define the recoil ion's trajectory through the magnet and hence its charge to momentum ratio, it is necessary to determine the ion's position as it leaves the magnet. Parallel plate avalanche counters (PPAC) are being built to measure this position to within 1 mm in two dimensions.[1] These counters will also provide a start signal for a time of flight measurement.

Following a 50 cm flight path, there will be a ΔE-E detector telescope for particle identification and energy measurement. This telescope will consist of a gas proportional counter (PC) for ΔE and a silicon strip array for E determination. For future experiments these telescope elements may be rearranged: for example, into a telescope of silicon followed by a scintillator. The proportional counter will have an active volume 65 cm long, 10 cm high and 1 cm deep. For the stopping detector, an array of silicon strips each 1 mm wide, 5 cm high and 300 μm thick covering the same area as the PC will be placed in the gas volume of the PC. In order to reduce the electronics required, the strip array will be installed as modules. Each module will consist of 50 strips connected to each other by resistors, and position information will be obtained by resistive division techniques. Modules that fail during an experiment can be readily replaced. These strips will be mounted in "transmission" rather than "stopping" mode so that they can be used as ΔE detectors in other experiments. Gates on the PC-Si histogram will separate out each mass group in the usual manner. Total energy may be obtained by summing these two detector signals. A velocity measurement will be obtained from PPAC start and Si stop signals. The goal for the timing resolution is 500 ps. Finally, a second position measurement will be made by the Si strip array. A combination of these signals will serve to identify the charge, mass, energy and angle of each ejectile passing through the system.

An elaborate raytracing program has been written that calculates the recoil ion trajectories and energy losses in each segment of the detection system. Simulations of each reaction of interest have been performed to optimize gas pressures, window thicknesses, detector placement and detector thicknesses. Examples of rays from the $^{12}C(p,\pi^0)^{13}N$ reaction at E_p=300 MeV are superimposed on the schematic of the detection system in Figure 1.

PROPOSED PHYSICS PROGRAM

The first recoil experiment approved by the IUCF PAC is CE-06, which involves detection of the mass-13 recoils from the reactions $^{12}C(p,\pi^0)^{13}N$, $^{12}C(p,\pi^+)^{13}C$, $^{12}C(p,\pi^-)^{13}O$, $^{12}C(p,\gamma)^{13}N$, and $^{12}C(p,\pi\pi)$ employing a 320 MeV proton beam and a carbon whisker target of thickness 5 $\mu g/cm^2$. The range of energies and angles of the recoils from these reactions is such that data for all of the reactions can be taken at the same time. Previous IUCF work has concentrated on charged pion production below 200 MeV bombarding energy,[2] and significant progress has been made in understanding positive pion production in terms of a two-nucleon model.[3] Employing the recoil technique on the Cooler, it will be possible to extend this work to higher energies while carrying out at the same time the first systematic studies of neutral pion production. Radiative capture experiments are expected to provide new data (especially analyzing powers, which are very difficult to measure by the inverse reaction) that will test the latest relativistic theories[4] of this process at intermediate energies. The recoil technique is ideal for studying double pion production near threshold, which is a process that is believed to proceed through channels different from those that dominate single pion production.[5]

With a rearrangement of detectors within the detector box and thicker stopping counters, light ejectiles may be detected. Light-ion reactions of interest include $d(p,{}^3He)\gamma$, $d(p,{}^3He)\pi^0$, $d(p,t)\pi^+$, ${}^3He(p,\alpha)\pi^+$, $d(p,{}^3He)\pi\pi$, and ${}^3He(p,\alpha)\pi\pi$. Data on these reactions very near threshold is sparse or nonexistent. Few nucleon systems provide the best testing ground for microscopic calculations of pion production because of the well-known nuclear wave functions and the small number of nucleons, which makes fully microscopic calculations tractable. It is anticipated that the Cooler luminosities will be large enough to allow measurements to be made with polarized beam.

REFERENCES

1. J. Homolka *et al.*, NIM **A260** (1987) 418.
2. B. Höistad *et al.*, Phys. Lett. **94B** (1980) 315, for example.
3. M. J. Iqbal and G. E. Walker, Phys. Rev. **C32** (1985) 556;
   P. W. F. Alons, R. D. Bent, and M. Dillig,
   Nuclear Physics **A493** (1989) 509.
4. M. Gari and H. Hebach, Physics Reports **72** (1981) 1.
5. N. Grion *et al.*, Phys. Rev. Lett. **59** (1987) 1080

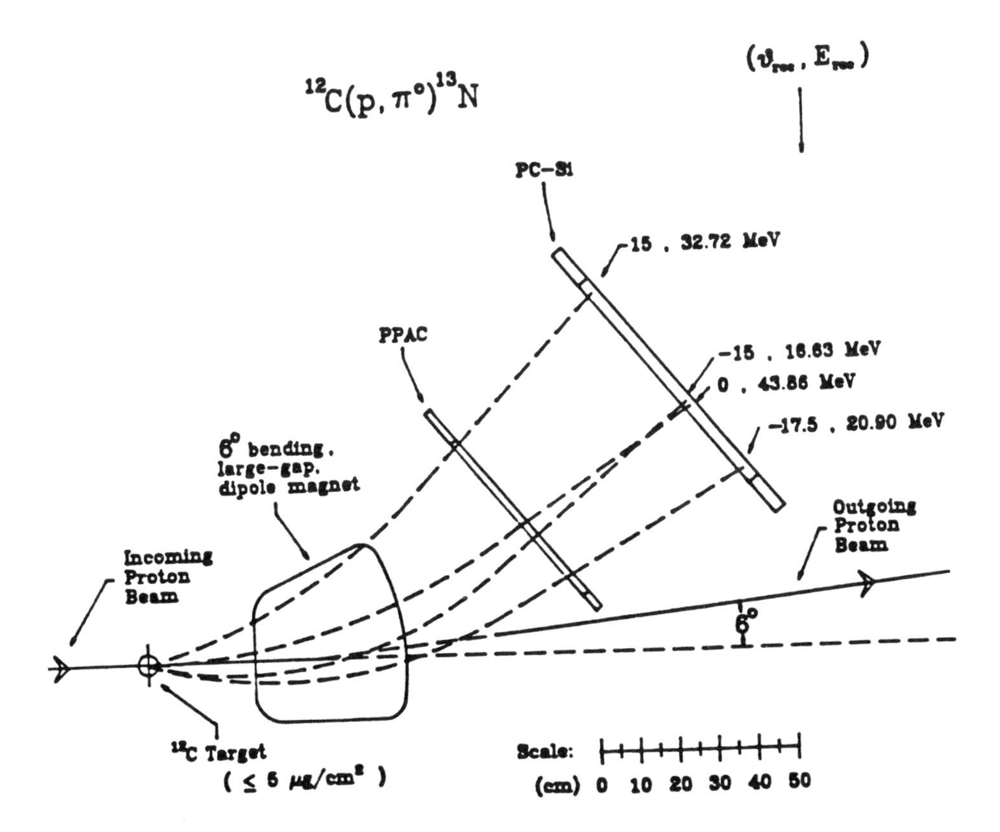

Figure 1: Schematic of detector configuration and representative rays for the $^{12}C(p,\pi^0)^{13}N$ reaction at $E_p = 300$ MeV.

PARITY VIOLATION IN PROTON-PROTON SCATTERING

P.D. Eversheim, F. Hinterberger
Institut für Strahlen- und Kernphysik, Universität Bonn, D-5300 Bonn 1, Germany

H. Paetz gen. Schieck
Institut für Kernphysik, Universität Köln, D-5000 Köln 41, Germany

W. Kretschmer
Physikalisches Institut, Universität Erlangen-Nürnberg, D-8520 Erlangen, Germany

ABSTRACT

Experimental results of parity violation in proton-proton scattering are reviewed. Since the parity violation in proton-proton scattering at 230 MeV is ρ-exchange dominated, this experiment gives additional independent information for the weak meson-nucleon coupling constants. With the cooler synchrotron COSY at Jülich, this aspect of weak interaction and the energy-dependence of A_z up to 2.5 GeV can be measured. Consequences for the experimental set-up are discussed.

INTRODUCTION

Parity violation in the N-N interaction is the only experimentally accessible signature of the flavor-conserving purely hadronic part of the weak interaction. Up to intermediate energies parity violation in N-N interaction is mediated by the exchange of mesons (π^+, π^-, ρ^+, ρ^0, ρ^- and ω) involving a weak and a strong vertex. The strong meson-nucleon couplings are derived from e.g. the Bonn potential. The weak coupling constants (f_π^1, h_ρ^0, h_ρ^1, h_ρ^2, h_ω^0, h_ω^1 - the superscripts refer to the isospin changes) are predicted[1] from the Glashow-Weinberg-Salam model with regard to W and Z exchanges between the quarks of the nucleon and meson.

The weak coupling constants have to be derived from 6 independent experiments, allowing a clear interpretation. Therefore, Simonius[2] proposed to measure the parity violation in N-N scattering. The quantity of interest is the longitudinal analyzing power A_z, which is the relative difference of the proton beam helicity dependent cross sections σ^+ and σ^-, normalized to the longitudinal mean polarization P_z.

$$A_z = 1/P_z\,(\sigma^+ - \sigma^-)\,/\,(\sigma^+ + \sigma^-) \tag{1}$$

It has been shown[3] that as a consequence of the well known strong phase-shifts at 230 MeV the weak interaction is dominantly described by ρ-exchange, giving independent information for a consistent set of the 6 weak meson-nucleon coupling constants.

In the high energy range above 1 GeV parity violation is described at the quark level[4]. The results are encouraging, but the large value for A_z at 5.3 GeV (cf. Fig. 1) is still not understood. Thus more and precise experiments are necessary. With the cooler synchrotron COSY experiments are planned at 230 MeV and just below the Λ-production threshold at 1.5 GeV, in order to avoid systematic errors from the parity violating Λ-decay.

Up to now A_z has been measured for $\vec{p}$-p scattering at only a few energies: at 15 MeV in Bonn[5] and Los Alamos[6], at 45 MeV e.g. at the PSI[7], at 800 MeV in Los Alamos[8] and for $\vec{p}$-H_2O scattering at 5.3 GeV in Argonne[9]. The challenge of these experiments is the reduction, measurement and correction of all relevant systematic errors. The most accurate result has been measured at 45 MeV at the PSI[7]. At Bonn, in addition to the procedures employed at the PSI, a multilinear regression of the noise of the data has been utilized. It provides besides most of the corrections also information on the sources of error reduction or points out how the experimental set-up can be improved.

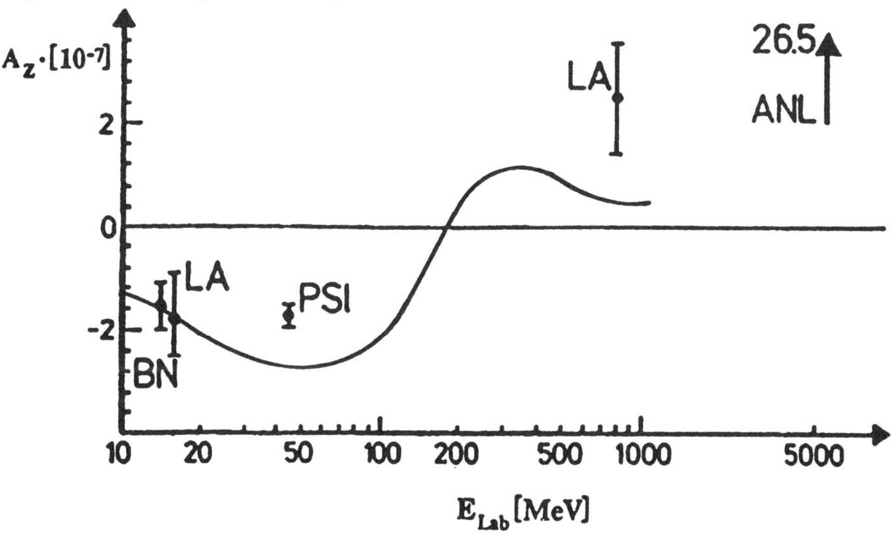

Fig. 1 Driscoll's and Miller's prediction[10] and measurements of A_z for $\vec{p}$-p scattering at Bonn (BN), Los Alamos (LA), and Paul Scherrer Institut (PSI), and a measurement for $\vec{p}$-H_2O scattering at the Argonne National Laboratory (ANL).

THE COSY EXPERIMENT

A polarized, pulsed proton beam will be produced in a colliding beams type ion-source. Having passed the injector cyclotron JULIC the normally polarized beam is injected into COSY. At the appropriate energy the beam is transferred to the experiment. In order to have longitudinal and sideways polarization at fixed geometry over a wide energy range, the polarization is prepared by a combination of solenoid - bending magnet - solenoid - bending magnet.

The experimental set-up reflects the experiences of the Bonn experiment. The accepted experiment proposal comprises a liquid hydrogen target and devices that measure and control permanently with high precision the phase-space distribution of the proton beam. In particular a x-y polarimeter is used to tune the beam for pure longitudinal polarization. As has been demonstrated at Bonn[5] and TRIUMF[11], the beam can be fixed to about one μm and one μrad transversally by an analog control-loop, comprised of two secondary electron monitors (SEM), fast steering magnets and power amplifiers. A third SEM should be implemented for cross-check purposes. Further, three x-y beam profile scanners provide intensity and polarization profiles at three positions in order to get sufficient information on the higher moments of the phase-space distribution.

A_z will be measured redundantly by a transmission as well as a scattering experiment. For the transmission experiment the target is placed between two ionization chambers measuring the amount of beam that is missing due to scattering in the target. Simultaneously, protons scattered off the target are detected by ionization chambers surrounding the end of the target. The ionization chambers are low noise transverse field chambers which are operated in current mode, in order to provide sufficient statistics.

The experiment is designed to give an accuracy of $2\ 10^{-8}$.

REFERENCES

1. B. Desplanques, J.F. Donoghue and B.R. Holstein, Ann Phys. **124**, 449 (1980)
2. M. Simonius, Nucl. Phys. B41, 415 (1972)
3. M. Simonius, Can. J. Phys. **66**, 245 (1988)
4. T. Goldman and D. Preston, Phys. Lett. **B168**, 415 (1986)
5. P.D. Eversheim W. Schmitt, F. Hinterberger, S. Kuhn, R. Gebel, U. Lahr, B. von Przewoski, P. von Rossen, I. Nies, P. Dresbach and M. Neuser, Journal de Physique, to be published
6. D.E. Nagle, J.D. Bowman, C. Hoffman, J. McKibben, R.E. Mischke, J.M. Potter, H. Frauenfelder and L. Sorenson, AIP Conf. Proc. **51**, 224 (1978)
7. S. Kistryn, J. Lang, J. Liechti, T. Maier, R. Müller, F. Nessi-Tedaldi, M. Simonius, J. Smyrski, J. Jaccard, W. Haeberli and J. Sromicki, Phys. Rev. Lett. **58**, 1616 (1987)
8. V. Yuan, H. Frauenfelder, R.W. Harper, J.D. Bowman, R. Carini, D.W. McArthur, R.E. Mischke, D.E. Nagle, R.L. Talaga and A.B. McDonald, Phys. Rev. Lett. **57**, 1680 (1986)
9. M. Lockyer et al., Phys. Rev. **D30**, 860 (1984)
10. D.E. Driscoll and G.A. Miller, Phys. Rev. **C39**, 1951 (1989)
11. S.A. Page, in:"Proceedings of the Symposium/Workshop on Spin and Symmetries", eds.W.D. Ramsay and W.T.H. van Oers, TRI-89-5, 2 (1989)

ASSOCIATED STRANGENESS PRODUCTION IN pp REACTIONS

K. Kilian, H. Machner, W. Oelert,
E. Roderburg, M. Rogge, O. Schult, P. Turek
Institut für Kernphysik, Forschungszentrum Jülich GmbH,
D–5170 Jülich, F.R.G.

W. Eyrich, M. Kirsch, R. Kraft, F. Stinzing
Phys. Institut der Universität Erlangen–Nürnberg,
D–8520 Erlangen, F.R.G.

ABSTRACT

For the associated strangeness production process pp → KYN, we want to measure total and differential cross sections, the mass spectra of all subsystems and the hyperon polarization. The latter is related to the polarization of the strange quark. The data should shed light on the dynamics of $s\bar{s}$ quarkpair creation. They should be compared with results from $p\bar{p} \to Y\bar{Y}$ and $\gamma p \to KY$ experiments, where the $s\bar{s}$ creation is embedded in different hadronic surroundings. The experimental setup, on an external COSY beam with a LH_2 target, will consist of a highly granulated scintillation counter time of flight spectrometer and wire chambers for precise track and decay vertex reconstruction. The on line trigger will make use of the delayed decay of $Y \to N\pi$ or $K \to K^+\pi^-$. Full solid angle coverage will be achieved. This will allow for complete kinematical reconstruction of each event. This experiment would further profit from a polarized beam.

INTRODUCTION

This experiment is proposed to investigate the complicated process of associated strangeness production in the "simple" case of nucleon–nucleon interaction, namely in the proton–proton reaction leading to a K–meson, a hyperon and a nucleon[1]

$$p + p \to K + Y + N.$$

In particular, we will investigate the process in which the hyperon is a Λ–hyperon and thus the nucleon a proton

$$p + p \rightarrow K^{+} + \Lambda + p.$$

The hyperon production process can be described either in the framework of meson exchange models leading to an intermediate resonante state R which decays into the K–meson and the Λ–hyperon channel or in terms of quark flow diagrams (Fig. 1), with creation of a $s\bar{s}$ quark pair by dissociation of the vacuum and the exchange of two up–quarks.

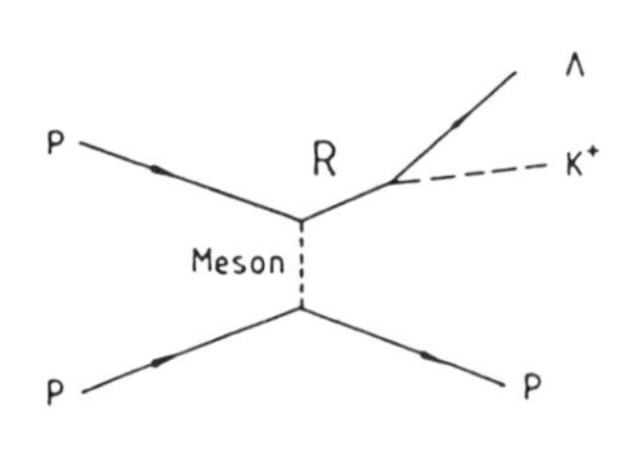

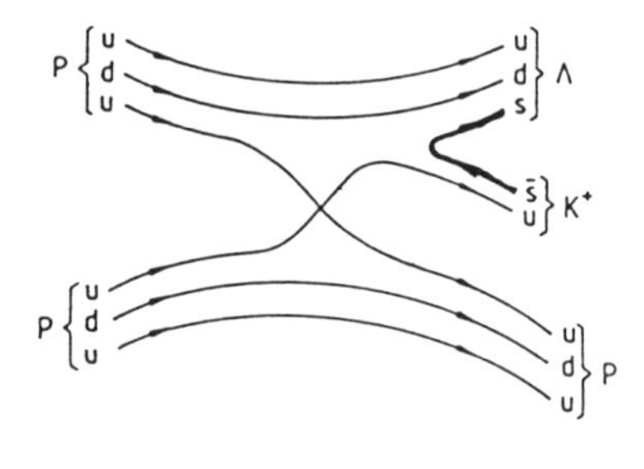

Fig. 1: Schematic diagrams describing the associated strangeness production process pp $\rightarrow K^{+}\Lambda p$.
upper part: The meson exchange process, leading to a resonante state R.
lower part: In the quark flow diagram the creation of an $s\bar{s}$ quark pair by dissociation of the vacuum and its coupling scheme.

In the energy range of COSY one can measure various hyperon production processes (Table 1) and by the excellent beam quality realized by electron– and stochastic–cooling processes all the tabulated reactions are measurable nearest to their thresholds.

Table 1: Threshold values

		$\sqrt{s}$ [GeV/c²]	beam momentum [GeV/c]	kin. energy [GeV]
pp	$\rightarrow K^{+}\Lambda p$	2.548	2.339	1.582
pp	$\rightarrow K^{+}\Sigma^{+}n$	2.622	2.560	1.789
pp	$\rightarrow K^{+}\Sigma^{0}p$	2.624	2.566	1.793
pp	$\rightarrow K^{0}\Sigma^{+}p$	2.625	2.569	1.796
pd	$\rightarrow K^{+}\Lambda d$	3.485	1.839	1.127
$p^{4}He$	$\rightarrow K^{+}\Lambda {}^{4}He$	5.338	1.581	0.900
pp	$\rightarrow ppK_{s}K_{s}$	2.872	3.327	2.518

The self–analyzing property of the parity violating weak decay of the produced hyperons enables to measure their polarization.

In the case of Λ– and Σ⁺–hyperons their weak decay modes into $p+\pi^-$ and $p+\pi^0$, respectively, are asymmetric with respect to the Λ– and Σ–spin direction. The proton distribution is described by : $I(\beta) = I_0\ (1+\alpha \cdot P \cdot \cos\beta)$, where β is the angle between the vector normal to the scattering plane of the Λ– or Σ–particle and the emitted proton, P is the hyperon polarization and α is the decay asymmetry parameter ($\alpha_\Lambda=0.64\pm0.013$; $\alpha_{\Sigma^+}= -0.98^{+0.019}_{-0.014}$).

Most interesting in this context is to study the spin dynamics of strange quarks: In the additive quark model the spin vector of the Λ is the same as that of the strange quark content in the Λ–hyperon, because the u– and d–quarks are coupled to spin zero and isospin zero. In the Σ–hyperon the corresponding light quarks are coupled to spin and isospin one. Comparing the two creation processes one has the possibility to study the interaction of strange quarks in two spin–isospin situations.

The comparison of results of the proposed COSY experiment with experiments performed at LEAR/CERN ($\bar{p}p \to \bar{\Lambda}\Lambda$)[2] and at the synchrotron in Bonn ($\gamma p \to K^+\Lambda$)[3] gives additional information about the creation of an $s\bar{s}$ quark pair in different hadronic surroundings.

Our proposed reaction: $pp \to K^+ p\Lambda$ has the lowest threshold energy for strangeness production in pp–collision (2339 MeV/c T = 1582 MeV). The total cross section in pp–reactions at 2.5 GeV is about 40 mb[4] and only 40 μb for the Λ–hyperon production cross section. The available data for this reaction from threshold up to 2.5 GeV are insufficient in quantity and statistics. Therefore we will measure the total and differential cross sections starting near threshold, the mass spectra of all subsystems and finally the polarization of the Λ–hyperon.

KINEMATICS

Close to the threshold all the reaction products are emitted in forward direction. At maximum COSY energy in the reaction $pp \to K^+\Lambda p$ the maximum angles for K^+–meson, proton and Λ–hyperon emission are 75^0, 35^0 and 30^0, respectively. This means that a relatively small forward detector corresponds to a 4π detector for this reaction in the c.m. system.

The decay distribution of the Λ along its flight path f follows an exponential function ($\exp - f/L$) with the decay length $L = c \cdot \tau_\Lambda \cdot p/m$. In the range of COSY energies the decay vertices are typically a few cm far from the target. This characteristic feature of the delayed decay of the Λ–hyperon ($\Lambda \to p + \pi^-$, life time $\tau_\Lambda = 2.6 \cdot 10^{-10}$s, branching ratio 64 %) in a vertex at a distance from the target gives an excellent trigger signature namely the increase of the number of charged particles from two (p, K^+) close to the target to four (p, K^+, p, π^-) far away from the target.

DETECTOR

The detector consists of an "inner" and an "outer" detector system. Both together are designed as a time of flight spectrometer as shown schematically in Fig. 2.

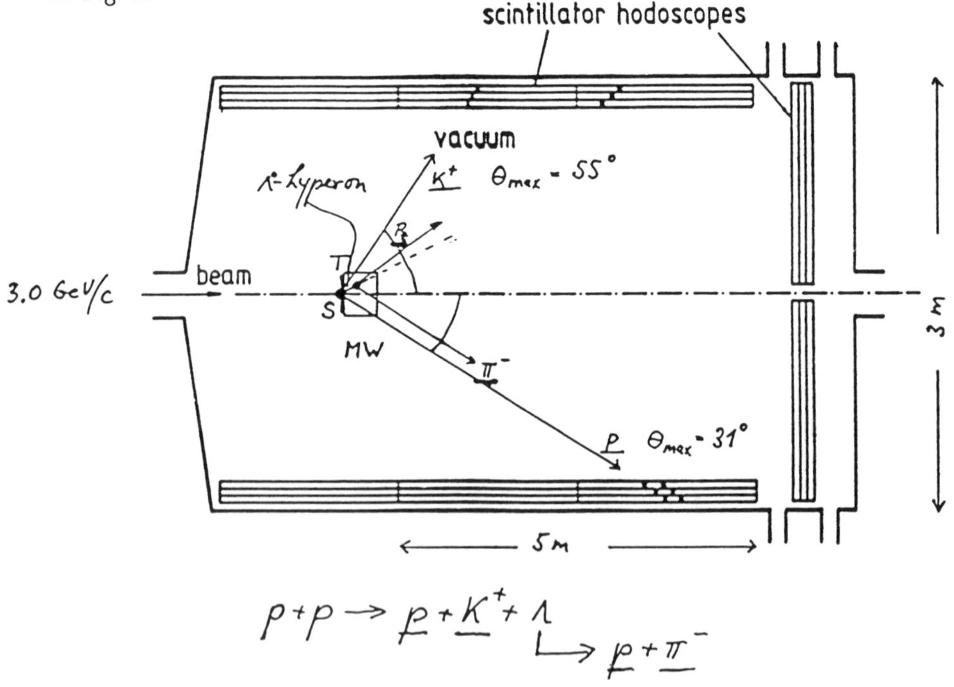

Fig. 2: The layout of the planned time of flight spectrometer. The target point T, the start detector S and the multiwire chamber MW are out of scale.

The target T is a 1 mm^3 cell of liquid hydrogen sealed by very thin windows of mylar foils[5].

The inner detector S works as start– and multiplicity–detector and consists of a granulated scintillator disk of 3 cm in diameter with a central hole of about 2 mm diameter and a silicon μ–strip disk with ring structure. Close behind it is placed a granulated cylindrical scintillator tube. The tube is 2 cm long, 3 mm in diameter with an inner opening of 2 mm for the passing zero degree beam.

Monte Carlo simulations for the transversal decay component of the Λ–decay show that about 6 % of the produced Λ–hyperons would decay inside a cylinder of 2 mm diameter. Such decay events would be lost in our one line trigger with the conditions: two hits in the inner detector and four hits in the outer one.

The multiwire chamber stack MW detects the tracks of the charged particles including the vertices of the Λ–decay particles p and π^-. These chambers may be of the type of "Induction Drift Chambers". They offer the best spatial resolution and rate capability[6].

The granulated cylindrical outer scintillator hodoscopes are used as stop– and multiplicity–detectors closed in forward direction by a granulated scintillator front plane. These hodoscopes provide enough pixels to localize the penetration points. The pixels are created as overlap regions between three segmented layers of straight, left and right tilted scintillator strips[7]. The outer hodoscope is hermetically closed for the wanted reaction partners (except some few of the decay pions). A size of > 1 m for these outer scintillators is enough for good trigger efficiency and for complete kinematical reconstruction of pp $\rightarrow$ KYN events. We prefer however to use the same large time of flight hodoscopes as proposed for the η'– and the bremsstrahlung experiment. This provides additional constraints and improves precision and background reduction. Details of this decector are given in ref. 7.

FUTURE ASPECTS

Other associated strangeness production processes can be measured with the same experimental set up. For example:

$$pp \to \Sigma^+ K_s p$$
$$pp \to K^+ K^- pp$$
$$pp \to K_s K_s pp$$

All these reactions offer the possibility to compare $s\bar{s}$ pair creation in different but well defined hadronic surroundings.

We are very interested to carry out such experiments with a polarized proton beam. This gives further information for quark spin dynamics and the "polarization transport" from entrance to final particles (Fig. 1).

REFERENCES

1. K. Kilian, COSY Note 62, KFA–IKP (1986).
2. P.D. Barnes et al., Proceedings of the IX European Symposium on Anti-proton–Proton Interactions, Mainz, Sept. 1988.
3. R. Haas et al., Nucl. Phys. B137 (1978) 261.
4. CERN–HERA 84–01, April 1984.
5. K. Kilian et al., A liquid hydrogen target with very thin windows, CERN EP 85–02 (1985).
6. E. Roderburg et al., The induction drift chamber, Nuclear Instruments & Methods A252 (1986) 285.
7. E. Roderburg, Proposal for the measurement of the cross section of η and η' production and inelastic scattering, E. Kuhlmann, H.P. Morsch, bremsstrahlung proposal.

ASSOCIATED STRANGENESS PRODUCTION ON LIGHT NUCLEI*

J. Ernst, J. Kingler, and C. Lippert
Institut für Strahlen- und Kernphysik der Universität Bonn,
5300 Bonn, Nussallee 14 - 16, Federal Republic of Germany.

ABSTRACT

The study of light hyper-nuclei via associated strangeness production in (p, K^+) reactions is discussed. Though the process is characterized by a very large momentum transfer the presence of short range correlations is expected to rise the cross section up to the order of nb/sr. Two approved proposals for high resolution studies of this reaction are discussed and respective detection limits are presented. The first is scheduled for October 1990 at the SPES4 spectrometer at the SATURNE accelerator (LNS Saclay). The second deals with the planned upgrading of the BIG KARL magnetic spectrograph at the cooled beam facility COSY being built at Forschungsanlage Jülich.

INTRODUCTION

The associated strangeness production of light hyper-nuclei via (p, K^+) reactions is hitherto completely unexploited. The reason for that lies in the inherent large momentum which has to be transferred to the final nucleus. Contrary to (K^-, π^-) reactions with small momentum transfer, in (p, K^+) reactions the transferred momentum is distributed at least on two interacting nucleons from beam and target which subsequently fuse with the residual target nucleons to the final hyper nucleus. As Shinmura [1] has pointed out, the presence of short range correlations will drastically enhance the cross section up to the order of nb/sr. Also other authors have estimated yields of this order [2], [3] partly by rescaling $A(p, \pi^+)$ processes. The (p, K^+) reaction mechanism is mainly a peripheral one, hence states of high spin should show up strongest. Hence, the reaction is a new tool in studying light hyper-nuclei with possibly better resolution than previous (K^-, π^-) or (π^+, K^+) investigations. Once experimental data exist it would allow to probe the strength of short range correlations in nuclei.

PROPOSALS FOR SATURNE AND COSY

The first experiment of this type is scheduled for the mid of October 1990 at SATURNE [4]. Measurements on D, ^3He and ^4He are being prepared for the 32 m

long SPES4 magnetic spectrometer covering forward angles up to 30°. An intermediate focus is situated 16 m behind the target. For TOF measurements an array of start and stop plastic scintillators is utilized in the intermediate and final focal planes of SPES4. Particles having the same momentum can be effectively separated from kaons by a newly developed TOF trigger providing preselectable time windows of an effective accuracy of 1 ns. In addition, pions, muons and electrons can be vetoed in the trigger by an array of 4 aerogel Čerenkov detectors. They consist of cubes of 10.8 cm height with an index of refraction of 1.03. The light of each cube is collected in a reflecting box containing two half-mirrors focussing the Čerenkov light to two 5 inch photo tubes above and below it. About 7% of the K^+ are vetoed by δ-electrons. This trigger has been successfully tested in previous SATURNE experiments where a ratio of K^+ to mainly π^+ background of 1 : 25 has been achieved on-line [4], [5]. The kaon trajectories are determined by the x- and y-coordinates measured from the hits in two sets of multi-hit drift chambers [4], [5]. Background events may be also removed by tracing the trajectories back to the entrance collimator of SPES4.

From 1993 onward, similar experiments are planned for the high quality proton beam of the new cooler synchrotron COSY at KFA Jülich [6]. The BIG KARL magnetic spectrograph will then be upgraded with entrance quadrupoles allowing for an opening of 10 msr. About 3 m behind the target there will be an intermediade horizontal focus where a thin laminated TOF start detector can be placed. The subsequent total TOF path length will be 13 m, the maximum particle momentum 1050 GeV/c. In the COSY experiments the kaons will be positively identified by a water Čerenkov detector behind the multi-hit focal plane detectors of Big Karl excluding protons while pions and faster particles will be vetoed by aerogel Čerenkov detectors similar to those at SPES4. In addition, particles of the same momentum will be effectively separated from kaons by the fast TOF techniques mentioned above. Drift chambers of 0.2 mm resolution FWHM in x- and y-direction are being developed in collaboration with people from Jülich and Cracow.

An alternative to (p, K^+) reactions is the associated hyper-nucleus production at rest via the $A(p, {}^4HeK^+)_\Lambda(A-3)$ process. In short, we mention the approved pilot experiment at COSY for Λ production at rest [7].

ESTIMATED DETECTION LIMITS

In order to estimate the detection limits in both experiments Monte-Carlo spectra have been generated for the (p, K^+) measurements on a ${}^4He$ target as a test case. Here, the g.s. of ${}^5_\Lambda He$ is 3.1 MeV separated from the ${}^4He + \Lambda$ breakup continuum. Due to final state interactions the continuum rises very sharply. In $pp \to K^+ p \Lambda$ a yield of up to 90 nb/sr/MeV has been found [8]. We assume an instant rise to twice this value (worst case assumption) and simulate the momentum smearing of the continuum to the region of the bound ground state (see Fig. 1). The reaction is calculated for an angle of 6° at an incoming kinetic energy of the protons of 1.344 GeV where the K^+ outgoing momentum is 1050 GeV/c. For optimum energy resolution a target thickness of 1 cm liquid ${}^4He$ is assumed together with a K^+ detection efficiency of 70 %. The input parameters taken to simulate the experiments at SPES4 and BIG KARL are condensed

Parameter	**SPES4**	**BIG KARL**
Solid angle	2.4 msr	10 msr
Beam intensity	$3.3 \times 10^{11}/s$	$10^{10}/s$
Beam resolution	$\pm 3 \times 10^{-4}$	$\pm 10^{-4}$
Momentum resolution	$\pm 5 \times 10^{-4}$	$\pm 10^{-4}$
Energy resolution FWHM	2.2 MeV	0.9 MeV
Counts/day per nb/sr	15.8	15.0

Table 1: Comparison of ${}^4He(p, K^+)^5_\Lambda He$ detection in SPES4 and BIG KARL.

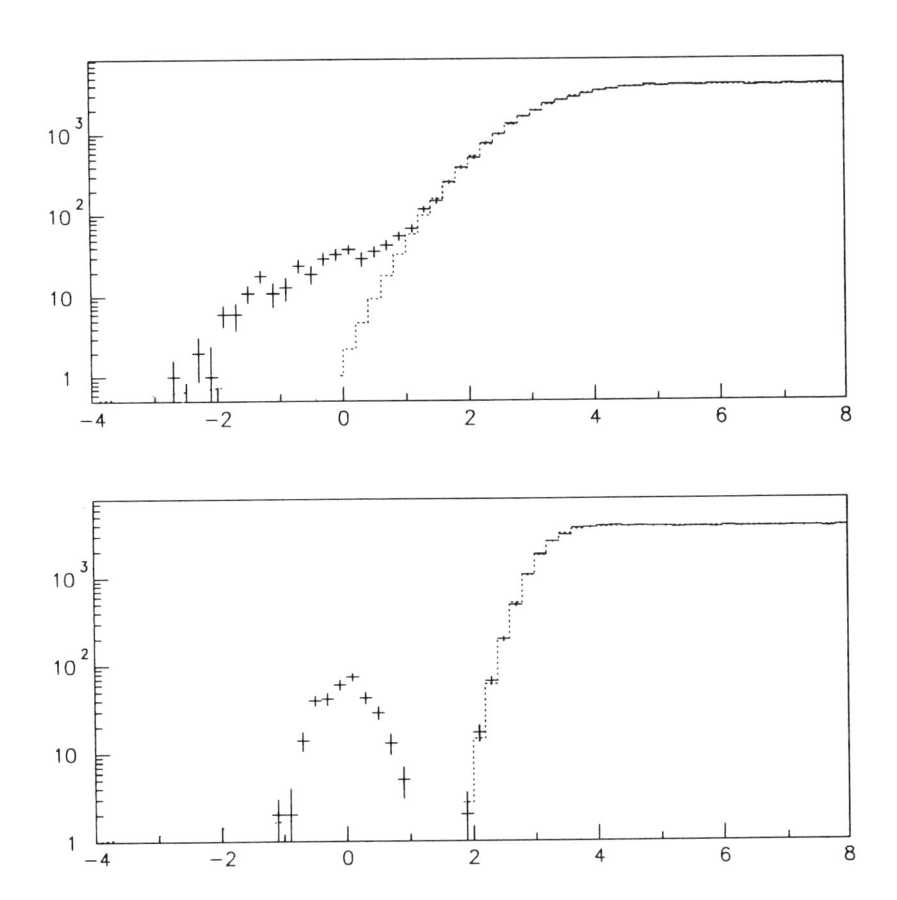

Fig. 1: Estimated missing mass spectra (counts/0.2 MeV) of the ${}^4He(p, K^+)^5_\Lambda He$ reaction with SPES4 (upper diagram) and BIG KARL (lower diagram). The deviation from the g.s. of ${}^5_\Lambda He$ is indicated in MeV.

in Table 1. The Fig. 1 shows the expected missing mass spectra for both detectors assuming cross sections of 3 nb/sr for the g.s. and a beam time of 1 week. At maximum resolution an upper detection limit of about 1 nb/sr is estimated for SPES4. For the upgraded BIG KARL spectrograph the detection limit can be improved to about 0.1 nb/sr at an energy resolution of about 1 MeV FWHM.

The most promising reaction is possibly the production of $^4_\Lambda He$ due to a lower momentum transfer than for $^5_\Lambda He$. Here, the separation from the breakup continuum is 2.42 MeV. On the other hand, the advantage of the still lower momentum transfer forming $^3_\Lambda H$ is compensated by the small short range correlations to be expected in the deuteron as well as in $^3_\Lambda H$ which is only bound by 0.13 MeV. Concluding we think that the proposed experiments will yield new and valuable information on the nucleon-nucleon and nucleon-hyperon interactions in nuclei.

References

[1] S. Shinmura et al., Nucl. Phys. **A450** (1986) 147, Prog. Theor. Physics **76**, 157 (1986), and private communication.

[2] Bo Höistad, "Some Few-Body Reactions at the High-Energy Limit of CELSIUS", Contribution to the workshop on the Physics Program at CELSIUS, CELSIUS-Note 83-15.

[3] M.G. Huber, B.C. Metsch and H.G. Hopf, "Quark Excitations in Nuclei", Lecture Notes in Physics, **242**, 499 (1986).

[4] G. Blanpied, M. Boivin, J.P. Didelez, M.A. Duval, J.P. Egger, J. Ernst, R. Frascaria, F. Hinterberger, R. Jahn, J. Kingler, C. Lippert, C.F. Perdrisat, B. Preedom, G. Rappenecker, T. Reposeur, B. Saghai, R. Siebert, J. Yonnet, E. Warde (updated list), Proposal CE-26 (1988) for the experiment # 182: "Experience Exploratoire de la Production de $^3_\Lambda H$ par Reaction $p + d \rightarrow K^+ + {^3_\Lambda H}$".

[5] C. Lippert, Thesis in preparation, University of Bonn.

[6] J.P. Didelez, J.Ernst, D. Eversheim, Ch. von Falkenhausen, R. Frascaria, F. Hinterberger, R. Jahn, J. Kingler, C. Lippert, O.W.B. Schult, H. Seyfarth, R. Siebert, D. Theis W. von Witsch, D. Witzke (updated list), COSY Proposal # 1, 1989: "Spectroscopy of Light Hyper-Nuclei with BIG KARL".

[7] J.Bisplinghoff, J.P. Didelez, J.Ernst, D. Eversheim, Ch. von Falkenhausen, R. Frascaria, F. Hinterberger, R. Jahn, J. Kingler, D. Kolev, C. Lippert, N. Nenov, R. Siebert, D. Theis, D. Witzke (updated list), COSY Proposal # 2, 1989: "Λ Production at Rest by means of the $^4He(p, {^4HeK^+})\Lambda$ Reaction at 1 GeV".

[8] R. Siebert, Doctoral Thesis, University of Bonn (1990); R. Siebert et al., Nucl. Phys. **A479**, 389 (1989), R. Frascaria et al., Nuovo Cimento **102A**, 561 (1989).

* Work supportd by the Bundesministerium für Forschung und Technologie

STUDY OF THE PRODUCTION OF HEAVY Λ HYPERNUCLEI IN THE (p,K+) REACTION BELOW THE N–N THRESHOLD AND OF THEIR DECAY

O.W.B. Schult, W. Borgs, D. Gotta, A. Hamacher, H.R. Koch, H. Ohm, R. Riepe, H. Seyfarth, K. Sistemich, V. Drüke, D. Filges
Institut für Kernphysik, Forschungszentrum Jülich,
D–5170 Jülich, F.R. Germany

L. Jarczyk, St. Kistryn, J. Smyrski, A. Strzalkowski, B. Styczen
Institute of Physics, Jagellonian Univ.,
PL–30–059 Cracow, Poland

P. von Brentano
Institut für Kernphysik, Universität zu Köln,
D–5000 Köln 41, F.R. Germany

ABSTRACT

An experiment is briefly described which we are preparing for the study of heavy Λ hypernuclei, their formation in the (p,K+)reaction, their Λ decay half life and other Λ decay features. The planned recoil shadow measurement of fission fragments is expected to yield the required background discrimination. The experiment uses the unique features of COSY–Jülich.

Through an experiment, to be performed at COSY–Jülich, we want to obtain information on the production of very heavy Λ–hypernuclei with $_{\Lambda}A \sim A_{target}$ in the $A_t(p,K^+)$ reaction[1] below the N–N threshold. We intend to investigate the reaction mechanism through the measurement of the momentum transfer as function of the bombarding energy T_p. We then plan to determine with higher accuracy the life time of the Λ particle in heavy nuclei by recording delayed fission induced by Λ decay in massive ($_{\Lambda}A \geq 180$) nuclei[2,3]. Further we want to measure the formation cross section of Λ–hyperfragments (with $_{\Lambda}A_f \sim 0.5\ A_t$) in the (p,K+ fast fission) process. The decay of such hyperfragments will be studied through the detection of heavy residues with $A_r \sim {_{\Lambda}A_f}$ which have been kicked out of the original direction of flight of the hyperfragment during its Λ decay.

Formation of heavy Λ–hypernuclei has been demonstrated[2,3] in the annihilation of stopped anti protons, where the $_{\Lambda}A$ recoil is isotropic. In reactions with fast projectiles, however, the $_{\Lambda}A$ recoil is expected mainly in the beam direction, which simplifies the use of the recoil shadow method. This is of advantage in our planned (p,K^+) study which is complementary to the (A_{beam},K^+) experiment[4] at the GSI, where high density achieved in heavy ion collisions might influence the production rate.

A heavy hypernucleus is formed through p bombardment if the Λ particle produced with cross section $\sigma(K^+)$ is trapped with probability $p(\Lambda)$ in the heavy system. Cross sections $\sigma(K^+)$ have been measured below 1 GeV[5], but $p(\Lambda)$ is unknown. Near the $A_t(p,K^+)$ threshold, cooperative processes become extreme. Thus p(Λ) is expected to be ~ 1, however $\sigma(K^+)$ is there very small. The large momentum transfer makes the $A_t(p,K^+)$ X reaction highly unlikely[6]. Below the N–N threshold for K^+ production $\sigma(K^+)$ is sizeable because of Fermi motion, but, in a single step $n(p,K^+)\Lambda$ reaction, the Λ momentum is large relative to the spectator nucleus so that p(Λ) will be small. However, K^+ production in the sub N–N threshold region is most likely[7] due to a two–step process: $p + N \rightarrow \pi + N + N$ and $\pi + N \rightarrow K^+ + \Lambda$, where Fermi motion contributes twice and where the Λ momentum is small enough for Λ capture by the reaction residue. It might lead to a target like Λ hypernucleus, if the nucleus does not undergo prompt fission. Otherwise a fission fragment with a built in Λ can be formed. In the first case A recoils out of the target with a momentum 0.5 GeV/c, while in the other case a hyperfragment and a fission fragment are formed in the target.

Detection of heavy Λ hypernuclei produced with cross sections below ~ 200 nb in the presence of an intense background of light particles should be possible through the recoil–shadow measurement of Λ delayed fission or of hyperfragments deflected during Λ decay. The geometry of the beam, target and detectors is sketched in Fig. 1. In a first measurement fission fragments will be recorded with the position sensitive shadow counters 1 and 2 which in

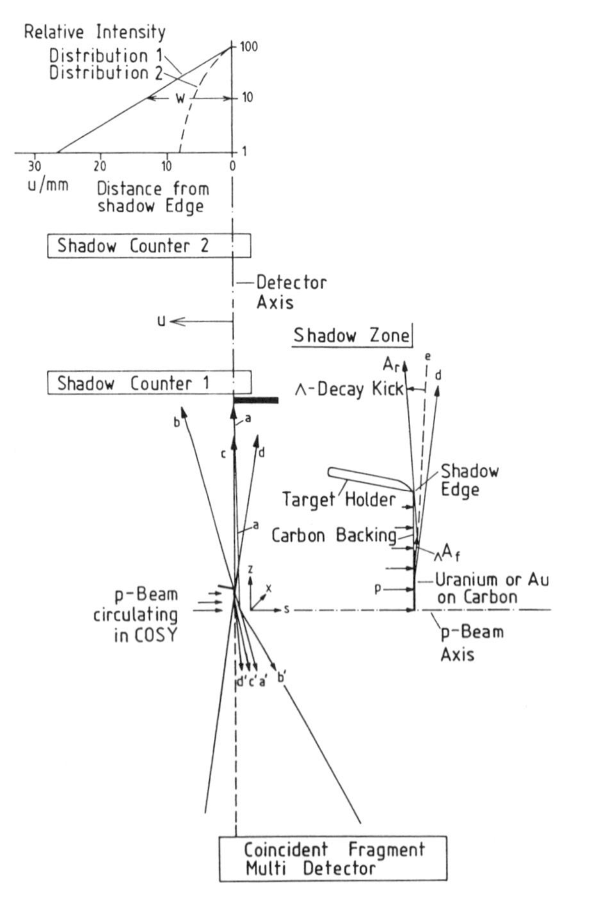

Fig. 1: Arrangement (not to scale) of the internal target with the shadow edge (enlarged in the right part of the figure), of the shadow counters and the coincident fragment multi–detector relative to the proton beam circulating in COSY. The fragments a (towards the shadow edge), b and the coincident fragments a', b' originate from positions where the Λ delayed fission of A occurs behind the target. The trajectories c (to the shadow edge) and d of hyperfragments A_f originate in the target (prompt fission). Their coincident partners are c', d'.

addition to x and u coordinates yield ΔE_1, ΔE_2 and TOF information leading to determination of the fragment mass and charge. The expected relative intensity of the Λ–delayed fission events in the shadow region of Counter 1 is depicted as distribution 1 in Fig. 1. The width W of this distribution is proportional to the product of the momentum transfer and the life time of the Λ in $_\Lambda A$. Superimposed on the indicated distribution 1 is a distribution 2 which results from hyperfragments that have started as $_\Lambda A_f$ e.g. in direction e and received a directional kick in Λ decay. Distribution 2 alone can be obtained (in s direction) by rotating the target and counters 1 and 2 through 180^0 around the detector axis.

In a second experiment events recorded by the shadow counters would be used as trigger for the measurement of the accompanying coincident fission fragment recorded in one element of the multi–detector counter. From its position x_ℓ and s and the x and u positions in the shadow counters the angle is found between the directions of flight of the fragments. In this way the momentum transfer and the velocity of $_\Lambda A$ is determined and hence, using the u coordinate, also the time between Λ formation and decay in a heavy nucleus.

COSY–Jülich will be a unique facility for the planned experiment for several reasons. The low cross section of the reaction requires high luminosity. Demand for suppression of the background by at least several orders of magnitude calls for a method which works with small and thin targets. These two conditions make it necessary to use an internal target in a circulating beam. In view of the expected beam life time rapid synchrotron acceleration and machine resetting is needed together with storage ring operation in less than ~ 3s in total without cooling. Projectile energies should be available with $0.65\ \text{GeV} \lesssim T_p \lesssim 1.5\ \text{GeV}$.

We have carried out simulation calculations which yield local source strengths on target and the dependence of the circulating p–beam intensity on time. With the assumption of $p(\Lambda) \sim 0.02$ we expect rates up to ~ 20/min for 10^{11} protons with $T_p \sim 1.3$ GeV in COSY.

We are very grateful to Prof. Polikanov and Dr. Schädel for very valuable hints.

REFERENCES

1. T. Yamazaki, Proc. KEK Int. Workshop on Nucl. Phys. in GeV Region, KEK, Nov. 13–15, 1984; KEK Report 84–20, p. 17
2. J.P. Bocquet, M. Epherre–Rey–Campagnoll, G. Ericsson et al., Phys. Lett. B182 (1986) 146
3. J.P. Bocquet, M. Epherre–Rey–Campagnoll, G. Ericsson et al., Phys. Lett. B192 (1987) 312
4. Proposal for an experiment at SIS "Subthreshold Production of Hypernuclei", K. Lützenkirchen, S. Polikanov, K. Sümmerer, M. Schädel, J.V. Kratz, and N. Trautmann
5. V.P. Koptev, S.M. Mikirtych'yants, M.M. Nesterov et al., JETP 67 (1988) 2177
6. S. Mrowczynski, Phys. Lett. B220 (1989) 337
7. W. Cassing, G. Batko, U. Mosel et al., Phys. Lett. B238 (1990) 25

SESSION F

The Production of Strange Particles

Investigation of the Reaction $\bar{p}p \to \bar{\Lambda}\Lambda$ Close to Threshold at LEAR

The PS185 Collaboration:
P.D. Barnes[1], P. Birien[4], W.H. Breunlich[8], G. Diebold[1], W. Dutty[4], R.A. Eisenstein[5], W. Eyrich[3], H. Fischer[4], R. von Frankenberg[6], G. Franklin[1], J. Franz[4], R. Geyer[8], N. Hamann[2], D. Hertzog[5], T. Johansson[7], K. Kilian[6], M. Kirsch[3], R. Kraft[3], C. Maher[1], D. Malz[3], W. Oelert[6], S. Ohlsson[7], B. Quinn[1], E. Rössle[4], H. Schledermann[4], H. Schmitt[4], R. Schumacher[1], T. Sefzick[6], G. Sehl[6], J. Seydoux[1], F. Stinzing[3], R. Tayloe[5], R. Todenhagen[4], and M. Ziolkowski[6]

1) Carnegie Mellon University, Pittsburgh, PA, USA
2) CERN, Geneva, Switzerland
3) University of Erlangen–Nürnberg, Erlangen, Fed.Rep.Germany
4) University of Freiburg, Freiburg, Fed. Rep. Germany
5) University of Illinois, Urbana IL, USA
6) Institut für Kernphysik der KFA, Jülich, Fed. Rep. Germany
7) Uppsala University, Sweden
8) Institut für Mittelenergiephysik der ÖAW, Vienna, Austria

presented by:

Walter Oelert
Institut für Kernphysik, Forschungszentrum Jülich
Postfach 1913, D–5170 Jülich, Germany

Abstract

The production of antihyperon–hyperon pairs out of the antiproton–proton collision has been studied by the PS185 collaboration using the Low Energy Antiproton Ring (LEAR) at CERN. The emphasis of the studies is to investigate the production threshold region. Results obtained for the reactions $\bar{p}p \to \bar{\Lambda}\Lambda$ and $\bar{p}p \to \bar{\Lambda}\Sigma^0$+c.c. are presented and discussed. Total– and differential cross sections were measured, polarizations, spin correlations and singlet fractions were extracted at a variety of incident antiproton momenta.

The investigations of the collaboration concentrated on the $\bar{\Lambda}\Lambda$–final channel. Data on $\bar{p}p \to \bar{\Lambda}\Lambda$ are a vailable for a rather wide energy range. A few results on the $\Lambda\Sigma$ channel were analyzed as well as some events for the $\bar{p}p \to \bar{\Sigma}^+\Sigma^+$ reaction. Further investigations on the charged sigma final hyperon–antihyperon pairs are planned and in progress.

Introduction

The exclusive production of hyperon–antihyperon pairs is an attractive channel to study the baryon–antibaryon reaction mechanisms. Measurements of the observables in the threshold region offer an excellent opportunity to examine the dynamics of strangeness creation. At LEAR/CERN the PS185 collaboration has initiated and performed a set of measurements of the process: $\bar{p}p \to \bar{\Lambda}\Lambda$ [1,2]. For the $\bar{p}p \to \bar{\Lambda}\Sigma^0$ + c.c. channel first results are available [3]; preliminary observations of $\bar{p}p \to \bar{\Sigma}^+\Sigma^+$ events have been extracted.

In the present contribution we concentrate to the final $\bar{\Lambda}\Lambda$ channel. The observables: total – and differential cross sections, polarization, and spin correlations have been determined in a range from the production threshold up to 1.92 GeV/c, covering a squared invariant mass S = 4.978 $(\mathrm{GeV}/c^2)^2$ to 5.77 $(\mathrm{GeV}/c^2)^2$. A track imaging decay spectrometer was used. Close to the reaction threshold the production mechanism should be less complicated due to only a few partial waves contributing to the strangeness production process or processes. Still the simultanious theoretical interpretation of the observables is not conclusive. The PS185 collaboration at LEAR will continue the study of antistrangeness– strangeness production in a variety of different exit channels. At other accelerators as CELSIUS, ELSA and COSY similar associated strangeness production experiments employing different environments are planned by other collaborations.

All theoretical interpretations [4–21] – concentrating especially on the $\bar{p}p \rightarrow \bar{\Lambda}\Lambda$ production reaction – emphasize the importance of initial and final state interaction. It leads to distortions (annihilation) preferentially for lower partial waves. Coupled channel effects are believed to contribute significantly, subthreshold resonance formation seems to contribute as well. Generally, the theoretical applications reflect a duality between constituent quark model considerations (Fig. 1a) and boson exchange (t–channel) (Fig. 1b).

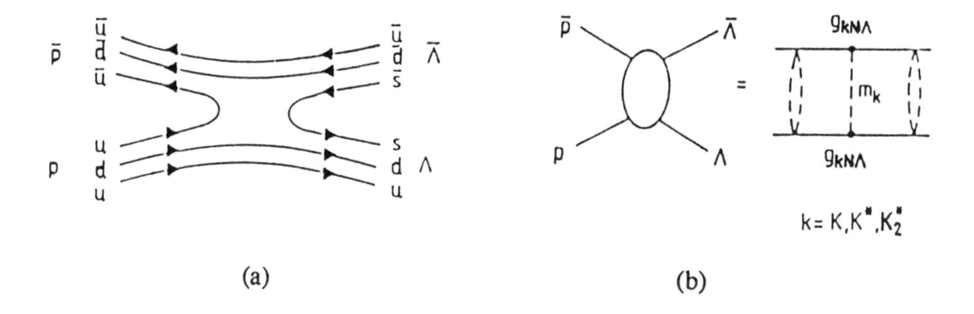

Figure 1: Simplified views of the pp → ΛΛ reaction as a) Simplest possible quark–line diagram, b) t–channel meson exchange.

In a simple constituent quark model the $\bar{p}p \rightarrow \bar{\Lambda}\Lambda$ reaction is related to the creation of an antistrange–strange quark–pair, since $s(\bar{s})$ quarks in $\Lambda(\bar{\Lambda})$, respectively, are coupled to a spin and isospin zero spectator di–quark which does not participate in the reaction process. Therefore, in this consideration the spin of the $\Lambda(\bar{\Lambda})$ is carried by the $s(\bar{s})$ quark and thus the polarization parameters of the hyperon (antihyperon) as well as the other observables reflecting the reaction dynamics, are those of the respective s–quark.
The measured angular momentum and spin correlation of the $\Lambda(\bar{\Lambda})$ could tell to which degree the $s\bar{s}$–pair is created with quantum numbers of an effective gluon field 1^- $(^3P_0)$ or an 0^+ vacuum state $(^3S_1)$.

Both theoretical approaches are rather successful in describing differential cross sections and polarizations. In fact, the quark and the boson exchange approaches give very similar results, perhaps reflecting the liberty in fitting the potential parameters. In this context the reaction $\bar{p}p \rightarrow \bar{\Lambda}\Sigma^o$+c.c. is of particular interest. The long ranged K(494) exchange is suppressed here: SU(3) relations give the ratio of the coupling constants $g^2 K N \Sigma^o / g^2 K N \Lambda = 1/27$. As a consequence the short ranged K*(892) exchange becomes more important for the $\bar{\Lambda}\Sigma^o$ reaction since $g^2 K^* N \Sigma^o / g^2 K^* N \Lambda \approx 0.8$, and, one might hope that the reaction dynamics are more sensitive to quark effects [8]. More details on the quark based model were given recently at the Stockholm meeting [22].

Experiment

The PS185 detector is a non–magnetic forward decay spectrometer that takes advantage of the two–body kinematics and the delayed decay signature of the $\bar{p}p \rightarrow \bar{Y}Y$ reactions using e.g. a neutral trigger in case of $Y = \Lambda^o$. The arrangement [1,2,23], shown in Fig. 2, consists of four principal components:

i) a trigger–active target system, which is sandwiched out of four CH_2 and one carbon target cells, each cylindrical with 2.5 mm diameter and 2.5 mm thickness surrounded by scintillators providing a neutral or charged event trigger;
ii) a track reconstructing system out of a stack of 10 multiwire proportional and 13 drift chamber planes.
iii) a scintillator hodoscope for triggering on charged decay particles of $\Lambda(\bar{\Lambda})$; and
iv) another three drift chambers, surrounded by a 0.1 T magnetic field solenoid for baryon number identification.

The laboratory coverage of $\pm$ 41 degree of the detector ensures a 100 % acceptance for the hyperon–antihyperon production at LEAR–momenta $\leq$ 2 GeV/c. The global efficiency in reconstructing a $\Lambda\bar{\Lambda}$ pair decaying into charged particles is about 60 % including geometrical and reconstruction efficiencies. The two–body kinematics allows for a complete kinematical reconstruction of the events in the non–magnetic detector; the information coming from the bending of the particles in the magnet is primarily used to assign the baryon number to the hyperons. The data evaluation is restricted to events having two neutral vees in the chamber stack and after applying geometrical and kinematical cuts, the event candidates are fed into a kinematic fitting procedure for the final analysis.

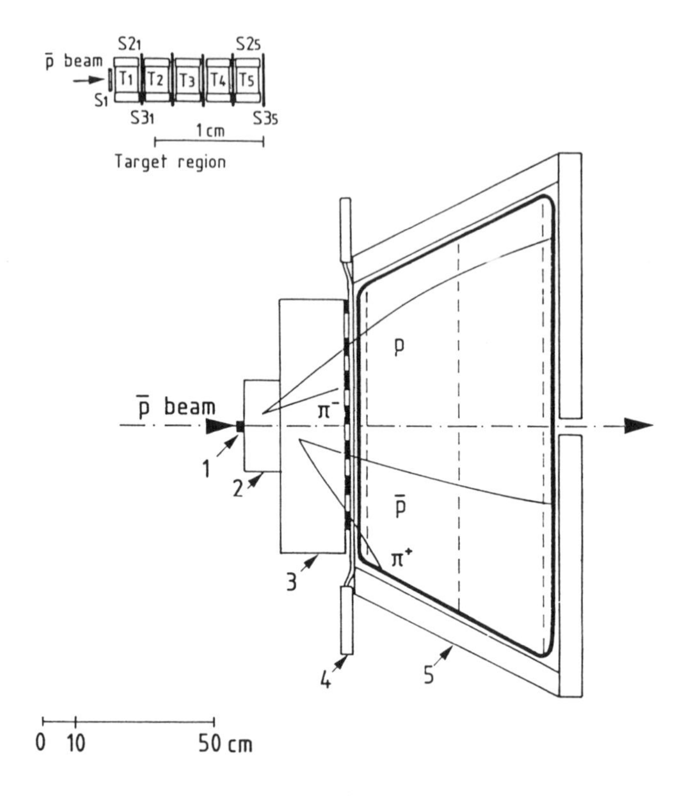

Figure 2: The PS185 experimental setup at the external beam line of LEAR, and expanded view of the target is shown: 1 = target, 2 = multiwire proportional chambers, 3 = drift chambers, 4 = scintillator hodoscope, 5 = magnetic solenoid with three drift chambers. The topology of a typical $\bar{p}p \to \bar{\Lambda}\Lambda \to \bar{p}\pi^+ p\pi^-$ event is indicated.

Results

a) cross sections

The measured excitation function for the $\bar{\Lambda}\Lambda$–production in the momentum range of LEAR is shown in Fig. 3 [1,2,24,25] together with other data points [26–30]. Cross sections for the $\bar{p}p \to \bar{\Lambda}\Sigma^0$ + c.c. channel are included as well. As can be seen, the threshold behaviour for the $\bar{\Lambda}\Lambda$ channel is well mapped out. The excitation function at the threshold region [2] shows that $l \geq 1$ partial waves contribute to the cross section already at very low $\bar{\Lambda}\Lambda$ relative energies.

In addition to the results on the total cross sections the differential cross sections confirm contributions of higher partial waves by their clear anisotropy in the angular distribution as can be seen in figure 4. Angular distributions at other momenta show a similar pattern, always – even as close to threshold as 800 keV excess energy – having a forward rise.

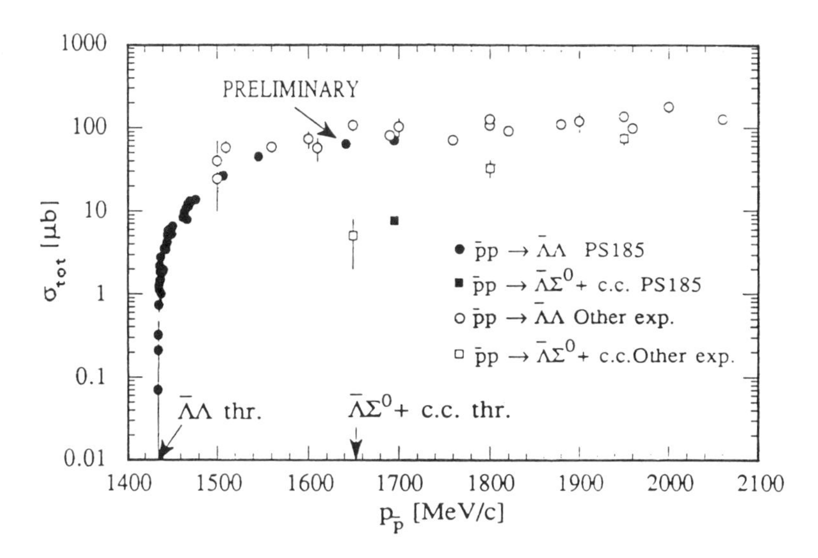

Figure 3: Compilation of available data for $\bar{p}p \to \bar{\Lambda}\Lambda$ and $\bar{p}p \to \bar{\Lambda}\Sigma^0$+c.c. in the momentum range of LEAR. The threshold momenta for the reactions are indicated by arrows.

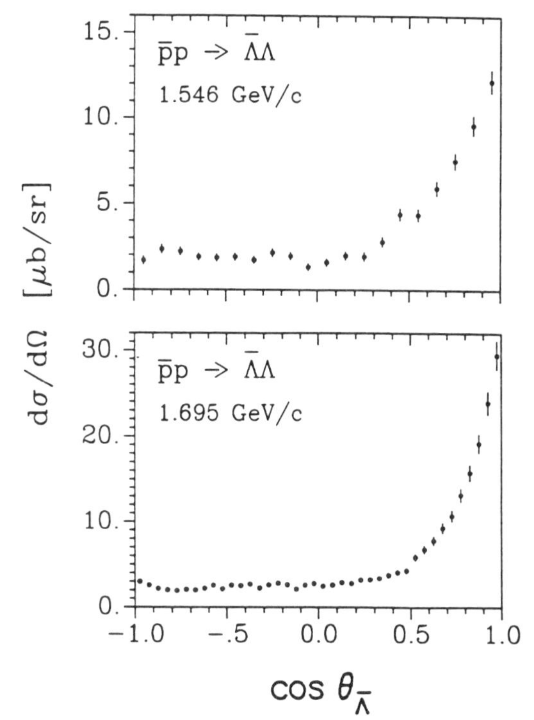

Figure 4: Differential cross sections at 1.546 GeV/c and 1.695 GeV/c via the center of mass angle for Λ.

The importance of the different partial waves can be deduced from the coefficients of a Legendre expansion $d\sigma/d\Omega = \Sigma a_n P_n(\cos\theta)$ of the data. For beam momenta ≤ 1.5 GeV/c the angular distributions could be described with $l < 2$; with increasing beam momentum an increasing d–wave contribution is obvious. The relative coefficients are shown in Fig. 5 together with other results [27].

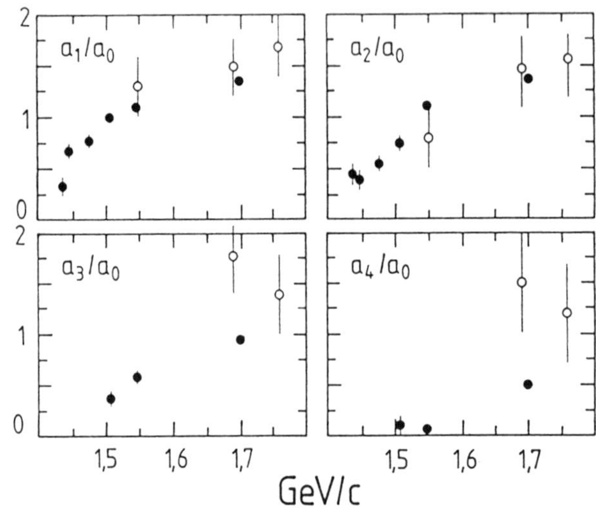

Figure 5: Results from legendre fits to the angular distribution of $\bar{p}p \to \bar{\Lambda}\Lambda$ at different beam momenta. The ratios a_1/a_0 to a_4/a_0 are shown, the open circles are from Ref. 27.

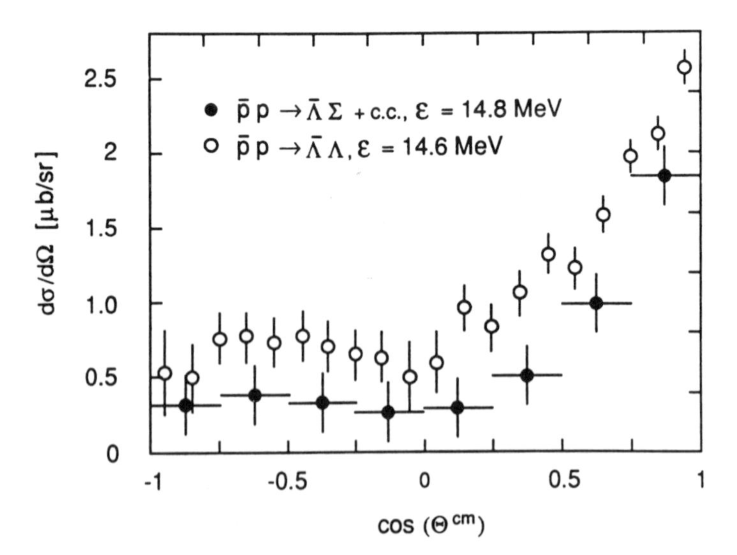

Figure 6: Comparison of the measured differential cross sections for the reactions $\bar{p}p \to \bar{\Lambda}\Sigma^0$+c.c. and $\bar{p}p \to \bar{\Lambda}\Lambda$.

The general features as observed for the $\bar{\Lambda}\Lambda$ angular distributions are also recognized in the $\bar{\Lambda}\Sigma^o$+c.c. channel: a forward rise and a flat distribution at backwards angles. Again in this channel strong $l > 0$ contributions govern the production mechanism. The PS185 collaboration measured both channels the $\bar{p}p \to \bar{\Lambda}\Lambda$ and the $\bar{p}p \to \bar{\Sigma}^o\Lambda$+c.c. at almost the same excess energy of 14.6 MeV and 14.8 MeV, respectively. The two differential cross sections are compared in Fig. 6.

We can compare the measured ratio of cross sections at equivalent excess energies:

$$R_\epsilon = \sigma(\bar{\Lambda}\Sigma^o)/\sigma(\bar{\Lambda}\Lambda) = 0.29 \pm 0.02.$$

This ratio is down to $R_p = 0.054 \pm 0.03$, however, when the comparison is made for the measurements at the same incident $\bar{p}$–momentum of $p_{\bar{p}} = 1.695$ GeV/c. Regarding the question which of these R–values would be the right one to compare to theoretical predictions it should be notified that a) initial state interactions should not be very different but b) final state interactions of the outgoing antihyperon–hyperon pair would more likely result in equivalent annihilation strength when the comparison is made for the same excess energy. On the basis of this argument the extracted experimental ratio R_ϵ agrees very well with theoretical averaged values [4,5,6,9] around 0.25 ± 0.01 from "quark–based" models.

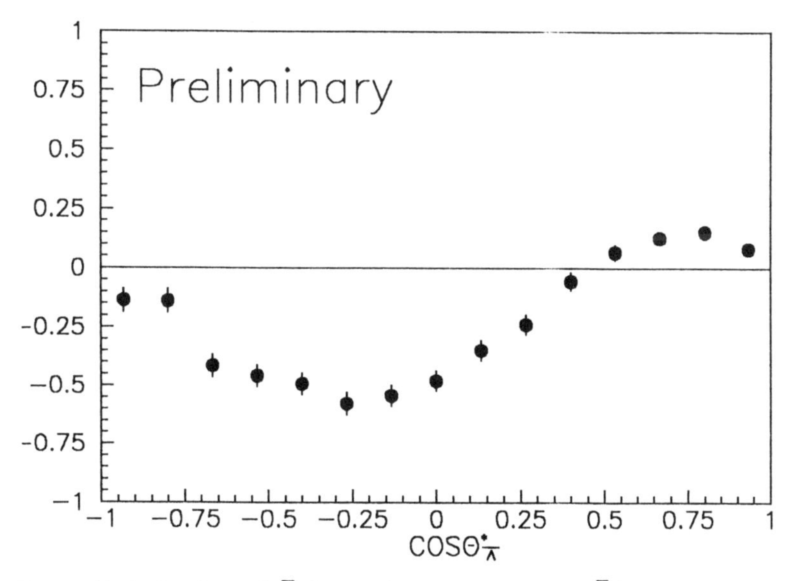

Figure 7: Polarization of $\bar{\Lambda}$ from the reaction $\bar{p}p \to \bar{\Lambda}\Lambda$ at 1.642 GeV/c incident momentum.

b) polarization
The self–analyzing weak decay of the $\Lambda(\bar{\Lambda})$–particles reveals a straight forward determination of the polarization with the decay assymmetry parameter $\alpha = 0.642 \pm 0.013$. Preliminary data of the highest statistic run with about 40.000 events analyzed at 1.642 GeV/c are shown in Figure 7 [31] as function of $\cos(\theta_{\bar{\Lambda}})$.
The polarization shows as a common feature a positive part at small angles (small momentum transfer) and a negative polarization for larger angles. Only for the preliminary data at an incident momentum of 1.922 GeV/c zero polarization seems to be measured for larger C.M. angles. These observations are summarized in Fig. 8 where the polarization data is shown versus the squared four–momentum transfer.

The zero crossing from the forward positive part of the polarization to the negative values appears to happen always at the same relative squared four–momentum transfer t' = t–t($\cos\theta_{\bar{\Lambda}}$ = 1) = 0.18 $\pm$ 0.02 $(\mathrm{GeV}/c)^2$. It should be noted here that the shape change from forward rise of the differential cross sections to the rather constant values for larger angles appear precisely at the same t' value.

In the PS185 experiment the polarization P has been analyzed independently in the $\bar{p}p \rightarrow \bar{\Lambda}\Lambda$ reaction for the kinematically correlated Λ's and $\bar{\Lambda}$'s. An apparent difference would signal a CP violating effect in the weak decay parameter α, since CP conservation requires zero for the relative quantity:

$$A = (\alpha_\Lambda P_\Lambda + \alpha_{\bar{\Lambda}} P_{\bar{\Lambda}})/(\alpha_\Lambda P_\Lambda - \alpha_{\bar{\Lambda}} P_{\bar{\Lambda}}).$$

The results obtained [24,25] are compatible with zero, the values being:

A = –0.07 $\pm$ 0.09 at 1.546 GeV/c and A = 0.006 $\pm$ 0.073 at 1.695 GeV/c

leading to a weighted average value for both incident momenta of A = –0.027 $\pm$ 0.057. These values are based on 4063 and 11362 analyzed events for the lower and the higher beam momenta, respectively.

Taking all data collected by the PS185 collaboration and being presently analyzed the uncertainty of the value A will decrease to about 1 %. Since the origin of CP–violation in the electro weak interaction is one of the fundamental questions, investigations have been started mainly by N. Hamann and D. Hertzog [32–35] to study the feasibility of CP–nonconservation measurements in antihyperon–hyperon productions. Here the asymmetry rate of decay angular distribution asymmetry parameters, and final baryon polarization parameters are regarded. The envisaged experiments will certainly be not easy to be performed and ask for a special effort for the high challenge to observe or not–observe CP–nonconservation other than the K^0–K^0 system [36–38].

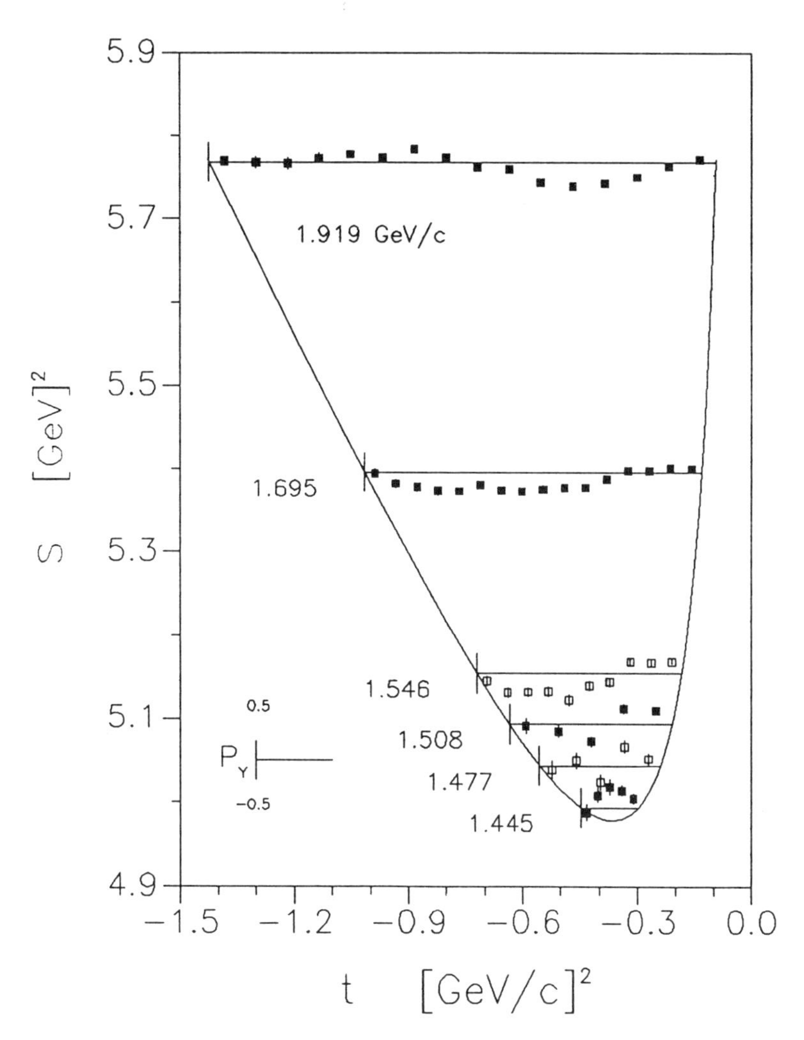

Figure 8: The polarization distributions as function of the four–momentum transfer squared t and the invariant mass squares s. The curve shows the kinematically allowed range of t for the S values.

c) spin correlations
Since in the present experiment Λ and $\bar{\Lambda}$ are always detected in pairs the spin correlations can also be extracted. Polarization and spin correlations are analyzed according to:

$$P_y(\theta) = 3/\alpha \quad 1/N \quad \Sigma \cos(\theta_p)$$

$$C_{ij}(\theta) = 9/\alpha\alpha \quad 1/N \quad \Sigma \cos(\theta_p) \cos(\theta p)$$

with θ_p being the angle of the decay proton $\Lambda \to p\pi^-$ in the Λ–rest frame, see Fig. 9 for the coordinate system used.

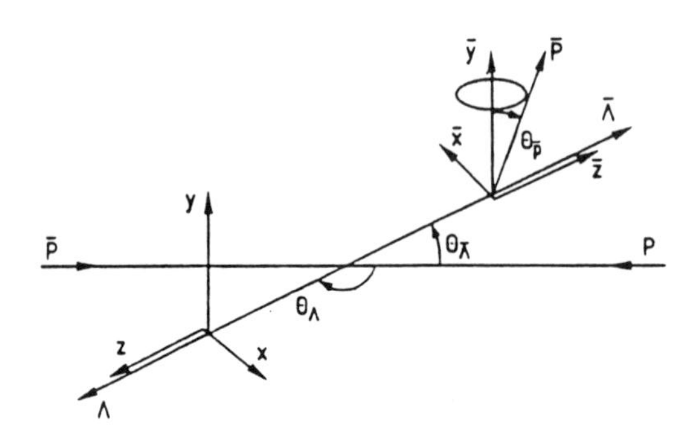

Figure 9: Used coordinate system for evaluating polarizations and spin correlations.

Symmetry laws require the following constrains to the polarization and spin correlation observables:

Parity conservation: $P_x(\theta) = P_z(\theta) = 0$; $C_{xy} = C_{yx} = C_{yz} = C_{zy} = 0$.

Charge conjugation in variance: $P_y(\theta)$; $C_{ij} = C_{ji}$.

CP invariance: $\alpha_\Lambda = \alpha_{\bar{\Lambda}}$.

Angular distributions of the $\bar{\Lambda}\Lambda$ spin correlations have been extracted. For the differential beam momenta the distributions coincide on a t' scale rather well with strong correlations in certain regions, as can be seen in Fig. 10 for two of the measured cases.

The singlet fraction $S = (1+C_{xx}-C_{yy}+C_{zz})/4 = 0.25\ (1-\sigma_{\bar{\Lambda}}\sigma_\Lambda)$ as function of the reduced four momentum transfer is shown in Fig. 11. This quantity is a measure of the spin alignment of the produced $\bar{\Lambda}$–Λ system, having S = 1 if the pair is produced purely in a relative singlet state and S=0 if produced purely in a relative triplet state. The averaged values for the singlet fractions are consistent with zero the numbers being:

$$S = -0.078 \pm 0.052 \quad \text{and} \quad S = -0.032 \pm 0.030$$

at beam momenta 1.546 GeV/c and 1.695 GeV/c, respectively.

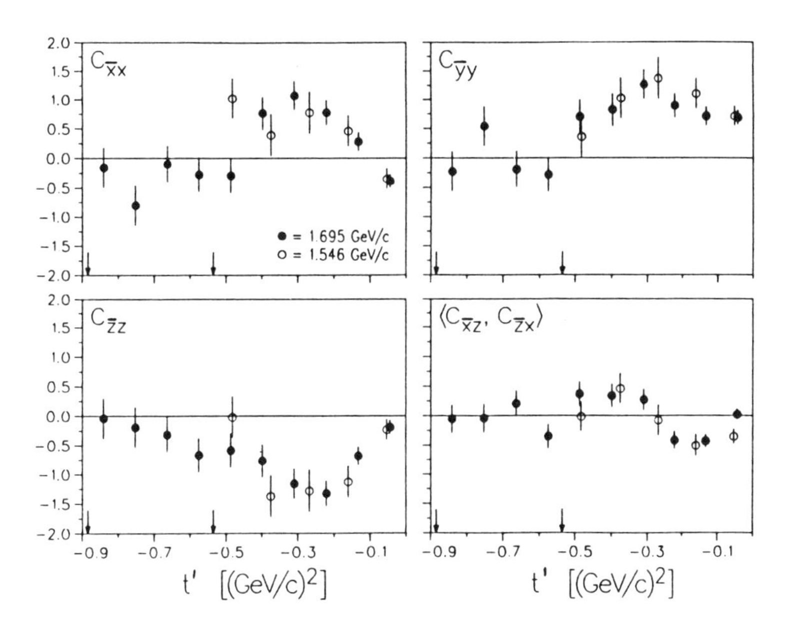

Figure 10: Relevant spin correlation coefficients and averaged values of $C_{\bar{x}z}$ and $C_{\bar{z}x}$ as function of t' at 1.546 GeV/c and 1.695 GeV/c. The arrows indicate the kinematically possible maximal values of t'.

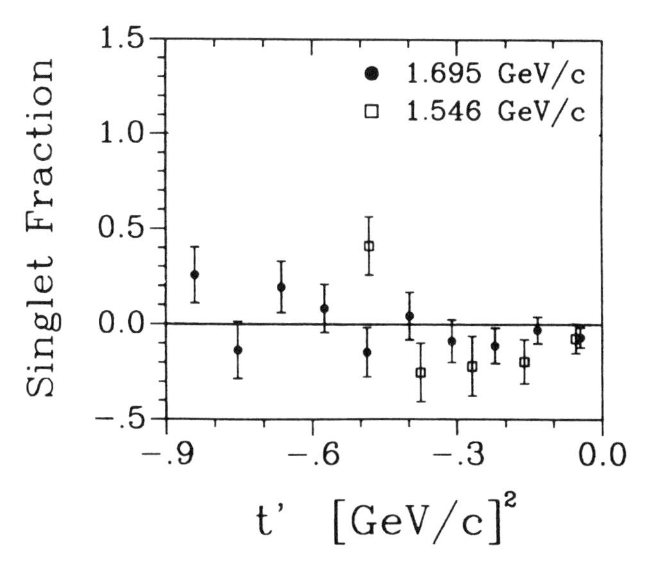

Figure 11: Differential singlet fractions for two beam momenta in the full kinematic range.

Conclusion

A global and coherent picture emerges from data taken by the PS185 collaboration on the reaction $\bar{p}p \to \bar{\Lambda}\Lambda$ at antiproton momenta < 2 GeV/c:

a) $\bar{\Lambda}\Lambda$ pairs are always produced in a relative triplet state.

b) Strong p–wave contribution is observed close to the production threshold, higher partial waves are noticeable in the momentum range studied.

c) The observables are to a large degree energy independent. At a four momentum transfer of $t' \approx -0.18$ $(GeV/c)^2$ there is a change observed in slope and sign of angular distributions and polarizations, respectively.

The entire body of precision data analyzed for several observables should allow to put constrains on the theoretical models. Further different reaction channels to be studied in addition will help to understand the nature of antistrange–strange associated production. Such experiments are in preparation for the charged Σ–channel at LEAR, after preliminary tests have been successful. Experiments using different entrance channels are planned at other accelerators.

Acknowledgements

These measurements reported here were only made possible by the excellent performance of the CERN antiproton complex, in particular the LEAR ring. We wish to thank the LEAR team for their enthusiasm and help.
The support from the Austrian Science Foundation, the German Federal Minister for Research and Technology and the Forschungszentrum Jülich, the Swedish Natural Science Research Council, the United States Department of Energy and the United States National Sciene Foundation is gratefull acknowledged.

References

1. P.D. Barnes et al., Phys. Lett. 189B (1987) 249
2. P.D. Barnes et al., Phys. Lett. 229B (1989) 432
3. P.D. Barnes et al., Phys. Lett. 246B (1990) 273
4. H. Genz and S. Tatur, Phys. Rev. D30 (1984) 63
5. H. Rubinstein and H. Snellman, Phys. Lett. 165B (1985) 187
6. M. Kohno and W. Weise, Preprint TPR–86–13, Contributed paper to 11th Int. Conf. on Few Body Systems in Particle and Nuclear Physics, Tokyo & Sendai, August 1986
7. M. Kohno and W. Weise, Phys. Lett. 176B (1986) 15
8. M. Kohno and W. Weise, Nucl. Phys. A479 (1988) 433c
9. S. Fururi and A. Faessler, Nucl. Phys. A508 (1987) 669
10. H. Burk and M. Dillig, Phys. Rev. C37 (1988) 1362
11. M.A. Alberg et al., Nucl. Phys. A508 (1990) 323c
12. W. Roberts, Preprint HUTP–90/A020
13. F. Tabakin and R.A. Eisenstein, Phys. Rev. C31 (1985) 1857
14. J.A. Niskanen, Preprint Helsinki Univ. HU–TFT–86–28
15. P. Lafrance, B. Loiseau and R. Vihn Mau, Phys. Lett. 214B (1988) 317
16. P. Lafrance and B. Loiseau, Preprint CRM–1673, 1990
17. T. Hippchen et al., Proc. 4 LEAR workshop, Villars, Harwood (1988)
18. R.G.E. Timmermans, T.A. Rijken and J.J. de Swart, Nucl. Phys. A479 (1988) 383c
19. M. Kohno and W. Weise, Phys. Lett. 206B (1988) 584
20. O.D. Dalkarov, K.V. Protasov and I.S. Shapiro, Preprint 37, Moscow, FIAN, 1988
21. A. Kudryatsev and V. Samilov, Mod. Phys. Lett. 4A (1989) 721
22. See the contributions from P. La France and M. Alberg to the First Biennial Conference on Low Energy Antiproton Physics, Stockholm, Sweden, 2–6 July 1990, proceedings to appear
23. P.D. Barnes et al., Phys. Lett. 246B (1990) 273
24. W. Dutty, Dissertation, Univ. of Freiburg (1988)
25. G. Sehl, Dissertation, Juel–Spez 535 (1990)
26. L. Durand and J. Sandweiss, Phys. Rev. 135 (1964) B540
27. B. Jayet et al., Il Nouvo Cimento 45A (1978)371
28. J.W. Cruz, Ph.D. thesis, Rutgers Univ. (1983)
29. B.Y. Oh et al., Nucl. Phys. B51 (1973) 57
30. J. Button et al., Phys. Rev. 121 (1961) 1788
31. H. Fischer, Dissertation in preparation
32. N.H. Hamann, Proceedings of the First Binnial Conference on Low Energy Antiproton Physics, Stockholm, Sweden, 2–6 July 1990
33. D.W. Hertzog, Proc. Symposium on Spin and Symmetries, TRIUMF, Vancouver 1989, Report TRI–89–5 (1989), p. 89
34. D.W. Hertzog, Proceedings of the TRIUMF workshop KAON, July 1990
35. N.H. Hamann, CERN–EP/89–66 (1989)
36. L. Wolfenstein, Ann. Rev. Nucl. Part. Sci. 36 (1986) 137
37. J.F. Donoghue et al., Phys. Rev. D34 (1986) 833
38. J.H. Christenson et al., Phys. Rev. Lett. 13 (1964) 138

HYPERON-NUCLEON INTERACTION STUDIES BY MEANS OF ASSOCIATED STRANGENESS PRODUCTION IN PROTON-PROTON COLLISIONS

R. Frascaria
Institut de Physique Nucléaire - 91406 ORSAY Cedex

ABSTRACT

The status of the results and the analyses of associated strangeness production in Nucleon-Nucleon collisions at Saturne (LNS, Saclay France) is presented. Plans for a new large acceptance spectrometer which are working out for spin observable measurements are exposed.

INTRODUCTION

It is now well established that the long range part of the nucleon-nucleon (NN) potential is described by the one pion exchange potential (OPEP). Differencies appear between potentials[1] for the description of the medium and short range part of the interaction. The $\sigma(2\pi)$ and ρ meson exchanges are responsible for the medium range part. At short distances, where the ω plays an important role a difficulty appear due to the finite size and the internal structure of the nucleon. In meson exchange potentials, the finite size of the nucleon is taken into account by the introduction of an ad-hoc form factor at the meson-nucleon vertex. The generalization of these studies to the baryon-baryon system allows us to get a deeper insight into the knowledge of the strong interaction between two baryons with different internal constitution. For instance if in NN interactions isospin 0 or 1 bosons can be exchanged, in ΛN interactions these are isospin 0 and 1/2 which can be exchanged. In a $\Lambda N \rightarrow \Lambda N$ elastic scattering neither π nor ρ are exchanged. In this case, the range must be shorter than in NN, with a weaker spin dependence.

Aside the boson exchange potentials, effective models based on QCD, describing rather well the hadrons are now used[2] to describe the strong interaction in its medium and short range parts.

Whatever is the theoretical approach used , these studies come up against the lack of hyperon-hyperon (YY) or hyperon-nucleon (YN) scattering data.

The YN scattering data obtained from bubble chamber experiments are scarce mainly due to the difficulty which exists in producing and making interact the short-living hyperons in the same detector volume. The associated production of an hyperon-nucleon pair (YN) in few body interactions is a simple way to study the baryon-baryon interaction at low relative energies. The simplest réactions where YN interaction at low relative energy can be studied through final state interactions (FSI) are of two types. The first one uses a deuterium target and the reactions $K^-d \rightarrow \pi YN$ and $\pi^-d \rightarrow KYN$ have been studied both in bubble chamber and in counter experiments[3]. New developments will probably emerge in the next future at CEBAF in studies of reactions such as $\gamma d \rightarrow K^+YN$ or $ed \rightarrow e'K^+YN$. The second one is the use of the $pp \rightarrow K^+YN$ reactions. These are rather simple reactions but nevertheless need a careful study of the associated strangeness production (ASP) mechanism. These 3-body final state reactions have never been exclusively measured in electronic experiments selecting the different channels $K^+\Lambda p$, $K^+ \Sigma_0 P$, $K^+\Sigma^+N$ when the $(\Sigma N)^+$ channels are open. The first $pp \rightarrow K^+X$ experiments were performed at Brookhaven National Laboratory[4] but these high statistic experiments suffered from poor K^+ momentum resolution and absolute momentum determination.

This contribution reports on a high resolution $pp \rightarrow K^+X$ experimental programme developped[5] in the last four years at Saturne National Laboratory (LNS , France) and presents the different analyses which have been developped further[6,7].

In a perspective for the next future at LNS, a new experimental apparatus designed to study the different ASP channels separately and allowing the measurements of spin observables is presented.

I - THE $pp \rightarrow K^+X$ EXPERIMENT AT LNS

The experiment utilizes the Saturne Synchrotron and the SPES4 beam line spectrometer[8]. The synchrotron delivers protons with energies ranging from .2 to 2.9 GeV with an intensity of 10^{12} p/c.

Scintillator telescopes viewing either a thin CH_2 film upstream from the target, or the liquid hydrogen target (280 mg/cm ^{2}) itself and a secondary emission monitor are used to monitor the relative proton flux. Absolute calibrations of the monitors were made by activation measurements from $^{12}C(p,pn)^{11}C$.

The SPES4 beam line is a 32 meter long spectrometer which allows the momentum analysis of particles up to 4 GeV/c. The (p,K^+) has to be done in a very high flux of protons and π^+. The K^+ discrimination is achieved by :

- Momentum analysis with SPES4. The solid angle of the spectrometer, defined by a lead collimator is $\Delta\Omega = 2.5\ 10^{-4}$, with momentum acceptance $\Delta p/p = \pm\ 3.5\ \%$, and momentum resolution 10^{-4}.

- Velocity measurement. Different time of flight (TOF) measurements are performed between the intermediate focal plane (IFP) and the region of the final focal plane (FFP) located 16 meters downstream. The "starts" are delivered by 12 scintillator counters, the stops are given either by one plane of 13 scintillator counters, or by Cerenkov counters or by two thin scintillator strips each one viewed by two photomultipliers at each end and covering the whole focal plane.

- Proton and pion rejections : the experiment has been done at two different bombarding energies T_p = 2.3 GeV and T_p = 2.7 GeV in two different runs. For the experiment at T_p = 2.3 GeV the proton and pion rejections has been obtained by 4 Total Reflection Cerenkov Detectors (TRCD) [8]. For the second part of the experiment for which the momentum range of the kaons is higher, aerogel counters have been used[9]. A very good proton and pion rejection has been obtained in both cases. For instance with the TRCD, a 99 % proton and a 92 % pion rejections have been achieved keeping a kaon efficiency as high as 90 %[8,9].A typical time spectrum is shown in figure 1 after applying more restrictive Cerenkov amplitude windows off line. The final good separation between π^+ and K^+ has a FWHM time resolution of typically 800 ps.

- Reconstruction of particle trajectories in FFP and missing mass spectrum at the target position. Two multidrift counters located close to FFP allow the reconstruction of the trajectories in both horizontal and vertical planes. Using the inverse matrix of SPES4, one can display the θ-momentum (or θ-missing mass) and θ-momentum planes at the target position. This allows the rejection of any particle that does not come from the target. The precision for the localisation of the particles in the FFP is around 1 mm. The total energy resolution is 1.5 MeV at T_p = 2.3 GeV and 2.5 at T_p = 2.7 GeV.

The results of this investigation at T_p = 2.3 GeV include cross sections for K^+ emission at θ_K = 0°, 6°, 8°, 10° and 12°[3]. The missing mass spectra are shown on Fig. 2, the sum of the spectra at θ = 8, 10 and 12° beeing shown on Fig. 3, since these 3 spectra have a similar behaviour.

The results of the experiment at T_p = 2.7 GeV are preliminary and show a similar general trend. The missing mass spectra have been obtained for θ_K = 8°, 13°, 16°, 20° and 23°. The θ_K = 13° spectrum is shown on Fig. 4.

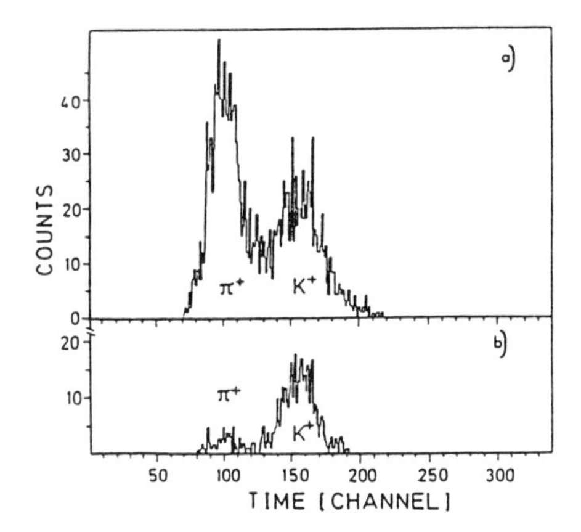

Fig. 1 a) Time of flight spectrum I-F

b) The same TOF-spectrum like in (a) ; in addition an "*amplitude window*" (see text) on the measured Cerenkov pulse height is used.

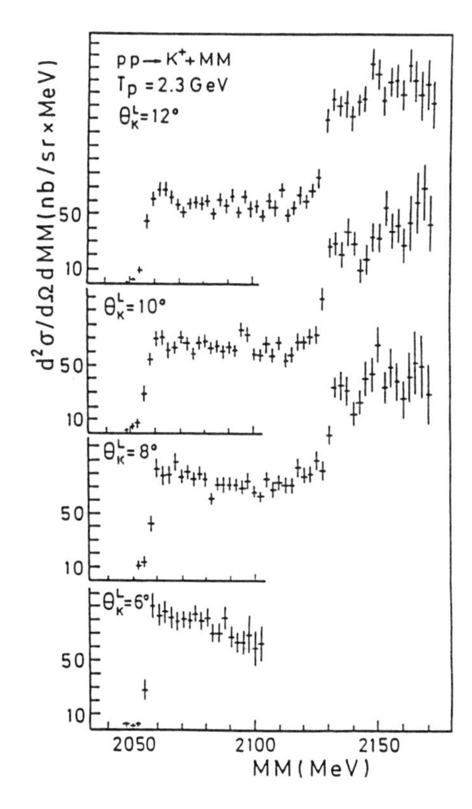

Fig. 2 Missing mass (MM) spectra of the reaction pp → K^+ + MM at T_p = 2.3 GeV for θ_K^{lab} = 6, 8, 10 and 12° ; the errors are stastistical, the bin width is 2.5 MeV.

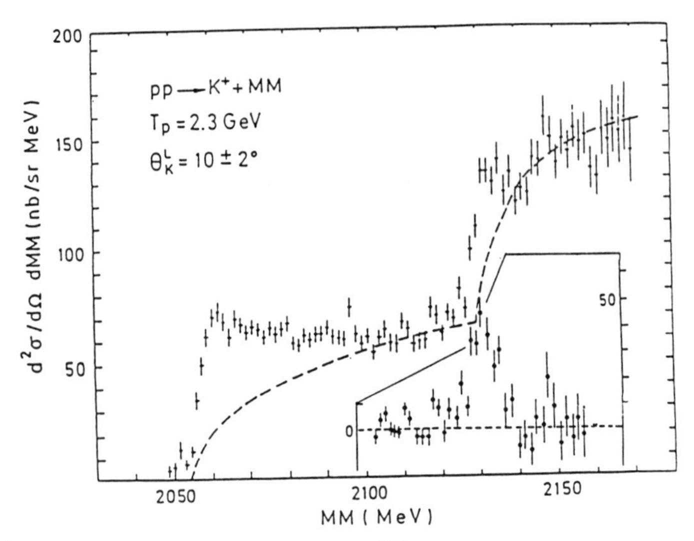

Fig. 3 Sum of the MM spectra at θ_K^{lab} = 8, 10 and 12° ; statistical errors are indicated. The dashed curve indicates the 3-body phase spaces (3-BPS). The data points shown at the bottom of the figure, are obtained by subtraction of these 3-BPS from the measured cross section in the vicinity of the $(\Sigma N)^+$ thresholds.

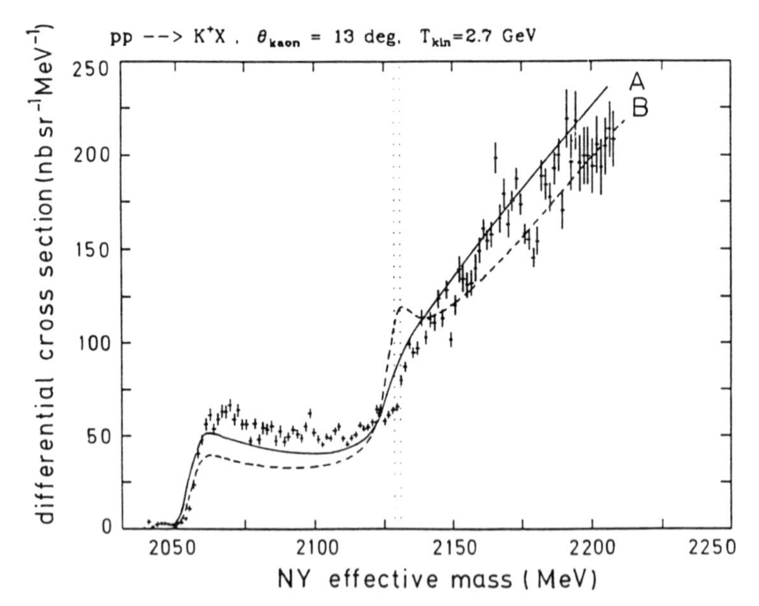

Fig. 4 New results on YN missing mass obtained at T_p = 2.7 GeV and $\theta_K^{lab.}$ = 13°. The curves A and B come from the Deloff calculation and are explained in the text.

II - RESULTS AND INTERPRETATION OF THE $pp \to K^+X$ EXPERIMENT

The missing mass spectra can be divided in three parts :

- The "Λp region" from 2056 MeV to 2129 MeV missing mass where the only open channel available is $K^+\Lambda p$ (Region I).
- The "$(\Sigma N)^+$ threshold region around the $K^+ \Sigma^+ n$ (2129.9 MeV) and the $K^+\Sigma^\circ p$ (2131.7 MeV) thresholds (Region II).
- The "(Λp) and $(\Sigma N)^+$" region above 2131 MeV but below the $K^+\Sigma N\pi$ channel (<2200 MeV) (Region III).

In region I a strong enhancement in the missing mass spectrum is observed in all the spectra at the opening of the $K^+ \Lambda p$ channel corresponding to the Λp FSI at low relative energies. The cross section of this effect decreases when the momentum transfer increases.

In region II, a peak is observed at $(\Sigma N)^+$ thresholds very well marked for $\bar{\theta}_K \approx 10^\circ$ and $T_p = 2.3$ GeV. This peak which is shown on Fig.3 after subtraction of the 3 body phase space is missing (or not evident) at other angles or at the other energy $T_p = 2.7$ GeV. Its position and width are respectively 2131 MeV ± 1 MeV and 9 ± 1 MeV.

In region III a strong increase of the cross section appear in all spectra, from $(\Sigma N)^+$ thresholds to the maximum of the explored missing mass.

In the basic meson production process traditionally a meson exchange model is employed and fair agreement with experiments has been achieved in the case of pion production where extensive data can be found. In the case of kaon production early calculations done in the 1960's showed that the kaon exchange mechanism dominates the $pp \to K^+\Lambda p$. The skeleton of this mechanism is shown in Fig. 5, graph a). The dominance of the K^+ exchange term over the π^+ exchange term (Fig. 5, graph b)) can be simply understood if a comparison between the products $G_{p\to k^+\Lambda} \times \sigma(K^+p \to K^+p)$ and $G_{p\to\pi N} \times \sigma(\pi^\circ p \to \Lambda K^+)$ is done. The two coupling constants $G_{NN\pi}/\sqrt{4\pi} = 3.8$ and $G_{NK\Lambda}/\sqrt{4\pi} = 4$ are near the same but $\sigma(K^+p \to K^+p)$ ($\simeq$ 10 mb) dominates over $\sigma(\pi^\circ p \to \Lambda K^+)$ ($\leqslant$ 1 mb) by a factor of ten in the kaon momentum range 0.1 GeV/c < p_k < 1 GeV/c. The YN final-state interaction (FSI) leads to triangle diagrams c) and d) of Fig. 5 respectively in K^+ and π° meson exchange, where Y is either the Σ or Λ hyperon.

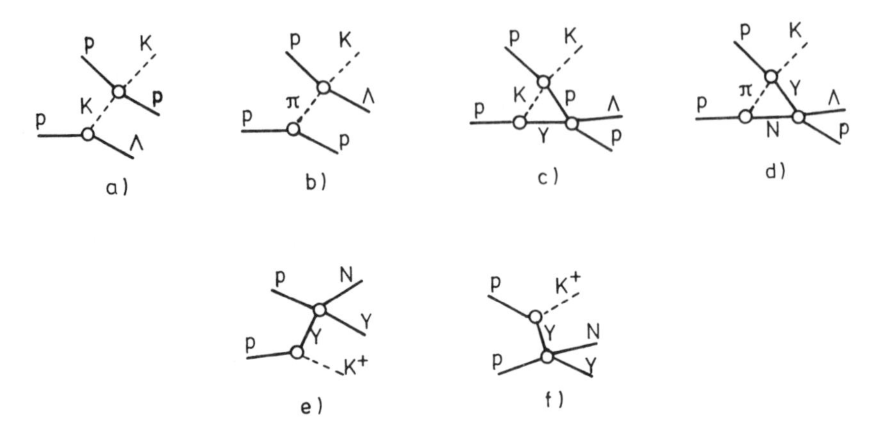

Fig. 5 Different possible graphs used in one meson exchange (graph a and b) with final state interactions (graphs c and d) or in hyperon exchange (graphs e and f) mechanisms.

When the $K^+p \rightarrow K^+p$ or $\pi^0p \rightarrow \Lambda K^+$ cross sections are compared with $NY \rightarrow NY$ cross sections which are of the order of hundreds of mb ($\Lambda p \rightarrow \Lambda p$ is about 500 mb at threshold), the idea comes out that, contrary to pion production, the ASP should be dominated by hyperon exchanges. The corresponding diagrams are shown in Fig. 5 e)f).

This remark served as a starting point to DELOFF[6)] to adopt the hyperon exchange (HE) mechanism as the dominant one in an impulse approximation calculation. In DELOFF model, the proton is allowed to dissociate into a kaon and an hyperon, the hyperon only being interacting with the second proton. The description of the overlap $N \rightarrow YK$ yields a mixing angle ϵ between the 3 different YK channels ($K^+\Lambda$, $K^+\Sigma^0$, $K^0\Sigma^+$) which is introduced as a free parameter to be determined from the $pp \rightarrow K^+X$ data. In fact ϵ can be expressed in terms of the coupling constants $g_{K\Lambda N}$ and $g_{K\Sigma N}$. There seems to be a general agreement that $g_{K\Lambda N}$ dominates over $g_{K\Sigma N}$, but both of them are not well determined. In particular the ratio $R = g^2_{K\Lambda N}/g^2_{K\Sigma N}$ which is expressed in terms of ϵ as :

$$R = g^2_{K\Lambda N}/g^2_{K\Sigma N} = 3\ \mathrm{cotg}^2\,\epsilon$$

has a value which depends strongly of the analyses. In a KN dispersion relation analysis, MARTIN gives $R = g^2_{K\Lambda N} / g^2_{K\Sigma N} = 4.2$, whereas the last analysis of kaon photoproduction reaction done by ADELSECK and SAGHAI[10] gives a value of about 12.5, in agreement with SU(3) predictions.

The Y-N scattering matrix is obtained from a s-wave separable potential. The parameters of the Y-N potentials named A and B come from a fit of the two-body YN scattering data. One of those fits is shown on Fig. 6 where one can see that A or B yield the same kind of agreement. These potentials were previously able to equally reproduce the $K^-d \to \pi YN$ and $\pi^-d \to KYN$ experiments[11].

The DELOFF calculation is shown on Fig. 7 in comparison to the $T_p = 2.3$ GeV, $\theta_K = 10 \pm 2^\circ$ missing mass spectrum for pot. A and pot. B. They appear to fit equally well the whole spectrum, except that for pot. B the value of R which comes out is 0.08, very far from all the expected values, whereas R = 1.6 for pot. A, closer to the Martin's value. The model A which results in a larger R value should be favoured. This model also yields the better χ^2 value. This result is confirmed by fitting the $T_p = 2.7$ GeV $\theta_K = 13^\circ$ missing mass spectrum in the new $pp \to K^+X$ experiment. This is shown on Fig. 4). The values for R have been kept the same than in the $T_p = 2.3$ GeV data fitting. One can see on this figure that the peak which was apparent at $T_p = 2.3$ GeV, $\theta_K = 10^\circ \pm 2^\circ$ (see Fig. 3) in region 2 disappears. Two reasons can be given for that : firstly, the energy resolution is poorer at $T_p = 2.7$ GeV (2.9 MeV FWHM) than at $T_p = 2.3$ GeV (1.9 MeV) ; secondly the energy and t-transfer are different in the two spectra. The t four-momentum transfer is close to zero in the T = 2.3 results for MM = 2130 MeV. This cusp effect must be very strongly t-dependent like in the $Kd \to \pi YN$ CERN experiment[12].

It is clear that this model should also include π and K meson exchanges because the supposed dominance of the hyperon exchange graph could be damped by a stronger off-shell suppression of the YN amplitudes relative to the KN amplitudes. DELOFF has done an estimation of the kaon exchange diagram and found a resulting cross section rather small, of the order of 5 % of the hyperon exchange cross section.

A meson exchange model has been developped by J.M. LAGET[7] including π and K exchanges with final state interactions in the coupled hyperon-nucleon channels. The result of the calculation is shown on Fig. 8 in comparison with the 2.3 GeV, $\theta = 10^\circ$ data. The different graphs which are calculated are shown on Fig. 9. For the region I, both π and K exchanges play the same role. For the region III the pion exchange diagram play the most important role exciting the S_{11}(1680), P_{13} (1690) and P_{11} (1730) resonances which dominate the $\pi N \to K\Lambda$ elementary subprocesses.

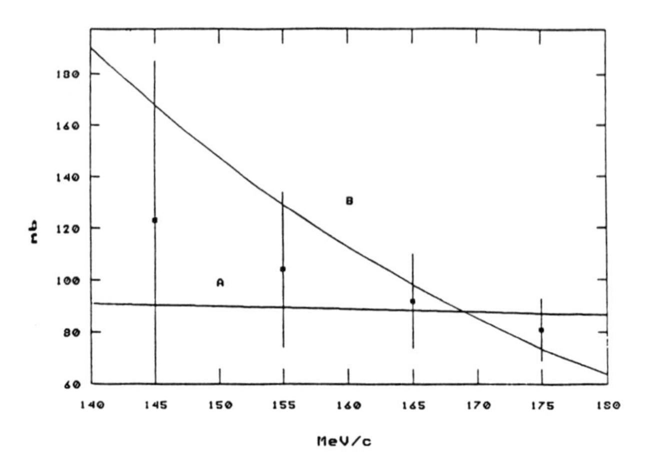

Fig. 6 Total $\Sigma^{+}p$ elastic cross sections versus incident momentums. The theoretical curves correspond to models A and B of Deloff (see text).

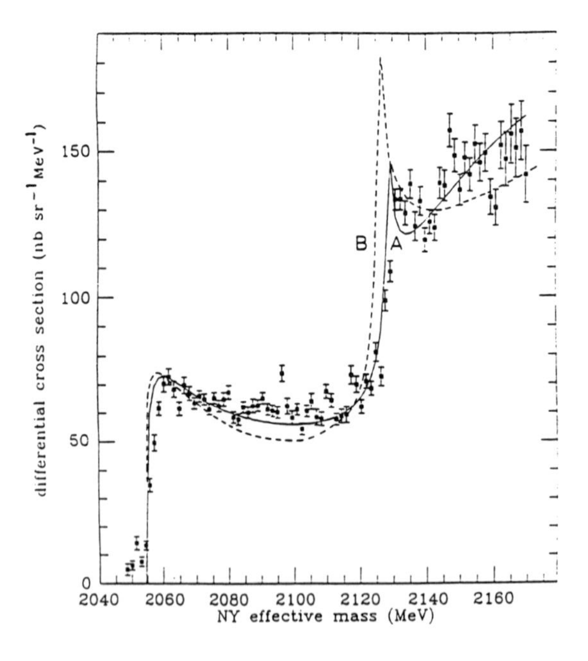

Fig. 7 Deloff calculation with pot. A or B in comparison to the experimental data at T_p = 2.3 GeV and θ_K^{lab} = 10 ± 2°.

The FSI dominates the Λp and ΣN threshold regions where the relative kinetic energies between the hyperon and the nucleon are small. This interaction is given through the on shell hyperon-nucleon Lorentz invariant scattering amplitude expanded in terms of partial waves. These partial waves are expressed with the phase shifts obtained by the Nijmegen group[13] in their coupled potential analysis of nucleon nucleon data and the available hyperon-nucleon data (potential D, Ref. 13). This is a check in fact of the $^3S_1$ amplitude part, which is dominant. A monopole factor is added to extrapolate off shell the scattering amplitude. Laget assumes a SU_3 relation between G^2_Λ and G^2_Σ , so $G^2_\Sigma \ll G^2_\Lambda$, such a way the conversion amplitude $\Sigma N \rightarrow \Lambda N$ is negligible. The ratio $R = g^2_{K\Lambda N}/g^2_{K\Sigma N}$ is taken equal to 14.

The graph II Fig. 9 corresponding to the kaon direct amplitude is directly proportional to the YN scattering amplitude and is responsible for the structure observed at ΣN threshold (region II). It also contributes in region I, distorting strongly the missing mass at ΛP threshold. If it gives the general shape it cannot alone reproduce the kaon spectrum.It is similar to the HE mechanism of Deloff (Fig. 5, graphs e) and f)).

From these two different analyses, some conclusions can be drawn. The HE graph seems to be the dominant mechanism at least in the ASP kinematical region studied here. Better values for $g_{K\Lambda N}$ and $g_{K\Sigma N}$ are needed. Taking the values coming from kaon photoproduction is the right way to proceed : extensive experimental investigations on those reactions are expected to be realized in the next future (Bates, Bonn, ESRF and CEBAF). That should allow a complete determination of the scattering amplitudes which are needed to define the $k\Lambda N$ and $K\Sigma N$ coupling constants precisely.

At this stage of the $pp \rightarrow k^+X$ analysis one is able to select some YN potentials (see for instance pot. A in Deloff analysis) but not yet able to refine the existing potentials, in particular in their spin dependent part. About the existence or not of a resonance in the ΛN amplitude, our observation of the $H_1(2130)$[5] at ΣN threshold is well reproduced by either pot. A of Deloff or pot. D of the Nijmegen group used in the Laget calculation. Both of them yield a pole close to ΣN threshold. This pole leads to an enhancement of the cusp effect at ΣN threshold. It does not correspond necessarily to a resonance.

From the experimental point of view, exclusive $pp \rightarrow K^+X$ experiments such as $pp \rightarrow K^+\Lambda p$ and $pp \rightarrow K^+\Sigma N$ reactions have to be studied. Asymetry measurements and transfer polarization to the hyperon can also be studied in polarized proton scattering, with or without Λ(or Σ) polarization measurements. These experiments could be done at LNS, so long as a new large acceptance detector is available.

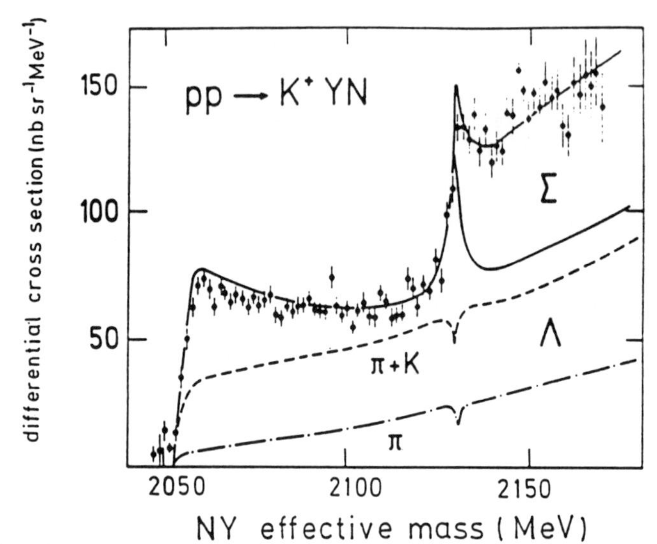

Fig. 8 Laget calculation with pot. D of the Nimegen group in comparison to the experimental data at T_p = 2.3 GeV.

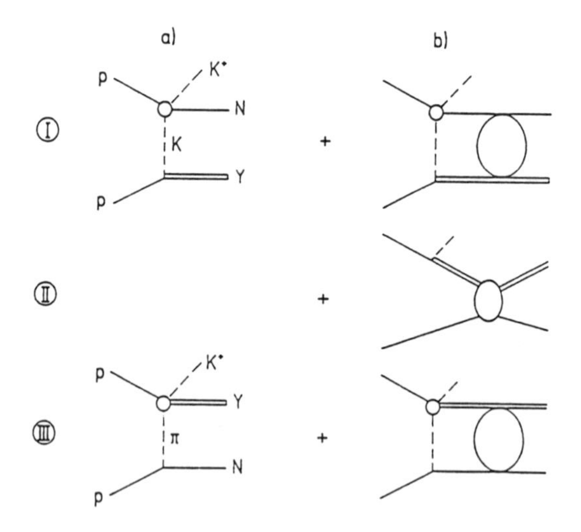

Fig. 9 The different graphs included in the Laget calculation.

III - PLANS FOR THE FUTURE AT LNS

It is clear that fully exclusive experiments are needed in particular to disentangle the different observations which are done at the opening of the $(\Sigma N)^+$ channels.

The detection of K^+ and p in correlation in the reactions $pp \to K^+pY$ is a good signature for the hyperon production. A good resolution on the $Y(\Lambda$ or $\Sigma)$ mass must be achieved. This can be obtained by a magnetic analysis of the K^+ and p with a large field path. The momentum acceptance and angular aperture of the spectrometer must be large, so as to get a good acceptance for two correlated particles. With precise measurements of the localization of these particles, the Y momentum can be reconstructed.

The experimental set up shown schematically on Fig. 10 should allow the measurement of the differential cross sections of $pp \to K^+p\Lambda$ and $pp \to K^+p\Sigma^\circ$ reactions just as beam asymetries with a proton polarized beam.

The position detectors located in front of the spectrometer which participate in the tracking of the K^+ and p can be used to determine the tracks of the products of the Λ decay $(\Lambda \to p\pi^-)$. This allows the determination of the Λ polarization in the $\vec{p}\,p \to K^+p\Lambda(\to p\pi^-)$ reaction. A programme of simulation based on Fowl has been developped at Orsay with intend to define and to optimize the detector shown on Fig. 10. The simulation of the event distributions in different configurations for a 4 particle large angle spectrometer yielded the following numbers for an optimized configuration.

- Mass resolution for the Λ — 12 MeV (FWHM)
- 2 particle (K^+p) acceptance — 4.10^{-3} of phase space
- 4 particle $(K^+p, p\pi^-)$ acceptance — 8.10^{-5} (with 40 x 10 and 80 x 20 cm^2 planes)

This acceptance can be enlarged with appropriate detectors.

- Angular resolution in the CM of the Λ for angle between the proton (or the pion) from the decay and the Λ reconstructed direction

$$\Delta\theta = 1^\circ \ (\theta \leqslant 90^\circ)$$
$$= 3^\circ \quad \theta \simeq 90^\circ.$$

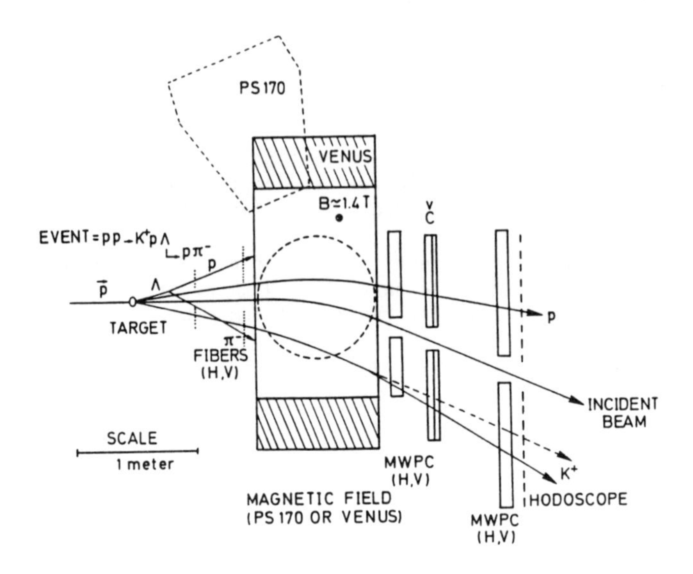

Fig. 10 Schematic drawing of the new proposed experimental set up at LNS for ASP studies.

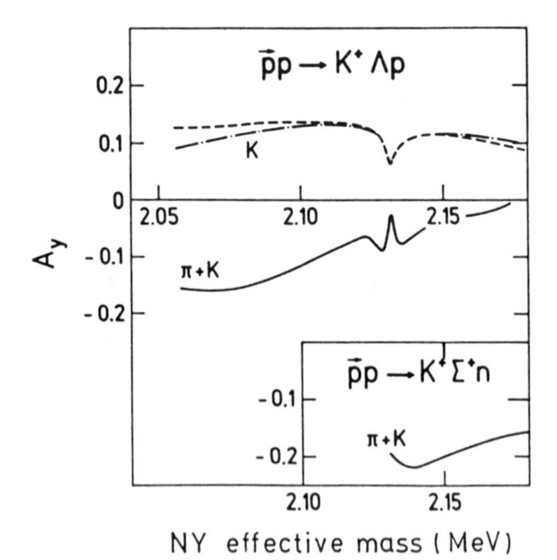

Fig. 11 Expected behaviour of beam asymmetry in ASP showing the contribution of π or K exchange in the pp collisions (Laget calculation, see text).

This value conditions partly the precision on the Λ polarization measurement.

This simulation is going on in three different laboratories (LNS, ORSAY and TRIUMF) and should end with a preconceptual design ready for december 90.

As a conclusion, the Fig. 11 presents the Laget calculation[7] in the same model presented above for the beam asymmetry Ay as a function of the missing mass ΛN or ΣN, showing the expected sensitivity of spin observables to the mechanisms involved in those experiments.

ACKNOWLEDGMENTS

I wish to thank my colleagues from the ASP collaboration at LNS[14]) -particularly C. Lippert and R. Siebert- to allow me to present a part of the preliminary results of our new data taken recently at 2.7 GeV.

REFERENCES

1. R. Machleidt, K. Holinde and Ch. Elster, Phys. Rep. 149, 1(1987)
2. K. Shimizu, Rep. Prog. Phys. 52, 1(1989)
   A. Faessler, Prog. in Part. and Nucl. Phys. 20, 151(1988)
3. T.H. Tan, Phys. Rev. Lett. 23, 395(1969) ; Phys. Rev. D17, 600(1972)
   O. Braun et al., Nucl. Phys. B124, 45(1977)
   G. D'Agostini et al., Phys. Lett. B104, 330(1981)
   C. Pigot et al., Nucl. Phys. B249, 172(1985)
4. J.T. Reed et al., Phys Rev. 168, 1495(1968)
   W.J. Hogan et al., Phys. Rev. Lett., 166,5,1472(1968)
5. R. Frascaria, R. Siebert, J.P. Didelez, G. Blanpied, M. Boivin, E. Bovet, J.P. Egger, J. Ernst, J.Y. Grossiord, T. Mayer-Kuckuck, Ch. Perdrisat, B. Preedom, T. Reposeur, B. Saghai, E. Warde, J. Yonnet, Nuovo Cimento, 102A, 2, 561(1989)
   R. Frascaria, R. Siebert, Few-Body Systems, Supp. 2, 425(1987) ;
   R. Siebert, R. Frascaria et al., Nucl. Phys. A479, 389C(1988)
6. A. Deloff, Nucl. Phys. A505, 583(1989)
7. J.M. Laget, submitted to Phys. Rev. Lett., Note CEA DPhN-Saclay 90-3 106 90(1990)
8. R. Siebert, PhD Thesis (IKSP, Bonn FRG)(1989) and references in there
9. R. Siebert, J.P. Didelez, R. Frascaria, E. Warde, C. Lippert, F. Hinterberger, Annual Report IPN(ORSAY, France), 130(1989)
10. Adelseck and B. Saghai, Phys. Rev. C, 42, 1, 108(1990)
11. A. Deloff, Nuovo Cim., 102A, 2(1989)
12. C. Pigot et al., Nucl. Phys., B249, 172(1985)
13. N.M. Nagels, T.A. Rijken and J.J. de Swart, Phys. Rev. D,

15, 9, 2547(1977)

14. The ASP collaboration at LNS
G. Blanpied[÷], M. Boivin[*], E. Bovet[■■], J.P. Didelez[■],
J.P. Egger[■■], J. Ernst[§], R. Frascaria[■], J.F. Germond[■■],
J.Y. Grossiord[**], F. Hinterberger[‡‡], C. Lippert[‡‡],
T. Mayer-Kuckuk[‡], Ch.F. Perdrisat[‴], B. Preedom[↠],
T. Reposeur[■], B. Saghai[˙], R. Siebert[■], E. Warde [■], J. Yonnet[*]

[■]IPN Orsay (France), [*]LNS Saclay (France), [■■]Univ. Neuchâtel (Suisse),

[§]Univ. Bonn (FRG), [**]IPN Lyon (France), [↠]Univ. South Carolina (USA),
[‴]Coll. William and Mary, Williamsburg (USA), [˙]DPNHE Saclay (France).

THE PRODUCTION AND DECAY OF HYPERNUCLEI

John J. Szymanski*
Los Alamos National Laboratory, Los Alamos, NM 87545

ABSTRACT

Hypernuclei have been studied for the last 35 years using several techniques. Since 1970, the ^{A}Z (K^-,π^-) $^{A}_{\Lambda}Z$ strangeness-exchange reaction has been used at CERN, BNL and KEK to produce hypernuclei and study their spectroscopic properties. These studies also include experiments where decay gamma rays and hypernuclear weak decay products are detected in coincidence with hypernuclear production. Recent experiments at BNL and KEK have proven the utility of the ^{A}Z (π^+,K^+) $^{A}_{\Lambda}Z$ reaction to study hypernuclear spectroscopy. Although I have no hope of covering this field in much detail, I write this paper as an introduction to the subject and highlight some of the recent experimental developments.

INTRODUCTION

The properties of hypernuclei reflect the characteristics of the hyperon-nucleon weak and strong interactions. The hyperon-nucleon strong interaction gives the level structure of hypernuclei. The hyperon is distinguishable from the nucleons because of the presence of a strange quark, so the hyperon is not Pauli-blocked from occupying low-lying states in the nucleus. The distinguishability of hyperons seems to hold true even for heavy hypernuclear systems.

A hyperon bound within a nucleus has a large probability of overlap with one of the constituent nucleons. Thus, a new weak-decay mode, $\Lambda+N\rightarrow N+N$ opens up in hypernuclei and is the dominant decay mode for all but the lightest hypernuclear species. These weak decays are interesting because of the modification of the weak force in the presence of strongly-interacting particles.

The Λ-N strong interaction differs from the N-N strong interaction in several ways[1]:

a) There is no 1-π exchange because of isospin conservation.

b) There is 1-K exchange, unlike the N-N interaction.

c) Because of a), the Λ-N interaction is generally weaker than the N-N interaction. In fact, the model of a Λ weakly coupled to the nuclear core is quite successfully applied to understand the structure of Λ-hypernuclei.

The Λ-N interaction may be written as

$$V_{\Lambda N}(r) = V_0(r) + V_\sigma(r)\,\bar{S}_N\bullet\bar{S}_\Lambda + V_\Lambda(r)\,\bar{l}_{N\Lambda}\bullet\bar{S}_\Lambda + V_N(r)\,\bar{l}_{N\Lambda}\bullet\bar{S}_N + V_T(r)\,S_{\Lambda N}\,,$$

where $S_{\Lambda N}$ is the usual tensor spin operator. The interaction can be expressed as a potential by integrating over harmonic oscillator wave functions. The radial integrals derived from the five terms of the above expression are designated as V, Δ, S_Λ, S_N

* Present address: Indiana University Cyclotron Facility, Bloomington, IN 47401.

and T, respectively. These are referred to as the spin-averaged central, the spin-spin, the spin-orbit, the induced spin-orbit and the tensor terms. Dalitz and Gal[2] have found the best fit to the existing p-shell data gives these parameters as 0.15, 0.57, -0.21 and 0.0 MeV for the Δ, S_Λ, S_N and T terms, respectively. More recently, Millener et al.[3] found 0.50, -0.04, -0.08 and 0.04 for these parameters. It is the goal of many hypernuclear production and decay experiments to better understand this two-body interaction.

HYPERNUCLEAR PRODUCTION REACTIONS

Several complimentary reaction mechanisms have been used to study hypernuclei. They fall under these broad categories:

a) Strangeness exchange, the general form of which is:

$$K^- + {}^{A}Z \rightarrow \pi^- + {}^{A}_{\Lambda}Z \ .$$

The elementary reaction is $K^- + n \rightarrow \pi^- + \Lambda$. Some of the targets that have been studied with the (K^-,π^-) reaction are: ^4He, $^{6,7}Li$, ^9Be, $^{12,13}C$, ^{16}O, ^{32}Si, ^{40}Ca and ^{209}Bi.

b) Associated production, where an outgoing (Strangeness=+1) K^+ tags the residual (S=-1) Λ. Examples are:

$$\pi^+ + {}^{A}Z \rightarrow K^+ + {}^{A}_{\Lambda}Z \qquad (\pi^+ + n \rightarrow K^+ + \Lambda)$$

$$\gamma + {}^{A}Z \rightarrow K^+ + {}^{A}_{\Lambda}Z\text{-}1 \qquad (\gamma + p \rightarrow K^+ + \Lambda)$$

$$p + {}^{A}Z \rightarrow K^+ + {}^{A+1}_{\Lambda}Z \qquad (p + n \rightarrow K^+ + \Lambda + n)$$

c) Strangeness and charge exchange:

$$K^- + {}^{6}Li \rightarrow \pi^+ + {}^{6}_{\Sigma}H$$

where the elementary process is $K^- + p \rightarrow \pi^+ + \Sigma^-$. The subject of Σ hypernuclei is under active study, but will not be pursued further here. See the paper by Tang et al.[4] and references therein.

d) Annihilation, where the two-step process

$$\bar{p} + N \rightarrow K\bar{K} + \pi \ ; \ \bar{K} + N \rightarrow \Lambda + \pi$$

is producing hypernuclei. This reaction has been used at LEAR to study the weak decay of heavy hypernuclei using the recoil-shadow technique.[5]

All of the above processes have been used to produce hypernuclei except (γ,K^+) and (p,K^+), although efforts to use these reactions are proceeding and will be covered in the section on new initiatives. Most of the existing data are for the (K^-,π^-) and (π^+,K^+) reactions leading to the formation of Λ hypernuclei. Thus, for the remainder of this paper, I switch to a discussion of the production and decay of Λ hypernuclei.

THE (K^-,π^-) AND (π^+,K^+) REACTIONS

The (K^-,π^-) reaction was first used at CERN in 1970.[6,7] The CERN-Heidelburg-Warsaw collaboration pioneered many techniques used in (K^-,π^-) studies.

An example of the typical level structure seen in hypernuclei is the high-resolution ^{12}C (K^-,π^-) $^{12}_{\Lambda}C$ missing-mass spectrum shown in Fig. 1.[8] The observed states are identified with the removal of a neutron from the $p_{3/2}$ shell and replaced by a Λ in the s shell (lower mass state) or by a Λ in the p shell (higher mass state). The (p^{-1},s) (p neutron-hole, s Λ-particle) state is lighter than the (p^{-1},p) state because the neutron is more strongly bound to the nuclear core than the Λ. Similar results are seen with the (π^+,K^+) reaction.

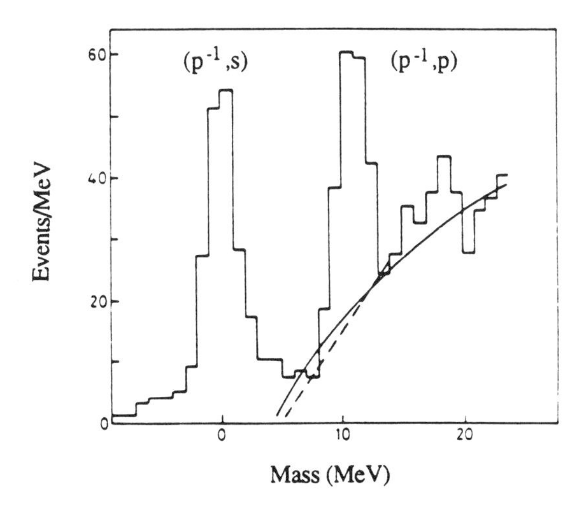

Fig. 1 High-resolution ^{12}C (K^-,π^-) $^{12}_{\Lambda}C$ mass spectrum, showing the neutron-hole Λ-particle states (p^{-1},s) and (p^{-1},p).

The (K^-,π^-) and (π^+,K^+) reactions are complementary in several ways. Fig. 2 contains a comparison of the momentum transfer, q, versus incoming beam momentum for the two reactions. The (π^+,K^+) reaction clearly yields a larger q than the (K^-,π^-)

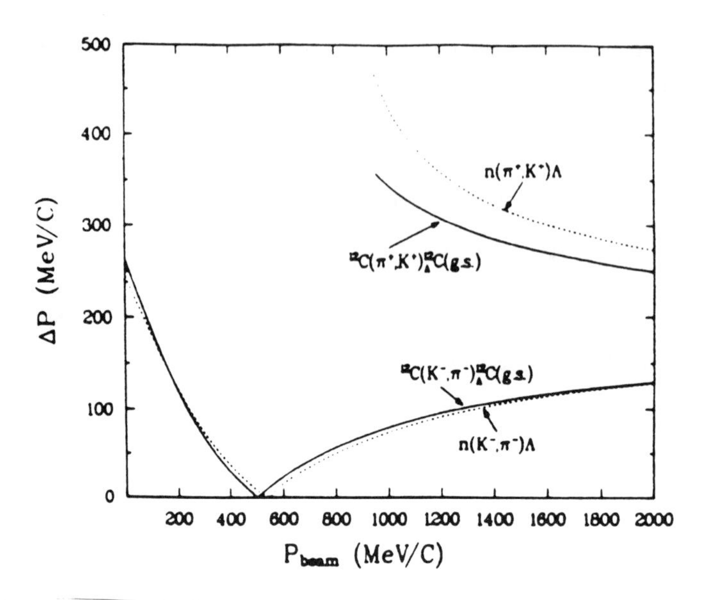

Fig. 2 Momentum transfer, q, for the (K^-,π^-) and (π^+,K^+) reactions versus beam momentum.

Table I

Comparison of the (K^-,π^-) and (π^+,K^+) reactions.[10] The listed cross section is the typical differential cross section to a particular hypernuclear state.

	(K^-,π^-)	(π^+,K^+)
q	<100 MeV/c	~300 MeV/c
Selectivity	low spin states	high spin states
Cross Section	0.3-1 mb/sr	10-20 μb/sr
Intensity	10^5	10^7
Beam Purity	K/all ~ 8-10%	π/all ~ 80%

reaction, which can actually produce a Λ with zero momentum (the 'magic momentum' of 530 MeV/c for the (K^-,π^-) reaction). The (π^+,K^+) reaction, on the other hand, has a typical q of 300 MeV/c and therefore populates higher spin states than (K^-,π^-). Table I compares the (K^-,π^-) and (π^+,K^+) reactions. The count rates for production of a particular hypernuclear state are comparable for the two techniques, because the larger available π^+ flux compensates for the smaller (π^+,K^+) cross section.

Fig. 3 shows the ^{89}Y mass spectrum obtained in BNL exp. 798, a recent (π^+,K^+) experiment.[9,10] The peaks are identified with Λ-particle-neutron-hole states (g^{-1},s), (g^{-1},p), (g^{-1},d) and (g^{-1},f). The high spin (g^{-1},s) and (g^{-1},s) states are seen.

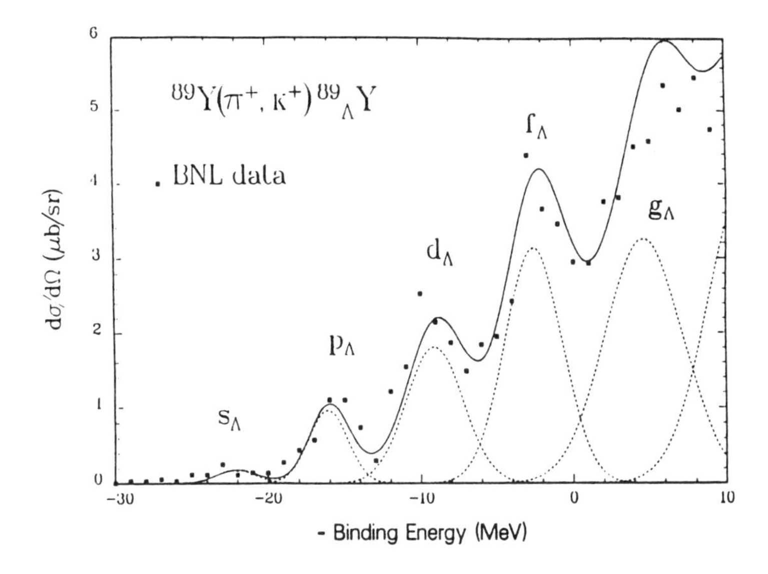

Fig. 3 A (π^+,K^+) mass spectrum obtained at BNL with a ^{89}Y target. The data points are compared to a DWBA calculation[11] using the experimental neutron-hole strength for ^{90}Zr.

The observation of these deeply bound states supports the view that the Λ hyperon is indeed distinguishable even within a heavy nucleus. The (π^+,K^+) program at KEK has also obtained good quality results on $^{12}_{\Lambda}C$ and $^{56}_{\Lambda}Fe$.[12]

Coincidence experiments to detect hypernuclear γ transitions have been carried out at BNL and CERN. The first hypernuclear γ rays were observed by stopping a K^- beam in Be and Li.[13,14] The observed γ rays at 1.04 MeV and 1.15 MeV for $^{4}_{\Lambda}H$ and $^{4}_{\Lambda}He$, respectively, fix the first excited 1^+ state at about 1 MeV above the ground state. The first in-flight experiment occured at BNL and utilized the (K^-,π^-) reaction to produce and tag hypernuclear states. The γ rays were detected in NaI detectors placed around the hypernuclear production target. This experiment resulted in the observation of hypernuclear transitions from the 2.034 MeV second excited state of $^{7}_{\Lambda}Li$ and from the 3.079 MeV $^{9}_{\Lambda}Be$ doublet. The doublet consists of $5/2^+$ and $3/2^+$ components and the fact that this spitting was not resolved puts an upper bound on the strength of the ΛN spin-orbit splitting of $|S_\Lambda| < 0.04$ MeV.[15]

A later experiment at BNL was designed to observe spin-flip transitions within the ground-state doublet of ${}^{10}_{\Lambda}B$. This experiment was sensitive to γ rays in the energy range of 100 keV to 500 keV, but saw no evidence for the spin-flip transition. The smallness of the ΛN spin-orbit term, S_{Λ}, and the absence of γ lines in the ${}^{10}_{\Lambda}B$ ground state implies the spin-spin splitting, Δ, is < 0.22 MeV.[16] This is in contrast to the value used by Millener et al.[3] of Δ=0.5. There is clearly a problem in fitting a consistent set of p-shell potential parameters.

HYPERNUCLEAR WEAK DECAY

In most cases, a particle-stable hypernucleus will decay electromagnetically to its ground state before undergoing a strangeness-changing weak decay to a normal nucleus. The predominant decay modes are of two types: mesonic and nonmesonic. The mesonic modes are $\Lambda \rightarrow p\,\pi^-$ (partial rate Γ_{π^-}) and $\Lambda \rightarrow n\,\pi^0$ (partial rate Γ_{π^0}) while the nonmesonic modes are $\Lambda\, p \rightarrow n\, p$ (proton stimulated partial rate Γ_p) and $\Lambda\, n \rightarrow n\, n$ (neutron stimulated partial rate Γ_n). The leptonic decay modes are orders of magnitude smaller and will be neglected in this discussion. The mesonic modes are analogous to free Λ decay, but are modified in hypernuclei because of phase-space changes and Pauli blocking of the final state nucleons[17]. Conversely, the nonmesonic modes become available in hypernuclei and are the dominant decay modes for all but the smallest hypernuclear systems. An experimental attraction of the nonmesonic modes is the ease of their identification because of the energetic nucleons produced in the final state. The total mesonic decay rate is denoted Γ_m where $\Gamma_m=\Gamma_{\pi^-}+\Gamma_{\pi^0}$ and the total nonmesonic decay rate is denoted Γ_{nm} where $\Gamma_{nm}=\Gamma_p+\Gamma_n$. The sum of all four partial rates gives the total decay rate $\Gamma_{total}=1/\tau$ for a particular hypernucleus.

At the quark level, the lowest-order Hamiltonian describing $\Delta S=1$ weak decay is the result of the combination of V-A theory with the Cabibbo hypothesis:[18]

$$H_{V\text{-}A} = \frac{G_F}{\sqrt{2}} \sin\theta_c \cos\theta_c \; Q_{V\text{-}A} + \text{c. c.}$$

where G_F is the Fermi coupling constant, θ_c is the Cabibbo angle and the operator, $Q_{V\text{-}A}$, is :

$$Q_{V\text{-}A} = \bar{u}\,\gamma_\mu\,(1-\gamma_5)\,s\;\bar{d}\,\gamma^\mu\,(1-\gamma_5)\,u.$$

This Hamiltonian leads to transitions of both $\Delta I = 1/2$ and $\Delta I = 3/2$ with comparable strength. It is well known,[19] however, that both kaon decay and free hyperon decay data suggest that the $\Delta I = 1/2$ amplitude is about a factor of twenty stronger than the $\Delta I = 3/2$ amplitude. The origin of this empirical "$\Delta I = 1/2$ Rule" is not understood. Some progress has been made using quark models which include strong interaction corrections to the weak Hamiltonian, but the full effect has yet to be explained.

The nonmesonic decay modes are described with the quark model and with meson exchange, as shown in Fig. 4.[20,21,22,23,24,25] The $\Delta I = 1/2$ rule is put into the meson exchange picture explicitly.

The experiment 759 and 788 collaborations at BNL have been studying the weak decay of ${}^{12}_{\Lambda}C$, ${}^{5}_{\Lambda}He$ and ${}^{4}_{\Lambda}He$.[26,27] The experiments utilize the (K^-,π^-) reaction to produce and tag hypernuclear states. Charged and neutral decay products are detected in plastic scintillator spectrometers located around the hypernuclear production target. Protons, pions and neutrons are identified and counted in coincidence with the formation of a particular hypernuclear state. A summary of some of the experimental results is contained in Table II.[27]

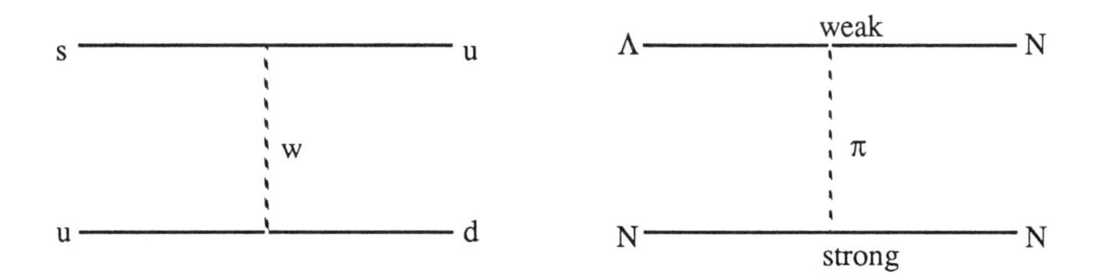

Fig. 4 Quark model and meson-exchange model of the nonmesonic weak decay $\Lambda N \to NN$.

Table II

${}^{12}_{\Lambda}C$ and ${}^{5}_{\Lambda}He$ weak decay experimental results. Γ_{nm} is quoted in units of the free Λ decay rate.

	Γ_{nm}	Γ_n/Γ_p	$\Gamma_{nm}/\Gamma_{\pi^-}$
${}^{12}_{\Lambda}C$	1.14 ± 0.20	$1.33^{+1.12}_{-0.81}$	22^{+43}_{-12}
${}^{5}_{\Lambda}He$	0.41 ± 0.14	0.93 ± 0.55	0.92 ± 0.31

The most striking feature of the results is the large drop in Γ_{π^-} in going from ${}^{5}_{\Lambda}He$ to ${}^{12}_{\Lambda}C$. The Γ_{π^-} rate is difficult to measure in ${}^{12}_{\Lambda}C$, because so few pions are observed in coincidence with the ${}^{12}_{\Lambda}C$ ground state. Γ_{nm} is increasing slowly and, in fact, roughly scales with the number of nucleons available for nonmesonic decay in these light hypernuclei. These experiments continued in the spring of 1990 with a run during which the ${}^{5}_{\Lambda}$He weak-decay measurements were repeated and ${}^{4}_{\Lambda}$He weak-decay data were taken.

The detection of decay products in coincidence with hypernuclear formation can give information on the structure of levels. For example, the ^{6}Li (K^{-},π^{-}) ${}^{6}_{\Lambda}$Li mass spectrum taken during the experiment 788 weak decay data run is shown in figure 5. This spectrum requires that a proton from nonmesonic decay (energy > 30 MeV) is observed in coincidence with the (K^{-},π^{-}) reaction. This is to be compared to the spectrum shown in figure 6, which was taken earlier by Bertini et al.[28] and which has

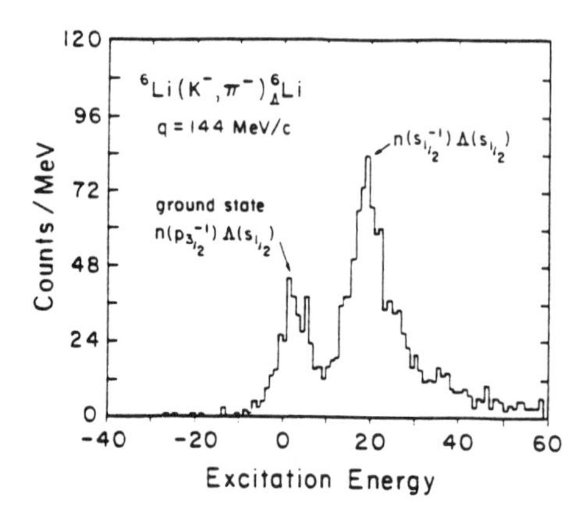

Fig. 5 ^{6}Li (K^{-},π^{-}) ${}^{6}_{\Lambda}$Li mass spectrum with an energetic (T>30 MeV) proton coincidence requirement.

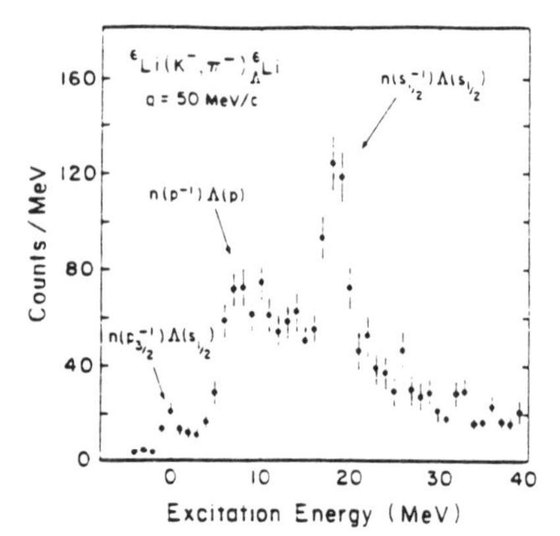

Fig. 6 ^{6}Li (K^{-},π^{-}) ${}^{6}_{\Lambda}$Li mass spectrum with no coincidence requirement.

no decay coincidence requirement. The (p^{-1},p) state is not seen when an energetic proton is required, indicating that this state must break up by Λ emission. That is, an energetic proton can only be produced by a bound Λ undergoing a nonmesonic decay, and cannot be produced by a free Λ (the proton in $\Lambda \rightarrow p\,\pi^{-}$ has less than 5 MeV). This observation confirms calculations by Majling et. al.[29], and Auerbach and Giai[30] in which the large width (Γ = 7 - 10 MeV) of the (p^{-1},p) state is related to its particle-unstable character.

NEW INITIATIVES IN HYPERNUCLEAR PHYSICS

There are four new programs that I wish to discuss. First, the ongoing, strong program at KEK needs to be mentioned in more detail. In the recent past, there have been experiments at KEK looking at Σ-hypernuclear states[31], mesonic weak decay[32] and (π^+,K^+)[12]. There are plans to construct a superconducting spectrometer for (π^+,K^+) studies[33] and to build a new, large-acceptance spectrometer[34]

There are plans for the study of hypernuclei using associted production with the (p,K^+) reaction. Specifically, the COSY ring is to be used to produce heavy hypernuclei and study their decay.[35] A new era in hypernuclear physics will open up if cooler rings can be successfully used for hypernuclear production.

The photoproduction of hypernuclei has been proposed for CEBAF.[36] These experiments are designed to reach an ultimate resolution of ~200 keV. The hope is to resolve the spin doublet splittings of p-shell hypernuclear ground states. The initial, lower resolution, stage of this experiment is expected to run soon after the CEBAF turn-on.

Another interesting new initiative is the PILAC project at LAMPF.[37] The PILAC is a superconducting RF linear accelerator, which accelerates pions from the P^3 channel at LAMPF to just over 1 GeV/c. There pions are above the (π^+,K^+) threshold. The preliminary design calls for 10^9 π^+/sec and 200 keV resolution. The construction of PILAC would represent a one- to two-order-of-magnitude increase in rate over existing facilities, with approximately a one-order-of-magnitude improvement in resolution.

SUMMARY

The study of hypernuclei has matured to the point where sophisticated coincidence experiments have been undertaken to better understand the Λ-N strong and weak interactions. This work will continue at the present facilities in the near future. Many experiments await new facilities with higher resolution and/or counting rates. I strongly encourage these efforts.

REFERENCES

1 C.B. Dover and G.E. Walker, Phys. Reports **89**, 1 (1982).

2 R.H. Dalitz and A. Gal, Ann. Phys. **116**, 167 (1978).

3 D.J. Millener, A. Gal, C.B. Dover and R.H. Dalitz, Phys. Rev. **C31**, 499 (1985).

4 L.G. Tang, E. Hungerford, T. Kishimoto, B. Mayes, L. Pinsky, S. Bart, R. Chrien, P. Pile, R. Sutter, P.Barnes, G. Diebold, G. Franklin, D. Hertzog, B. Quinn, J. Seydoux, J. Szymanski, T. Fukuda, R. Stearns, Phys. Rev. **C38**, 846 (1988).

5 M. Rey-Campagnolle, Nuovo Cimento **102A**, 653 (1989); J.P. Bocquet, M. Epherre-Rey-Campagnolle, G. Ericsson, T. Johansson, J. Konijn, T. Krogulski, M. Maurel, E. Monnand, J. Mougey, H. Nifenecker, P. Perrin, S. Polikanov, C. Ristori and G. Tibell, Phys. Lett. **B182**, 146 (1986); J.P. Bocquet, M. Epherre-Rey-Campagnolle, G. Ericsson, T. Johansson, J. Konijn, T. Krogulski, M. Maurel, E. Monnand, J. Mougey, H. Nifenecker, P. Perrin, S. Polikanov, C. Ristori and G. Tibell, Phys. Lett. **B192**, 312 (1987).

6 CERN-Heidelberg-Warsaw Collaboration (M.A. Faessler et al.), Phys. Lett. **46B**, 468 (1973).

7 W. Bruckner, M.A. Faessler, K. Kilian, U. Lynen, B. Pietrzyk, B. Povh, H.G. Ritter, B. Schurlen, H. Schoder and A. H. Walenta, Phys. Lett. **55B**, 107 (1975);

W. Bruckner, W.B. Granz, D. Ingham, K. Kilian, U. Lynen, J. Niewisch, B. Pietrzyk, B. Povh, H.G. Ritter and H. Schoder, Phys. Lett. **62B**, 481 (1976)

8 R.E. Chrien et al., Phys. Lett. **89B**, 31 (1979).

9 P.H. Pile, Nuovo Cimento **102A**, 413 (1989);

10 J.C. Peng, Intersections Between Particle and Nuclear Physics, AIP Conference Proceedings Series **176**, 39 (1988)

11 D.J. Millener, private communication (1990).

12 M. Akei, H. Ejiri, M. Fukuda, T. Irie, Y. Iseki, A. Kashitani, T. Kishimoto, H. Nagasawa, H. Noumi. H. Ohsumi, K. Okuda, Y. Umeda, T. Fukuda, T. Shibata, O. Hashimoto, S. Homma, Y. Matsuyama, T. Nagae, C. Nagoshi, K. Omata, F. Soga, S. Toyama, N. Yoshikawa, Y. Yamanoi, J.F. Amann, J.A. McGill, H.A. Thiessen, J.Chiba, M. Nomachi, O. Sasaki, K.H. Tanaka, S. Kato, K.Kimura, F. Takeuchi and K. Maeda, Nuovo Cimento **102A**, 457 (1989).

13 A. Bamburger, M. Faessler, U. Lynn, H. Piekarz, J. Piekarz, J. Pniewski, B. Povh, H. Ritter and V. Soergel, Nucl. Phys. **B60**, 1 (1973).

14 M. Bedjidian, A. Filipowski, J. Grossiord, A. Guichard, M . Gusakow, S. Majewski, H. Piekarz, J. Piekarz and J.R. Pizzi, Phys. Lett. **62B**, 467 (1976);
M. Bedjidian, E. Descroix, J. Grossiord, A. Guichard, M . Gusakow, M. Jacquin, M. Kudla, H. Piekarz, J. Piekarz, J. Pizzi and J. Pniewski,Phys. Lett. **83B**, 252 (1979)

15 M. May, S. Bart, S. Chen, R. Chrien, D. Maurizio, P. Pile, Y. Xu, R. Hackenburg, E.Hungerford, H. Piekarz, Y. Xue, M. Deutsch, J. Piekarz, P. Barnes, G. Franklin, R. Grace, C. Maher, R. Rieder, J. Szymanski, W. Wharton, R.L. Stearns, B. Bassaleck and B. Budick, Phys. Rev. Lett. **51**, 2085 (1983).

16 R.E. Chrien, S. Bart, M. May, P.H. Pile, R.J. Sutter, P. Barnes, B. Bassalleck, R. Eisenstein, G. Franklin, R. Grace, D. Marlow, R. Rieder, J. Seydoux, J. Szymanski, W. Wharton, J. Derderian, Y. Civelekoglu, M. Deutsch, J. Prater, C. Chu, R. Hackenburg, E. Hungerford, T. Kishimoto, T. Fukuda, M. Barlett, G. Hoffman, E.C. Milner, R.L. Stearns, Phys. Rev. **C41**, 1062 (1990).

17 C.B. Dover and G.E. Walker, Physics Reports **89**, 148 (1982).

18 N. Cabibbo, Phys. Rev. Lett. **10**, 531 (1963).

19 D. Bailin, "Weak Interactions", Sussex University Press, Sussex, 1977;
E.D. Commins and P.H. Bucksbaum, "Weak Interactions of Leptons and Quarks", Cambridge University Press, Cambridge 1983.

20 J.B. Adams, Phys. Rev. **156**, 1611 (1967).

21 B.H.J. McKeller and B.F. Gibson, Phys. Rev. **C30**, 322 (1984).

22 J. Dubach, Nucl. Phys. **A450**, 71c (1986)

23 K. Takeuchi, H. Takaki and H. Bando, Prog. Theor. Phys. **73**, 841 (1985).

24 E. Oset and L.L. Salcedo, Nucl. Phys. **A443**, 704 (1985).

25 C.-Y. Cheung, D.P. Heddle and L.S. Kisslinger, Phys. Rev. **C27**, 1277 (1983);
D.P. Heddle and L.S. Kisslinger, Phys. Rev. **C33**, 608 (1985);
L.S. Kisslinger, Second Int. Conf. on the Intersections of Particle and Nuclear Physics, AIP Conference Proceedings Series **150**, 940 (1986).

26 R. Grace, P.D. Barnes, R.A. Eisenstein, G.B. Franklin, C. Maher, R. Rieder, J. Seydoux, J. Szymanski, W. Wharton, S.Bart, R.E. Chrien, P. Pile, Y. Xu, R. Hackenburg, E. Hungerford, B. Bassalleck, M. Barlett, E.C. Milner and R.L. Stearns, Phys. Rev. Lett. **55**, 1055 (1985);

27 J.J. Szymanski, P.D. Barnes, G.E. Diebold, R.A. Eisenstein, G.B. Franklin, R. Grace, D.W. Hertzog, C.J. Maher, B. Quinn, R. Rieder, J. Seydoux, W.R. Wharton, S. Bart, R.E. Chrien, P. Pile, R. Sutter, Y. Xu, R. Hackenburg, E.V. Hungerford, T. Kishimoto, L.G. Tang, B. Bassalleck, R.L. Stearns, Submitted to Physical Review C.
P.D. Barnes Nucl. Phys. **A450**, 43c (1986).
J.J. Szymanski, Carnegie Mellon University PhD Dissertation, 1987.

28 R. Bertini, O. Bing, P. Birien, K. Braune, W. Bruckner, A. Chaumeaux, M.A. Faessler,

R.W. Frey, D. Garreta, T.J. Ketel, K.Kilian, B. Mayer, J. Niewisch, B. Pietrzyk, B. Povh, H.G. Ritter and M. Uhrmacher, Nuclear Phys. **A368** 365, (1981).

29 L. Majling, M. Sotona, J. Zofka, V.N. Fetisov and R.A. Eramzhyan, Phys. Lett. **92B**, 256 (1980).

30 N. Auerbach and N. Van Giai, Phys. Lett. **90B**, 354 (1980).

31 S. Paul, W. Bruckner, B.Povh and H. Dobbeling, Nuovo Cimento **102A**, 379 (1989).; T. Yamazaki, T. Ishikawa, K.H. Tanaka, Y. Akiba, M. Iwasaki, S. Ohtake, H. Tamura, M. Nakajima, Y. Yamanaka, I. Arai, T. Suzuki, F. Naito and R.S. Hayano, Phys. Rev. Lett. **54**, 102 (1985).

32 A. Sakaguchi, W. Bruckner, S. Paul, R. Schussler, B. Povh, H. Dobbeling, H. Tamura, T. Yamazaki, M. Aoki, R.S. Hayano, T. Ishikawa, M. Iwasaki, T. Motoki, H. Outa, E. Takada and K.-H. Tanaka, Nuovo Cimento **102A**, 511 (1989).

33 O. Hashimoto, T. Nagae, T. Fukuda, S. Homma, T. Shibata, Y. Yamanoi, T. Shintomi, J. Imazato, Y. Makida, T. Mito and S. Kato, Nuovo Cimento **102A**, 679 (1989).

34 T. Yamazaki, M. Aoki, Y. Fujita, R.S. Hayano, H. Tamura, T. Ishikawa, E. Takada, M. Iwasaki, H. Outa, J. Imazato and A. Sakaguchi, Nuovo Cimento **102A**, 695 (1989).

35 O.W.B. Schult, W. Borgs, D. Gotta, A. Hamacher, H.R. Koch, H.Ohm, R. Riepe, H. Seyfarth, K. Sistemich, V. Druke, D. Filges, L. Jarczyk, S. Kistryn, J. Smyrski, A. Strzalkowski, B. Styczen, and P. von Brentano, these proceedings.

36 E.V. Hungerford, B.W. Mayes, R. Phelps, L.S. Pinsky, L.G. Tang, S. Bart, R.E. Chrien, P. Pile, R. Sutter, L.C. Dennis, K.W. Kemper, S.R. Cotanch, J.J. Reidy and J. Wise, CEBAF Hall C Proposal, Oct. 1989.

37 Proceedings of the PILAC User's Meeting October, 1990.

ELECTROMAGNETIC PRODUCTION OF STRANGE PARTICLES

Reinhard A. Schumacher
Department of Physics, Carnegie Mellon University, Pittsburgh, PA 15213

ABSTRACT

A brief overview of the experimental status of electro- and photo-production of hyperons from a proton target is given; with some discussion of likely directions for future work.

INTRODUCTION

Electro- and photo-production of kaons and hyperons will soon enter a era of renewed experimental investigation with the advent of new facilities at new, high duty-factor electron beam accelerators, such as CEBAF. We will discuss measurements of elementary strangeness electro-production, p(e,e'K)Y, and real photoproduction, p(γ,K)Y. The physics motivation is focused on the production of strange quarks via the well-understood electromagnetic interaction. In contrast to hadronically induced reactions such as (π,K) or (p,K), the photon is a relatively weakly interacting probe, allowing sensible first-order calculations to be done. These studies typically aim to extract values for the fundamental KYN coupling constants and to understand the dynamics of the strangeness-producing interaction. Good understanding of the elementary interaction is a necessary step towards embedding it in the nuclear medium for the calculation of electromagnetic hypernuclear production.

While this talk will outline our knowledge of these reactions, it must be said at the outset that many open questions remain in this field because the data are rather sparse. Also, present models typically evaluate Lorentz invariant amplitudes in terms of Feynman diagrams with exchange of mesons and baryons; contact with quark models is only indirect, through the calculation of coupling constants. The information in this talk is based on experiments done typically 15 to 25 years ago, some theoretical work done in recent years, and a discussion of what kinds of new results may be expected in the future.

Three related strangeness producing reactions on a proton target which lead to hyperons in the $J^{\pi}=1/2^{+}$ baryon octet can be considered:

$$\gamma_{(v)} + p \rightarrow K^{+} + \Lambda \quad (E_{\gamma}^{Th} = 0.911 \text{ GeV}) \quad (1a)$$
$$\rightarrow K^{+} + \Sigma^{o} \quad (E_{\gamma}^{Th} = 1.046 \text{ GeV}) \quad (1b)$$
$$\rightarrow K^{o} + \Sigma^{+} \quad (E_{\gamma}^{Th} = 1.048 \text{ GeV}) \quad (1c)$$

where E_{γ}^{Th} are the real photon threshold energies. The first two have been studied via both real and virtual photon experiments, while the third one has not yet been measured. Analogous reactions on the neutron and the production of excited hyperons have not been extensively studied.

In each of these reactions the final state hyperon can be polarized. Using the parity violating weak decay of these particles, this polarization can be measured without the need to do double scattering experiments. In unraveling the production mechanism this is an attractive feature in comparison to pion electro- and photo-production. For example, in the rest frame of the Λ, the weak decay $\Lambda \rightarrow \pi^- p$ emits the nucleon preferentially along the direction of the Λ spin. For Λ's with polarization P, the decay yields a distribution $I(\theta_p) = I_o(1 + \alpha P \cos\theta_p)$, where the weak decay asymmetry parameter is $\alpha = 0.642$, and θ_p is the angle between the nucleon momentum and the normal to the $K\Lambda$ plane. For the $\Sigma^+ \rightarrow p\pi^o$ decay the weak asymmetry parameter is -0.98. Because the Σ^o decays 100% via an M1 electromagnetic transition to the Λ, a measurement of the decay Λ polarization in Reaction 2 also measures the polarization of the Σ^o; the relationship is that $P_\Lambda = -1/3\, P_{\Sigma^o}$.

ELECTROPRODUCTION EXPERIMENTS

In electroproduction experiments the relevant quantities are the 4-momentum of the virtual photon $q=(\vec{q},\omega)$ and the direction (θ_x,ϕ) of the hyperon relative to $\vec{q}$ and the electron scattering plane, as shown in Figure 1. In terms of longitudinal and transverse photon polarization, the cross section can be written down most generally in terms of nine structure functions[1]; this corresponds to knowledge of the electron's initial helicity state, the proton polarization, and the polarization of the produced hyperon. Some of the most ambitious experiments undertaken to date[2],[3] used unpolarized electrons on unpolarized protons, and detected the electrons and produced kaons but not the polarization of the hyperon. When initial state helicities, proton polarization, and hyperon polarization are averaged or summed over, only four of these nine structure functions are left. The experimental cross section takes the form [2],[3],[4]

$$\frac{d^4\sigma}{dW\, dq^2\, dt\, d\phi} = \Gamma\left(\frac{d\sigma_U}{dt} + \varepsilon\frac{d\sigma_L}{dt} + \varepsilon\frac{d\sigma_P}{dt}\cos(2\phi) + \sqrt{2\varepsilon(\varepsilon+1)}\,\frac{d\sigma_I}{dt}\cos\phi\right) \quad (2)$$

where the virtual photon flux is

$$\Gamma = \frac{\alpha}{4\pi^2}\,\frac{E_e'}{E_e}\,\frac{W^2-M^2}{MQ^2}\,\frac{1}{1-\varepsilon},$$

and the virtual photon polarization parameter is

$$\varepsilon = \left(1+2\frac{\vec{q}^{\,2}}{Q^2}\tan^2\frac{\theta_e}{2}\right)^{-1}$$

Here W is the total center of mass energy, t the square of the 4-momentum transfer to the hyperon, $Q^2 = -q^2$ the electron 4-momentum transfer, M the proton mass, and E_e, E_e', and θ_e are the electron energies and angle in the lab frame. The first two terms are the (e,e') inclusive (unpolarized) transverse and longitudinal photon cross sections, separable by the Rosenbluth method or by 180° scattering. When the direction of the kaon is detected at the azimuthal angle ϕ out of the (e,e') scattering plane, the transverse photons can be polarized parallel or perpendicular to the hyperon production plane. The term $\varepsilon\, d\sigma_P/dt \cos(2\phi)$ is due to the interference of transverse components of this polarization. The term $(2\varepsilon(\varepsilon+1))^{1/2}\, d\sigma_I/dt \cos\phi$ is due to the interference between the transverse part of the transverse polarization and the longitudinal photon polarizations. The cross sections are functions of W, q^2, and t.

In the one-photon exchange shown in Figure 1 the interaction depends on the product of a leptonic current, $J_\mu(l) = \bar{u}(e')\gamma_\mu u(e)$, and a hadronic current, $J^\mu(h)$. The hadronic current is written in the CGLN formalism as a sum over six possible gauge and Lorentz invariant amplitudes A_i and matrices M_i^μ:

$$J^\mu(h) = \bar{u}(p_\Lambda)\,\gamma_5 \sum_{i=1}^{6} A_i\, M_i^\mu\, u(p_p)). \tag{3}$$

Expressions for the M_i^μ and the connection between the measured cross sections in Eq(2) and the currents are given in Refs 5], 6], and 7].

The hadronic part of electromagnetic strangeness production is therefore reduced to the computation of the six complex amplitudes A_i. In principle this should be possible starting from QCD and the quark model. In practice this has not been done, especially in the threshold region; instead models have been developed using hadronic and mesonic degrees of freedom. Figure 2 shows the diagrams which are needed. The lowest order Born terms involve proton, lambda, and kaon exchange in the s, u, and t-channels, respectively. Other important terms involve the exchange of the Σ, the spin-1 kaon resonances, the spin 1/2 and spin 3/2 nucleon resonances, and spin 1/2 hyperon resonances. The coupling constants at the vertices can be extracted from the data and compared either with values predicted by SU(3) or SU(6), or with values obtained from other strangeness-producing reactions. For virtual photon interactions the A_i also contain proton and kaon form factors [5].

We can consider the DESY experiment by Azemoon *et al.*[3], which measured $e + p \rightarrow e' + K^+ + Y$ for kinematics $-0.1 > q^2 > -0.6$ GeV^2/c^2, and 1.9<W<2.8 GeV (Λ threshold is at W=1.6 GeV). This experiment used two magnetic spectrometers at forward angles to detect the e' and K^+, with optical spark chambers, Cerenkov counters, and time-of-flight hodoscopes. Figure 3 shows the missing mass distribution $M = \sqrt{(\gamma+p-K)^{1/2}}$ where γ, p, and K are the 4-vectors for the virtual photon, target proton, and kaon, from which it is clear that several hyperons could be isolated for study. The photon polarization varied in the fairly narrow range $0.5<\varepsilon<0.8$, and so no separation of the contribution of longitudinal and transverse virtual photons was made. Figure 4 shows the result for the transverse-transverse interference cross section and the longitudinal-transverse interference cross sections defined in Eq.(2) as a function of W, q^2, and t, compared to the sum of the longitudinal and unpolarized cross sections. One can either say that the interference terms are consistent with zero, or remark that individual points show interference cross sections roughly 10% as large as the inclusive cross sections. Without theoretical guidance one cannot extract much physics from these data, but they suggest that new experiments may have sizeable interference cross sections for which to look.

In an similar experiment done at Cornell [8], the ϕ dependence was averaged, leaving only the first two terms in Eq.(2). Figure 5 shows the Q^2 dependence of the electroproduction of Λ and Σ^0 for a fixed value of the total c.m. energy W. $d\sigma/d\Omega^*$ is the sum of longitudinal and unpolarized transverse cross sections in the c.m. of the photon and proton. For Λ production the cross section first rises from the real photon

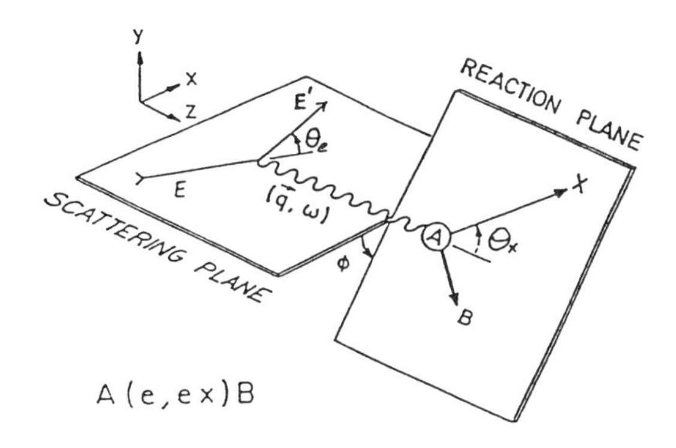

Figure 1) Kinematic variables for electroproduction. From Ref 1].

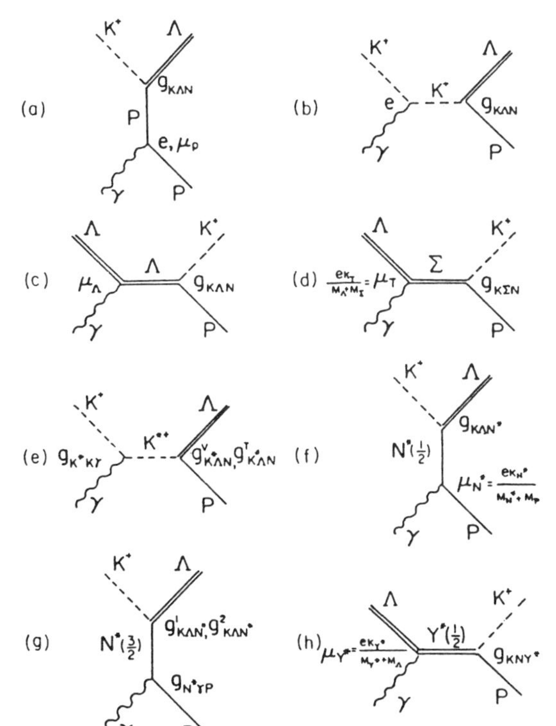

Figure 2) Diagrams for process $\gamma + p \rightarrow K^+ + \Lambda$. Born terms (a-e), and resonant terms (f-h). From Ref. 15].

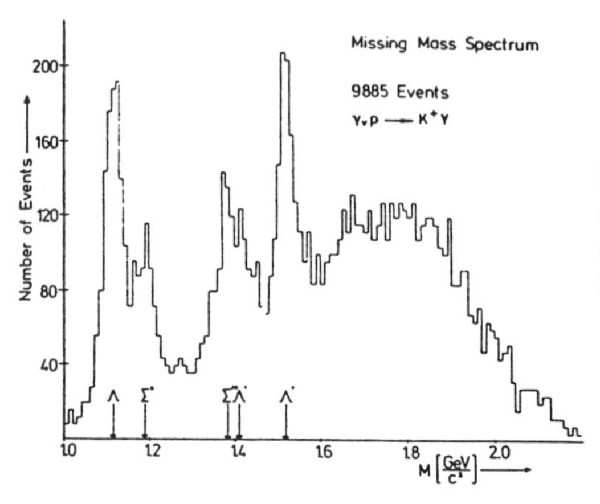

Figure 3) Missing mass spectrum for $\gamma_{(v)} + p \rightarrow K^+ + X$ for electon energy of 4.9 GeV, showing hyperons created. From Ref 3].

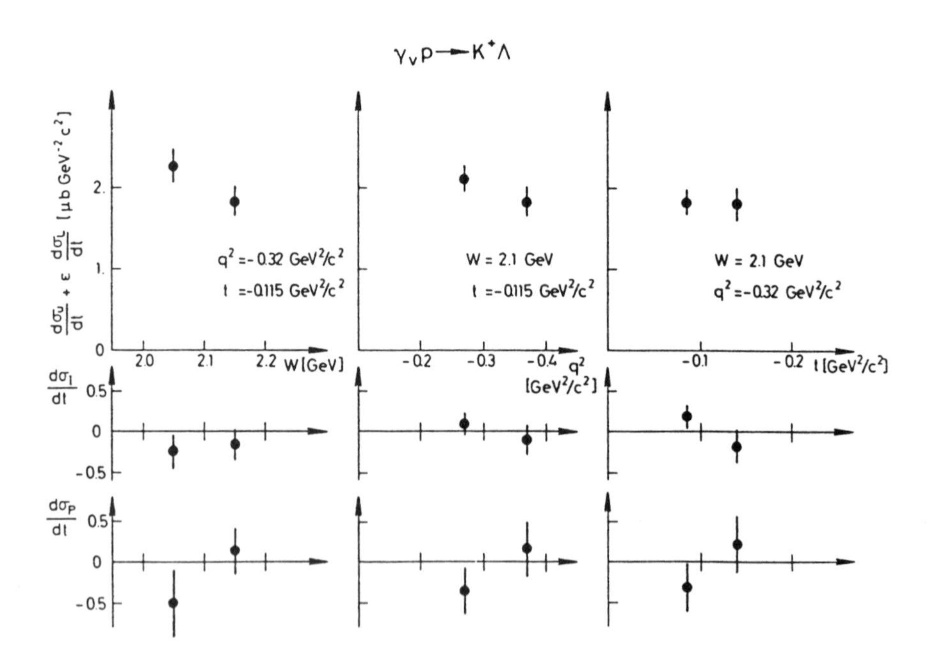

Figure 4) The cross section components in Eq. (2) for $\gamma_{(v)} + p \rightarrow K^+ + \Lambda$ as a function of W, q^2, and t for 4.0 GeV primary electron energy. From Ref 3].

point at Q^2=0, while for Σ^o production the drop is monotonic with Q^2. This is interpreted as an indication that longitudinal photon contributions are important for Λ production but not Σ production. A rough longitudinal-transverse separation was obtained [9] by measuring at different ε for the same W and Q^2, supporting the conclusion that a large longitudinal component contributes to Λ production only. A suggested explanation [3] is that longitudinal photons contribute strongly to kaon exchange in the t-channel, but this process is expected to be larger for Λ production than for Σ production since $g_{K\Lambda N} > g_{K\Sigma N}$, and hence the difference in Q dependence.

The Q^2 dependence of Σ^o production is much steeper than that for Λ production. In quark-parton models [10],[11],[12] this feature was interpreted as a consequence of the decrease of the ratio $F_1^{\gamma n}/F_1^{\gamma p}$, the deep inelastic electron-nucleon structure functions of the neutron and proton, as Bjorken $x = Q^2/2M\nu$ goes to 1 . In this limit the production of forward-going kaons off u-quarks tends to leave behind an isospin 0 pair u-d quarks, from which the production of I=1 baryons (Σ) is suppressed in favor of I=0 baryons (Λ).

The Q^2 dependence can also be interpreted in terms of the kaon form factor. In the vector dominance picture the photon can convert to a ϕ meson which decays to K^+K^- before interacting with the proton. Using a vector dominance form factor, Cotanch and Hsiao [5], [13] found an rms kaon radius of 0.3 to 0.4 fm. They used a set of coupling parameters (Fig. 2) derived from early photoproduction data[6]. However, an analysis by Adelseck and Wright [7] of the complete electro- and photoproduction data set gave a kaon rms radius of 0.51 fm, which agrees with more direct measurements where kaons are scattered from atomic electrons [14].

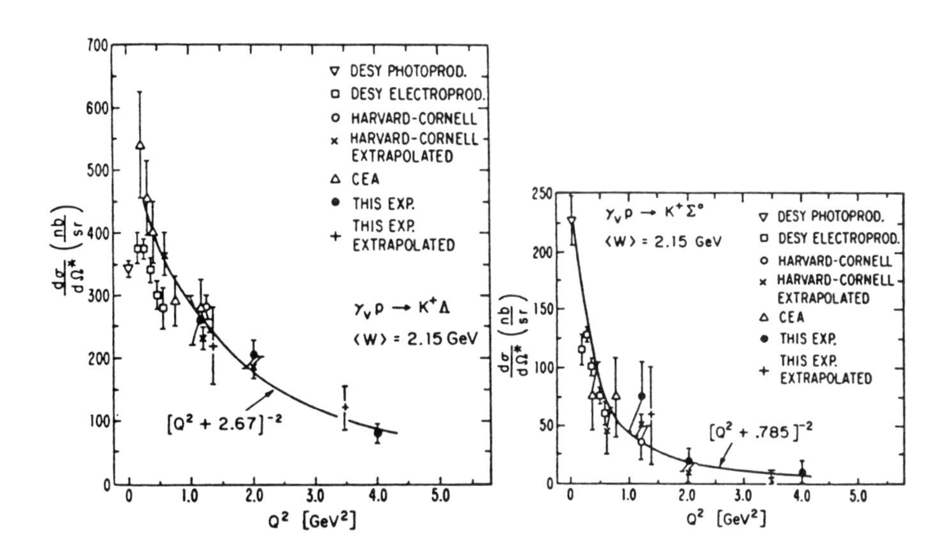

Figure 5) Q^2 dependence for $\gamma_{(v)} + p \rightarrow K^+ + \{\Lambda,\Sigma^o\}$. The trends away from the real photon point at $Q^2 = 0$ differ strongly. Data are from Ref. 8] and others cited therein.

PHOTOPRODUCTION EXPERIMENTS

In strangeness production by real photons, the longitudinal components of the cross sections vanish and the computational situation therefore simplifies somewhat. The complex invariant amplitudes are reduced from six to four in number and are given explicitly in Adelseck and Wright [7], for example.

Real photon experiments using tagged or untagged bremsstrahlung beams have been more numerous than electron scattering experiments. While being restricted to $q^2=0$, the relative simplicity of these experiments has led to more efforts to look for polarization effects as a way to pin down to amplitudes and couplings. Figures 6 and 7 give overviews of the existing Λ differential cross section data and polarization data[15],[16],[17]. The polarization data in Figure 7, which are for a kaon c.m. angle of $90^\circ \pm 5^\circ$, represent well over half of the data points ever measured. While the differential cross section is moderately well established, little more than the sign of the polarization is known.

Table I compiles the principal couplings $g_{K\Lambda N}$ and $g_{K\Sigma N}$ (see Fig. 2) which have been extracted from the data over the years. The sparseness of the data is such that a rather wide range of values for these constants has been obtained, combined with longstanding uncertainty about which resonances are really important in the production mechanism. The older photoproduction measurements tended to result in systematically smaller values for $g_{K\Lambda N}$ than the hadronically inferred values. For comparison, the table gives the SU(3) prediction of the couplings from Refs. 20] and 24], based on the experimental πNN coupling, and the values of Bozoian *et al*.[21], who use a quark-pairing model to describe meson exchange between baryons in an MIT bag model. Also listed are some values obtained from hadronic data. Workman has emphasized the instability of the couplings against addition of many resonances on top of the Born terms [18],[19], and hence the importance of trying to assess model uncertainties; his uncertainty estimates for several of the models are shown in brackets in Table I. Besides the values of the basic couplings, models differ in prescriptions for including resonance exchange terms to the Born amplitudes.

Recent authors have attempted to improve the early models by exploring more combinations of resonances in their fits to data. Adelseck and Wright [7] found that including $K_1(1280)$ exchange in the t-channel could increase $g_{\Lambda KN}$ to the hadronic value. Tanabe, Kohno, and Bennhold [22], on the other hand, pointed out the necessity of including K^++Λ final state correlations explicitly; by including a partial-wave dependent absorptive factor on top of the usual Born and resonance terms, they fit the cross section data at all energies, including beyond the typical 1.4 GeV cutoff in these studies (see Figure 8), and claim to obtain the hadronic-reaction value for $g_{K\Lambda N}$. Cohen[23] has found that taking the couplings derived from photoproduction and recalculating hadronic reactions such as low energy KN scattering unexpectedly produces *better* agreement with data than the "standard" hadronic values. He questions, however, the validity of the all diagrammatic models used to obtain elementary coupling constants from photo-kaon reactions on the grounds that the kaon mass is large on the scale of the nucleon mass, resulting in non-applicability of the Kroll-Ruderman theorem.

Adelseck and Saghai [24] have recently re-examined all the available data on Λ photoproduction. They find that data from just one of the dozen separate old experiments is "internally inconsistent" when compared to a simple model involving only a few exchanged particles. Removing this one data set from their analysis then leads them to a model for Λ production which involves exchanges of only five particles: $K^*(892)$, K1(1280), Σ^o, $N^*(1440)$, and $\Lambda^*(1670)$; the resulting coupling constants are in agreement with the SU(3) predictions. Unfortunately, the excluded data set happens to provide most of the cross section data for kaon angles larger than 90°, so it may be that the simple mechanism is an artifact of a now incomplete data set [25].

Numerous authors [15],[16],[18],[22],[24] have pointed out that better hyperon polarization data are the necessary next step for progress in the field. Figure 7a, for example, shows the older analysis of Renard [16] which illustrates the sensitivity of his model to Λ polarization data: the hatched region shows the range of predictions due to reasonable variations of the couplings $g_{\Lambda KN}$ and $g_{\Sigma KN}$ (compare to Figure 6). Figure 7b is from the newer analysis of Adelseck and Wright [7] (same P data with minus sign), showing again the inability of present data to discriminate strongly among fits. Future experiments are expected to emphasize polarization measurements. For example, at CEBAF there are plans [26] to measure the Λ polarization in detail using the toroidal spectrometer CLAS [27].

One can itemize [28],[24] sixteen observables corresponding to the combinations of photon, target, and lambda polarization that one can measure with single or double polarization measurements. Adelseck and Saghai have explored [24] the sensitivity of these observables to parameters in their model. There is sensitivity comparable to or better than that of the Λ polarization, P, in several observables, essentially no data exist with which to compare. Only one non-P polarization measurement exists [29] in which the polarized target asymmetry was measured one kaon angle and three photon energies. Workman found [30] even these few crude data points useful in comparing models.

Photoproduction data of Σ^+ and Σ^o (Reactions 1b, and 1c) are supplemental to the Λ data, since all three reactions are described within the same framework. A useful feature of Σ^o production is the absence of the t-channel diagrams because the photon does not couple to the K^o. This results in a backward peak in the predicted differential cross section, which would be easy to find experimentally [31]. By detecting $K^o \rightarrow \pi^+ \pi^-$ this reaction will be accessible at CEBAF [26].

Photoproduction of spin 3/2 hyperons is exemplified by a Daresbury/NINA experiment [32] which examined the reaction $\gamma + p \rightarrow K^+ + \Lambda(1520)$ between 2.8 and 4.8 GeV. Λ(1520) production was found to be identical to Λ(1115) production, having the same total cross section, same s- and t- dependencies, and the same spin polarization.

CONCLUSIONS

The basic features of electromagnetic hyperon production are experimentally known and understood in terms of conventional particle production models. Theoretical progress is still possible, however, after more detailed experiments are done. Cross sections for two of the three "elementary" reactions on the proton (Eq(1)) have been determined at the 10% to 20% level. But only a very crude longitudinal/transverse separation been performed for electroproduction, and no adequate measurements of the out-of-plane interference cross sections exist. Persistent difficulties with interpretation of the photoproduction data (values of coupling constants, importance of the non-Born terms) have left the picture rather incomplete. The future lies in improved coincidence measurements and the measurement of polarization observables. We can expect such progress at new facilities, such as those under construction at CEBAF and Bonn.

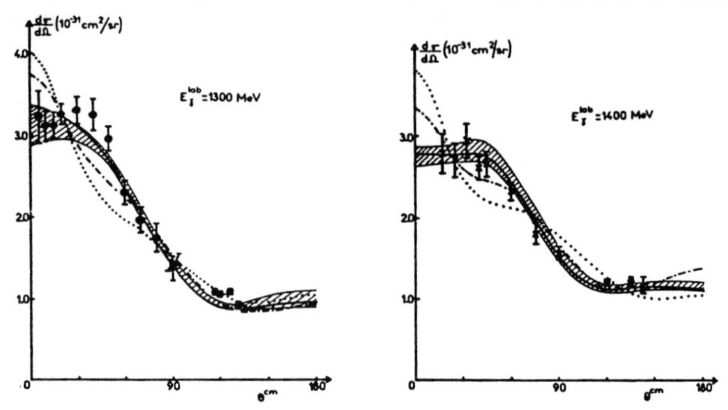

Figure 6) Differential cross sections for $\gamma + p \rightarrow K^+ + \Lambda$ plotted as a function of kaon c.m. angle, from Ref 16]. The shaded region corresponds to $g_{\Lambda KN}/\sqrt{4\pi}$ varying from 1.1 to 2.8.

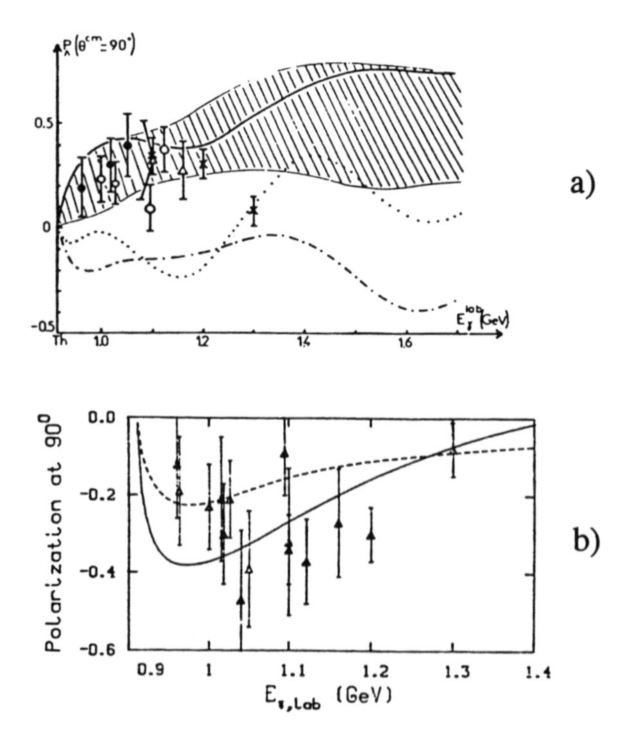

Figure 7) Λ polarization data for $p(\gamma,K^+)\Lambda$ for a kaon c.m. angles of $90^0 \pm 5^0$. (a) From Renard (Ref. 16]), with curves corresponding to those in Figure 6. Note the sensitivity of this model to the polarization data. (b) From by Adelseck, Bennhold, and Wright. (Ref 15]). Polarization has opposite sign definition from (a).

Table I

Values of $g_{K\Lambda N}/\sqrt{4\pi}$ and $g_{K\Sigma N}/.\sqrt{4\pi}$ obtained by various authors. Uncertainties in parenthesis have been computed by Workman 20] 27].

	$\frac{g_{K\Lambda N}}{\sqrt{4\pi}}$	$\frac{g_{K\Sigma N}}{\sqrt{4\pi}}$	Reference
Quark model or SU(3) Prediction:			
deSwart '63	−3.0 to −4.4	0.9 to 1.3	20,24
Bozoian *et al* '83	−4.13	+0.82	21
From Λ photoproduction data:			
Kuo '63	−2.0		33
Thom '66	−1.1 to −2.6	−0.9 to +1.0	6
Renard '72	−1.1 to −2.8	+0.4	16
Pickering '73	−2.8 to −3.4	+.9 to +1.0	34
Adelseck, Bennhold, and Wright '85	−1.3 (±.1)	+2.0	15
Rosenthal *et al* '88	−0.9	+0.6	35
Adelseck and Wright '88	−4.3	−1.8	7
(including (e,e'K$^+$))	−3.2	−1.7	
Cohen '89	−2.0	−0.8	36
Adelseck and Saghai '90	−4.2 (±1.6)	1.2 (±2.3)	24
From Σ photoproduction data:			
Renard and Renard '71	−2.4 to −3.6	+0.6	16
Bennhold '89	−1.9 to 2.1	+2.7	37
From K and Y data (signs undetermined):			
Granovskii and Starikov '68	2.43	–	38
Knudsen and Pietarinen '73	3.5±2.5	–	39
Martin '81	3.73±.3	<1.82	40
Antolin '86	3.53	1.53	41
Dover and Walker '82	4.62	0.9	42

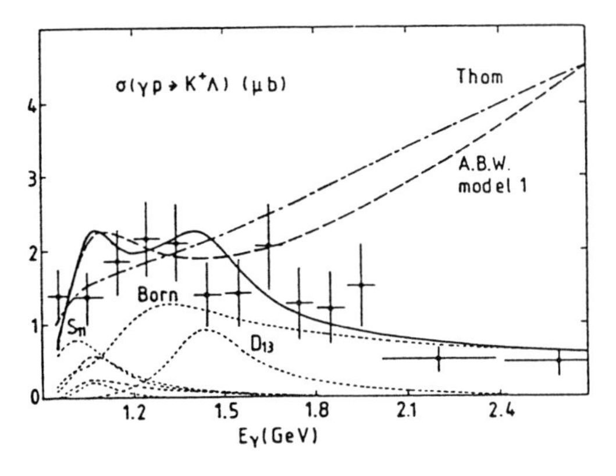

Figure 8) Total cross section. Fits are from Ref. 37] (solid and dotted lines), Ref. 6] (dot-dash), and Ref. 15] (dashed).

1] Report of the 1987 Summer Study Group, CEBAF, VA, p. 185 (1987) .
2] C.J. Bebek *et al.*, Phys Rev Lett **32**, 21 (1974).
3] T. Azemoon *et al* ., Nucl. Phys. **B95**, 77 (1975).
4] S. Boffi, C. Giusti, and F.D.Pacati, Nucl. Phys. **A435**, 697 (1985).
5] Stephen Cotanch and Shian Hsiao, Nucl. Phys. **A450**, 419c (1986).
6] H. Thom, Phys. Rev. **151** , 1322 (1966).
7] R.A.Adelseck and L.E.Wright, Phys. Rev. **C38** 1965 (1988); Ralf Anton Adelseck, PhD thesis, Ohio University (1988).
8] C. J. Bebek *et al* ., Phys Rev **D15**, 594 (1977).
9] C. J. Bebek *et al* ., Phys Rev **D15**, 3082 (1977).
10] F.E. Close, Nucl. Phys. **B73**, 410 (1974).
11] O. Nachtmann, Nucl. Phys. **B74**, 422 (1974).
12] J. Cleymans and F.E.Close, Nucl. Phys. **B85**, 429 (1975).
13] Shian Hsiao and Stephen Cotanch, Phys. Lett. **28**, 200 (1985).
14] E.B.Dally *et al*.., Phys Rev Lett **45**, 232 (1980).
15]Available data are summarized in R.A.Adelseck, C. Bennhold, and L.E.Wright, Phys. Rev. **C32** , 1681 (1985).
16] Y.Renard Nucl. Phys. **B40** , 499 (1972); F. M. Renard and Y. Renard, Nucl. Phys. **B25**, 491 (1971).
17] T. Fujii *et al*.., Phys. Rev. D **2** , 439 (1970).
18] R.L.Workman, Phys Rev **C39** , 2076 (1989); reply to comment by Workman in R.A.Adelseck and L.E.Wright, Phys Rev **C39** , 2078 (1989).
19] R.L. Workman, Proc. of "Few Body XII", Vancouver, ed. B.K.Jennings (1989).
20] J.J.deSwart, Rev. Mod. Phys. **35**, 916 (1963).
21] M. Bozoian, J.C.H. van Doremalen, and H.J.Weber, Phys. Lett **122B**, 637 (1983).
22] H. Tanabe, M. Kohno, and C. Bennhold, Phys Rev **C39**, 741 (1989).
23] Joseph Cohen, Phys. Lett. B **192** 291 (1987); Joseph Cohen, Phys. Rev. C **37** , 187 (1988).
24] R.A.Adelseck and B. Saghai, Phys. Rev. **C42**, 108 (1990).
25] R. Workman, preprint and private communication.
26] CEBAF proposal 89-04, Carnegie Mellon, Catholic U., V.P.I., CEBAF, Florida State, Los Alamos; R. Schumacher, Spokesman (1990); CEBAF proposal 89-43, CEBAF N* Collaboration, L.Dennis and H. Funsten, Spoeksmen (1990).
27] CEBAF Conceptual Design Report, April, (1990).
28] I.S.Barker, A. Donnachie, and J.K.Storrow, Nucl. Phys. **B95**, 347 (1975).
29] K.H.Althoff *et al.*, Nucl. Phys. **B137** , 269 (1978).
30] R.L Workman, Phys. Rev. **C42**, 781 (1990).
31] C. Bennhold, private communication.
32] D.P.Barber *et al.*, Z. Physik **C7**, 17 (1980).
33] T.K.Kuo, Phys. Rev. **129**, 2264 (1963).
34] A. R. Pickering, Nucl. Phys. **B66** , 493 (1973).
35] A. S. Rosenthal, Dean Halderson, Kimberly Hodgkinson, and Frank Tabakin, Annals of Physics **184** , 33 (1988).
36] Joseph Cohen, Phys. Rev. **C39**, 2285 (1989).
37] C. Bennhold, Phys. Rev. **C39**, 1944 (1989).
38] Ya. I. Granovskii and V.N. Starikov, Sov. J. Nucl. Phys. **6**, 444 (1968).
39] C.P.Knudsen and E. Pietarinen, Nucl. Phys. **B57** , 637 (1973).
40] A.D.Martin, Nucl. Phys. **B179** , 33 (1981).
41] J. Antolin Z. Physik **C31** , 417 (1986).
42] Carl B. Dover and George E. Walker, Phys. Rep. **89**, 1 (1982).

THE JETSET EXPERIMENT AT LEAR

R. A. Eisenstein
Nuclear Physics Laboratory
University of Illinois
Champaign, IL 61820

Representing the Jetset Collaboration:

R. Armenteros[1], D. Bassi[3], P. Birien[2], R. Bock[1], A. Buzzo[3], E. Chesi[1], P. Debevec[4], D. Drijard[1], W. Dutty[2], S. Easo[3], R. A. Eisenstein[4], T. Fearnley[1], M. Ferro-Luzzi[1], J. Franz[2], R. Greene[4], N. Hamann[1], R. Harfield[1], P. Harris[4], D. Hertzog[4], S. Hughes[4], T. Johansson[7], R. Jones[1], K. Kilian[5], K. Kirsebom[3], A. Klett[2], H. Korsmo[6], M. Lovetere[3], A. Lundby[6], M. Macri[3], M. Marinelli[3], L. Mattera[3], B. Mouëllic[1], W. Oelert[5], S. Ohlsson[7], J.-M. Perreau[1], M. G. Pia[3], A. Pozzo[3], M. Price[1], P. Reimer[4], K. Röhrich[5], E. Rössle[2], A. Santroni[3], A. Scalisi[3], H. Schmitt[2], O. Steinkamp[5], B. Stugu[5], R. Tayloe[4], S. Terreni[3], H.-J. Urban[2], H. Zipse[2]

(1) CERN (2) University of Freiburg (3) University of Genoa (4) University of Illinois (5) IKP at KFA Jülich (6) University of Oslo (7) University of Uppsala

ABSTRACT

The Jetset experiment (PS 202) at LEAR will search for gluonic hadrons and other exotics in the interaction of in-flight antiprotons with protons at rest. The mass range to be covered extends from 1.96 to 2.43 GeV. Our experiment uses a molecular hydrogen cluster jet target which is inserted in the LEAR ring and is surrounded by a general-purpose detector of advanced design. In "Phase I" the detector has been constructed to provide selection at the trigger level of four kaon events, allowing a search for resonances in the OZI-forbidden, gluon-rich reaction $\bar{p}p \rightarrow \phi\,\phi \rightarrow 4K$. The combination of LEAR with our detector is unique in that it allows scanning over a large momentum range with excellent momentum resolution, implying correspondingly excellent mass resolution. The physics interest in this process is outlined briefly, and the design of the detector elements is described in some detail. A brief discussion of possible upgrade paths ("Phase II) for the detector is given at the end.

INTRODUCTION

The last few years has seen a growing interest in the search for gluonic matter and exotic quark-gluon formations. The importance of these objects lies simply in the fact that, being allowed and expected states of matter in the standard model, evidence of their existence should be forthcoming. In fact, the lack of a clear signature of such states could be a genuine problem for QCD. Because of this, there has been recently a heightened interest in experimental searches for such exotica, and indeed some of the data point to the possible existence of interesting new physics at masses above 2 GeV. A number of recent review articles and conferences[1-6] have stressed the importance of these studies and the need to continue work in this area in both the experimental and theoretical domains.

We have begun a new experimental program (PS202) at CERN/LEAR that is intended to examine these questions in more detail. The general purpose is to study rare or "OZI-forbidden" formation reactions of the type $\bar{p}p \rightarrow M_1M_2$, where the new forms

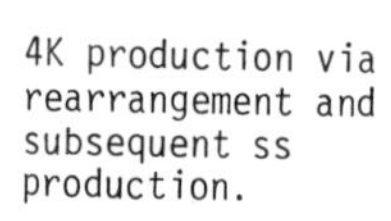
4K production via rearrangement and subsequent ss production.

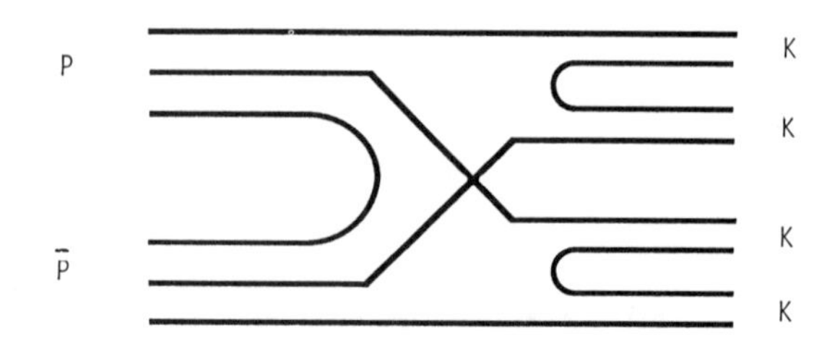

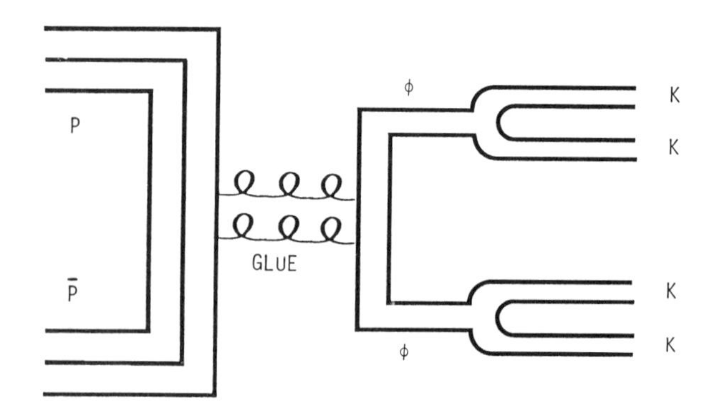

4K production via φφ intermediate state.

Resonance appearing on top of the 4K and φφ production cross sections. The relative sizes are exaggerated for the purpose of illustration.

FIGURE 1

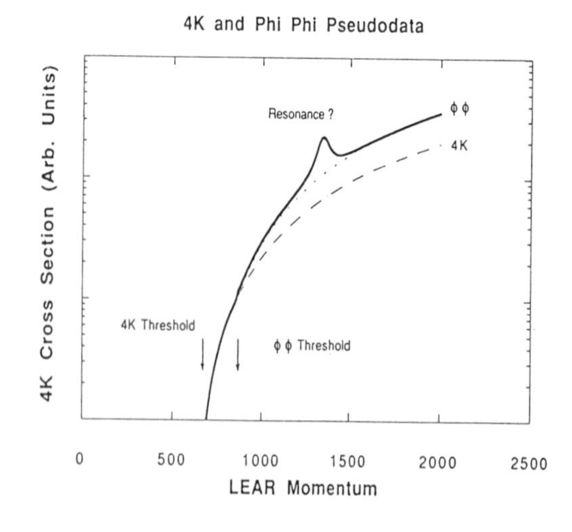

HYDROGEN CLUSTER JET TARGET

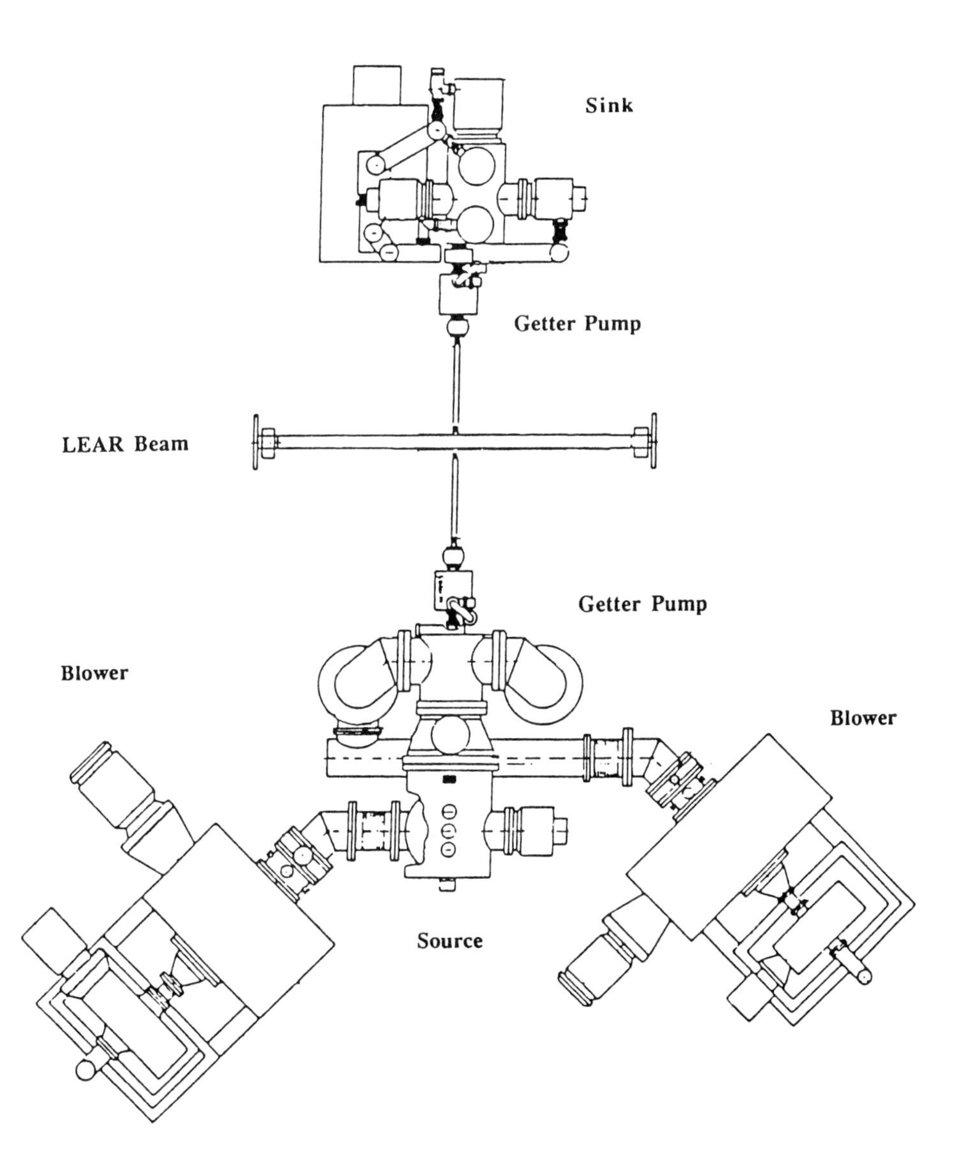

Plan view of the hydrogen cluster jet target, now installed on the LEAR facility.

FIGURE 2

of matter are most likely to be visible. We have divided our study into two phases; the focus of "Phase I" is to study in detail the in-flight production of $K^+K^-K^+K^-$, including ϕKK and most importantly $\phi\phi$. In Fig. 1 we indicate two possible ways of reaching the 4K final state -- one process proceeding via an intermediate state that is "pure glue" to form the $\phi\phi$ state that subsequently decays to 4K with a 25% branch; the other possibility proceeds directly to 4K by a combination of rearrangement and subsequent $s\bar{s}$ pair creations. (Other paths to formation of $\phi\phi$ exist; these include the possibility of freeing strange sea quarks from the proton, connections to d or u quark admixtures in the ϕ, or $\phi-\omega$ mixing.) The bottom part of the figure is a schematic drawing of what we hope to see: the onset of 4K production at its threshold of 646.8 MeV/c, followed by an increase in cross section at the $\phi\phi$ threshold (866.6 MeV/c), followed by the appearance of a resonance in $\phi\phi$ at some undetermined point. The drawing is intended to be illustrative only. In "Phase II" we intend to improve the detector in ways that will make possible the study of a much wider variety of related physics channels. Here we describe the design of "Phase I" and indicate the direction that we envision for "Phase II".

THE "PHASE I" JETSET EXPERIMENT

The central feature of the Jetset apparatus is the **molecular hydrogen cluster jet target** which is installed in the south straight section (SD2) of the LEAR ring. First tests show that the jet presently has a density of about $2\text{x}10^{13}$ atoms/cm^2; we expect another factor two improvement. The jet provides a "massless" pure gaseous hydrogen cylindrical target of about 1 cm diameter (see Fig. 2).

For a stored beam containing $4\text{x}10^{10}$ antiprotons moving with a circulation frequency of $3.2\text{x}10^6$ Hz, the peak luminosity will be about $5\text{x}10^{30}$/cm^2/sec in an interaction volume of about 1 cm^3 and will exceed the most optimistic external target experiment by about a factor of 20. It thus offers the possibility of searching efficiently for rare physics channels such as $\bar{p}p \rightarrow \phi\ \phi$. Such luminosities provide rates of about $5\text{x}10^5$ events/μb/day, so that we will double the world's existing sample of $\phi\phi$ events in a very short running time.

Unfortunately, the desired signal is small compared to a number of other physics channels of much less interest to us. Fig. 3 shows some of the relevant total cross sections for these processes in the LEAR energy range. The high luminosity of the jet target apparatus also means that the detector must be equipped to allow extraction of good events from a background rate of about 1 MHz. To do this we use the characteristics of the events we wish to study: they have (only) 4 charged prongs; they are forward of 90° in the lab frame; they each have a moderate β value. At least 3 of the 4 particles are almost always forward of 45°. Thus we will apply cuts at the trigger level that focus on the multiplicity, the event geometry and the β values of the outgoing particles.

With a trigger based on the above ideas, the very large "purely pionic" background can be reduced to a small fraction of the good event rate. Other, more troublesome, reactions are $\pi^+\pi^-K^+K^-$ and $\bar{p}p\pi^+\pi^-$; these provide signatures looking much more like good events. However, the "Phase I" detector is designed to distinguish these in both the on-line and off-line analysis.

A plan view of the "Phase I" detector is given in Fig. 4. It is a compact ($\approx$ 1m^3) device that surrounds the jet with 85% of 4π solid angle. Although the LEAR beam pipe (an oval shape with major and minor half-axes 6.8 and 3.0 cm resp.) limits the very forward acceptance below about $\theta = 8°$, it will be possible to measure the $\phi\phi$ production cross section near threshold with reduced efficiency.

The detector includes a number of advanced features, as described in [7]. A major constraint in the design is the need to be able to dismount it rapidly in case of needed maintenance on either the detector itself or the LEAR facility.

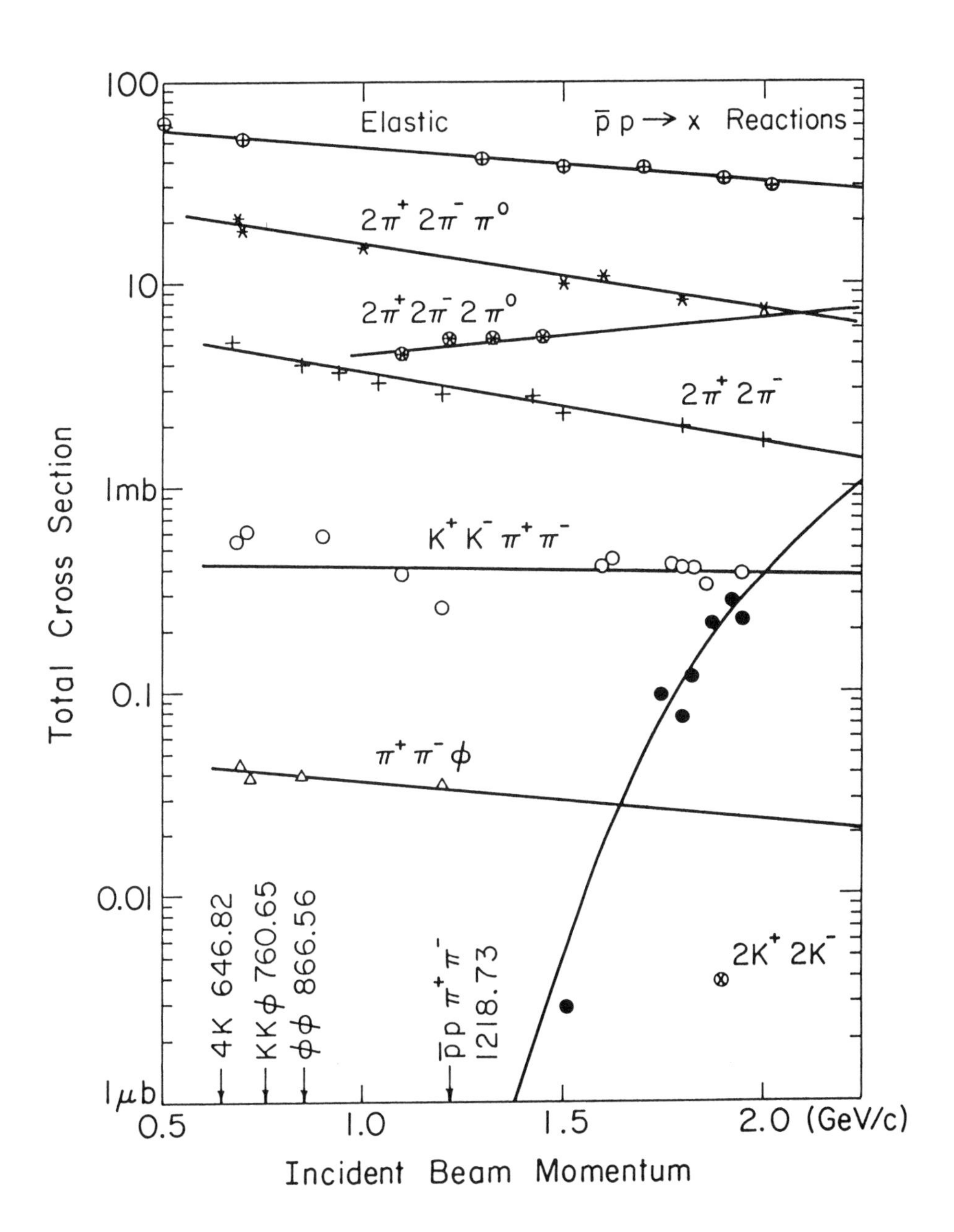
Elastic
p̄ p → x Reactions
2π+ 2π- π0
2π+ 2π- 2π0
2π+ 2π-
K+ K- π+ π-
π+ π- φ
2K+ 2K-
4K 646.82
KKφ 760.65
φφ 866.56
p̄ p π+ π-
1218.73
Total Cross Section
100
10
1mb
0.1
0.01
1μb
0.5
1.0
1.5
2.0 (GeV/c)
Incident Beam Momentum

FIGURE 3

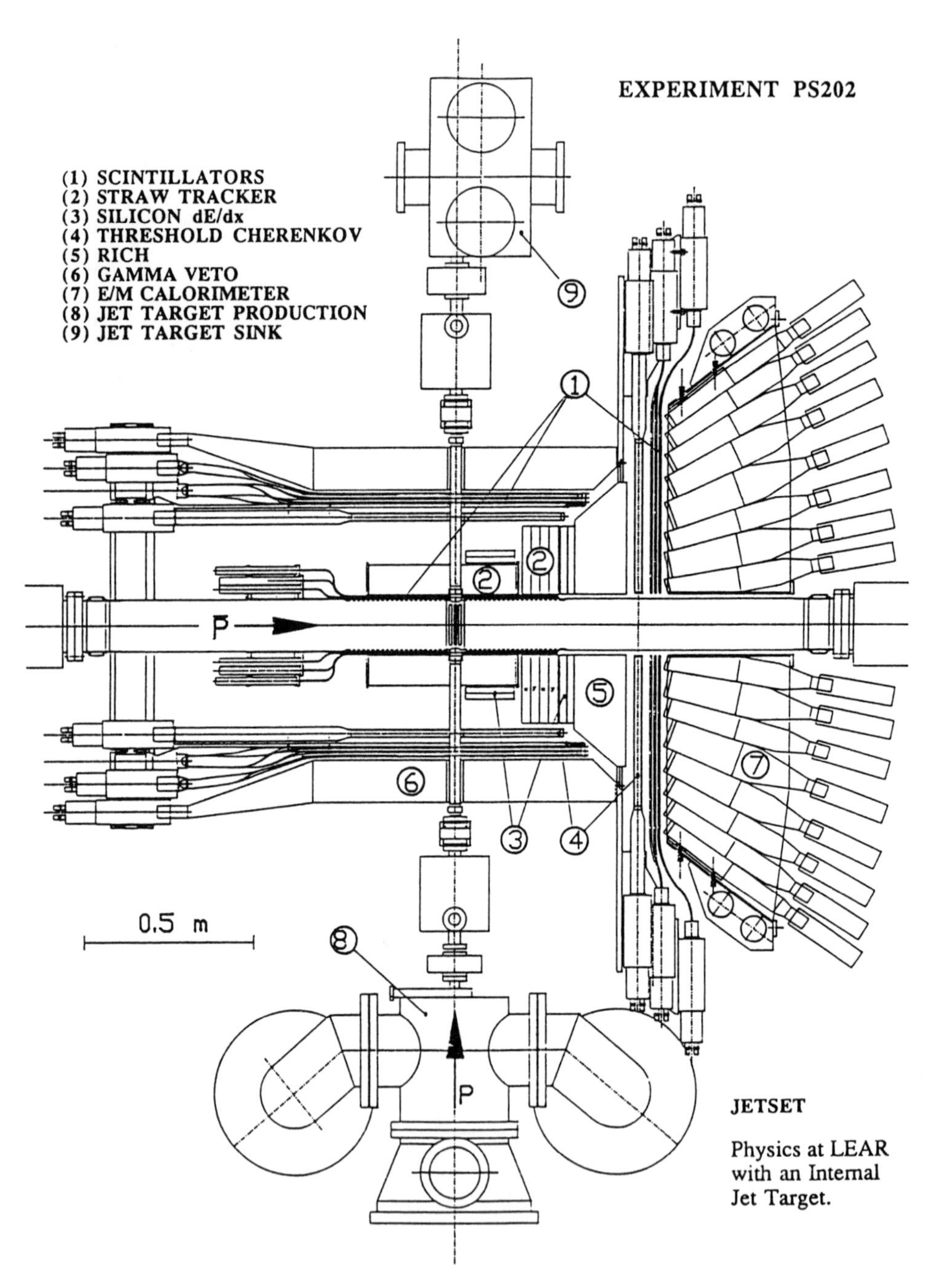

Plan view of Jetset.

FIGURE 4

The **pipe trigger scintillators**, which lie directly on the LEAR beam pipe, are arranged in two layers, each of which consists of scintillator strips 2 mm thick: the first set (40 scintillators, each 10 mm wide) covers the θ range from 15° to 45°. Because of the oval shape of the beam pipe the length of these strips varies in order to keep fixed the angular range covered by each strip. A second set of 20 scintillators (each 20 mm wide) covers the θ range from 45° to 65°.

The purpose of these scintillators is to provide a fast multiplicity and time-zero trigger, in conjunction with the "outer" trigger counters (discussed below). These have the advantage of being close to the interaction region so that they will be less affected by kaon decay, secondary interactions, or multiple scattering. At low momentum (near threshold), where these effects are large, they play a major role in keeping the acceptance at a reasonable level. This is done by exploiting the fact that at low momentum the four outgoing kaons are mostly forward of 45°, whereas the contaminant pion reactions are more uniformly spread out in θ.

The precision barrel tracker provides both azimuthal and z-direction position information for off-line analysis of the charged tracks in the barrel region ($\theta > 45°$). It consists of about 1500 individual "straw" drift-tube counters running parallel to the beam direction and glued together into a self-supporting unit (see Fig. 5). The straws fill about 85% of the available volume. Such a construction reduces the amount of supporting material needed in the forward direction and thus minimizes multiple scattering. For this reason all of the electronics connections and gas handling fittings are mounted on the barrel upstream endplate. Each wire is equipped separately with drift-time readout in order to measure the radial distance of impact. In order to measure the longitudinal coordinate by means of charge division, two wires are connected to each other using resistors (in SMD technology) that are located at the forward end of the assembly. The combined information from TDC's (drift time) and ADC's (charge division) will provide unambiguous three-dimensional information for off-line track finding and fitting.

To allow for perpendicular entry of the jet, about 400 of the straws are split roughly at their midpoint. In order to disassemble the tracker easily the entire unit is split along its horizontal midplane into two halves.

Each "straw counter" consists of a tube that is 8 mm in diameter and 400 mm in length with a resistive anode wire (30 μm diameter stainless steel, $R \approx 1k\Omega/m$) running down its center at a tension of about 100 g. The tubes are pure aluminum extruded tubing with a wall thickness of 60 μm (about 0.001 radiation length).

Using a gas mixture of $Ar/CO_2 = 1/1$ at atmospheric pressure we have obtained drift time resolutions corresponding to about 160 μm.. For this mixture a charge division resolution of σ/L of about 1% has also been obtained. These figures are easily maintained at the rates characteristic of this experiment and will provide excellent tracking.

A monolithic preamplifier (Fujitsu MB 43468) is used for each of the drift tubes. The preamplified signals are then fed into a receiver card which contains a differential amplifier (μA 733), analog drivers for the connection to the ADC's, and fast discriminators (LeCroy MVL 407) for the connection to the TDC's. This system comprises 2560 channels because it serves both the barrel and forward trackers.

The precision forward tracker provides tracking information for the forward-going ($\theta < 45°$) charged particles. It is constructed from the same aluminum tubes and electronics as described above. These straws, numbering 1020, are mounted in 12 layers perpendicular to the beam axis, three each for the x-, y-, and u-, v- coordinates. We have achieved essentially the same resolution figures as mentioned above for the barrel. As is

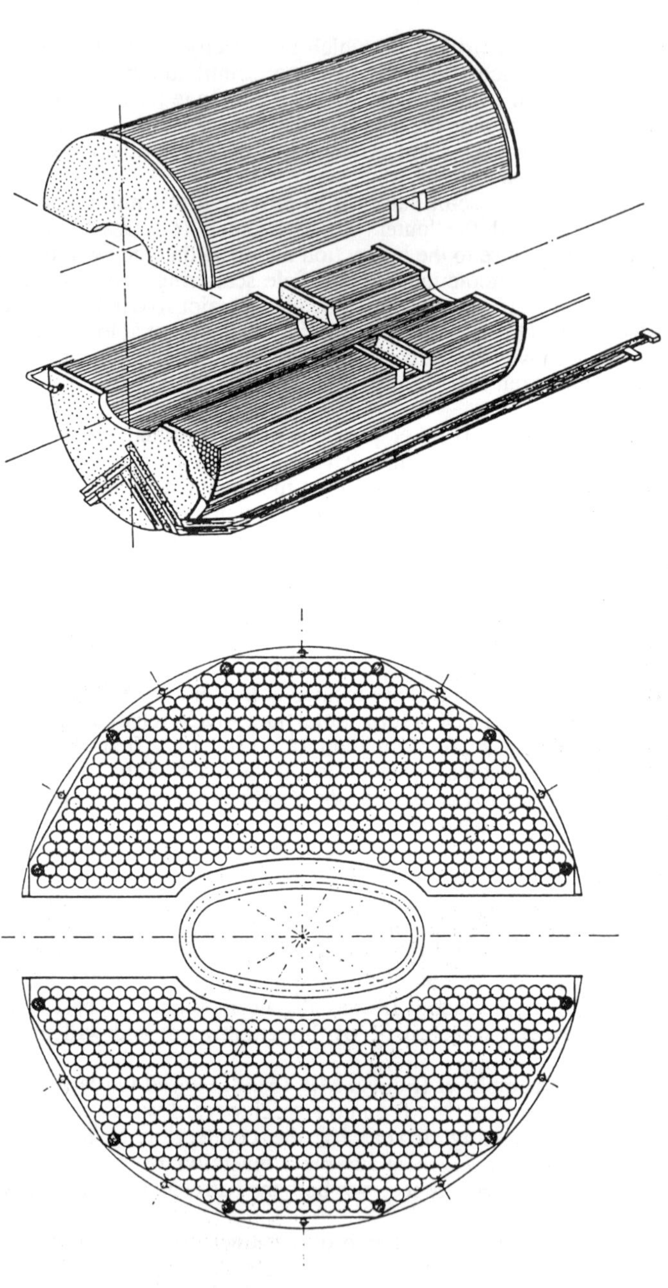

(Top) Exploded view of the barrel "straw" tracker. Note the rear entry of the gas and electronics connections, and the side entry of the jet. (Bottom) End view of the tracker showing the cutout for the jet. This region will not be instrumented in Phase I.

FIGURE 5

the case with the barrel tracker, the independence of each straw insures that there will be no track ambiguities from this source in the off-line analysis.

The silicon dE/dx forward counters measure β of the charged tracks and thus greatly help to distinguish 4K events from the $\bar{p}p\pi^+\pi^-$ (and other) backgrounds off line. The silicon complements the threshold Cerenkov and RICH counters, as it measures β best at low energies. Cuts on dE/dx, used in conjunction with the trackers, will reduce backgrounds to a fraction of a percent without loss of the desired signal.

The forward silicon will be mounted in two modules, each of which yields an (x,y) measurement for a passing charged particle. Half of one module is displayed in Fig. 6. These will thus provide two energy loss measurements as well as geometric information about each particle track. The silicon planes consist of units 1.95 x 2.40 cm^2 in area, each made up of four pads of 280 μm thickness. There are 924 such detectors in the forward counter covering an area of 0.43 m^2. The amount of material at normal incidence is estimated to be $\approx$ 3.5% of a radiation length with the silicon itself representing about 0.6%. The silicon is supplied by SI/AME, Norway. The electronics are mounted on the same printed circuit board as the detector elements. The front-end VLSI electronics are based on the Amplex chip developed at CERN/EF Division for UA2. An elaborate multiplexing scheme will be used to read out the 3696 pads in the forward region.

Tests have been made at the CERN T_{11} test beam which clearly show the ability to measure dE/dx adequately for our purposes.

The silicon dE/dx barrel counters provide two energy loss measurements in the barrel. We use the same silicon pads here as in the forward counter, in this case arranging them in a cylindrical geometry. The barrel Si counters extend to $\theta = 78^o$, which corresponds to the maximum value of θ for the process $\bar{p}p \rightarrow \phi\,\phi \rightarrow K^+K^-K^+K^-$. This counter will be installed in 1991.

The threshold Cerenkov counters are used to reject the fast charged pion background at the trigger level. The detectors consist of "liquid freon" (C_6F_{14}) filled plexiglass wedges (3 mm wall thickness) in the barrel and foward regions (see Fig. 7). The "liquid freon" material (3M Corporation FC-72) has an index of refraction n = 1.26, making the detector sensitive to particles of $\beta > 0.8$.

In the barrel counter the wedges run along the beam direction. They have a trapezoidal cross section of thickness 3.5 cm in the radial direction and a length of about 60 cm. The barrel counter (29 cm radius) is formed of 24 such wedges. Light is collected at the upstream end of each wedge due to lack of space downstream and our wish to avoid putting more material there.

The forward threshold Cerenkov counter is formed of 24 pie-shaped wedges of outer radius about 30 cm and thickness in the beam direction of 5 cm. The segmentation matches that of the trigger scintillator and the gamma calorimeter/veto.

The Ring-Imaging Cherenkov (RICH) counter will provide additional (off line) information about the momentum of the charged tracks. This device is in the R&D stage. It is presently conceived to have a 1 cm thick CaF_2 or quartz (n = 1.56, $\beta_{threshold}$ = 0.64) radiator and a 6 cm empty "drift space" whose purpose is to allow the Cherenkov light cone to broaden as much as is practical before detection. We estimate that for particles of $\beta > 0.64$, measurement of the cone under these conditions will yield β with about 10% uncertainty. This device therefore complements very well the measurements we will obtain from the Si layers.

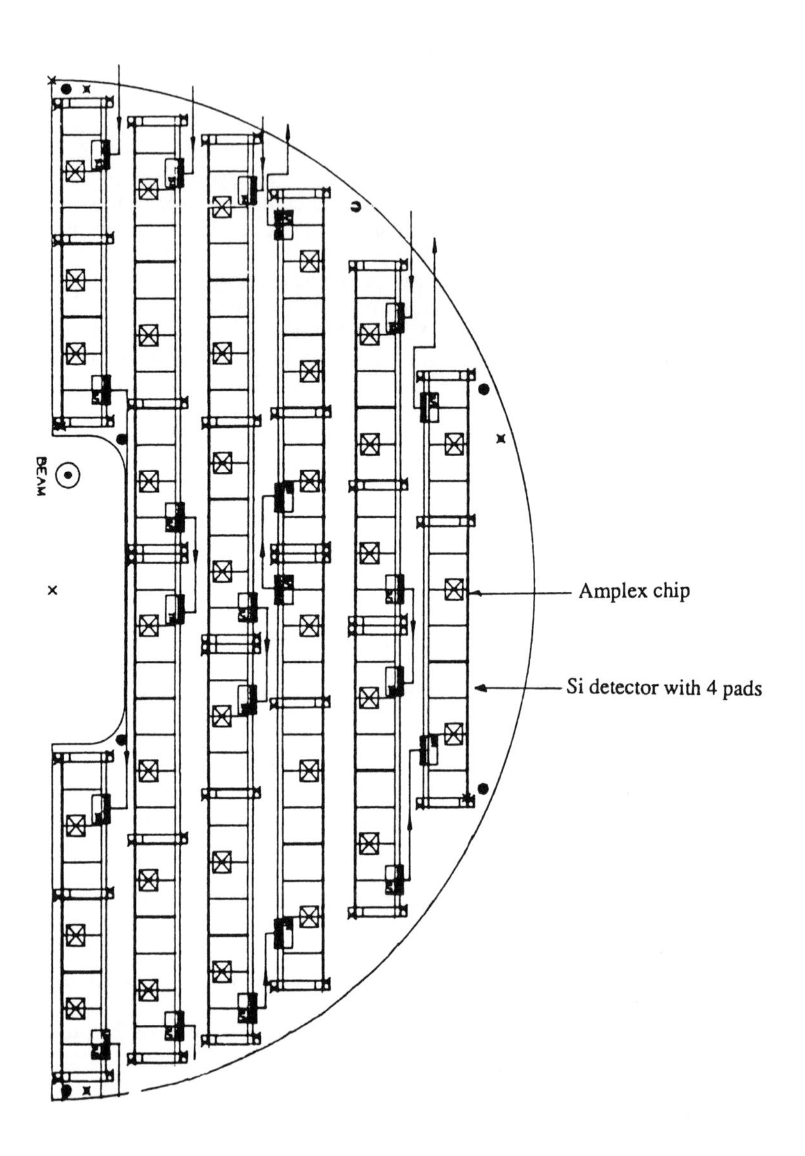

Plan view of mounting for one of the four silicon half-planes for the silicon dE/dx forward counter. The vertical strips are rows of silicon detectors and associated Amplex electronics. Two half-planes have this orientation; the other two have the silicon running perpendicular to what is shown here. This arrangement gives full coverage for the dE/dx counter.

FIGURE 6

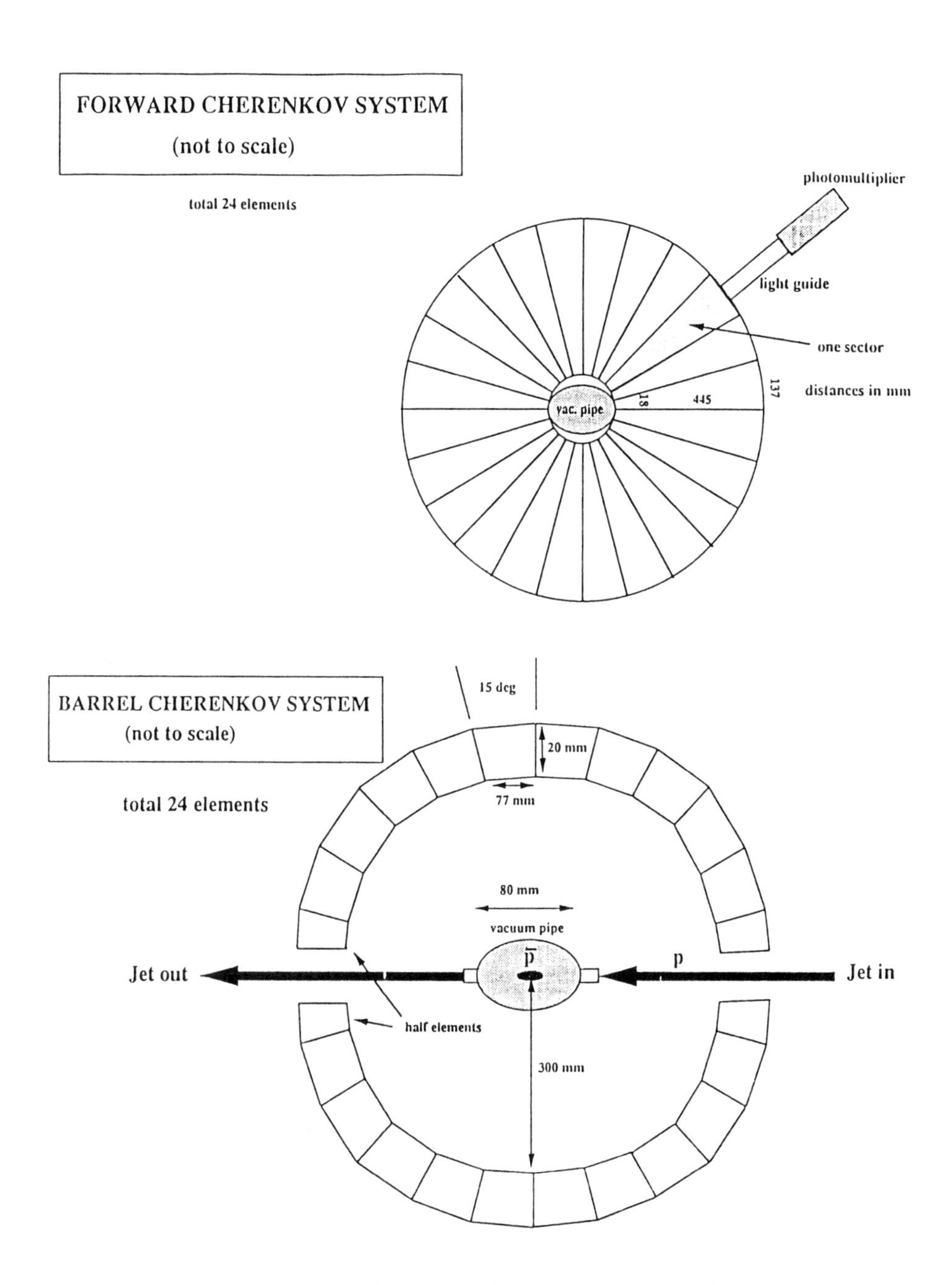

(Top) Schematic view of the forward Cherenkov system. (Bottom) Cross section of the Cherenkov counters for the barrel region. There are 24 counters in each.

FIGURE 7

Photons in the arc will be converted into photoelectrons by striking TMAE gas carried in He-ethane at room temperature. This mixture is contained in a two-dimensional "honeycomb" module consisting of 64 miniature wire chambers, each operating independently. The module is 64 mm on a side, and each wire chamber cell has dimensions of 8 mm by 8 mm. The pattern of struck cells provides the information necessary to determine the β value. The wires of the honeycomb are perpendicular to the radiator surface and are cloisonné. An assembly of such modules make up the RICH photosensitive detector.

The fast outer trigger counters in the barrel and forward regions (Fig. 8) perform several functions. In conjunction with the "pipe trigger scintillators", these counters define the time reference for the trackers and measure the event multiplicity. In addition, they will provide: a charged particle veto shield for the forward and barrel gamma detectors; crude first-level kinematic filters (e.g. rudimentary momentum balance); additional dE/dx information (especially for the non-annihilation background in the forward region); and fast θ–ϕ coordinates for possible use in the next level trigger, where additional constraints characteristic of true 4K events can be applied.

In both the barrel and the foward regions the fast charged trigger counters are constructed by overlapping 3 scintillator layers to form a high-granularity charged multiplicity detector. In the barrel region the counter is constructed of one straight (ie parallel to the beam axis) and two helical layers, the each of the latter being "wound" with opposite screw sense. This construction gives rise to 288 cells. In the forward detector the 3 layers are planar; one consists of 48 wedge-shaped segments, while the other two are each made from 24 curved segments, each layer again having opposite sense of rotation. Overlaying these segments provides 2304 sensitive regions. All scintillator layers are made of 5 mm NE-110 equivalent material. Tests of the trigger counter segments at the CERN T11 test beam, and with sources, have shown that light collection from these somewhat unorthodox counters is not a problem. Collection efficiencies remain high regardless of the source position along the counter. This enables measurement of the multiplicity with very high reliability (>99%).

The γ calorimeter in the forward region (see Fig. 9) will measure the energy of γ rays coming from the decay of neutral mesons. This detector is a part of the full-scale γ calorimeter which will be completed only for "Phase II"; at present it is used as a part of the overall γ veto scheme in the first-level trigger and to measure the charged particle multiplicity. The forward calorimeter is segmented in such a way as to match the segmentation of the forward threshold Cerenkov counter and the trigger scintillators. Hence it contains 300 individual "Pb-Sci/Fi" modules grouped in 8 theta rings with either 12, 24 or 48 modules per ring in the phi direction. The individual detectors are referred to as "towers" and are mounted in a geometry which "points" toward the interaction region but slightly upstream to preserve hermeticity. Each tower is machined from a basic "element" that consists of a block of Pb-scintillator fiber mixture. The block consists of scintillator fibers (1 mm diameter) embedded in a matrix of Pb made of sheets of corrugated Pb/Sb alloy laid on top of one another; overall the block is mixed in the volume ratio 50%/40%/10% : scintillator/Pb/glue. In this construction each scintillator fiber is completely surrounded by Pb. Because of this an excellent resolution has been achieved with a Pb/SciFi element of dimensions $(7.5\ \text{cm})^2 \times 22$ cm (14 radiation lengths). In tests with electrons ranging in energy from 88 to 5000 MeV, nearly constant energy resolutions ($\sigma(E)/E$) of about $6.3\%/\sqrt{E}$ (E in GeV) have been achieved.

The γ veto counter in the barrel region is intended to detect the γ rays from the decay of neutral mesons possibly produced in $\bar{p}p$ collisions, and then veto those events.

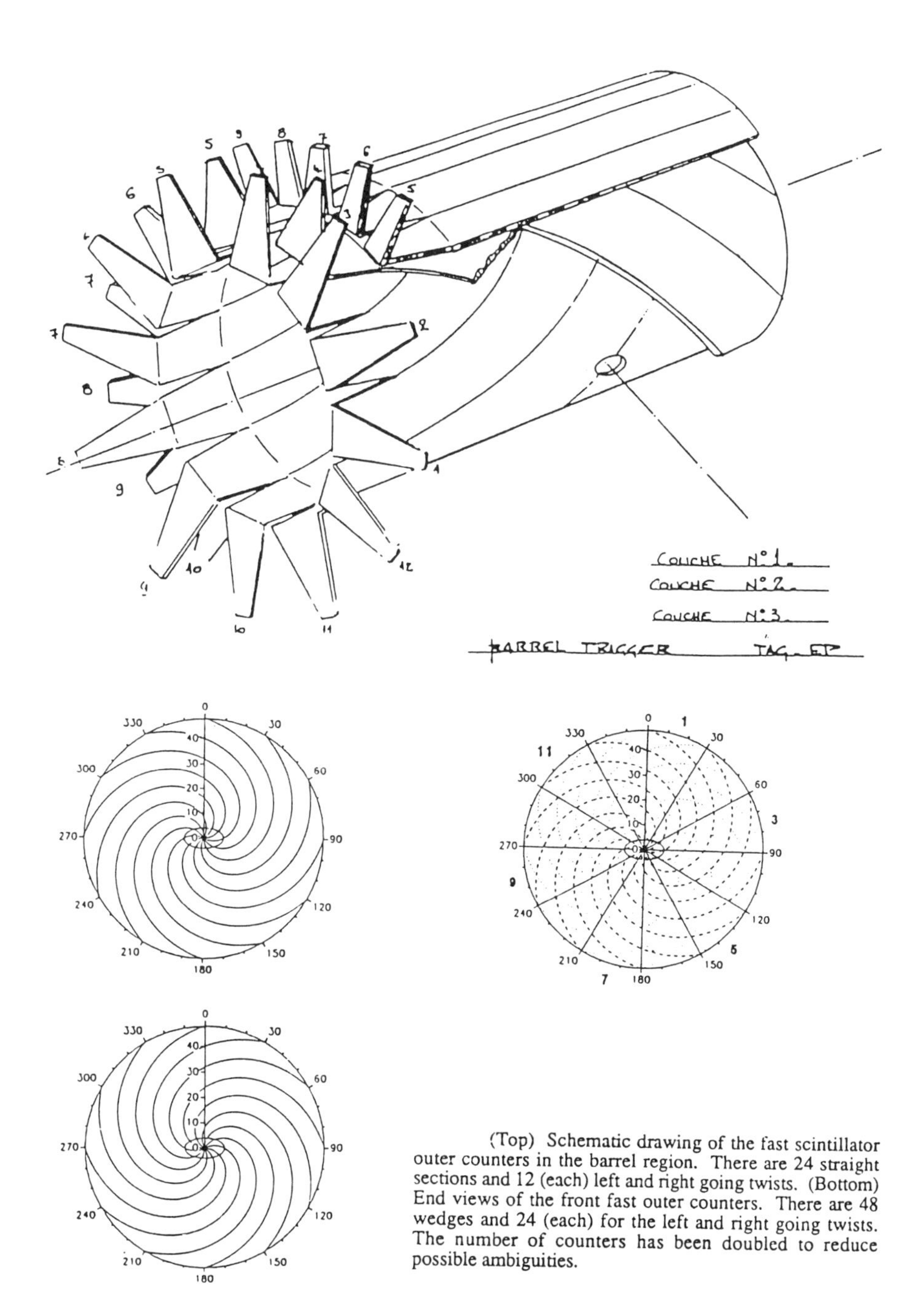

(Top) Schematic drawing of the fast scintillator outer counters in the barrel region. There are 24 straight sections and 12 (each) left and right going twists. (Bottom) End views of the front fast outer counters. There are 48 wedges and 24 (each) for the left and right going twists. The number of counters has been doubled to reduce possible ambiguities.

FIGURE 8

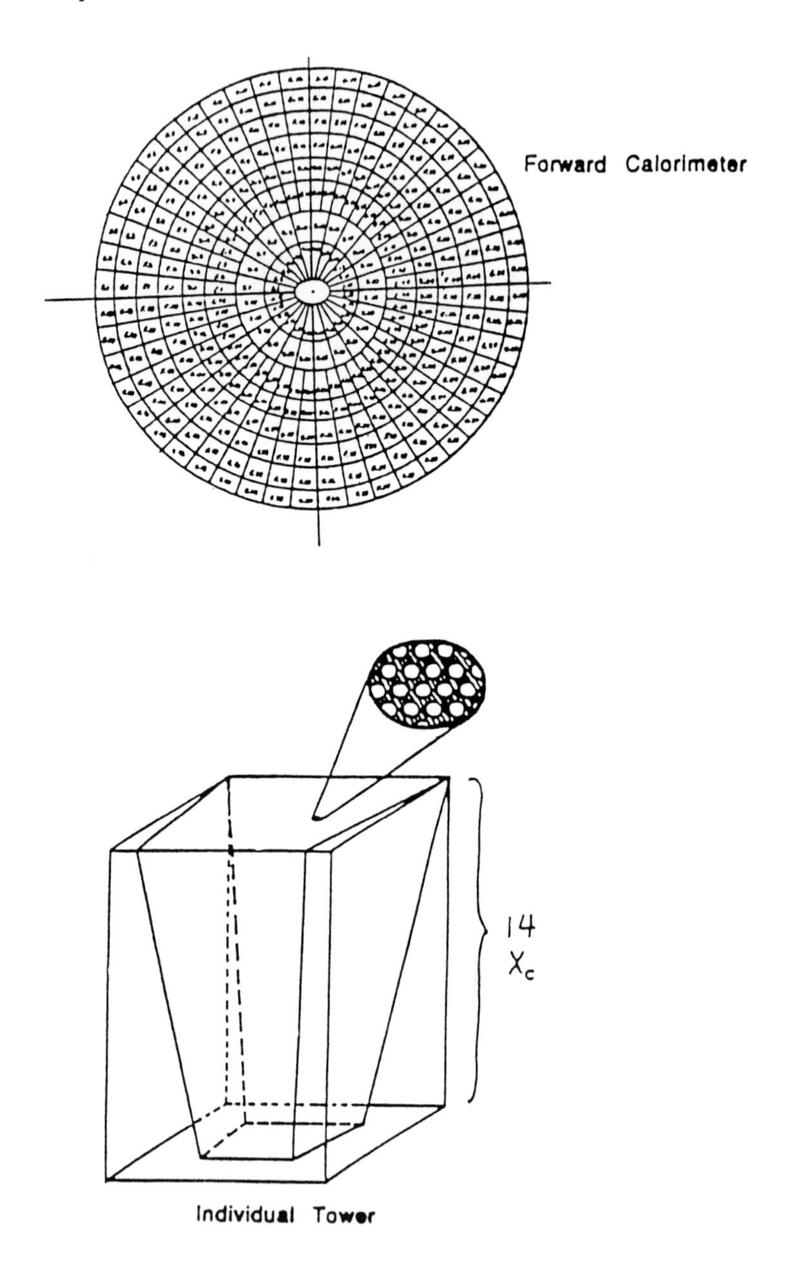

(Top) Front view of the 300-element electromagnetic SciFi calorimeter. (Bottom) Schematic view of an individual element showing the Pb-scintillator fiber matrix. The element contains 50% Pb, 40% fiber, and 10% glue by volume.

FIGURE 9

It is intended to "fill the gap" in the barrel region for the "Phase I" program while the full barrel calorimeter is under construction. The geometry of the barrel gamma veto will be similar to that of the barrel Cerenkov counter; there are 24 wedge-shaped elements in the azimuthal direction, each 1 m long in the beam direction and 9.5 cm (6 radiation lengths) thick. Each wedge is constructed using the same techniques described above, but in this case the fibers are arranged to lie along the beam direction.

The data acquisition computer array is based on the CERN Valet-Plus (VME-based) system with a fast Transputer interconnect . Seven Valets will be used to control the experiment: one for the dE/dx silicon counters; one for the barrel and forward straw trackers; one for the γ calorimeter and γ veto; one for the outer trigger scintillators; one for the pipe scintillators and threshold Cherenkovs; and one for the slow controls. The results from this first-level processing are assembled by the Transputer event builder in the last Valet into an "event" that can be further analyzed and written to tape. Analysis takes place in a system of μVAX computers which monitor the experiment.

TRIGGERING AND EVENT RECOGNITION IN "PHASE I"

The triggering for Jetset must be capable of bringing a 1 MHz rate for all events (the vast majority of which are due to unwanted interactions) down to a rate of about 100 Hz that can be handled by the data acquisition system with tolerable deadtimes. In doing this, as few good events as possible should be sacrificed.

The fast trigger will be produced for each event within 200 ns by the pipe trigger scintillators (PS), the outer trigger scintillators (OS), the electromagnetic calorimeter (EC) and γ veto (GV), and the threshold Cherenkov counters (TC). Events with the wrong charged/neutral multiplicity are discarded, and some decisions based on β are made.

At this time the readout system ADC's are latched and TDC's stopped, beginning the various event digitizations. For the $\phi\phi$ (4K) events of Phase I, this involves the following rather stringent but simple conditions: (1) a multiplicity of 4 charged particles, not more than 1 backward of 45°; (2) no gammas; (3) not more than 1 of these charged particles with $\beta > 0.8$ (the threshold of the fast Cherenkov counter). A crude test of momentum conservation ("balance") can be applied by simply checking that not all tracks emerge on one side of the detector. At this level the potentially troublesome $\bar{p}p\pi^+\pi^-$ reaction can be substantially reduced by its topology and possibly by the detection of the antiproton in the EC. The multiplicity is determined by the PS, the segmented OS systems, and the forward EC.

The pipe trigger scintillators can be used with the outer trigger scintillators and the calorimeter elements to establish hit correlations. Ideally, this will clean up the trigger sample significantly, leaving for further processing in a second level trigger four (θ,ϕ) prongs pointing to the vertex.

Neutrals are rejected by the EC and GV layers; this is important because these processes are quite prominent (eg, $\sigma(\bar{p}p \rightarrow 4\pi^{\pm} + n\pi^o) > 20\text{mb}$). The GV efficiency has been studied for the proposed detectors, including all known effects such as construction cracks, thresholds, and charged particle "blinding" (from true charged particles entering the same ϕ segment as a potential gamma). It is found to be essentially 95% effective for events where a single π^0 accompanies 4 charged particles.

Since the threshold Cherenkov counters give a reasonable signal for charged fast pions, we do not expect any problems in keeping the event sample which passes these first-level trigger quite pure. Thus the level-one trigger will be a series of "yes" signals from all of the detector subsystems and can be thought of as a number of loose cuts on the event sample.

We hope to reduce the initial 1 MHz rate to less than one kHz after the first level without allowing the deadtime to exceed 20%. During the commissioning of Jetset the luminosity will be lower than the design figure, and the trigger will at first be somewhat loose in order to be as bias-free as possible. Multiplicities will be "windowed" around the desired 4 charged particle value, and at least 1 gamma will be allowed in the event. We will run above and below the $\bar{p}p\pi^+\pi^-$ threshold to test the effects of that unwanted interaction on our event sample. As always, we must accumulate "luminosity" events in order to measure not only the total number of $\bar{p}p$ interactions (to normalize the data), but also to determine the performance of our detector and thus to develop more stringent on-line triggers based on hardware cuts.

THE "PHASE II" JETSET EXPERIMENT

"Phase II" in JETSET is intended to broaden our study considerably, searching for other interesting physics channels in which exotica may appear. An example in which one might expect to observe new phenomena is the channel $\bar{p}p \rightarrow \phi\,\omega$, which is also an OZI-forbidden decay from an ordinary $\bar{q}q$ state. To make such searches requires a more advanced detector than will exist in "Phase I". Thus, we plan to augment the "Phase I" detector by the addition of (1) a complete electromagnetic calorimeter in the barrel region; (2) Si dE/dx counters in the barrel; (3) the RICH counter; and (4) for the more distant future, a magnetic field. These devices will provide the necessary means to permit a complete identification of the momentum and sign of the outgoing charged particles, and will also allow the investigation of channels containing neutral particles. With such a detector, a broad attack on the important, basic questions involving gluonic and exotic quark matter can be undertaken with great confidence. Another promising approach is to build an atomic beam target capable of providing polarized protons. The ability to explore spin degrees of freedom will enhance our studies even further.

This work is supported by CERN, the Italian Istituto Nazionale di Fisica Nucleare, the German Bundesministerium für Forschung und Technologie, the Norwegian Research Council, the Swedish Natural Science Research Council, and the US National Science Foundation.

REFERENCES

1. F. E. Close, Rpts. Prog. Phys. **51** (1988) 833.
2. M. S. Chanowitz and S. R. Sharpe, Phys. Let. **132B** (1983) 412.
3. The Second International Conference on Hadron Spectroscopy, KEK, Tsukuba, Japan, April 1987. Proceedings edited by Y. Oyanagi, K. Takamatsu, and T. Tsuru.
4. The Third Conference on the Intersections between Particle and Nuclear Physics, Rockport, ME, May 1988.
5. Proceedings of the BNL Workshop on Glueballs, Hybrids, and Exotic Hadrons, Brookhaven, NY, August 1988, ed. by S.-U. Chung, AIP Conference Proceedings **185**, American Institute of Physics, New York (1989).
6. D. W. Hertzog, in [4].
7. Jetset collaboration reports to the PSCC, CERN/PSCC/88-34 and 89-33.

SESSION G

Future Machine Developments

POLARIZED BEAM AT COSY

K. Bongardt and R. Maier
Forschungszentrum Jülich GmbH, Postfach 1913, D–5170 Jülich, Germany

ABSTRACT

At present the cooler synchrotron COSY, a synchrotron and storage ring for medium energy physics is built at Jülich. The cooler ring will deliver protons and light ions in the momentum range of 275 to 3300 MeV/c. To increase the phase space density of the circulating protons electron and stochastic cooling will be used. The cooled beams will be used for experiments with internal and after slow extraction with external targets.

The facility consists of different ion sources, the cyclotron JULIC as injector, the 100 m long injection beam line, the ring with a circumference of 184 m and the extraction beam lines to the external areas. The start of the users operations is provided to be in autumn 1992. A short overview of the machine, beam parameters and the status is given. Special emphasis is given to the polarized beam facility.

INTRODUCTION

The COSY ring, see Fig. 1, is composed of two 180 degree bending arcs and two straight sections. The two arcs consist of six mechanical identical periods. The straights, bridged by four optical triplets, provide free space for internal target areas, for the RF stations and for the phase space cooling devices [1].

The magnetic lattice is based on a six fold symmetry. Each of the mirror symmetric half cell has a QF–bend QD–bend structure. Additional flexibility of the tune is given by interchanging the focusing and defocusing quadrupoles. In the straights the momentum dispersion can be suppressed. The straight sections are built as 1:1 telescopes. The phase advance in both telescopes can be either π or 2π. The machine and beam parameters are summarized in Table I.

ACCELERATOR COMPONENTS

MAGNETS AND POWER SUPPLIES

After successful testing and field measurements of the prototyp dipole the series manufacturing was started. The eleventh magnet is under construction now. Sextupole magnets are ready to be assembled for the first prototypes of the long and short version. Steering magnets are under construction. The design of the different injection and extraction septum magnets are finished and all components are under construction [2].

For the pulsed operation COSY needs an overall power between 3 MVA at injection energy and 15 MVA at final energy of 2.5 GeV. Thus very stable power converters with a wide dynamic operation area and nominal output powers in the range of about 100 kW to several MW are required for most of the magnets in the ring. The first converter of this type with an output current change from 200 A to 5000 A within 2 s has been installed mid '89 as main power converter for all field measurements.

momentum (kinetic energy) range		275–3300 MeV/c (40 MeV–2500 MeV)
injection momentum (kinetic energy)		275 MeV/c (40 MeV)
max. number of stored protons		$\lesssim 2 \cdot 10^{11}$
circumference		184 m
typical cycle	injection	$\lesssim$ 10 ms
	ramp up/down	1.5s/1.5s
	e–cooling	1–4s
	s–cooling	10–100s
internal target areas		4/6 m
bending magnets, No, radius, field at 3.5 GeV/c, max. fieldrate		24, 7 m, 1.67 T, 1 T/s
arc quadrupoles, No, No of families, magnetic length, max. grad. at 3.5 GeV/c		24, 6, 0.29 m, 7.5 T/m
telescope quadrupoles, No, No of families, magnetic lenght, max. gard. at 3.5 GeV/c		32, 8, 0.65 m, 7.65 T/m
focusing structure in the bending section		6 periods, seperated function FoBoDoBooBoDoBoF
betatron wave numbers		3.38, 3.38
γ_{tr}, transition energy		2.22, 1150 MeV
Aperture limitation		$a_h = \pm$ 70 mm, $a_v = \pm$ 27.5 mm
nat. chromaticity hor./ver.		$-5.2/-4.5$
geometrical acceptances	horizontal	130 π mm mard
	vertical	35 π mm mrad
		$\frac{\Delta p}{p} = \pm 0.5$ %
vacuum system pressure		10^{-10}–10^{-11} hPa
bake out temperature		300 °C
pump down time		70 h
RF system		
frequency range (h=1)		0.462 – 1.572 MHz
gap voltage (at duty cycle)		5 kV (100%) 8 kV (50%)

Table I: COSY Basic Parameters

ACCELERATING STATION

The radio frequency acceleration system for COSY will employ a ferrite–loaded cavity scheme of the symmetric push–pull type [3]. CW gap voltages in excess of 5 kV will be achieved by a pair of power amplifiers with total nominal output power of 50 kW. A provision for amplitude modulation will permit adiabatic trapping of the beam bunches after injection. During acceleration, the frequency has to be ramped from 450 kHz to 1.7 MHz.

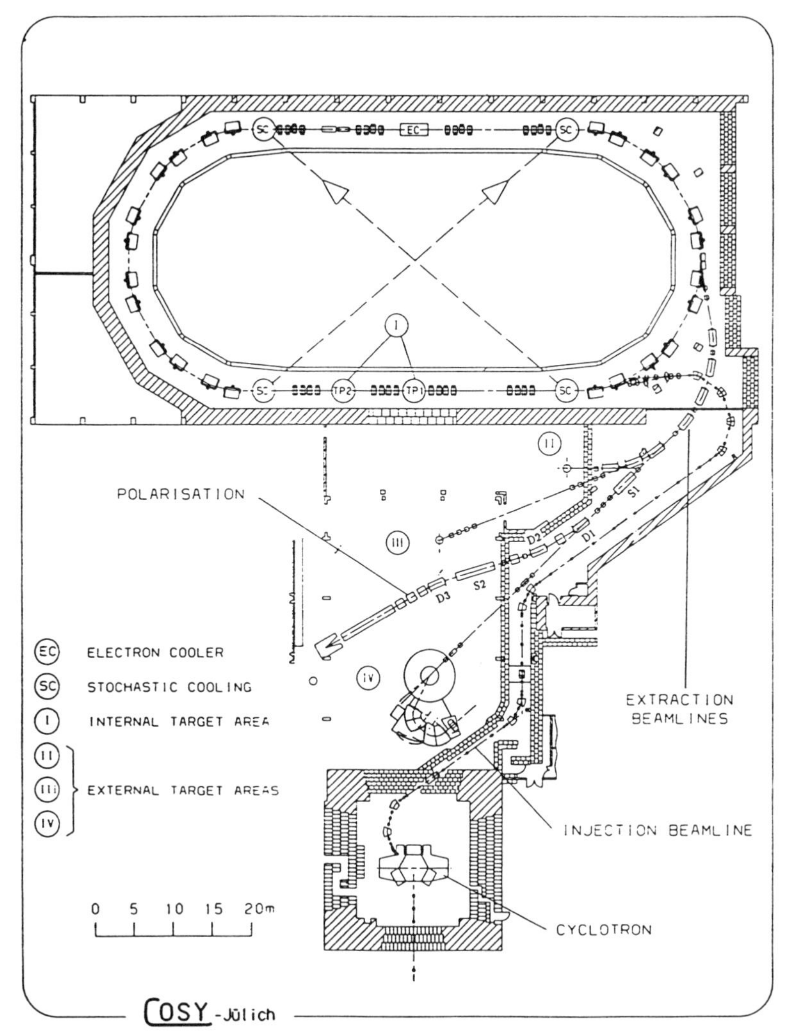

Figure 1: Layout of the COSY accelerator facility

SLOW EXTRACTION

Slow beam extraction is planned to take place via a 1/3–integer type, i.e. sextupole–driven resonance. A two–channel longitudinal drive system is fed by a common noise source. One (higher power) channel has a narrow noise spectrum with center frequency around the resonance. It serves for the rapid "sweep–out" of particles at resonance. The other channel serves at first to flatten the beam profile into a near–rectangle; thereafter, it generates a noise distribution (that is sweeping in center frequency) in between the original beam profile and the resonance to transport beam particles to the resonance. In the present design, the system will operate at a center frequency of 10 times the revolution frequency, with a minimum extraction time of 5 sec imposed by rf and kicker power limitations.

ELECTRON COOLER

An electron cooler of the active length of 2 m will be installed in the 7.2 m long free section of the cooler telescope (Fig. 1). The electron cooling will serve as a tool for beam preparation before acceleration. This offers highest storable phase space density in a fast cycling operation. The phase space condensation leads to a reduction of the transverse emittance of the injected beam down to the order of 1 π mm mrad, to the decreasing of the momentum spread (rms) to the order of 10^{-4}, and for bunch beam to shortening of the bunch length. In stage 1 the electron cooling will operate during the injection, using a 22 keV electron beam, in stage 2 after injection and acceleration to an intermediate flat top, using an 100 keV electron beam. The simulated cooling times for stage 1 and stage 2 operation for chosen lattice parameters are summarized in Table II [4]. τ^* is the time to cool a beam of 15 π mm mrad emittance (both horizontal and vertical) down to 5 π mm mrad.

E_e[keV]	E_p[MeV]	I_e[A]	B[mT]	τ^*[s]
22	40	1.8	84	1
100	184	4.0	150	4

Table II: Simulated cooling times for stage 1 and stage 2 operation of the COSY electron cooler

Local compensation of the solenoid and the toroid fields by two 'anti'–solenoids are foreseen to avoid depolarisation and coupling between the transverse planes. For a proton beam at 40 MeV, only from the solenoid the spin is rotated by 30^0, whereas the two transverse planes are rotated by 11^0.

STOCHASTIC COOLING

The transverse and longitudinal cooling of protons in the range $N = 10^8$ ppp up to 10^{10} ppp and energies above $T = 0.85$ GeV will be done with a 2 GHz bandwidth system splitted into two sub–systems consisting of band I and band II with a frequency range (1 – 1.8) GHz and (1.8 – 3) GHz, resp.

The number of pickup loops is 48 and 64 for band I and band II respectively. The number of kicker loops is half that of the pickups. The pickup and kicker have a width of 3 cm and an impedance of 50 Ω. Pickup and kicker plates are moveable up to maximal gap height of 15 cm. The cooling system will be installed diagonally in the telescopic section of the ring (see Fig. 1) in order to ensure cooling above and below γ_{tr}. The total power of the cooling system including losses is about 250 W per plane and band [5].

In Table III results are given at two different energies for the transverse and the longitudinal cooling system.

transverse:						
	Band I (1-1.8 GHz)		Band II (1.8-3 GHz)		Band I+II (1-3 GHz)	
	N	t [sec]	N	t [sec]	N	t [sec]
E_{kin} = 850 MeV	10^{10}	79	10^{10}	53	10^{10}	32
ε = 10 π mm mrad → 1 π mm mrad	10^{8}	5	10^{8}	3	10^{8}	2
E_{kin} = 2.5 GeV	10^{10}	88	10^{10}	36	10^{10}	24
ε = 5 π mm mrad → 1 π mm mrad	10^{8}	4	10^{8}	2	10^{8}	1
longitudinal:						
	N	τ_0 [sec]	N	τ_0 [sec]	N	τ_0 [sec]
E_{kin} = 850 MeV	10^{10}	20	10^{10}	15	10^{10}	9
	10^{8}	4	10^{8}	2	10^{8}	1
E_{kin} = 2.5 GeV	10^{10}	49	10^{10}	38	10^{10}	21
	10^{8}	6	10^{8}	4	10^{8}	2

Table III: Stochastic cooling times for COSY
transverse: total cooling time t
longitudinal: inverse of the cooling rate

EXPERIMENTAL AREAS

For high luminosity experiments the ring will run with thin internal targets in the recircular mode. Two internal target stations will be available, they are located in the center of the target telescope (TP1) and TP2 between the two quadrupole quadrublets 7.3 m further downstream of TP1 (Fig. 1). The free length for experimental installations will be 4 m for TP1 and 6 m for TP2.

The extracted beam will be directed to three different target areas (Fig. 1). The target area II and III will be fed by direct beamlines without special ion optical request. The beamline IV is a high resolution beamline to the spectrometer BIG KARL, which allows for small spot sizes at the target. Here target spot sizes of $\lesssim$ 1 mm for an emittance of 5 π mm mrad will be available.

The transfer line 'Polarisation' to the external target area allows to rotate the vertical polarization into a sideways or a longitudinal one.

In Fig. 1 the layout of this beam line including the spinrotating elements are shown.

The geometrical parameters of the spin rotation elements are given in Table IV:

Solenoids	length [m]	field [T]
S1	3.25	1.94
S2	5.00	2.48

Dipols	length [m]	field [T]	field index	deflection angle [degree]
D1	2.169	0.83	12.9	13.7
D2	2.169	0.83	12.9	13.7
D3	2.169	0.83	12.9	–13.7

Table IV: Spin Rotation Elements in the 'Polarisation'Beam Line

With the proposed arrangement, sideways and longitudinal polarization up to T = 1.35 GeV can be adjusted, whereas up to maximal energy of T = 2.5 GeV only longitudinal polarization can be tuned at the target station.

POLARIZED BEAM FACILITY

In order to make use of the full scientific potential of COSY, the machine has to be equiped with a polarized beam facility. The main components, the negative polarized ion source, and the methods in order to keep the polarisation in COSY are discussed.

POLARIZED ION SOURCE

At the universities of Bonn, Köln and Erlangen, a negative source for polarized H^- and D^- is under construction.

In this source of the colliding beam type neutral hydrogen atoms are polarized in a ground–state atomic beam source and then ionized via the charge exchange reaction $Cs^0 + \vec{H}^0 \rightarrow Cs^+ + \vec{H}^-$

Only the Cs^0–beam is pulsed. A negative polarized source is preferred due to the stripping injection process into the COSY synchrotron. The injection energy into the existing JULIC cyclotron is 4.5 keV/N.

The main source parameters are given in Table V:

Particles:	H^-, D^-
pulse current:	30 μA (at 4,5 keV/N)
polarisation in degree:	$\geq$ 75%
pulse length:	20 m sec
repetion rate:	1 Hz
geometrical emittance:	60 π mm mrad (at 4,5 keV/N)
extraction energy:	$\leq$ 30 keV

Table V: Parameter of the Polarized Ion Source

The polarized ion source will be mounted in the existing beamline behind the unpolarized negative ion source. In order to keep the polarization in the JULIC cyclotron the polarization vector will be 90^0 turned around by a Wien filter. In the cyclotron by itself no depolarizing effects are expected.

At 40 MeV, in the beamline between the cyclotron and the ring the spin orientation should be controlled. To determine the polarization, a scattering chamber inside the beamline is foreseen. With two symmetric 3–counter scintillator telescopes the elastically scattered protons from a carbon target at 52.5^0 laboratory angle can be measured.

ACCELERATION OF POLARIZED IONS IN COSY

Acceleration of polarized ions in a circular accelerator is possible, since the (vertical) guide field does not change the component of the polarization vector parallel to it. However, the existing horizontal field components can lead to partial or total depolarization of the beam when their effect adds up coherently over many cells and/or turns. These depolarizing resonances are due to interference of the spin precession frequency or "spin tune" with certain harmonic components of the horizontal fields.

On an ideal orbit in a "flat" ring (vertical fields only) this leads to a spin precession frequency

$$\nu_p = \gamma\, G$$

where G = (g–2)/2 is the magnetic anomaly, 1.7928 for protons and –0.1430 for deuterons.

In the presence of betatron oscillations and field imperfections, depolarizing fields occur. These contain frequencies in units of revolution frequency.

(a) k (integer) due to orbit imperfections.
(b) k P ± ν_z due to linear betatron oscillations.
(c) k P ± $(a\nu_z + b\nu_x + c\nu_s)$ due to nonlinear betatron and synchrotron oscillations. Here P is the number of identical superperiods in the lattice, and k, a, b, c are integers.

Resonant depolarization occurs wherever ν_p equals one of these frequencies. Since ν_p is proportional to energy, the resonance condition occurs at many energies (spaced 520 MeV apart for protons and imperfection resonances); therefore a particle being accelerated will traverse many resonances.

Due to the optical flexibility of COSY, both, the superperiodicity P and the vertical tune ν_z can vary. The two telescopic straight sections have a phase advance of 2 π each, which changes the intrinsic resonance condition to

$$\gamma\, G = kP \pm (\nu_z - 2)$$

In order to avoid intrinsic resonances each arc should be made out of 3 identical cells: P = 6. This implies a non–vanishing dispersion in the telescopic straight sections. For zero dispersion the two central quadrupoles in the arc break the sixfold symmetry of the machine and reduces it to a superperiod of two.

For such a case with $\nu_x = \nu_z = 3.38$ in Table VI the resonance strength ϵ_R of all kinematically allowed resonances are given for a proton beam with $\epsilon_0 = 60\ \pi$ mm mrad. ϵ_0 is the geometrical total emittance at 40 MeV injection energy. The resonance strength ϵ_R is calculated by the Program DEPOL, written by E.D. Courant.

$k \pm \nu_z$	T[MeV]	ϵ_R
$-1 + \nu_z$	292	$< 10^{-4}$
$6 - \nu_z$	448	$8 \cdot 10^{-4}$
$0 + \nu_z$	815	$8 \cdot 10^{-4}$
$7 - \nu_z$	972	$< 10^{-4}$
$1 + \nu_z$	1338	$< 10^{-4}$
$8 - \nu_z$	1495	$5.3 \cdot 10^{-3}$
$2 + \nu_z$	1861	$1.8 \cdot 10^{-3}$
$9 - \nu_z$	2018	$< 10^{-4}$
$3 + \nu_z$	2384	$< 10^{-4}$
$10 - \nu_z$	2532	10^{-4}

Table VI: Intrinsic Resonance and their strength ϵ_R for $\epsilon_0 = 60\ \pi$ mm mrad

It is evident that 5 even intrinsic resonances have to be crossed for accelerating the beam up to 2.5 GeV. The odd ones are excited by gradient errors. For a lattice with periodicity P = 6, corresponding to non–vanishing dispersion in the straights, only the $\gamma G = 8 - \nu_z$ resonance is excited with almost the same strength.

Using the formula of Froissart and Stora [6] for crossing an isolated resonance, the resulting polarization for a 'pencil' beam ($\epsilon_0 = 10\ \pi$ mm mrad) and for a high intensity beam ($\epsilon_0 = 60\ \pi$ mm mrad) are shown in Fig. 2 and Fig. 3. The 'pencil' beam will contain about 10^9 ppp, whereas the high intensity beam have up to 10^{11} ppp. The solid line belongs to the maximal crossing speed α_0, corresponding to the fastest magnetic field ramp of $\dot{B}$ = 1T/sec or a maximal energy gain of 1.3 keV/turn. The broken line belongs to $\alpha = 1/3\ \alpha_0$.

For the 'pencil' beam with a uniform particle distribution the polarization will be 75% after crossing the $(6 - \nu_z, 0 + \nu_z)$ – resonances with the maximal crossing–speed. Reducing the speed to $1/3\ \alpha_0$ for the $8 - \nu_z$ resonance, the spin completely flips, resulting in an average beam polarization of about 85%.

After crossing the $(2 + \nu_z)$ – resonance no significant beam polarization remains for a uniform distribution.

This results in a final polarization of about $P_f = 50\% \cdot P_i$ at $T \sim 1{,}5$ GeV, where P_i is the polarization value of the at 40 MeV injected beam. By using the standard power supplies for the quadrupoles, which allow to change the tune by about 10^{-6}/turn, a higher polarization degree should be possible by creating a more suitable spin distribution.

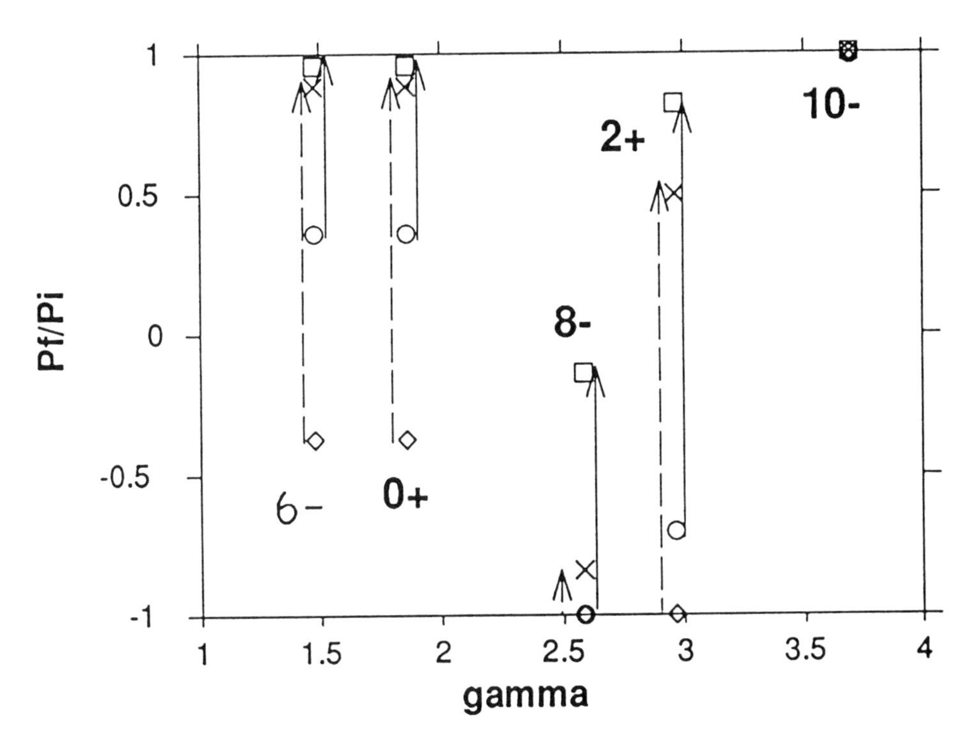

Figure 2: Polarization values after crossing the intrinsic resonances for a 'pencil'beam: $\epsilon_0 = 10\ \pi$ mm mrad
The arrows are pointing from the 100% emittance value to the 5% one. The solid line belongs to the maximal crossing speed, the broken line to 1/3 of it.

For the high intensity beam no significant beam polarization remains after crossing the $(6 - \nu_z,\ 0 + \nu_z)$ – resonance for a uniform particle distribution.

In order to keep the polarization and to accelerate the beam up to 2.5 GeV, fast pulsed quadrupoles should be added later on.

The applied tune jump should be larger than 10^{-2}/turn in order to get for the high intensity beam more than 90% beam polarization after crossing each resonance except for the $8 - \nu_z$ one. Their adiabatic spin–flip is unavoidable. The value for the needed tune jump was calculated by using the formulae of Courant and Ruth [7].

For achieving high polarization at full energy also the 5 odd intrinsic resonances, see Table VI, and the imperfection resonances have to be considered. The odd intrinsic resonances will be cured, if necessary, by using the fast pulsed quadrupoles. The kinematically allowed imperfection resonances are listed in Table VII together with their resonance strength ϵ_R and the resulting polarization for an average closed orbit distortion of 1 mm.

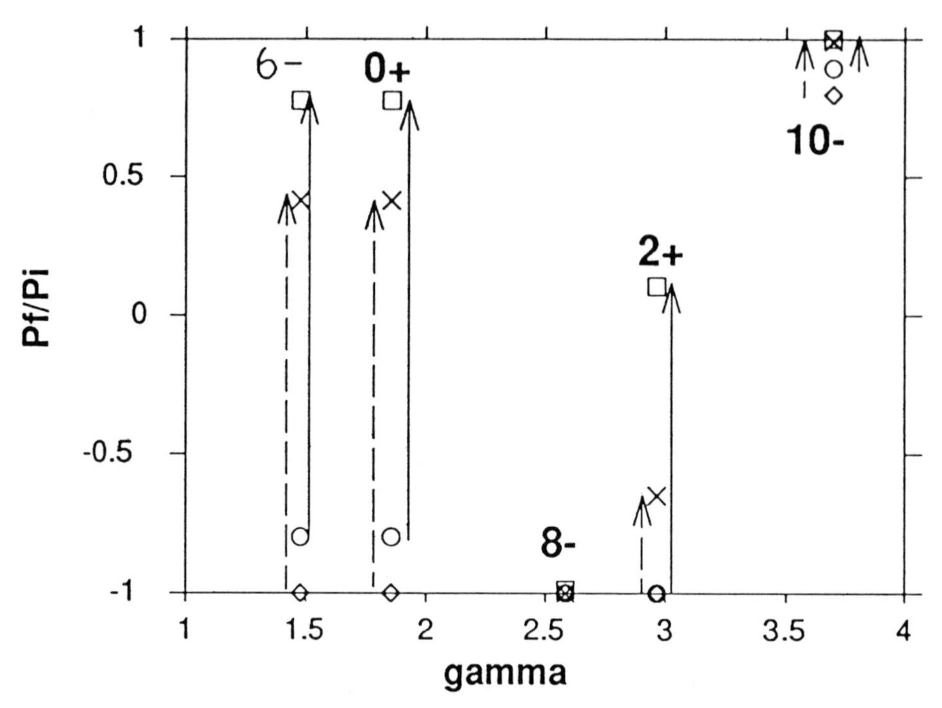

Figure 3: Same as in Fig. 2, but for $\epsilon_0 = 60\ \pi$ mm mrad

k	T[MeV]	ϵ_R	P_f/P_i
2	108	$4 \cdot 10^{-4}$	+ 0.25
3	631	$3 \cdot 10^{-4}$	+ 0.53
4	1155	$4.3 \cdot 10^{-4}$	+ 0.15
5	1678	$1 \cdot 10^{-3}$	− 0.87
6	2201	$2.7 \cdot 10^{-4}$	+ 0.62

Table VII: Imperfection resonance and (ϵ_R, P_f/P_i) for an average closed orbit distortion of 1 mm

In order to get no polarization loss harmonic spin matching is needed. This will be done by the 27 vertical pickups and steering elements for the closed orbit control.

For tuning up the ring should be equiped with an internal polarimeter.

All above considerations only apply for protons. For deuterons there is only one odd intrinsic resonance $3 - \nu_z$ at T = 1.17 GeV, but no imperfection resonances in the energy range of COSY because the magnetic anomaly is to small.

SUMMARY

The accelerator building had been finished in May 1990. More than 70% of the equipment has been ordered and delivery has been envisaged within the foreseen timeschedule.

A negative polarized ion source for H^- and D^- is under construction. In the COSY synchrotron by itself, for a low intensity proton beam a relatively small polarization degree of about 50% at T = 1.5 GeV can be expected with the existing components. For getting, espescially at high intensities, large polarization at the maximal energy fast pulsed quadrupoles should be added later on.

ACKNOWLEDGEMENT

We are indebted to our colleague D. Prasuhn for helping with the program DEPOL. Special thanks go to P. Blüm, University of Karlsruhe, for getting his version of DEPOL.

REFERENCES

[1] R. Maier, U. Pfister, R. Theenhaus, The COSY–Jülich Project, EPAC Nice (France) 1990.

[2] U. Bechstedt et al., Injection and Extraction Magnets of COSY–Jülich, same conference.

[3] C. Fougeron, Ph. Guidee, and K.C. NGuyen, RF System for SATURNE II, PAC, San Francisco, 1979.

[4] R. Maier, B. Seligmann, H.J. Stein, Electron Cooling in COSY–Jülich, same as ref. 1.

[5] K. Bongardt, S. Martin, D. Prasuhn, H. Stockhorst, Theoretical Analysis of Transverse Stochastic Cooling in the Cooler Synchrotron COSY, same as ref. 1.

[6] M. Froissart and F. Stora, Nucl. Instrum. Meth. 7, 297 (1960).

[7] E.D. Courant and R.D. Ruth, BNL Report 51270, Brookhaven National Laboratory, Upton, NY 11973 (1980).

Limits on Electron Cooling

D. Reistad
The Svedberg Laboratory, 751 21 Uppsala, Sweden

T.J.P. Ellison
Indiana University Cyclotron Facility, Bloomington, Indiana 47405, U.S.A.

ABSTRACT

A simple discussion of electron cooling theory, with emphasis on relativistic scaling relations, is given. This discussion is used to predict the behaviour of electron cooled beams, subjected also to an internal target and to intrabeam scattering. Technical considerations relating to the alignment and the confinement of electron beams for cooling indicate that the longitudinal magnetic field will be abandoned in favour of focusing by discrete lenses at electron energies above about 1 MeV. The experiments on acceleration, transport, and collection of prototype electron beams made at Novosibirsk and Wisconsin are discussed. It is concluded that intermediate energy electron cooling is within the capability of present technology, and although its most important application may be for improving the luminosity of colliders, nuclear and particle physics applications requiring high energy, spatial, or time resolution also may benefit.

INTRODUCTION

Electron cooling [1-5] is a way to shrink the emittance and the momentum spread of a stored ion beam without removing ions from the beam. This is accomplished by mixing the ion beam over at fraction, η_c, of the storage ring circumference with a "cold" electron beam of the same average velocity, see Fig. 1. In the moving frame the system appears as a hot ion gas mixed with a cold electron gas. Temperature relaxation occurs through Coulomb collisions between the electrons and the ions.

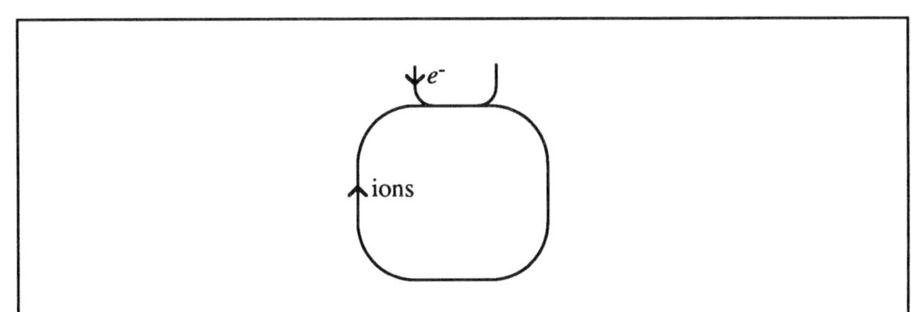

Fig. 1. Electron cooler can be seen as a heat exchanger, with cold electrons as the cooling medium.

Electron cooling was proposed by Budker in the early sixties, and published in 1966 [6] as a means for accumulating antiprotons. The high energy at which it is favourable to produce antiprotons, and the large emittance and momentum spread of the antiproton beam emerging from a production target, favour the use of another cooling method, stochastic cooling, invented in 1972 by van der Meer [7] for this purpose.

Stochastic cooling makes use of signals from a very high bandwidth pickup in order to control a kicker placed at a suitable distance downstream to the pickup, in order to kick the ions in the stored beam towards the centre of the phase space volume.

The ultimate temperature of a stochastically cooled ion beam becomes determined by the signal-to-noise ratio at the pickup pre-amplifiers, and stochastic cooling works best for "hot" ion beams. The ultimate temperature of an electron cooled ion beam becomes determined by the electron temperature and by target, rest gas, and intrabeam scattering. Electron cooling can cool ion beams to a considerably higher phase space density, than stochastic cooling can. On the other hand, electron cooling works best when the ion velocity spread is less than the electron velocity spread, so electron cooling works best for beams that are not too hot to begin with. Stochastic and electron cooling are therefore complementary [8], and it has been discussed to use electron cooling as a way to prepare the beam for colliders, by increasing the phase space density after accumulation with stochastic cooling at an intermediate energy [9,10] or even at the full energy [11].

The first practical realization of electron cooling was achieved at Novosibirsk in 1974 [12]. The second generation of electron coolers was subsequently built at CERN [13] and FNAL [14]. Third generation electron cooling systems have since become routine tools in a number of low-energy rings [15,16]. In these rings electron cooling is used:

— to shrink the phase space volume occupied by the beam in order to allow *accumulation*.

- to increase the *lifetime* of stored beams, in particular when the beams are exposed to an internal target.
- to improve the experimental *resolution*, both in *spatial* coordinates and in *energy* of the beam. The high energy resolution also leads to a high *time* resolution if rf is applied to the beam; a time resolution of about one nanosecond can be achieved without using excessive rf voltages.

The highest energy electron cooling to date has been obtained at IUCF, Bloomington, Indiana [17], where 170 keV electrons have been used to cool 310 MeV protons.

INTRODUCTION TO THE ELECTRON COOLING THEORY

The theory of electron cooling is discussed in detail in a number of papers [1-3,18-23]. It is not our intention here to further this discussion, but instead to show that with a very simple model, it is possible to get a fair idea of how the cooling process works, and to understand the relevant scaling relations.

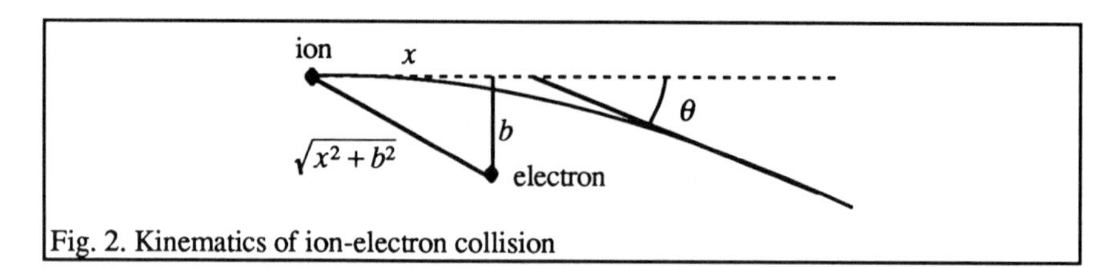

Fig. 2. Kinematics of ion-electron collision

Consider an ion which is passing a stationary electron (Fig. 2). The force acting between the ion and the electron is

$$F = \frac{1}{4\pi\varepsilon_0}\frac{Ze^2}{x^2+b^2} \tag{1}$$

We wish to evaluate the momentum, which is transferred from the ion to the electron. The proton is much heavier than the electron, so its trajectory is not appreciably affected (θ is small). The collision is lasting a short time, so in first order the electron acquires transverse momentum, but does not have time to change its position. Thus the integrated longitudinal momentum transfer.vanishes, and we can neglect the longitudinal force. The transverse part of the force is

$$F_\perp = F\frac{b}{\sqrt{x^2+b^2}} \tag{2}$$

The momentum, which is transferred from the ion to the electron, is then

$$\Delta p = \int_{-\infty}^{\infty} F_{\perp}\, dt = \frac{1}{4\pi\varepsilon_0} \int_{-\infty}^{\infty} \frac{Ze^2}{x^2 + b^2} \cdot \frac{b}{\sqrt{x^2 + b^2}} dt \tag{3}$$

Using $v_i = dx/dt$ we get

$$\Delta p = \frac{2Ze^2}{4\pi\varepsilon_0 v_i b} \tag{4}$$

and the energy, which is transferred from the ion to the electron is

$$\frac{(\Delta p)^2}{2m_e} = \frac{2Z^2 e^4}{(4\pi\varepsilon_0)^2 m_e b^2 v_i^2} \tag{5}$$

The energy, which is lost by the ion when it travels through the electron cloud, is obtained by integrating over all possible impact parameters (Fig. 3).

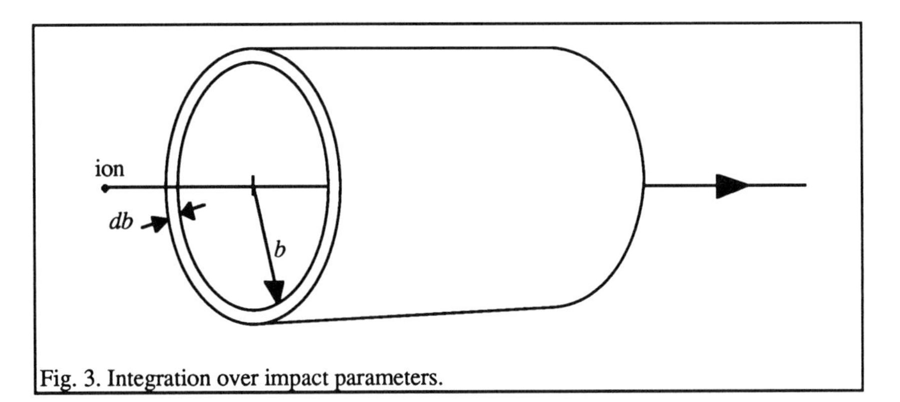

Fig. 3. Integration over impact parameters.

$$\frac{dE}{dx} = \frac{4\pi Z^2 e^4 n_e}{(4\pi\varepsilon_0)^2 m_e v_i^2} \int_0^{\infty} \frac{db}{b} \tag{6}$$

We have to restrict the integration to an impact parameter range.

$$\int_0^{\infty} \frac{db}{b} \rightarrow \int_{b_{\min}}^{b_{\max}} \frac{db}{b} = L_c \tag{7}$$

The logarithmic ratio between the maximum and the minimum impact parameter, L_c , is called the Coulomb logarithm. The minimum impact parameter is determined by the maximum classical momentum transfer $\Delta p_{\max} = 2m_e v_i$

$$b_{\min} = \frac{Zr_e c^2}{v_i^2} \tag{8}$$

The maximum impact parameter is determined by the screening radius in the electron gas, usually the Debye length

$$b_{\max} = \lambda_D = \sqrt{\frac{\varepsilon_0 kT_e}{e^2 n_e}} \tag{9}$$

In practical cases, the Coulomb logarithm assumes a value between 5 and 15:

$$L_C \approx 10$$

We rewrite eq. (6)

$$\mathbf{F}(\mathbf{v}_i) = -F_0 L_C c^2 \frac{\mathbf{v}_i}{v_i^3} \tag{10}$$

where

$$F_0 = 4\pi Z^2 n_e r_e^2 m_e c^2 \tag{11}$$

where $r_e = e^2/(4\pi\varepsilon_0 m_e c^2)$ is the classical electron radius, and

$$n_e = \frac{I_e}{\pi r_0^2 \beta\gamma ec} \tag{12}$$

is the electron density in the rest frame. r_0 is the radius of the electron beam. Note the $\frac{1}{\beta\gamma}$ scaling in F_0.

So far, we have neglected the motion of the electrons. To account for this, replace the ion velocity $\mathbf{v}_i$ with the ion-electron velocity difference $\mathbf{u} = \mathbf{v}_i - \mathbf{v}_e$, and obtain the force by integrating over the electron velocity distribution

$$\mathbf{F}(\mathbf{v}_i) = -F_0 c^2 \int L_C(u) f(v_e) \frac{\mathbf{u}}{u^3} d^3 v_e \tag{13}$$

L_C varies slowly with u, and can therefore be brought out of the integral:

$$\mathbf{F}(\mathbf{v}_i) = -F_0 c^2 L_C \int f(v_e) \frac{\mathbf{u}}{u^3} d^3 v_e \tag{14}$$

The expression for the frictional force $\mathbf{F}(\mathbf{v}_i)$ resembles the expression for the Coulomb force $\mathbf{F}(\mathbf{r})$ of a charge distribution $f(r_e)$ acting on a test charge at **r**. We call this *the electrostatic analogy*. It is useful for understanding the behaviour of the cooling force as a function of the electron velocity distribution and the ion velocity.

ELECTRON VELOCITY DISTRIBUTION

Several sources contribute to the electron energy spread. The most obvious is maybe the cathode temperature $\frac{1}{2}kT_C = 0.06$ eV. Another source of electron energy spread is the ripple of the electron cooler high voltage power supply. The present state of the art for high voltage power supply regulation for operating voltages up to 300 kV is about 2×10^{-5} times the maximum output voltage. As is discussed below, this performance should be achievable also at higher voltages. The least understood but sometimes most important source of electron energy spread is intrabeam scattering among the electrons. Both transverse-longitudinal and longitudinal-longitudinal scattering are discussed in the literature [23-25]. The transverse-longitudinal intrabeam scattering among the electrons seems to be the most important effect, except in the case of a very strong longitudinal magnetic field in the electron cooler. It is shown in ref. 22 that for a beam with a current density of $0.5\ \mathrm{A/cm^2}$, transverse-longitudinal intrabeam scattering contributes of the order of 10 eV to the (laboratory frame) electron energy spread.

So the cathode temperature will in fact contribute insignificantly to the electron energy spread compared to the contributions from power supply ripple and electron intrabeam scattering.

The longitudinal electron velocity distribution will be deformed due to the acceleration of the electrons. We will illustrate this effect by a non-relativistic example.

Consider an initially flat electron velocity distribution

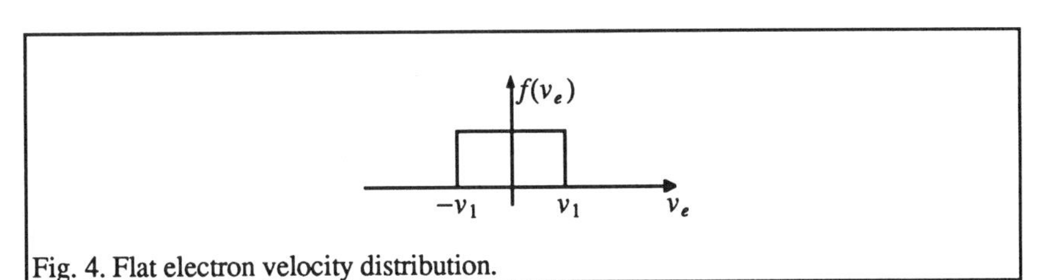

Fig. 4. Flat electron velocity distribution.

$$f(v) = \frac{1}{2v_1}; \qquad -v_1 < v < v_1$$

The maximum kinetic energy of the electrons is then $T_1 = \dfrac{mv_1^2}{2}$

Suppose these electrons are accelerated, so that an energy $W = \dfrac{mV^2}{2}$, where $V >> v_1$, is added to the kinetic energy of each electron. Let the total velocity of an electron with initial velocity v be $(V+u)$. Then, obviously,

$$v^2 + V^2 = (V+u)^2$$

It is an interesting observation that $u \geq 0$, also for electrons which before acceleration had $v < 0$ i.e. were moving backwards.

Making use of $V >> v_1$ we calculate

$$u = \frac{v^2}{2V}$$

The electrons that had the initial velocities $\pm v_1$ will have the highest lab frame velocity after the acceleration $V + u_1 = V + \dfrac{v_1^2}{2V}$. The electrons that were initially without kinetic energy will of course have the lowest velocity after acceleration, V. So the longitudinal velocity distribution gets folded over, and its width gets very much reduced. The maximum kinetic energy after acceleration in the moving frame becomes $\frac{T_1^2}{4W}$, i.e. much less than T_1. The electron velocity distribution becomes

$$f(u) = \frac{1}{v_1}\sqrt{\frac{V}{2u}}; \quad 0 < u < \frac{v_1^2}{2V} \quad \text{see Fig. 5.}$$

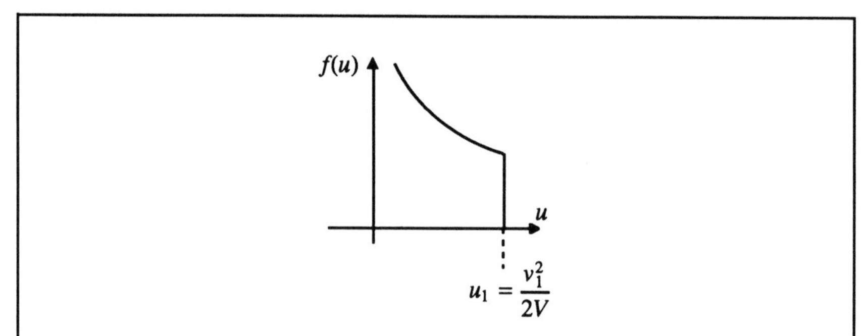

Fig. 5. Electron velocity distribution of Fig. 4, seen in moving frame after acceleration to energy $\frac{m_e V^2}{2}$.

The transverse electron velocity distribution on the other hand, is (ideally) not influenced by the acceleration, and remains determined by the cathode temperature $kT_C = 0.12$ eV.

A relativistic calculation [5] shows that the factor by which the longitudinal velocity spread gets reduced by the acceleration is $\frac{T_1}{2\beta^2\gamma^2 mc^2}$. This is a large factor, so even though there are other contributions to the longitudinal velocity than from the cathode temperature, the longitudinal velocity spread of the accelerated electron beam becomes much less than the transverse velocity spread.

The rest frame electron velocity distribution can be seen as a flat disk in velocity space, Fig. 6.

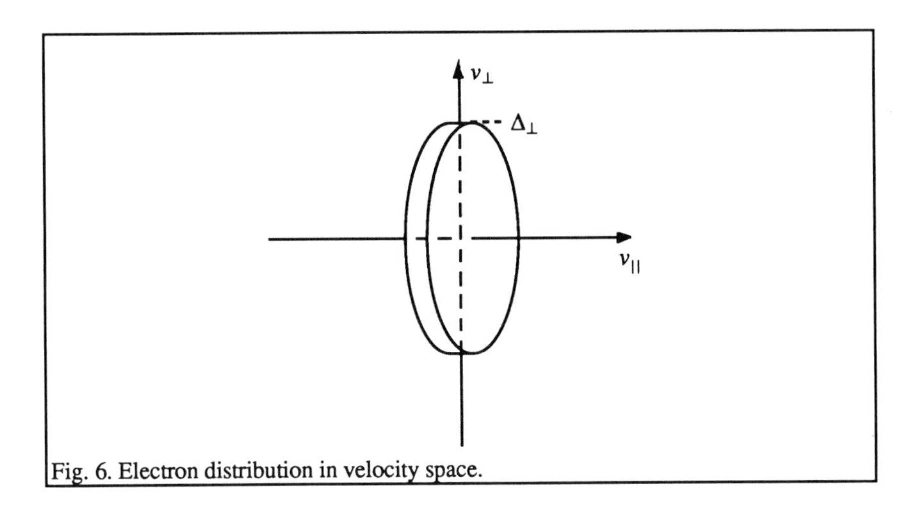

Fig. 6. Electron distribution in velocity space.

FRICTION FORCES

Since the transverse and longitudinal electron velocity distributions become very different, also the friction forces become very different in the transverse and longitudinal cases. We will discuss the behaviour of the friction forces under the simple assumptions that the transverse electron velocity distribution is flat until the edge of the distribution at $\Delta_\perp$, and that the longitudinal electron velocity distribution is completely flattened.

<u>Transverse</u>

The electrostatic analogy tells us that the transverse friction force will grow linearly with the transverse ion velocity $v_\perp$ until the ion reaches the edge of the electron velocity distribution at $\Delta_\perp$ (Fig. 6) and decrease as $v_\perp^{-2}$ when $v_\perp >> \Delta_\perp$.

We define the electron transverse "temperature" as $kT_e = \frac{m_e \Delta_\perp^2}{2}$; where $\Delta_\perp^2 = \Delta_x^2 + \Delta_y^2$

We wish to relate the friction force to a familiar lab frame quantity, and choose the transverse emittance, which at a symmetry point in a ring will be represented by an upright ellipse in the (x,x') – plane .

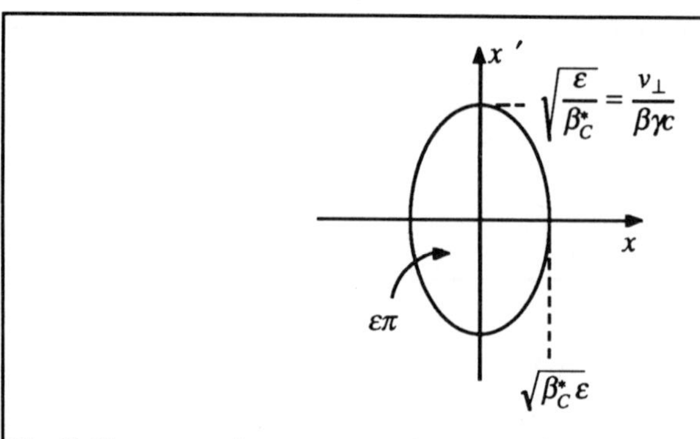

Fig. 7. Transverse phase space at electron cooler. Area of ellipse in (x, x') – plane is $\varepsilon\pi$. Ratio of x and x' half axii is the Twiss β-function, which we designate β_C^* to distinguish from the relativistic β. Ion beam half angular spread in the lab frame is $\sqrt{\frac{\varepsilon}{\beta_C^*}} = \frac{v_\perp}{\beta\gamma c}$, where $v_\perp$ is transverse ion velocity in moving frame. Ion beam half size is $\sqrt{\beta_C^* \varepsilon}$.

The maximum transverse force is

$$F_{\perp\max} = F(\Delta_\perp) = F_0 c^2 L_C \frac{1}{\Delta_\perp^2} \propto \frac{1}{\beta\gamma} \tag{15}$$

The scaling relations for the transverse friction force are shown in Fig. 8.

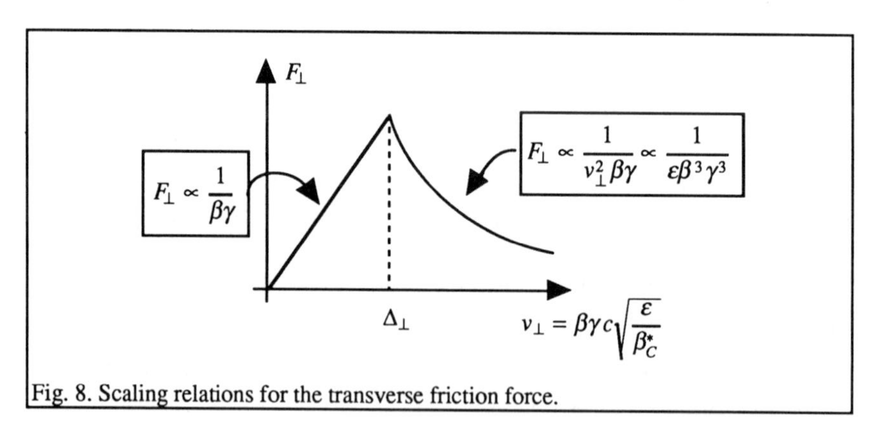

Fig. 8. Scaling relations for the transverse friction force.

Longitudinal

The electrostatic analogy and Fig. 6 show that the longitudinal friction force will be constant as long as $v_{||} << \Delta_\perp$, but decrease as $v_{||}^{-2}$ when $v_{||} >> \Delta_\perp$.

We wish to relate also the longitudinal friction force to a familiar lab frame quantity, and choose the relative momentum spread $\delta = \frac{\Delta p}{p}$. The relation between relative momentum spread and longitudinal rest frame velocity is $v_{||} = \beta\delta c$.

The maximum longitudinal force is like the maximum transverse force

$$F_{||\,max} = F(\Delta_\perp) = F_0 c^2 L_C \frac{1}{\Delta_\perp^2} \propto \frac{1}{\beta\gamma} \tag{16}$$

The scaling relations for the longitudinal friction force are shown in Fig. 9.

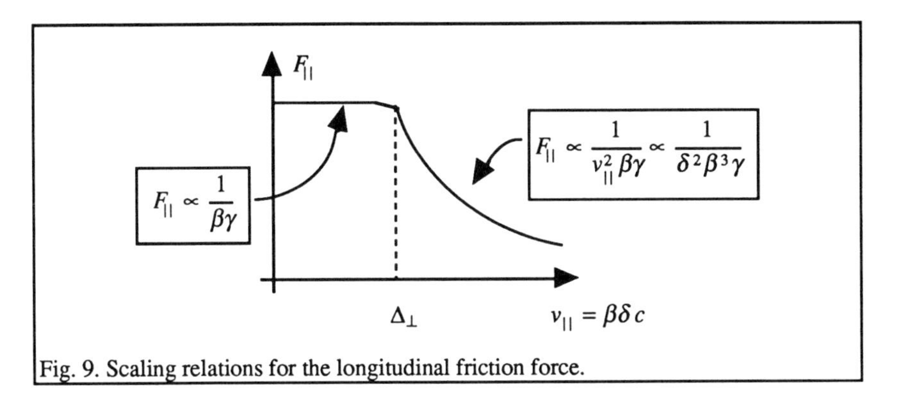

Fig. 9. Scaling relations for the longitudinal friction force.

RANGE OF COOLING

We can (cf. Figs. 8-9) define the *range of cooling* as the regime where the transverse and longitudinal ion velocities are below the transverse electron velocity $\Delta_\perp$. We find

$$\text{transverse:} \qquad v_\perp(\text{range}) = \Delta_\perp \qquad \Rightarrow \varepsilon_{range} = \frac{\Delta_\perp^2 \beta_C^*}{\beta^2\gamma^2 c^2} \propto \frac{\beta_C^*}{\beta^2\gamma^2} \tag{17}$$

$$\text{longitudinal:} \qquad v_{||}(\text{range}) = \Delta_\perp \qquad \Rightarrow \delta_{range} = \frac{\Delta_\perp}{\beta c} \propto \frac{1}{\beta} \tag{18}$$

COOLING TIMES

We can define the *cooling time* as the momentum divided by the time-average friction force, taking into account the fraction η_c of the ring circumference occupied by the cooler:

$$\tau_\perp = \frac{\gamma m_i v_\perp}{F_\perp \eta_C} \quad \text{and} \quad \tau_\| = \frac{\gamma m_i v_\|}{F_\| \eta_C} \tag{19}$$

We find the scaling relations for the transverse and longitudinal cooling times, which are indicated in Fig. 10. Note that it is only for the case of the transverse force, for ion velocities below $\Delta_\perp$, that the cooling time is independent of the ion velocity, and that an exponential decrement is obtained. Then, in the absence of heating mechanisms,

$$\varepsilon(t) = \varepsilon(0)\, e^{-2t/\tau_\perp} \tag{20}$$

$$\dot{\varepsilon}_{\text{cooler}} = \frac{-2\varepsilon}{\tau_\perp} \tag{21}$$

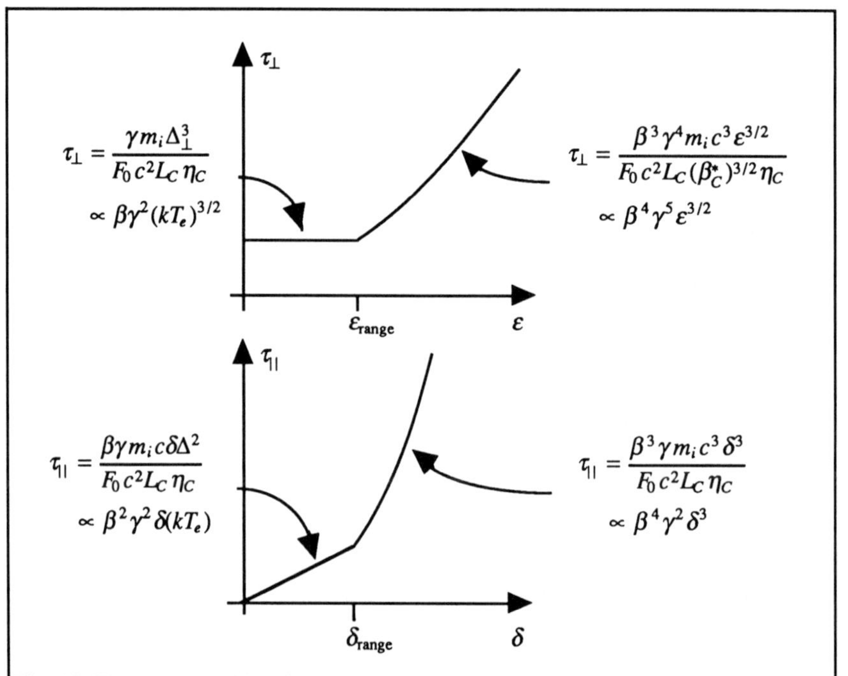

Fig. 10. Transverse and longitudinal cooling time vs. transverse emittance ε and relative momentum spread δ. In the regime where the transverse and longitudinal ion beam velocity spreads exceed the electron beam transverse velocity spread the relativistic scaling (assuming constant ion beam emittance and momentum spread) are as γ^5 and γ^2 respectively. In the regime where the ion beam velocity spreads are less than the electron beam transverse velocity spread, the transverse cooling time has a much less severe relativistic scaling relation, increasing as a mere γ^2.

INTERNAL TARGET OPERATION

Transverse

Consider a storage ring with a thin internal target. It is well known [26], that the emittance growth due to the multiple scattering in the target is

$$\dot{\varepsilon}_{\text{m.s.}} = \frac{f}{2}\beta_T^* \xi^2 \tag{22}$$

where f is the revolution frequency of the ions in the ring, β_T^* is value of the Twiss β -function at the target location, and ξ is the rms. multiple scattering angle in the target,

$$\xi = \frac{E_S}{\beta^2 \gamma E_0}\sqrt{\frac{l}{l_R}} \tag{23}$$

where $E_S \approx 14$ MeV , E_0 is the rest energy of the beam ion, l is the target thickness, and l_R is the radiation length in the target material.

The equilibrium between heating of the beam by multiple scattering in the target and cooling is obtained by requiring that $\dot{\varepsilon}_{\text{cooler}} + \dot{\varepsilon}_{\text{m.s.}} = 0$.

$$\varepsilon_{\text{eq}} = \frac{\dot{\varepsilon}_{\text{m.s.}}\,\tau_\perp}{2} \tag{24}$$

The luminosity is $L = f N \rho l N_A$, where N is the number of stored ions in the ring, and $\rho l N_A$ is the number of target atoms per unit cross section, ρ being the density of the target material and N_A being Avogadro's number. So we can relate the luminosity achieved in internal target operation to the equilibrium emittance

$$L = \frac{4F_0 \beta^4 \gamma N \varepsilon_{\text{eq}}}{\Delta_\perp^3}\,\frac{\rho m_i c^6 L_C \eta_C l_R N_A}{\beta_T^* E_S^2} \propto \frac{\beta^3 N \varepsilon_{\text{eq}}}{(kT_e)^{3/2}} \tag{25}$$

We use eq. (17) to write the luminosity which is corresponding to the range of cooling

$$L_{\text{range}} = \frac{4F_0 \beta^2 N}{\gamma \Delta_\perp}\,\frac{\beta_C^*}{\beta_T^*}\,\frac{\rho m_i c^4 L_C l_R \eta_C N_A}{E_S^2} \propto \frac{\beta N \beta_C^*}{\gamma^2 \beta_T^*} \tag{26}$$

Let us now introduce some approximate engineering scaling laws:

First, lets assume that the circumference of the ring will be proportional to γ. This is reasonable, because for large γ the ring circumference will become proportional to the rigidity of the ions, but the circumference will level off for small γ.

Also, lets assume that the number of stored ions is proportional to its circumference. This assumption is not self-evident, but seems acceptable when considering that the injection energy of a large and high-energy ring will be higher than that of a small and low-energy ring.

$$N \propto \text{circumference} \propto \gamma \tag{27}$$

Then we find

$$L_{\text{range}} \propto \frac{\beta}{\gamma}\frac{\beta_C^*}{\beta_T^*}\eta_C \tag{28}$$

Now introduce another scaling law, that the length of the electron cooling section scales with the ring circumference, and also that the value of the β-function at the cooler scales with the circumference.

$$\beta_C^* \propto \gamma; \quad \eta_C = \text{constant} \tag{29}$$

Then

$$L_{\text{range}} \propto \beta \tag{30}$$

i.e. as soon as $\beta \approx 1$ the luminosity which is compatible with the range of cooling becomes essentially independent of the energy.

Fig. 11 shows the calculated equilibrium emittance, using the formalism and the engineering scaling laws outlined above.

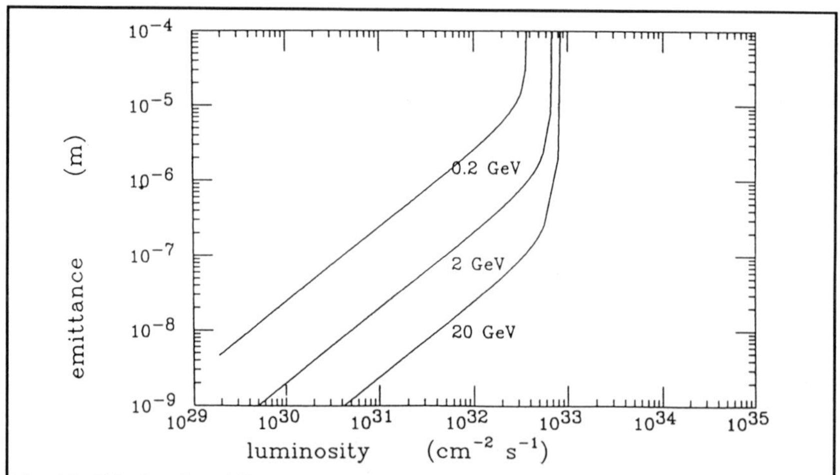

Fig. 11. Calculated equilibrium emittance as a function of luminosity with $\beta_C^* = 4\gamma$ m, circumference = 30 γ m, 10^8 ions/m, $I_e = 1$ A, $r_0 = 10$ mm, and $kT_e = 0.5$ eV.

Longitudinal

The longitudinal situation is quite different from the transverse one. In the longitudinal plane the strong cooling force tends to bring all ions into a very narrow peak. Occasional large losses, associated with the emission of $\delta-$rays, produce a long tail [27].

The average energy loss per target passage scales with energy as β^{-2}. With our first scaling law (27) we have $f \propto \beta/\gamma$, and therefore the energy loss per unit time in the target scales as $1/\beta\gamma$. The cooling drag rate, on the other hand, scales as $\beta F_{||} \propto 1/\gamma$. So, since we know that that the drag rate can keep up with the average energy loss of useful internal targets at low energies, we see that it will also be able to do so at high energies.

The maximum energy loss in the target, associated with δ-ray production, is $\beta^2\gamma^2 m_e c^2$. The associated momentum loss, $\beta\gamma^2 m_e c$, scales much faster with energy than both the range of cooling (eq. 18) $\Delta p_{\text{range}} = \dfrac{p\Delta_\perp}{\beta c} = \gamma m_i \sqrt{\dfrac{2kT_e}{m_e}}$ and the maximum momentum error for a constant relative momentum acceptance. This is shown in Fig. 12.

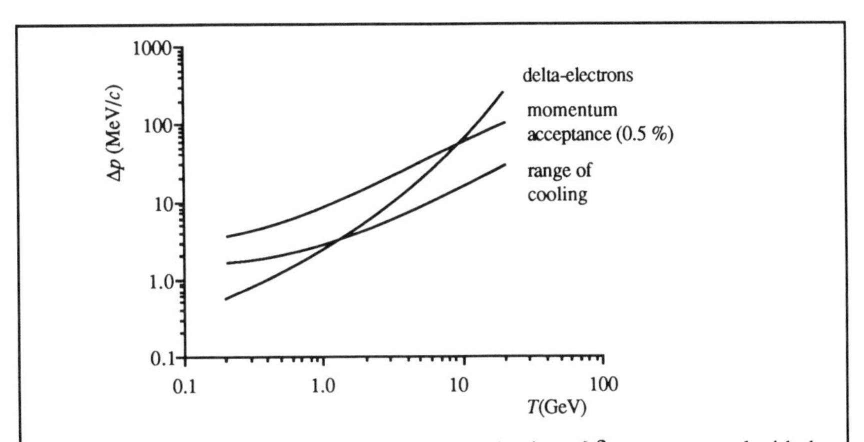

Fig. 12. Momentum loss associated with the production of δ-rays compared with the range of cooling and with momentum deviation inside 0.5 % momentum acceptance.

The probability for an energy loss ΔE is approximately [28]

$$P(\Delta E) \approx \frac{\xi^2}{(\Delta E)^2}, \quad \text{where } \xi = \frac{2\pi N_A \rho l e^4}{(4\pi\varepsilon_0)^2 m_e \beta^2 c^2} \frac{Z_T Z_P^2}{A_T} \tag{31}$$

A calculation shows that the probability for large momentum loss becomes small enough that the loss rate due to momentum loss larger than the acceptance of a ring does not become a serious loss mechanism if the momentum acceptance is greater than about ±0.2 % (Fig. 13), and that the maximum target thickness consistent with a requirement that 90 % of the beam should be in the core is about 5×10^{15} atoms/cm^2 for high energies (Fig. 14).

The width of the narrow peak will be determined by the electron beam energy spread as discussed above (of the order of 2×10^{-5}), and by intrabeam scattering among the ions.

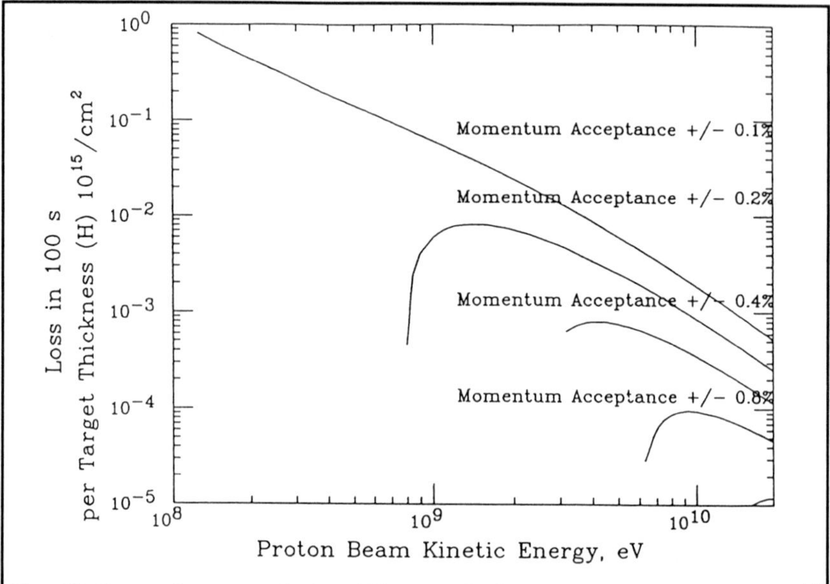

Fig. 13. Loss of protons during 100 s per hydrogen target thickness of 10^{15} atoms/cm^2 as a function of proton beam kinetic energy.

INTRA-BEAM SCATTERING

Coulomb scattering between beam particles can exchange energy spread from one phase plane to another, and also transfer energy from the common longitudinal motion into longitudinal or transverse energy spread.

One may write [29]

$$\frac{1}{\tau_{\text{ibs}}} = \frac{N c r_p^2 \left(\frac{Z^2}{A}\right)^2}{\frac{32}{\sqrt{\pi}} \beta^3 \gamma^4 \varepsilon_h \varepsilon_v \delta \, 2\pi r} F \tag{32}$$

where F is a form factor, $-10 < F < 10$ (intrabeam scattering can cool in the plane where the beam temperature is highest), r_p is the classical proton radius, and $2\pi r$ is the ring circumference. We will follow the example of ref. 28 and put $F = 1$ in order to evaluate the scaling laws.

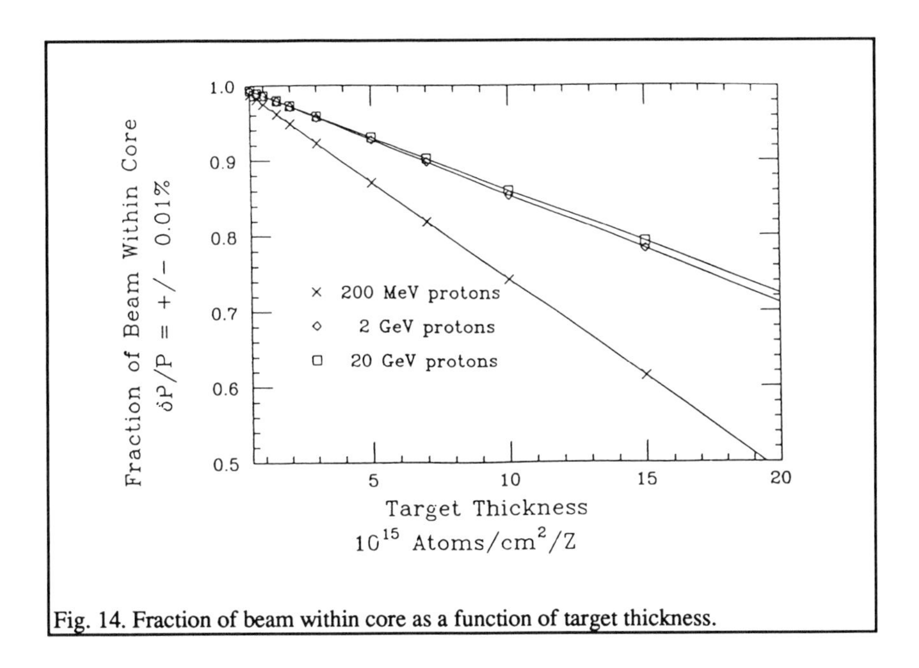

Fig. 14. Fraction of beam within core as a function of target thickness.

Since an electron cooled ion beam will have the smallest rest frame velocity spread in the longitudinal phase plane, it is in this phase plane that intrabeam scattering will tend to increase the longitudinal rest frame velocity of the beam ions. Equilibrium between intrabeam scattering and the longitudinal cooling force will obey a scaling relation

$$\delta \propto \frac{(\frac{Z^2}{A})^2 N \tau_{\parallel}}{\beta^3 \gamma^4 \varepsilon_h \varepsilon_v \, 2\pi r} \tag{33}$$

If $\tau_{\parallel} \propto \beta^2\gamma^2$ (fig. 10), and $2\pi r \propto \gamma$ (eq. 27), then

$$\delta \propto \frac{(\frac{Z^2}{A})^2 N}{\varepsilon_h \varepsilon_v \beta \gamma^3} \tag{34}$$

if ε_h and ε_v are made to scale as $\dfrac{1}{\beta\gamma}$ then

$$\delta \propto \frac{N\beta}{\gamma} \tag{35}$$

It should be pointed out, however, that intrabeam scattering is a very non-linear and complicated function of the machine and beam parameters, and requires detailed study for each case.

ALIGNMENT

The apparent "temperature" of a misaligned electron beam is

$$kT_e = \frac{\beta^2\gamma^2 m_e c^2\theta^2}{2} \tag{36}$$

So, in order to get a contribution to the electron beam temperature which is less than kT_e , we have to align the ion and electron beams better than $\theta \le \frac{1}{\beta\gamma}\sqrt{\frac{2kT_e}{m_e c^2}}$.

Alignment of the electron and the ion beams is done with pickup electrodes at each end of the electron cooling section. If the length of the cooler scales as γ (eq. 29), the required precision in the pickup electrodes remains essentially constant with energy. On the other hand, it does get more and more difficult to shape the magnetic field with the required straightness $\theta < \frac{0.6\text{ mrad}}{\beta\gamma}$ corresponding to $kT_e < 0.1$ eV .

MAGNETIC FIELD

At low energies, a longitudinal magnetic field is the way to keep the electron beam together against its own space charge force. In order to limit the transverse electron temperature to the value which is determined by the cathode, the electrons must be transmitted adiabatically through unavoidable imperfections in the magnetic field. Therefore the ratio between the "wavelength" of the imperfections and the gyro-wavelength of the electrons has to be maintained. The approximate scaling laws are [25]:

$$B\, r_{\text{toroid}} \ge 0.08\beta\gamma \text{ Tm} \tag{37}$$

$$B\, d_{\text{solenoid}} \ge 0.03\beta\gamma \text{ Tm} \tag{38}$$

Magnetic confinement becomes unpractical for energies above approximately 1 MeV. Focusing in discrete lenses has to be employed. These lenses will be quadrupole magnets, arranged in doublets or triplets. It is fortunate, that the acceptable distance between lenses, that makes the divergence of the electron beam corresponding to an electron temperature of 0.1 eV, becomes manageable at about that energy, see fig. 15.

$$s < \frac{4\pi\varepsilon_0 r_0 \beta^2\gamma^2 c}{eI_e}\sqrt{2m_e c^2 kT_e} \tag{39}$$

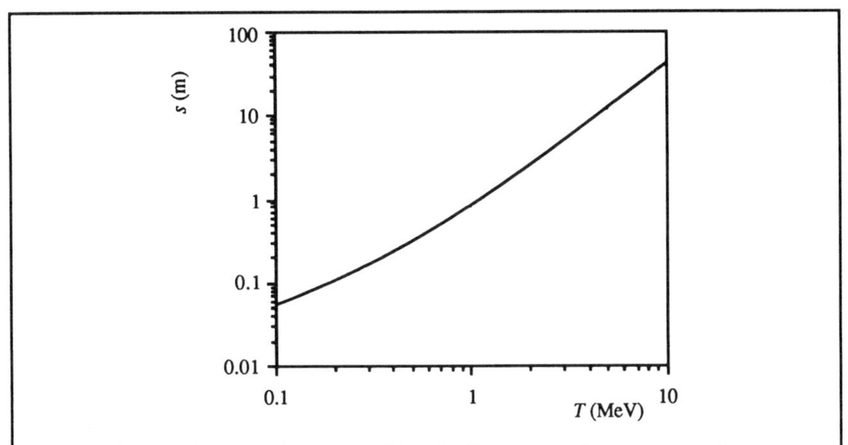

Fig. 15. Distance between lenses resulting in divergence due to space charge corresponding to an electron temperature of 0.1 eV, $I_e = 1$ A, and $r_0 = 0.01$ eV.

EXPERIMENTS RELATING TO HIGH ENERGY COOLING

Novosibirsk

At Novosibirsk, an experiment to accelerate, transport and collect an MeV electron beam in a magnetic confinement system has been concluded [30]. In this device, the adiabatic scaling laws discussed above (eq. 37-38) were not respected, and it was only at discrete values of the longitudinal magnetic field that transport of the electron beam was possible. Yet, in spite of poor vacuum and slow regulation of power supplies (which turned out to be a problem) reasonably stable operation with 1 MeV and 1 A was achieved.

Frascati

These problems should be overcome in a project to build a magnetically confined prototype electron cooler of 750 keV and 3.5 A at Frascati [31]. The system is expected to have a collected electron beam during the first half of 1991 [32].

Wisconsin

The transverse temperature of a 3 MeV electron beam from an electrostatic accelerator (Pelletron of National Electrostatic Corporation of Middleton, Wisconsin) has been measured [9,33] to correspond to about three times the cathode temperature. This encouraged a collaboration between the University of Wisconsin, the National Electrostatics Corporation, and FNAL to embark on a project to recirculate a 3 MeV ampere-intensity beam, intended as a prototype for electron cooling of antiprotons.

A non-magnetized recirculation system was built and succeeded in recirculation of a 2 MeV, 105 mA electron beam [34]. It was foreseen that beam losses would have to be limited to the current capacity of the electrostatic accelerator, which was 250 μA in this experiment. It is important to note however, that the available charging current never became a limitation. It was other difficulties related to beam losses which limited operation whenever the losses approached 50 μA. The beam current during a particular run appeared to be limited by outgassing from the collector, with each day's run reaching a new maximum beam current until system pressure rose, collector efficiency deteriorated, and the recirculation ended.

This project represents a source of very valuable experience which will be a basis for new designs of multi-MeV electron coolers. There is a great number of ideas among the experimenters about things they would do differently if they would be building such a device again [35,36]. Several of these ideas aim to improve the degree of vacuum and ion clearing in the equipment, the collection efficiency by collecting at a higher energy and have better focusing near the collector, better stability in power supplies, better high voltage performance, etc.

A particularly interesting idea [35] of Larson is to put a magnetic dipole in front of the collector, in order to bend any secondaries that are leaving the collector away from the deceleration tube.

STATE OF THE ART

Electron temperature

It has been shown by measurements [9,33] that an electron temperature corresponding to three times the cathode temperature can be achieved from an electrostatic accelerator. This is encouraging, but further work should be made to limit the electron beam temperature to that determined by the cathode. One way to further reduce the electron temperature would be to adiabatically expand the electron beam size.

High voltage regulation

The commercial state of the art for regulation of electrostatic accelerators is about 10^{-5}.

Un-controlled beam loss, leading to unstable operation

Both the experiments at Novosibirsk and in Wisconsin have been limited by un-controlled beam loss, leading to unstable operation. This problem area requires further work, based on the experience gained. It appears as if better power supply regulation against transients and better vacuum would help. Larson's idea of a dipole magnet before the collector should be tried.

CONCLUSIONS

We conclude that if we scale the circumference of the ring, the length of the cooling section, and the β-value at the cooler with γ, then the cooling force will scale as $1/\gamma$, the equilibrium emittance for constant luminosity will also scale as $1/\gamma$, the equilibrium relative momentum spread will scale as N/γ (but not below 2×10^{-5} due to electron cooler power supply regulation and intrabeam scattering among the electrons).

The transverse cooling range ε_{range} will scale as $1/\gamma$ and the corresponding luminosity will be independent of γ (if $\beta \approx 1$). The longitudinal cooling range δ_{range} will be independent of γ.

If the ion velocity distribution is within the range of cooling, the relativistic scaling for the transverse and longitudinal cooling times will be as γ^2. If the ion velocity distribution is outside of the range of cooling, on the other hand, the transverse cooling time will scale as γ^5. The relativistic scaling for the longitudinal cooling time will remain as γ^2.

Among the benefits achieved with electron cooling at low energies (improving the lifetime, shrinking the beam size to allow stacking, and improving the resolution) the improvement of the resolution may be the one that becomes most important at intermediate and high energies. This is true both for the energy resolution, which can be used to also obtain a high time resolution by applying rf. to the beam, and also for the transverse beam size. A small beam size will also be essential for obtaining high luminosity in colliders, and it may be this last point which will be the most important case for high energy electron cooling in the near future.

ACKNOWLEDGEMENTS

Much of the enthusiasm for electron cooling around the world is a result of the work and the devotion of Helmut Poth. This paper is dedicated to his memory.

Useful discussions with A. Johansson, B. Karlsson, S.Y. Lee, and M. Sedlacek are acknowledged.

REFERENCES

[1] H. Poth, *Electron Cooling: Theory, Experiment, Application.* CERN-EP/90-04 (1990).

[2] H. Poth, *Electron Cooling.* Proc. CERN Accelerator School, Advanced Accelerator Physics, The Queen's College, Oxford, U.K. 16-27 September 1985, ed. S. Turner, CERN 87-03 (1987).

[3] W. Kells, *Electron Cooling*, Proc. Fermilab Summer School, Batavia, Illinois, July 13-24, 1981, ed. R.A. Carrigan, F.R. Huson, M. Month. AIP Conference Proceedings **87** (1982) 656.

[4] H. Poth, *Applications of Electron Cooling in Atomic, Nuclear and High-Energy Physics.* Nature **345** (1990) 399.

[5] Y. Derbenev, I. Meshkov, *Studies on Electron Cooling of Heavy Particle Beams made by the VAPP-NAP Group at the Nuclear Physics Institute of the Siberian Branch of the USSR Academy of Science at Novosibirsk.* CERN 77-08 (1977)

[6] G.I. Budker, Proc. Int. Symp. on Electron and Positron Storage Rings, Saclay, France, 1966, ed. H. Zyngier and E. Crémieux-Alcan (PUF, Paris 1967), p. II-1-1.

[7] S. van der Meer, *Stochastic Damping of Betatron Oscillations in the ISR*, CERN/ISR-PO/72-1 (1972).

[8] D. Möhl, *A Comparison between Electron Cooling and Stochastic Cooling.* Proc. ECOOL 1984, Karlsruhe, Germany, Sept. 24-26, 1984. Ed. H. Poth, KfK 3846, Kernforschungszentrum Karlsruhe (1984).

[9] D.B. Cline, D.J. Larson, W. Kells, F.E. Mills, J. Adney, J. Ferry, M. Sundquist, *Intermediate Electron Cooling for Antiproton Sources using a Pelletron Accelerator.* IEEE Trans. Nucl. Sci. **NS-30** (1983) 2370.

[10] D.J. Larson, *Intermediate energy electron cooling for antiproton sources,* thesis, University of Wisconsin, Madison, 1986.

[11] H. Herr, C. Rubbia, *High Energy Cooling of Protons and Antiprotons for the SPS Collider*, Proc. 11th Int. Conf. on High Energy Accelerators, Geneva, July 1980.

[12] G.I. Budker, N.S. Dikansky, V.I. Kudelainen, I.N. Meshkov, V.V. Parchomchuk, D.V. Pestrikov, A.N. Skrinsky, B.N. Sukhina, *Experimental Studies of Electron Cooling.* Particle Accelerators **7** (1976) 197.

[13] M. Bell, J. Chaney, H. Herr, F. Krienen, P. Møller-Pedersen, G. Petrucci, *Electron Cooling in ICE at CERN.* Nucl. Instr. and Meth., **190** (1981) 27.

[14] T. Ellison, W. Kells, V. Kerner, F. Mills, R. Peters, T. Rathbun, D. Young, P.M. McIntyre, *Electron Cooling and Accumulation of 200 MeV Protons at Fermilab*. IEEE Trans. Nucl. Sci., **NS-30** (1983) 2636.

[15] E. Jaeschke, *Heavy Ion Storage Rings*, to be published in Proc. 2nd European Particle Accelerator Conference, Nice, France, June 12-16, 1990.

[16] S.T. Belyaev (editor), *Heavy Ion Storage Rings (USSR proposals and projects collection)*, I.V. Kurchatov Institute of Atomic Energy, Moscow, 1990.

[17] T. Ellison, *Electron Cooling at IUCF*. To be published in Proc. ECOOL90, Legnaro, Italy, May 15-17, 1990.

[18] Ya. S. Derbenev, A.N. Skrinsky, *The kinetics of electron cooling of beams in heavy particle storage rings*. Particle Accelerators **8** (1977) 1.

[19] Ya. S. Derbenev, A.N. Skrinsky, *The effect of an accompanying magnetic field on electron cooling*. Particle Accelerators **8** (1978) 235.

[20] Ya.S. Derbenev, A.N. Skrinskii, *The Physics of Electron Cooling*, Physics reviews, Harwood Academic Publishers, Chur, **3** (1981) 165.

[21] A.H. Sørensen, E. Bonderup, *Electron Cooling*, Nucl. Instr. Meth. **215** (1983) 27.

[22] M. Bell, *Electron Cooling with Magnetic Field*. Particle Accelerators **10** (1980) 101.

[23] N.S. Dikansky, V.I. Kudelainen, V.A. Lebedev, I.N. Meshkov, V.V. Parkhomchuk, A.A. Sery, A.N. Skrinsky, B.N. Sukhina, *Ultimate possibilities in Electron Cooling*, Institute of Nuclear Physics, Novosibirsk, Preprint 88-61 (1988).

[24] V.I. Kudelainen, V.A. Lebedev, I.N. Meshkov, V.V. Parkhomchuk, B.N. Sukhina. *Temperature Relaxation in Magnetized Electron Flux*. Sov. Phys. JETP **56** (1982) 1191.

[25] T. Ellison. *Electron Cooling*. Thesis, Department of Physics, Indiana University, To be published 1990.

[26] F. Hinterberger, D. Prasuhn. *Analysis of Internal Target Effects in Light Ion Storage Rings*. Nucl. Instr. and Meth. in Phys. Res. **A279** (1989) 413.

[27] H.O. Meyer. *Beam properties in Storage Rings with Heating and Cooling*. Nucl. Instr. and Meth. in Phys. Res. **B10/11** (1985) 342

[28] T. Ellison. *Longitudinal Equilibrium Distributions of Ion Beams in Storage Rings with Internal Targets and Electron Cooling*. IEEE Trans. Nucl. Sci. **NS-32** (1985) 2418

[29] D. Möhl. *Limitations and Instabilities in Cooled Beams*. Proc. Workshop on Crystalline Ion Beams, Wertheim, Germany, 4-7 October 1988, 108.

[30] M.E. Veis, B.M. Korabelnikov, N.K. Kuksanov, P.I. Nemytov, R.A. Salimov, A.N. Skrinsky, *Electron Beam with an Energy 1 MeV in the Recuperation Regime*. Proc. European Part. Acc. Conf., Rome, Italy, June 7-11, 1988, 1361.

[31] M.E. Biagini, U. Bizzarri, R. Calabrese, M. Conte, S. Guiducci, F. Petrucci, L. Picardi, C. Ronsivalle, C. Salvetti, M. Savrié, S. Tazzari, L. Tecchio, A. Vignati. *Perspectives for a High Energy Electron Cooling at LEAR. An Experimental Test*. IEEE Trans. Nucl. Sci. **NS-32** (1985) 2409.

[32] R. Calabrese, L. Tecchio. Private communication October 1990.

[33] D.J. Larson. *Design of an Electrostatic Accelerator for use in Intermediate-Energy Electron Cooling*. Particle Accelerators, **23** (1988) 239.

[34] J.R. Adney, M.L. Sundquist, D.R. Anderson, D.J. Larson, F.E. Mills, *Successful DC Recirculation of a 2 MeV Electron Beam at Currents More Than 0.1 Ampere*. Proc. 1989 IEEE Part. Acc. Conf., Chicago, Illinois, March 20-23 1989, 348.

[35] D.J. Larson. Private communication, September 1990.

[36] M.L. Sundquist. Private communication, September 1990.

CLOSING

Conference Summary

ON THE THRESHOLD OF PARTICLE PHYSICS

D.F. Measday

University of British Columbia, Vancouver, B.C., Canada V6T 2A6

INTRODUCTION

It is worth stopping to ask ourselves why we are here; why hesitate on the threshold of particle physics before plunging in. In fact there is an excellent reason, because on the edge you can take your bearings and establish your reference system, but once you have jumped into the morass of the strong interactions, you can lose your sense of direction very easily.

The easiest place to start is exotic atoms such as muonic or pionic atoms. Historically studies of these systems established many basic properties of the particles such as mass, spin, and magnetic moment, even whether the muon was a hadron. Because the electromagnetic interaction is well understood and the atomic states limit the angular momentum, other properties can be determined with confidence. Today there is still interest in exotic atoms, but the questions are getting quite sophisticated. At this conference H. Toki discussed the exciting possibility of studying the 1s level in pionic lead which normally is impossible to reach because the pion is absorbed higher in the cascade. So far no experiment has detected this exotic state of an exotic atom, but reactions such as (n,p), (d,^{2}He) and (t,^{3}He) have been suggested as useful probes.[1] If such a state can be identified, it will help our understanding of pion propagation in nuclei.

SCATTERING LENGTHS

A popular way to describe low energy scattering is via the effective range approximation. The total s-wave scattering cross-section is then given by

$$\sigma = \frac{4\pi}{k^2 + \left[\frac{1}{2}k^2 r_o - \frac{1}{a}\right]^2}$$

where k is the wave number, r_o is the effective range of the interaction and a is the scattering length.

The perfect case in which this type of analysis is used is neutron-neutron scattering, one of our most fundamental interactions. This classic example even uses an exotic atom as the initial state in order to further simplify the theoretical and experimental situation. The reaction $\pi^- d \rightarrow nn\gamma$ can be studied either via the γ–ray only or via the neutron-neutron pair. Both techniques have been applied at PSI and give consistent results viz.

$$a_{nn} = \ -18.5 \pm 0.4\ fm \qquad r_{nn} = 2.80 \pm 0.11 \qquad \text{[ref. 2]}$$

$$a_{nn} = -18.7 \pm 0.6 \; fm \qquad \text{[ref. 3]}$$

The gamma-ray spectrum is illustrated in Fig.1 and includes the resolution function obtained from the reaction $\pi^- p \rightarrow \gamma n$, which provides a convenient monochromatic line at 129.4 MeV. The gamma-ray detector was a pair spectrometer which obtained a resolution of 720 keV which, although excellent, was still sufficient to distort considerably the theoretical spectrum, depicted by the dash-dot line.

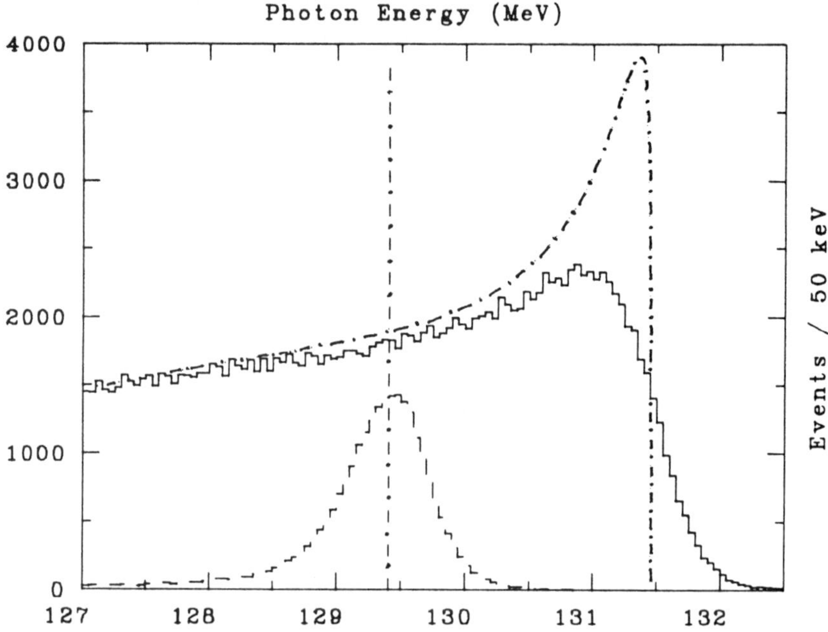

Figure 1. Experimental photon spectrum from the reaction $\pi^- d \rightarrow nn\gamma$ which is used to deduce the neutron-neutron scattering length of -18.50 ± 0.4 fm. The monochromatic line at 129.4 MeV is a calibration line using the reaction $\pi^- p \rightarrow \gamma n$.

This can be exploited in other examples. Thus there was a recent attempt to study the reaction $K^- d \rightarrow \Lambda n \gamma$ in order to determine the Λn scattering length[4] using the theory of Fearing and Workman.[5] In this case the branching ratio is much lower (1.9×10^{-3} instead of 26%) and kaon beams are much poorer, so the experiment was unable to improve on measurements from bubble chambers of the Λp interaction which gave[6]:-

$$a = -1.80 \; fm \qquad r_o = 3.16 \; fm$$

However new detectors at a Kaon Factory should improve on this situation.

It is worth noting at this juncture that the ΛN interaction has been suspected of harbouring a dibaryon. Claims have been made by Piekarz[7] that an S=-1 dibaryon has been uncovered in the reaction $K^-d \rightarrow \pi^-\Lambda p$. However the data can also be interpreted as a kinematic broadening similar to that observed in reactions such as $d(n,p)nn$ [ref. 8] or $d(p,n)pp$ [ref. 9]. A related reaction $pp \rightarrow K^+YN$ has been discussed here at this conference by Frascaria and no dibaryon is required.

Another scattering length that is of great interest is that of the $\pi\pi$ system. Experimentally it is very difficult to measure and the best data come from the rare K_{e4} decay, viz. $K^+ \rightarrow e^+\nu\pi^+\pi^-$, which has a branching ratio of only 3.9×10^{-5}. Nevertheless Rosselet et al.[10] obtained 30,318 events using a 2.8 GeV/c K^+ beam and obtained $a_0^0 = 0.28 \pm 0.05\ m_\pi^{-1}$, which can be compared to the calculated value of $0.192\ m_\pi^{-1}$. Their data are the solid circles in Fig. 2 and the scatter amongst the

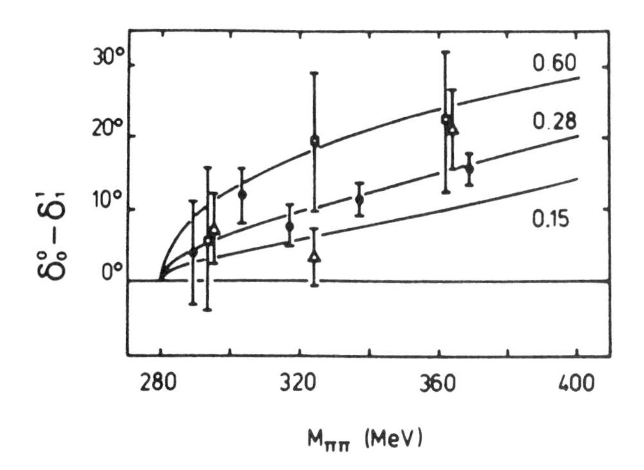

Figure 2. Phase shift difference $\delta_0^0 - \delta_1^1$ for $\pi\pi$ scattering obtained from K_{e4} experiments (viz. $K^+ \rightarrow e^+\nu\pi^+\pi^-$).

points speaks for itself, so another experiment is clearly needed. A recent attempt has been made by Lowe et al. to study this interaction indirectly via the reaction $\pi^-p \rightarrow \pi^0\pi^0n$ using the Crystal Box at Brookhaven.[11] They have followed the type of analysis carried out by Manley[12] catalyzed by the results from $\pi^-p \rightarrow \pi^+\pi^-n$ obtained at LAMPF. Lowe et al. have also included the Omicron data presented to us here by Zavrtanik. The extraction of $\pi\pi$ information from these reactions is very difficult, as is best illustrated by the calculation of Oset and Vicente-Vacas.[13] Further data are expected from TRIUMF on the reaction $\pi^+p \rightarrow \pi^+\pi^+n$ and from LAMPF on the reaction $\pi^+p \rightarrow \pi^+\pi^0p$. It may then be possible to sort out the various contributions to the reaction amplitudes and thereby to extract information on the $\pi\pi$ scattering length.

The $\pi\pi$ interaction was extensively discussed at this conference and some beautiful calculations were presented to us. Weinstein has a straightforward quark molecule approach, using constituent quarks.[14] Speth has a more complex approach, but both obtained unbelievably good fits to the $\pi\pi$ phase-shifts. They also agree that the $S^*(975)$ and $\delta(980)$ are both $K\overline{K}$ quasi-bound states. The calculations also explain long-standing problems in the decay $J/\psi \to \omega\pi^+\pi^-$. A large peak is observed at 400 MeV/c^2 which is now understood as resulting from the $\pi\pi$ attraction in the final state interaction. These data were presented here by W. Toki but the effect has also been observed by the DM2 collaboration at Orsay[15]. See Fig. 3.

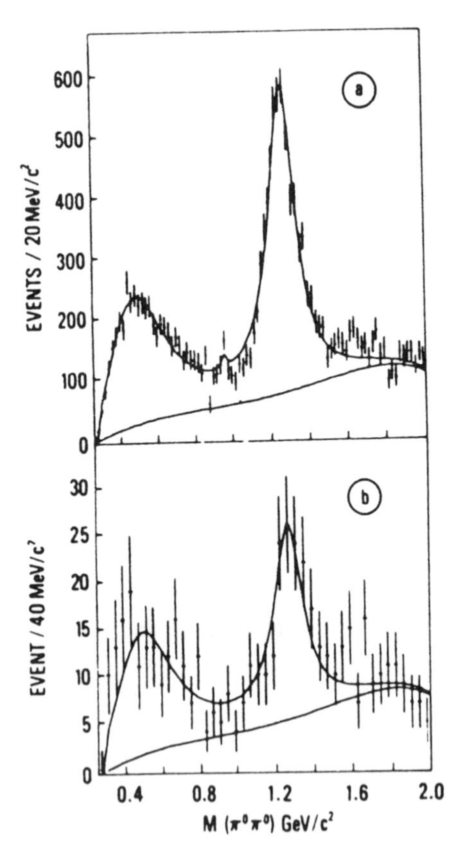

Figure 3. The $\pi\pi$ distribution: (a) for $J/\psi \to \omega\pi^+\pi^-$ events; (b) for $J/\psi \to \omega\pi^0\pi^0$ events, from the DM2 collaboration.[15]

The low energy $\pi\pi$ interaction reminds us of the ABC, an effect first observed in the reaction $pd \to {}^3He\ X$, where X had a mass of about 280 MeV/c^2. The apparent final state interaction was too strong to be explained by a large $\pi\pi$ scattering length, but Brody[16] showed that it could be explained satisfactorily by including the contribution from the ^{3}He form factor. We shall come back to the ABC effect when we consider the new publication (but old data) on $np \to dX$.

PION PRODUCTION

There were several discussions of pion production near threshold. Hutcheon described the new TRIUMF results on $np \to \pi^0 d$. [ref. 17] This difficult experiment is the first to get so near threshold and supplants the old experiment of Rose[18] on $\pi^+ d \to pp$, which used the Chicago bubble chamber. The low energy cross-sections are normally described in terms of two contributions,

$$\sigma(pp \to \pi^+ d) = 2\sigma(np \to \pi^0 d) = \alpha\eta + \beta\eta^3$$

where η is the pion momentum in the centre of mass in units of $(m_\pi c)^{-1}$, α is the s-wave pion production and β is the p-wave pion production. Hutcheon et al. obtain $\alpha = 184 \pm 5$ μb. (Coulomb corrections have to be made for $pp \to \pi^+ d$.) This s-wave contribution can be related to photo-meson production cross-sections using branching ratios for $\pi^- p$ and $\pi^- d$ atoms.[19] The new value for α re-opens the old problem of inconsistency between the various cross-sections. Maybe it is time to look again at some of the photo-meson production data near threshold, and a proposal has been made to study $\gamma p \to \pi^+ n$ at Saskatoon.

There is every reason to check these data because new experiments on the reaction $\gamma p \to \pi^0 p$ have revealed very interesting anomalies. The initial unearthing came from Saclay[20], but recent data from Mainz were presented to us by Beck.[21] The measurements agree and demonstrate that the cross-section is much lower than was originally thought. it is complicated by the need to include in the analysis, a cusp from the threshold of the reaction $\gamma p \to \pi^+ n$. [refs. 22,23]

Perhaps the most dramatic new results in this field, however, came from the IUCF cooler itself.[24] Some new results were presented by Meyer on the reaction $pp \to pp\pi^0$. This experiment approached close to threshold, taking advantage of the characteristics of the cooler ring. By obtaining data within a few MeV of threshold at 279.6 MeV they have shown that previous phenomenological analyses need to be improved by adding Coulomb effects and the restriction in phase-space of the pp final-state interaction.[25] This reaction is very important as this isospin transition (σ_{11}) is a major contributor to the neutron induced reactions $np \to \pi^+ nn$ and $np \to \pi^- pp$. These in their turn are used in the analysis of pion absorption experiments to describe the reaction $\pi^- +$"pp"$\to np$.

RESONANCE BELOW THRESHOLD

The data which we have discussed so far have been relatively straightforward. The only corrections that need to be made are for phase-space and Coulomb effects.

However a nearby resonance can severely complicate the analysis and there are many examples. We can take the K^-p atom as an example (m = 1431.9 MeV/c^2). Just below threshold is the $\Lambda(1405)$ which has a full width of about 55 MeV/c^2. It is so close to threshold that the K^-p atom can be used to study some of its properties via the reactions $K^-p \rightarrow \Lambda\gamma$ and $K^-p \rightarrow \Sigma\gamma$. [ref. 26]

The K^-–nucleon system is interesting because it is basically attractive, but the presence of the $\Lambda(1405)$ makes it appear repulsive above the resonance. Thus in K^- atoms the 2p levels are less bound. However one residual problem is that the experiments on the K^-p atom find that the 1s level is more bound. This problem has been around for many years and has been pursued with dedication by Landau. In recent papers Schnick and Landau[27,28] have shown that it is possible to encompass the opposing results in a relativistic theory, but the alternative possibility is that the experimental results were incorrect.

RESONANCE ABOVE THRESHOLD

An example of the opposite effect is also easy to find where a resonance is just above threshold. In fact there are a plethora of cases in nuclear physics, but let us focus on the S_{11} baryon which has received a lot of attention at this conference.

The η meson is a hermit which keeps to itself. In proton-proton collisions, it is rarely produced, and the majority of excited baryons have a vanishingly small branching ratio to decay into this meson. The one exception is the S_{11} (1535), which also happens to coincide approximately with the ηN threshold at 1490. This fact alone has engendered much speculation that thresholds are somehow connected to baryons.

The S_{11} (1535) affects all the η–production reactions and must be included if a reasonable description is to be obtained. Mukhopadhyay discussed the reaction $\gamma p \rightarrow \eta p$ and demonstrated the dramatic impact of this resonance, which boosts the cross-section near threshold by an order of magnitude. Similarly the reaction $\pi^-p \rightarrow \eta n$ is magnified by the presence of this resonance.

In nucleon-nucleon collisions the η is not produced prolifically, except for the reaction $pd \rightarrow \eta^3He$, which is being advertised as the source for an η factory. This inconsistency may now have been explained in the recent paper by Plouin et al.[29], who resurrected old data on the reaction $np \rightarrow dX$, which were taken to study the ABC effect. The ABC was in fact clearly visible on their data, but in addition there was another clear peak in their data at 1.5°, which they now attribute to the reaction $np \rightarrow \eta d$ (see Fig. 4). It is curious because it is apparently an order of magnitude stronger than the reaction $np \rightarrow \pi^0 d$ near its own threshold, but it does explain why the η is seen so clearly in the reaction $pd \rightarrow \eta^3He$. Unfortunately the experiment of Plouin was done right at threshold (p_n = 1880 MeV/c : T_n = 1162 MeV), so a more complete measurement is desperately needed.

The effect of the η is also seen as a cusp in π^-p elastic scattering at 690 MeV/c. Bekrenev discussed this effect and showed that the existing phase-shift analyses are

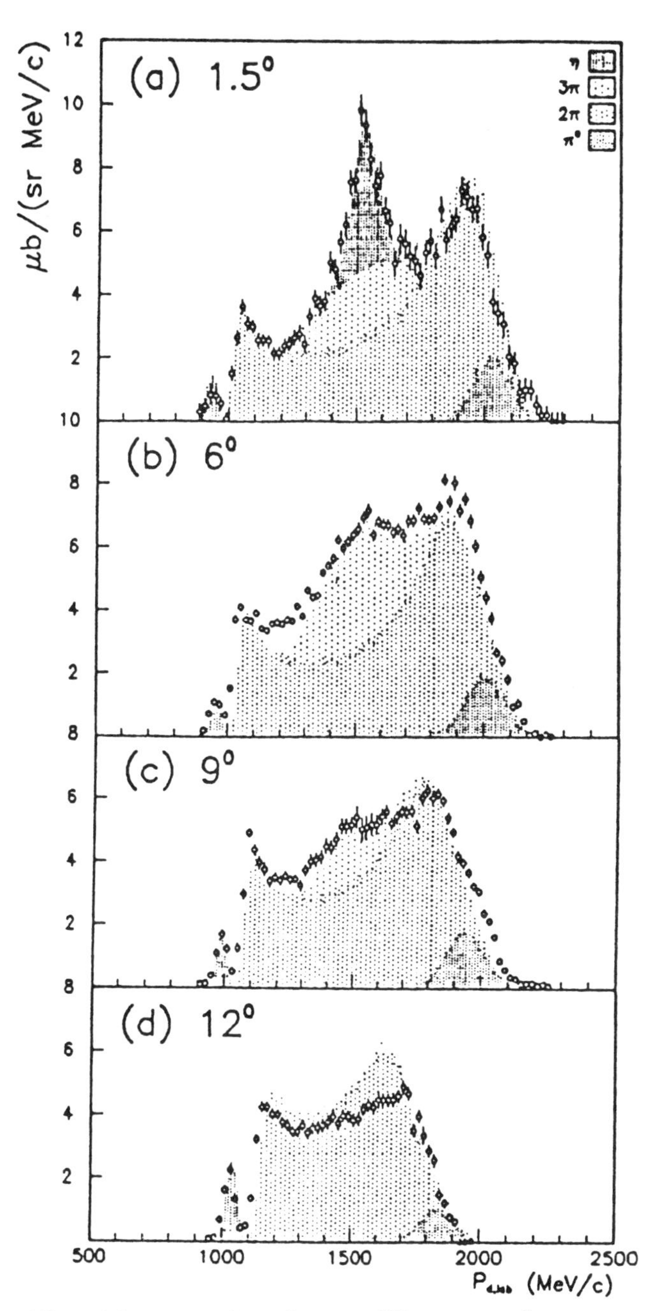

Figure 4. Differential cross-section of $np \rightarrow dX$ at a central neutron momentum of 1.88 GeV/c. The shaded areas correspond to the Monte Carlo simulations of the η, $\pi\pi$, and $\pi\pi\pi$.

not compatible with the observation. A measurement of the charge exchange reaction $\pi^- p \rightarrow \pi^0 n$ should help to clarify this situation.

ANTI-PROTON REACTIONS

Although it does not fit neatly into the tapestry that I have woven, it would be unforgivable to omit the beautiful results that have been obtained in experiment PS 185. This topic was reviewed in a scholarly manner by Oelert. In this LEAR experiment, they have studied a variety of reactions but the most elegant data come from $p\overline{p} \rightarrow \overline{\Lambda}\Lambda$. This reaction rises abruptly from threshold at 1440 MeV/c and quickly exhibits the need for p-waves as well as the expected s-waves.[30] A tantalizing dip has also been observed just above threshold and could be evidence for interference from a mesonic state. The data are still being analyzed but the interest is obvious.

These experiments are important because they can be used as information to design new experiments to study the properties of the Λ or $\overline{\Lambda}$. The reaction $p\overline{p} \rightarrow \overline{\Lambda}\Lambda$ could also be used as a source, so that one could study Λp or $\overline{\Lambda}p$ interactions, in fact the opportunities are limitless and Eisenstein laid out a smorgasbord of delicacies that the jet-set of LEAR physicists will enjoy in the next few years.

CONCLUSION

In conclusion I would like to commend everyone for an enjoyable and collegial atmosphere. The talks were also excellent, and gave us all an opportunity to learn about work quite distant from our normal daily toil. Special thanks go to Ed Stephenson and his colleagues at IUCF for making this conference possible and for ensuring that everything ran so smoothly.

REFERENCES

1. T. Yamazaki, R.S. Hayano, H. Toki, and P. Kienle, Nucl. Instr. Methods *A292*, 619 (1990).
2. B. Gabioud, J.-C. Alder, C. Joseph, J.-F. Loude, N. Morel, A. Perrenoud, J.-P. Perroud, M.T. Tran, E. Winkelmann, W. Dahme, H. Panke, D. Renker, G. Strassner, P. Truöl, and G.F. de Téramond, Nucl. Phys. *A420*, 496 (1984).
3. O. Schori, B. Gabioud, C. Joseph, J.-P. Perroud, D. Rüegger, M.T. Tran, P. Truöl, E. Winkelmann, and W. Dahme, Phys. Rev. *C35*, 2252 (1987).
4. K.P. Gall, E.C. Booth, W.J. Fickinger, M.D. Hasinoff, N.P. Hessey, D. Horváth, J. Lowe, E.K. McIntyre, D.F. Measday, J.P. Miller, A.J. Noble, B.L. Roberts, D.K. Robinson, M. Sakitt, M. Salomon, and D.A. Whitehouse, Phys. Rev. *C42*, R475 (1990).
5. R. Workman and H. Fearing, Phys. Rev. *C41*, 1688 (1990).
6. B. Povh, Prog. Part. Nucl. Phys. *18*, 183 (1987).
7. H. Piekarz, Nucl. Phys. *A479*, 263c (1988).

8. D.F. Measday, Phys. Lett. *21*, 66 (1966).
9. B. Murdock, M.Sc. Thesis, M.I.T. (1983), unpublished.
10. L. Rosselet, P. Extermann, J. Fischer, O. Guisan, R. Mermod, R. Sachot, A.M. Diamant-Berger, P. Bloch, G. Bunce, B. Devaux, N. Do-Duc, G. Marel, and R. Turlay, Phys. Rev. *D15*, 574 (1977).
11. J. Lowe et al., Exp 811 at BNL, private communication.
12. D.M. Manley, Phys. Rev. *D30*, 536 (1984).
13. E. Oset and M.J. Vicente-Vacas, Nucl. Phys. *A446*, 584 (1985).
14. J. Weinstein and N. Isgur, Phys. Rev. *D41*, 2236 (1990).
15. J.E. Augustin et al., Nucl. Phys. *B320*, 1 (1989).
16. H. Brody et al., Phys. Rev. Lett. *28*, 1215, 1217 (1972).
17. D.A. Hutcheon, E. Korkmaz, G.A. Moss, R. Abegg, N.E. Davison, G.W.R. Edwards, L.G. Greeniaus, D. Mack, C.A. Miller, W.C. Olsen, I.J. van Heerden, and Ye Yanlin, Phys. Rev. Lett. *64*, 176 (1990).
18. C.M. Rose, Phys. Rev. *154*, 1305 (1967).
19. J. Spuller and D.F. Measday, Phys. Rev. *D12*, 3550 (1975).
20. R. Beck, F. Kalleicher, B. Schoch, J. Vogt, G. Koch, H. Ströher, V. Metag, J.C. McGeorge, J.D. Kellie, and S.J. Hall, Phys. Rev. Lett. *65*, 1841 (1990).
21. E. Mazzucato, P. Argan, G. Audit, A. Bloch, N. de Botton, N. d'Hose, J-L. Faure, M.L. Ghedira, C. Guerra, J. Martin, C. Schuhl, G. Tamas, and E. Vincent, Phys. Rev. Lett. *57*, 3144 (1986).
22. A.N. Kamal, Phys. Rev. Lett. *63*, 2346 (1989).
23. L. Tiator and D. Drechsel, Nucl. Phys. *A508*, 541c (1990).
24. H-O. Meyer, M.A. Ross, R.E. Pollock, A. Berdoz, F. Dohrmann, J.E. Goodwin, M.G. Minty, H. Nann, P.V. Pancella, S.F. Pate, B. v. Przewoski, T. Rinckel, and F. Sperisen, Phys. Rev. Lett., to be published.
25. S. Stanislaus, D. Horváth, D.F. Measday, A.J. Noble, and M. Salomon, Phys. Rev. *C41*, R1913 (1990).
26. R.L. Workman and H.W. Fearing, Phys. Rev. *D37*, 3117 (1988).
27. J.W. Schnick and R.H. Landau, Phys. Rev. Lett. *58*, 1719 (1987) and Phys. Rev., to be published.
28. P.J. Fink, G. He, R.H. Landau, and J.W. Schnick, Phys. Rev. *41C*, 2720 (1990).
29. F. Plouin, P. Fleury, and C. Wilkin, Phys. Rev. Lett. *65*, 690 (1990).
30. P.D. Barnes et al., Nucl. Phys. *A508*, 311c (1990).

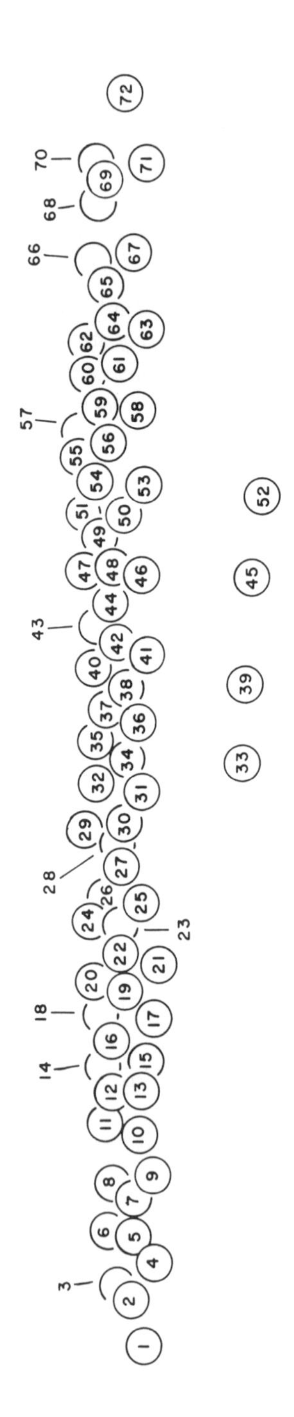

Particle Production Near Threshold

1. Danilo Zavrtanik
2. Boris Blankleider
3. Jeremy Brown
4. Edward Stephenson
5. Dieter Eversheim
6. Heiko Rohdjess
7. Herbert Funsten
8. Jens Bisplinghoff
9. Peter Herczeg
10. Elie Korkmaz
11. Theo Mayer-Kuckuk
12. Wilfried Daehnick
13. Robert Frascaria
14. David Hutcheon
15. Jen-Chieh Peng
16. Eberhard Kuhlmann
17. Paul Debevec
18. Ben Nefkens
19. Erwin Rössle
20. Frank Hinterberger
21. Ralph Segel
22. David Measday
23. Wolfgang Eyrich
24. Kurt Kilian
25. Hermann Nann
26. Reinhard Beck
27. Paul Hoyer
28. Matthias Kirsch
29. Rudolf Maier
30. Sven Kullander
31. Hiroshi Toki
32. John Cameron
33. Guy Emery
34. Wolfgang Kretschmer
35. Hartmut Machner
36. Eulogio Oset
37. Sergio Lemaître
38. Tim Ellison
39. Otto Schult
40. Shunzo Kumano
41. Lon-chang Liu
42. Josef Speth
43. Walter Toki
44. Dieter Drechsel
45. Nimai Mukhopadhyay
46. Benjamin Mayer
47. Dan Miller
48. Franz Sperisen
49. Paul Pancella
50. Michiko Minty
51. Chuck Bloch
52. Igor Strakovsky
53. Eric Jacobsen
54. Alan Ross
55. Walter Oelert
56. Doug Fields
57. Scott Wissink
58. Yun Wang
59. John Weinstein
60. Bob Pollock
61. Ted Bowyer
62. Curtis Whiddon
63. Irwan The
64. Andreas Engel
65. Will Brooks
66. Jerry Lisantti
67. Sonya Bowyer
68. Josef Homolka
69. Heinrich v. Geramb
70. John Szymanski
71. Chengwei Sun
72. Valdimir Bekrenev

PROGRAM FOR A TOPICAL CONFERENCE ON

Particle Production Near Threshold

Sunday, 30 September

OPENING

14:00 Welcome from Indiana University
GEORGE WALKER, Vice-President
Welcome from the Indiana University Cyclotron Facility
JOHN CAMERON, Director
Welcome from the Forschungszentrum Jülich
KURT KILIAN, Director

SESSION A: Rare Decays

PETER HERCZEG, session chairman

14:30 Rare and Forbidden Decays of Mesons
JOHN NG, *TRIUMF*

15:15 π^0-Decay
SVEN KULLANDER, *Svedberg Laboratory*

15:45 η-Decay
BENJAMIN MAYER, *CEN-Saclay*

SESSION B: Electroproduction and Photon-Induced Reactions

BEN NEFKENS, session chairman

16:45 J/ψ and Υ Production
WALTER TOKI, *SLAC*

17:30 Nuclear Bremsstrahlung near One- and Two-Pion Thresholds
EBERHARD KUHLMAN, *Ruhr-Universität Bochum*

17:45 Study of $\gamma p \rightarrow n\pi^+$ Very Close to Threshold
ELIE KORKMAZ, *University of Alberta*

18:00 η Photoproduction Near Threshold
NIMAI MUKHOPADHYAY, *Rensselaer Polytechnic Institute*

18:15 Reception (Woodland Room)

Monday, 1 October

SESSION B (cont.): Electroproduction and Photon-Induced Reactions

8:30 $\gamma p \rightarrow \pi^0 p$ Near Threshold
REINHARD BECK, *MIT*

9:00 Pion Photoproduction at Threshold
DIETER DRECHSEL, *University of Mainz*

SESSION C: Meson Production with Hadronic Beams
LON-CHENG LIU, session chairman

10:15 Precision Measurements of $NN \rightarrow \pi d$ Cross Section and Analyzing Power
DAVID HUTCHEON, *TRIUMF*

11:00 Near Threshold $NN \rightarrow NN\pi$
H.-O. MEYER, *IUCF*

11:45 Pion Production in *NN* Collisions
BORIS BLANKLEIDER, *Universität Basel*

14:00 Hadronic η-Production
WILL JACOBS, *IUCF*

14:45 Coherent Meson Production and Velocity Matching
KURT KILIAN, *Jülich*

15:15 Meson Physics at LNPI
VLADIMIR BEKRENEV, *Leningrad Nuclear Physics Institute*

16:00 From Continuum to Bound States
GUY EMERY, *Bowdoin College*

16:45 The $pp \rightarrow pn\pi^+$ Reaction Near Threshold
WILFRIED DAEHNICK, *University of Pittsburgh*

17:00 Measurements of Pion Production in π^+–p Interactions Near Threshold
DANILO ZAVRTANIK, *CERN*

18:00 Pig Roast

Tuesday, 2 October

SESSION D: Meson–Meson and Meson-Baryon Interactions
NIMAI MUKHOPADHYAY, session chairman

8:30 Production of Pionic Atoms
HIROSHI TOKI, *Tokyo Metropolitan University*

9:15 Recoil-less Production of π^- Atoms
HARTMUT MACHNER, *Jülich*

9:30 New States of the Strong Interaction
PAUL HOYER, *Stanford University and Helsinki University*

10:45 Low Energy Meson–Meson Interactions in the Quark Model
JOHN WEINSTEIN, *University of Tennessee and ORNL*

11:15 Threshold Effects in Meson–Meson and Meson–Baryon Interactions
JOSEPH SPETH, *Jülich*

11:45 The Width of the Bound η in Nuclei
EULOGIO OSET, *University of Valencia*

SESSION E: Future Experimental Developments
THEO MAYER-KUCKUK, session chairman

19:00 0° Facility at COSY
KORNELIUS SISTEMICH, *Jülich*

19:15 Chicane/K300 Spectrometer at the IUCF Cooler
DAN MILLER, *IUCF*

19:30 Near Threshold Two-Meson Production
RAINER JAHN, *Universität Bonn*

19:45 Excitation Function Measurements with Internal Targets
JENS BISPLINGHOFF, *Universität Bonn*

20:00 Recoil Ion Detection System for the IUCF Cooler
JEREMY BROWN, *Princeton University*

20:15 Parity Violation in *pp* Scattering
DIETER EVERSHEIM, *Universität Bonn*

21:00 Associated Strangeness Production in *pp* Reactions
PETER TUREK, *Jülich*

21:15 Associated Strangeness Production in Light Nuclei
JÜRGEN ERNST, *Universität Bonn*

21:30 Production on Heavy Hypernuclei
OTTO SCHULT, *Jülich*

Wednesday, 3 October

SESSION F: The Production of Strange Particles

PETER BARNES, session chairman

8:15 Hyperon–Antihyperon Production in the Antiproton–Proton Interaction
WALTER OELERT, *Jülich*

9:00 Associated Production of Kaons and Hyperons
ROBERT FRASCARIA, *Institut de Physique Nucléaire*

9:35 Hypernucleus Production
JOHN SZYMANSKI, *LANL*

10:50 Electromagnetic Production of Strange Particles
REINHARD SCHUMACHER, *Carnegie Mellon University*

11:20 Polarization Transfer in Associated Hyperon Production
RAIMONDO BERTINI, *LNS and CEN Saclay*

11:50 The Jetset Experiment
ROBERT EISENSTEIN, *University of Illinois*

SESSION G: Future Machine Developments

S.-Y. LEE, session chairman

14:00 Development of Polarized Beams at COSY
RUDOLF MAIER, *Jülich*

14:35 Limits on Electron Cooling
DAG REISTAD, *Svedberg Laboratory*

CLOSING

15:45 Conference Summary
DAVID MEASDAY, *University of British Columbia*

16:30 Closing Remarks

ROSTER OF PARTICIPANTS

PARTICLE PRODUCTION NEAR THRESHOLD
The Brown County Inn September 30 - October 3, 1990

Dr. Andrew D. Bacher
Indiana University
IUCF
2401 Milo B. Sampson Ln
Bloomington, IN 47405
812-855-2675
Fax: 812-855-6645
BACHER@IUCF

Dr. Peter D. Barnes
Carnegie Mellon University
Department of Physics
Medium Energy Physics
Pittsburgh, PA 15213
412-268-2745
Fax: 412-681-0648
BARNES@CMPHYSME

Dr. Reinhard Beck
M.I.T.
Department of Physics, MIT 26-456
77 Massachusetts Avenue
Cambridge, MA
617-258-5450
Fax: 617-258-6923
BECK@MITLNS

Dr. Vladimir S. Bekrenev
Leningrad Nuclear Physics Institute
Nuclear Physics Institute
Gatchina
Leningrad Dist. 188350
USSR
Fax: 812-71-37196

Dr. Robert D. Bent
Indiana University
IUCF
2401 Milo B. Sampson Lane
Bloomington, IN 47405
812-855-9365
Fax: 812-855-6645
BENT@IUCF

Dr. Raimondo Bertini
LNS/CEN Saclay
11191 Gif-Sur-Yvette
France
33 (1) 69 08 56 45
Fax: 33 (1) 69 08 29 70
BERTINI@FRCPN11

Dr. Jens Bisplinghoff
University of Bonn
Institüt für Strahlen-
und Kernphysik
Nussallee 14-16
D-5300 Bonn West Germany
49-228-732543
Fax: 49-228-733728
JENS@DBNISKP5

Dr. Boris Blankleider
Universitüt Basel
Physicalisches Institüt
CH-4056 Basel
Switzerland

Dr. Chuck Bloch
Indiana University
IUCF
2401 Milo B. Sampson Ln
Bloomington, IN 47405
812-855-9365
Fax: 812-855-6645
CBLOCH@IUCF

Ms. Sonya Bowyer
Indiana University
IUCF
2401 Milo B. Sampson Ln
Bloomington, IN 47405
812-855-9365
Fax: 812-855-6645
SMFRAY@IUCF

Mr. Ted W. Bowyer
Indiana University
IUCF
2401 Milo B. Sampson Ln
Bloomington, IN 47405
812-855-9365
Fax: 812-855-6645
BOWYER@IUCF

Dr. William K. Brooks
University of Pittsburgh
Physics Department
3941 O'Hara St.
100 Allen Hall
Pittsburgh, PA 15260
412-624-1566
Fax: 412-624-9163
WBROOKS@PITTVMS

Dr. Jeremy Brown
Princeton University
Physics Department
Princeton, NJ 08854

Dr. John Cameron
Indiana University
IUCF
2401 Milo B. Sampson Ln
Bloomington, IN 47405
812-855-9365
Fax: 812-855-6645
CAMERON@IUCF

Dr. Wilfried W. Daehnick
University of Pittsburgh
Physics Department
Pittsburgh, PA 15260
412-624-9236
DAEHNICK@PITTYMS

Dr. Paul T. Debevec
University of Illinois
Nuclear Physics Laboratory
23 Stadium Drive
Champaign, IL 61820
217-333-0285
Fax: 217-333-1215
DEBEVEC@UIUCNPL

Dr. Dieter Drechsel
University of Mainz
Institüt für Kernphysik
Soarstr.
D-6500 Mainz
W. Germany
TUBA@DMZNAT51.BITNET

Dr. Robert Eisenstein
University of Illinois
Department of Physics
1110 W. Green Street
Urbana, IL 61801

Dr. Tim Ellison
Indiana University
IUCF
2401 Milo B. Sampson Lane
Bloomington, IN 47405
812-855-9365
Fax: 812-855-6645
TIME@IUCF

Dr. Guy T. Emery
Bowdoin College
Physics Department
Brunswick, ME 04011
207-725-3708
Fax: 207-725-3405
GEMERY@BOWDOIN

Mr. Andreas Engel
University of Giessen
Institüt für Theoretische Physik
Heinrich-Buff-Ring 16
6300 Giessen West Germany
0641 70222828
UGO9@DDAGSI3

Prof. Jürgen Ernst
der Universität Bonn
Institüt fuer Strahlen-
und Kernphysik
Nussallee 14-16
D-5300 Bonn 1 West Germany
49-228 732364
Fax: 49-228 733728
ERNST@DBNISKP5

Dr. Dieter Eversheim
der Universität Bonn
Institüt für Strahlen-
und Kernphysik
Nussallee 14-16
D-5300 Bonn 1 W. Germany
49-288-735299
Fax: 49-288-733728
EVERSHEI@DBNISKP5

Prof. Wolfgang Eyrich
Universitat Nurnberg
Physikalisches Institüt
Erwin-Rommel-Str. 1
D-8520 Erlangen West Germany
9131/85 7078
Fax: 9131/15249
MPPI29@DERRZE1

Dr. Charles C. Foster
Indiana University
IUCF
2401 Milo B. Sampson Lane
Bloomington, IN 47405
812-855-9365
Fax: 812-855-6645
FOSTER@IUCF

Prof. Robert Frascaria
Institut de Physique Nucleaire
Orsay/Essonne/91406 France
69416302
Fax: 69416470
FRASCARI@FRIPN51

Dr. Herbert Funsten
College of William and Mary
Department of Physics
Williamsburg, VA 23185
804-221-3515

Dr. Charles D. Goodman
Indiana University
IUCF
2401 Milo B. Sampson Lane
Bloomington, IN 47405
812-855-9365
Fax: 812-855-6645
GOODMAN@IUCF

Dr. Peter Herczeg
Los Alamos National Laboratory
Group T-5
Los Alamos, NM 87545

Prof. Frank Hinterberger
der Universität Bonn
Institüt für Strahlen-
und Kernphysik
Nussallee 14-16
D-5300 Bonn 1 W. Germany
49-228-73-2251
Fax: 49-228-73-3728
FH@DBNISKP5

Dr. Josef Homolka
Technische Universität München
Physik Department E12
James Franck Str.
D-8046 Garching West Germany
0891 32092439
Fax: 0891 3206780
HOMOLKA@D6ABLGSP

Dr. Paul Hoyer
Stanford University/
Helsinki University
SLAC
SLAC Bin 81, P.O. Box 4349
Stanford, CA 93409
415-926-2650
HOYER@SLACVM

Dr. David A. Hutcheon
TRIUMF
4004 Wesbrook Mall
Vancouver, BC V6T2A3
CANADA
604-222-1047
Fax: 603-222-1074
SMURF@TRIUMFCL

Dr. Will W. Jacobs
Indiana University
IUCF
2401 Milo B. Sampson Lane
Bloomington, IN 47405
812-855-9365
Fax: 812-855-6645
JACOBS@IUCF

Mr. Erik Jacobsen
Princeton University
Physics Department
Princeton, NJ 08544
GR.ERJ@PUPCYC.PRINCETO
N.EDU

Prof. Rainer Jahn
der Universität Bonn
Institüt fuer Strahlen-
und Kernphysik
Nussallee 14-16
D-5300 Bonn 1 W. Germany
49-228-732364
JAHN@DBNISKP5

Prof. Kurt Kilian
Institüt für Kernphysik
Forschungszentrum Jülich GmbH
Postfach 1913
D-5170 Jülich 1
W. Germany
02461-615943
Fax: 02461-613930

Mr. Matthias Kirsch
Universität Erlangen-Nurnberg
Physicalisches Institüt
Erwin-Rommel-Str. 1
D-8520 Erlangen
W. Germany
9131/85 7063
Fax: 9131/15249
MPPI29@DERRZE1

Dr. Elie Korkmaz
University of Alberta
Nuclear Research Centre
Edmonton, Alberta T6G2N5
CANADA
403-492-3637
Fax: 403-492-3408
KORKMAZ@TRIUMFCL

Prof. Wolfgang Kretschmer
der Universitat Erlangen-Nurnberg
Physikalisches Institüt
Erwin-Rommel-Str. 1
D-8520 Erlangen
West Germany
9131/85 7075
Fax: 9131/15249
MPPI67@Cnve.Rrze.Uni-Erlange
n.Dbp.De

Dr. Eberhard Kuhlmann
Ruhr-Universität Bochum
Institüt für Kernphysik I
D-4630 Bochum
West Germany
02461 61 2381
Fax: 02461 61 3930
KPH049@DJUKFA11

Dr. Sven Kullander
The Svedberg Laboratory
Box 533
S-75121 Uppsala 1
Sweden

Dr. Shunzo Kumano
Indiana University
IUNTC
2401 Milo B. Sampson Ln
Bloomington, IN 47405
812-855-9365
Fax: 812-855-6645
KUMANO@IUCF

Dr. T.-S.Harry Lee
Argonne National Lab
Physics Division
Argonne, IL 60439
708-972-4094
Fax: 708-972-3903
LEE@ANLPHY

Dr. Sergio Lemaitre
Institut für Kernphysik
Universität zu Köln
Zülpicherstr 77
Köln D5000 GERMANY
0221/4703459
(Internet)
Lemaitre@IKP.UNI_KOELN.DE

Dr. Jerry Lisantti
Indiana University
IUCF
2401 Milo B. Sampson Lane
Bloomington, IN 47405
812-855-9365
Fax: 812-855-6645
LISANTTI@IUCF

Dr. Lon-chang Liu
Los Alamos National Laboratory
Group T-2
Mail Stop B243
Los Alamos, NM 87545
505-667-7669
Fax: 505-667-9671
LIU@LAMPF.BITNET

Dr. Tim Londergan
Indiana University
IUNTC
2401 Milo B. Sampson Lane
Bloomington, IN 47405
812-855-9365
Fax: 812-855-6645
LONDERGAN@IUCF

Dr. Malcolm Macfarlane
Indiana University
IUNTC
2401 Milo B. Sampson Ln
Bloomington, IN 47405
812-855-9365
Fax: 812-855-6645
MACFARLANE@IUCF

Dr. Hartmut Machner
KFA Jülich
Institüt f. Kernphysik
Postfach 1913
D-5170 Jülich
West Germany

Dr. Rudolf Maier
Kernforschungsanlage Jülich
Institüt f. Kernphysik
Postfach 1913
D-5170 Jülich West Germany

Dr. Benjamin Mayer
SEPN, CEN-Saclay
91191 Gif-sur-yvette Cedex
FRANCE
1 69 08 32 36
Fax: 1 69 08 86 43
MAYER@FRSAC11

Prof. Theo Mayer-Kuckuk
der Universität Bonn
Institüt für Strahlen-
und Kernphysik
Nusallee 14-16 D-5300 Bonn 1
West Germany
49 228-732201
Fax: 49 228-733728
EMKA@DBNISKP5

Dr. David F. Measday
University of British Columbia
Physics Department
2075 Wesbrook Mall
Vancouver, BC V6T 2A3 Canada

Dr. Hans-Otto Meyer
Indiana University
IUCF
2401 Milo B. Sampson Ln
Bloomington, IN 47405
812-855-9365
Fax: 812-855-6645
MEYER@IUCF

Dr. Dan W. Miller
Indiana University
IUCF
2401 Milo B. Sampson Lane
Bloomington, IN 47405
812-855-5465
Fax: 812-855-6645
MILLER@IUCF

Ms. Michiko G. Minty
Indiana University
IUCF
2401 Milo B. Sampson Ln
Bloomington, IN 47405
812-855-9365
Fax: 812-855-6645
MICHIKO@IUCF

Dr. Nimai C. Mukhopadhyay
Rensselaer Polytechnic Institute
Department of Physics
Troy, NY 12180

Dr. Hermann Nann
Indiana University
IUCF
2401 Milo B. Sampson Lane
Bloomington, IN 47405
812-855-9365
Fax: 812-855-6645
NANN@IUCF

Dr. Ben Nefkens
UCLA
Physics Department
Los Angeles, CA 90024
213-825-4970
Fax: 213-206-4397
NEFKENS@UCLAPH

Dr. John N. Ng
TRIUMF
4004 Wesbrook Mall
Vancouver, B. C. V6T2A3 CANADA
604-222-1047
Fax: 604-222-1074

Dr. Benwen Ni
Indiana University
IUCF
2401 Milo B. Sampson Ln
Bloomington, IN 47405
812-855-8918
Fax: 812-855-6645
BENWEN@IUCF

Dr. Walter Oelert
KFA Jülich
Institüt f. Kernphysik
Postfach 1913
D-5170 Jülich West Germany
492461-
Fax: 492461-61 3930
KPH037@DJUKFA11

Ms. Allena K. Opper
Indiana University
IUCF
2401 Milo B. Sampson Lane
Bloomington, IN 47405
812-855-8918
Fax: 812-855-6645
OPPER@IUCF

Dr. Eulogio Oset
University of Valencia
Departamento De Fisica Teorica
Facultad De Fisica
Burjassot, Valencia SPAIN 46100
34 (6) 3864551
Fax: 34 (6) 3642345
OSET@EVALVX.DECNET.CE
RN.CH

Dr. Paul Pancella
Western Michigan University
Physics Department
Kalamazoo, MI 49008
616-387-4941
PANCELLA@GW.WMICH.EDU

Dr. Stephen Pate
Indiana University
IUCF
2401 Milo B. Sampson Lane
Bloomington, IN 47405
812-855-9365
Fax: 812-855-6645
PATE@IUCF

Dr. Jen-Chieh Peng
Fermi Lab, EN89
Batavia, IL 60510
708-840-4605
Fax: 708-840-4343
PENG@FNAL

Dr. Robert E. Pollock
Indiana University
IUCF
2401 Milo B. Sampson Ln
Bloomington, IN 47405
812-855-9365
Fax: 812-855-6645
POLLOCK@IUCF

Dr. Dag Reistad
The Svedberg Lab
Box 533
S751 21 Uppsala
Sweden
46 18 18 31 77
Fax: 46 18 10 85 42
REISTAD@TSL.UU.SE

Mr. Heiko Rohdjess
Bereich Teilchenphysik
Universitaet Hamburg
Luruper Chaussee 149
2000 Hamburg 50
West Germany
49-40-8998-2168
Fax: 49-40-8998-2101
I04ROH@DHHDESY3

Mr. Alan Ross
Indiana University
IUCF
2401 Milo B. Sampson Lane
Bloomington, IN 47405
812-855-9365
Fax: 812-855-6645
ROSSO@IUCF

Prof. Erwin Rössle
der Universität Freiburg
Fakultat für Physik
Hermann-Herder-Str. 3
D 7800 Freiburg
West Germany
00497612033713
Fax: 00497612034527

Prof. Otto Schult
IKP-KFA Jülich
Postfach 1913
D-5170 Jülich West Germany
2461-614408
Fax: 2461-613930
KPH200@DJUKFA

Prof. Reinhard Schumacher
Carnegie Mellon University
Department of Physics
Pittsburgh, PA 15213
412-268-5177
Fax: 412-681-0648
REINHARD@CMPHYSME

Dr. Peter Schwandt
Indiana University
IUCF
2401 Milo B. Sampson Lane
Bloomington, IN 47405
812-855-9365
Fax: 812-855-6645
SCHWANDT@IUCF

Dr. Ralph Segel
Northwestern University
Physics Department
Evanston, IL 60204
708-491-5459
RALPH@ANLPHY

Prof. Kornelius Sistemich
Institüt für Kernphysik
Forschungszentrum Jülich GmbH
Postfach 1913
D-5170 Jülich 1 West Germany
492461613100
Fax: 492461613930
KPH208@DJUKFA11

Mr. Adam Smith
Indiana University
IUCF
2401 Milo B. Sampson Lane
Bloomington, IN 47405
812-855-9365
Fax: 812-855-6645
ASMITH@IUCF

Dr. James Sowinski
Indiana University
IUCF
2401 Milo B. Sampson Ln
Bloomington, IN 47405
812-855-9365
Fax: 812-855-6645
SOWINSKI@IUCF

Dr. Franz Sperisen
Indiana University
IUCF
2401 Milo B. Sampson Ln
Bloomington, IN 47405
812-855-2948
Fax: 812-855-6645
SPERISEN@IUCF

Prof. Joseph Speth
Institüt für Kernphysik
Forschungszentrum Jülich GmbH
Postfach 1913
D-5170 Jülich 1
West Germany
0 2461-61 4401
Fax: 0 2461-61 3930
KPH119@DJUKFA11

Mr. Mark Spraker
Indiana University
IUCF
2401 Milo B. Sampson Ln
Bloomington, IN 47405
812-855-8215
Fax: 812-855-6645
SPRAKER@IUCF

Dr. Edward J. Stephenson
Indiana University
IUCF
2401 Milo B. Sampson Lane
Bloomington, IN 47405
812-855-9365
Fax: 812-855-6645
STEPHENSON@IUCF

Dr. Igor Strakovsky
Leningrad Nuclear Physics Institute
Nuclear Physics Institute
Gatchina
Leningrad Dist. 188350
USSR
604-222-1047
IGOR@TRIUMFCL

Dr. Evan Sugarbaker
Ohio State University
Physics Department
1302 Kinnear Rd.
Columbus, OH 43212
614-292-4775
SUGARBAK@OHSTPY

Dr. Chengwei W. Sun
Indiana University
IUCF
2401 Milo B. Sampson Ln
Bloomington, IN 47405
812-855-9365
Fax: 812-855-6645
CHENGWEI@IUCF

Prof. John Szymanski
Indiana University
IUCF
2401 Milo B. Sampson Ln
Bloomington, IN 47405
812-855-9365
Fax: 812-855-6645
SZYMANSKI@IUCF

Mr. Jeff Templon
Indiana University
IUCF
2401 Milo B. Sampson Ln
Bloomington, IN 47405
812-855-0932
Fax: 812-855-6645
TEMPLON@IUCF

Dr. Hiroshi Toki
Tokyo Metropolitan University
Dept. of Physics
Setagaya Tokyo 158
Japan

Dr. Walter Toki
Stanford Linear Accelerator
SLAC Bin 65
P.O. Box 4349
Stanford, CA 94309
Fax: 415-926-2923
TOKI@SLACVM

Dr. Peter Turek
KFA Jülich
IKP
D-5170 Jülich
West Germany

Dr. Tamotsu Ueda
Osaka University
Faculty of Engineering Science
Machikaneyama 1-1
Toyonaka, Osaka 560
Japan
06-844-1151-ext.4862
Fax: 06-857-0253

Dr. Barbara v. Przewoski
Indiana University
IUCF
2401 Milo B. Sampson Lane
Bloomington, IN 47405
812-855-9365
Fax: 812-855-6645
PRZEWOSKI@IUCF

Dr. Vic E. Viola
Indiana University
IUCF
2401 Milo B. Sampson Lane
Bloomington, IN 47405
812-855-2878
Fax: 812-855-6645
VICV@IUCF

Dr. Heinrich V. von Geramb
University of Hamburg
Theoretical Nuclear Physics
Luruper Chaussee 149
D-200 Hamburg 50,
West Germany
40-8998-2131
Fax: 40-8998-2196
DHHDESY3@I04GER

Dr. George Walker
Assoc. Vice President for Research
and Dean of The Graduate School
Bryan Hall
Bloomington, IN 47405
812-855-6153
GEWALKER@IUCF

Dr. Yun Wang
Indiana University
IUCF
2401 Milo B. Sampson Lane
Bloomington, IN 47405
812-855-9365
Fax: 812-855-6645
YWANG@IUCF

Dr. John Weinstein
University of Tennessee
and ORNL
Department of Physics
203 South College
Knoxville, TN 37996
615-974-0771
Fax: 615-974-2667
JOHNW@UTKVX.BITNET

Mr. Curtis Whiddon
Indiana University
IUCF
2401 Milo B. Sampson Ln
Bloomington, IN 47405
812-855-8214
Fax: 812-855-6645
WHIDDON@IUCF

Dr. Scott W. Wissink
Indiana University
IUCF
2401 Milo B. Sampson Ln
Bloomington, IN 47405
812-855-9365
Fax: 812-855-6645
WISSINK@IUCF

Ms. Sherry Yennello
Indiana University
IUCF
2401 Milo B. Sampson Ln
Bloomington, IN 47405
812-855-9365
Fax: 812-855-6645
YENNELLO@IUCF

Dr. Zhou Yu
Northwestern University
Physics Department
Evanston, IL 60204

Dr. Danilo Zavrtanik
CERN
CERN-PPE
CH-1211 Geneva 23
Switzerland
22-767-5953
Fax: 22-783-0564
DANILO@CERNVM

AUTHOR INDEX

AIP Conference Proceedings

		L.C. Number	ISBN
No. 68	High Energy Physics – 1980 (XX International Conference, Madison, WI)	81-65032	0-88318-167-3
No. 69	Polarization Phenomena in Nuclear Physics – 1980 (Fifth International Symposium, Santa Fe, NM)	81-65107	0-88318-168-1
No. 70	Chemistry and Physics of Coal Utilization – 1980 (APS, Morgantown)	81-65106	0-88318-169-X
No. 71	Group Theory and its Applications in Physics – 1980 (Latin American School of Physics, Mexico City)	81-66132	0-88318-170-3
No. 72	Weak Interactions as a Probe of Unification (Virginia Polytechnic Institute – 1980)	81-67184	0-88318-171-1
No. 73	Tetrahedrally Bonded Amorphous Semiconductors (Carefree, AZ, 1981)	81-67419	0-88318-172-X
No. 74	Perturbative Quantum Chromodynamics (Tallahassee, FL, 1981)	81-70372	0-88318-173-8
No. 75	Low Energy X-Ray Diagnostics – 1981 (Monterey, CA)	81-69841	0-88318-174-6
No. 76	Nonlinear Properties of Internal Waves (La Jolla Institute, 1981)	81-71062	0-88318-175-4
No. 77	Gamma Ray Transients and Related Astrophysical Phenomena (La Jolla Institute, 1981)	81-71543	0-88318-176-2
No. 78	Shock Waves in Condensed Matter – 1981 (Menlo Park, NJ)	82-70014	0-88318-177-0
No. 79	Pion Production and Absorption in Nuclei – 1981 (Indiana University Cyclotron Facility)	82-70678	0-88318-178-9
No. 80	Polarized Proton Ion Sources (Ann Arbor, MI, 1981)	82-71025	0-88318-179-7
No. 81	Particles and Fields –1981: Testing the Standard Model (APS/DPF, Santa Cruz, CA)	82-71156	0-88318-180-0
No. 82	Interpretation of Climate and Photochemical Models, Ozone and Temperature Measurements (La Jolla Institute, 1981)	82-71345	0-88318-181-9
No. 83	The Galactic Center (Cal. Inst. of Tech., 1982)	82-71635	0-88318-182-7
No. 84	Physics in the Steel Industry (APS/AISI, Lehigh University, 1981)	82-72033	0-88318-183-5
No. 85	Proton-Antiproton Collider Physics –1981 (Madison, WI)	82-72141	0-88318-184-3
No. 86	Momentum Wave Functions – 1982 (Adelaide, Australia)	82-72375	0-88318-185-1
No. 87	Physics of High Energy Particle Accelerators (Fermilab Summer School, 1981)	82-72421	0-88318-186-X
No. 88	Mathematical Methods in Hydrodynamics and Integrability in Dynamical Systems (La Jolla Institute, 1981)	82-72462	0-88318-187-8

No. 89	Neutron Scattering – 1981 (Argonne National Laboratory)	82-73094	0-88318-188-6
No. 90	Laser Techniques for Extreme Ultraviolet Spectroscopy (Boulder, CO, 1982)	82-73205	0-88318-189-4
No. 91	Laser Acceleration of Particles (Los Alamos, NM, 1982)	82-73361	0-88318-190-8
No. 92	The State of Particle Accelerators and High Energy Physics (Fermilab, 1981)	82-73861	0-88318-191-6
No. 93	Novel Results in Particle Physics (Vanderbilt, 1982)	82-73954	0-88318-192-4
No. 94	X-Ray and Atomic Inner-Shell Physics – 1982 (International Conference, U. of Oregon)	82-74075	0-88318-193-2
No. 95	High Energy Spin Physics – 1982 (Brookhaven National Laboratory)	83-70154	0-88318-194-0
No. 96	Science Underground (Los Alamos, NM, 1982)	83-70377	0-88318-195-9
No. 97	The Interaction Between Medium Energy Nucleons in Nuclei – 1982 (Indiana University)	83-70649	0-88318-196-7
No. 98	Particles and Fields – 1982 (APS/DPF University of Maryland)	83-70807	0-88318-197-5
No. 99	Neutrino Mass and Gauge Structure of Weak Interactions (Telemark, 1982)	83-71072	0-88318-198-3
No. 100	Excimer Lasers – 1983 (OSA, Lake Tahoe, NV)	83-71437	0-88318-199-1
No. 101	Positron-Electron Pairs in Astrophysics (Goddard Space Flight Center, 1983)	83-71926	0-88318-200-9
No. 102	Intense Medium Energy Sources of Strangeness (UC-Santa Cruz, CA, 1983)	83-72261	0-88318-201-7
No. 103	Quantum Fluids and Solids – 1983 (Sanibel Island, FL)	83-72440	0-88318-202-5
No. 104	Physics, Technology and the Nuclear Arms Race (APS, Baltimore, MD, 1983)	83-72533	0-88318-203-3
No. 105	Physics of High Energy Particle Accelerators (SLAC Summer School, 1982)	83-72986	0-88318-304-8
No. 106	Predictability of Fluid Motions (La Jolla Institute, 1983)	83-73641	0-88318-305-6
No. 107	Physics and Chemistry of Porous Media (Schlumberger-Doll Research, 1983)	83-73640	0-88318-306-4
No. 108	The Time Projection Chamber (TRIUMF, Vancouver, 1983)	83-83445	0-88318-307-2
No. 109	Random Walks and Their Applications in the Physical and Biological Sciences (NBS/La Jolla Institute, 1982)	84-70208	0-88318-308-0
No. 110	Hadron Substructure in Nuclear Physics (Indiana University, 1983)	84-70165	0-88318-309-9

No. 111	Production and Neutralization of Negative Ions and Beams (3rd Int'l Symposium) (Brookhaven, NY, 1983)	84-70379	0-88318-310-2
No. 112	Particles and Fields – 1983 (APS/DPF, Blacksburg, VA)	84-70378	0-88318-311-0
No. 113	Experimental Meson Spectroscopy – 1983 (7th International Conference, Brookhaven, NY)	84-70910	0-88318-312-9
No. 114	Low Energy Tests of Conservation Laws in Particle Physics (Blacksburg, VA, 1983)	84-71157	0-88318-313-7
No. 115	High Energy Transients in Astrophysics (Santa Cruz, CA, 1983)	84-71205	0-88318-314-5
No. 116	Problems in Unification and Supergravity (La Jolla Institute, 1983)	84-71246	0-88318-315-3
No. 117	Polarized Proton Ion Sources (TRIUMF, Vancouver, 1983)	84-71235	0-88318-316-1
No. 118	Free Electron Generation of Extreme Ultraviolet Coherent Radiation (Brookhaven/OSA, 1983)	84-71539	0-88318-317-X
No. 119	Laser Techniques in the Extreme Ultraviolet (OSA, Boulder, CO, 1984)	84-72128	0-88318-318-8
No. 120	Optical Effects in Amorphous Semiconductors (Snowbird, UT, 1984)	84-72419	0-88318-319-6
No. 121	High Energy e^+e^- Interactions (Vanderbilt, 1984)	84-72632	0-88318-320-X
No. 122	The Physics of VLSI (Xerox, Palo Alto, CA, 1984)	84-72729	0-88318-321-8
No. 123	Intersections Between Particle and Nuclear Physics (Steamboat Springs, CO, 1984)	84-72790	0-88318-322-6
No. 124	Neutron-Nucleus Collisions: A Probe of Nuclear Structure (Burr Oak State Park, 1984)	84-73216	0-88318-323-4
No. 125	Capture Gamma-Ray Spectroscopy and Related Topics – 1984 (Int'l Symposium, Knoxville, TN)	84-73303	0-88318-324-2
No. 126	Solar Neutrinos and Neutrino Astronomy (Homestake, 1984)	84-63143	0-88318-325-0
No. 127	Physics of High Energy Particle Accelerators (BNL/SUNY Summer School, 1983)	85-70057	0-88318-326-9
No. 128	Nuclear Physics with Stored, Cooled Beams (McCormick's Creek State Park, IN, 1984)	85-71167	0-88318-327-7
No. 129	Radiofrequency Plasma Heating (Sixth Topical Conference) (Callaway Gardens, GA, 1985)	85-48027	0-88318-328-5
No. 130	Laser Acceleration of Particles (Malibu, CA, 1985)	85-48028	0-88318-329-3
No. 131	Workshop on Polarized ^{3}He Beams and Targets (Princeton, NJ, 1984)	85-48026	0-88318-330-7
No. 132	Hadron Spectroscopy–1985 (International Conference, Univ. of Maryland)	85-72537	0-88318-331-5
No. 133	Hadronic Probes and Nuclear Interactions (Arizona State University, 1985)	85-72638	0-88318-332-3

No. 134	The State of High Energy Physics (BNL/SUNY Summer School, 1983)	85-73170	0-88318-333-1
No. 135	Energy Sources: Conservation and Renewables (APS, Washington, DC, 1985)	85-73019	0-88318-334-X
No. 136	Atomic Theory Workshop on Relativistic and QED Effects in Heavy Atoms (Gaithersburg, MD, 1985)	85-73790	0-88318-335-8
No. 137	Polymer-Flow Interaction (La Jolla Institute, 1985)	85-73915	0-88318-336-6
No. 138	Frontiers in Electronic Materials and Processing (Houston, TX, 1985)	86-70108	0-88318-337-4
No. 139	High-Current, High-Brightness, and High-Duty Factor Ion Injectors (La Jolla Institute, 1985)	86-70245	0-88318-338-2
No. 140	Boron-Rich Solids (Albuquerque, NM, 1985)	86-70246	0-88318-339-0
No. 141	Gamma-Ray Bursts (Stanford, CA, 1984)	86-70761	0-88318-340-4
No. 142	Nuclear Structure at High Spin, Excitation, and Momentum Transfer (Indiana University, 1985)	86-70837	0-88318-341-2
No. 143	Mexican School of Particles and Fields (Oaxtepec, México, 1984)	86-81187	0-88318-342-0
No. 144	Magnetospheric Phenomena in Astrophysics (Los Alamos, NM, 1984)	86-71149	0-88318-343-9
No. 145	Polarized Beams at SSC & Polarized Antiprotons (Ann Arbor, MI & Bodega Bay, CA, 1985)	86-71343	0-88318-344-7
No. 146	Advances in Laser Science–I (Dallas, TX, 1985)	86-71536	0-88318-345-5
No. 147	Short Wavelength Coherent Radiation: Generation and Applications (Monterey, CA, 1986)	86-71674	0-88318-346-3
No. 148	Space Colonization: Technology and The Liberal Arts (Geneva, NY, 1985)	86-71675	0-88318-347-1
No. 149	Physics and Chemistry of Protective Coatings (Universal City, CA, 1985)	86-72019	0-88318-348-X
No. 150	Intersections Between Particle and Nuclear Physics (Lake Louise, Canada, 1986)	86-72018	0-88318-349-8
No. 151	Neural Networks for Computing (Snowbird, UT, 1986)	86-72481	0-88318-351-X
No. 152	Heavy Ion Inertial Fusion (Washington, DC, 1986)	86-73185	0-88318-352-8
No. 153	Physics of Particle Accelerators (SLAC Summer School, 1985) (Fermilab Summer School, 1984)	87-70103	0-88318-353-6
No. 154	Physics and Chemistry of Porous Media—II (Ridge Field, CT, 1986)	83-73640	0-88318-354-4
No. 155	The Galactic Center: Proceedings of the Symposium Honoring C. H. Townes (Berkeley, CA, 1986)	86-73186	0-88318-355-2

No. 156	Advanced Accelerator Concepts (Madison, WI, 1986)	87-70635	0-88318-358-0
No. 157	Stability of Amorphous Silicon Alloy Materials and Devices (Palo Alto, CA, 1987)	87-70990	0-88318-359-9
No. 158	Production and Neutralization of Negative Ions and Beams (Brookhaven, NY, 1986)	87-71695	0-88318-358-7
No. 159	Applications of Radio-Frequency Power to Plasma: Seventh Topical Conference (Kissimmee, FL, 1987)	87-71812	0-88318-359-5
No. 160	Advances in Laser Science–II (Seattle, WA, 1986)	87-71962	0-88318-360-9
No. 161	Electron Scattering in Nuclear and Particle Science: In Commemoration of the 35th Anniversary of the Lyman-Hanson-Scott Experiment (Urbana, IL, 1986)	87-72403	0-88318-361-7
No. 162	Few-Body Systems and Multiparticle Dynamics (Crystal City, VA, 1987)	87-72594	0-88318-362-5
No. 163	Pion–Nucleus Physics: Future Directions and New Facilities at LAMPF (Los Alamos, NM, 1987)	87-72961	0-88318-363-3
No. 164	Nuclei Far from Stability: Fifth International Conference (Rosseau Lake, ON, 1987)	87-73214	0-88318-364-1
No. 165	Thin Film Processing and Characterization of High-Temperature Superconductors (Anaheim, CA, 1987)	87-73420	0-88318-365-X
No. 166	Photovoltaic Safety (Denver, CO, 1988)	88-42854	0-88318-366-8
No. 167	Deposition and Growth: Limits for Microelectronics (Anaheim, CA, 1987)	88-71432	0-88318-367-6
No. 168	Atomic Processes in Plasmas (Santa Fe, NM, 1987)	88-71273	0-88318-368-4
No. 169	Modern Physics in America: A Michelson-Morley Centennial Symposium (Cleveland, OH, 1987)	88-71348	0-88318-369-2
No. 170	Nuclear Spectroscopy of Astrophysical Sources (Washington, DC, 1987)	88-71625	0-88318-370-6
No. 171	Vacuum Design of Advanced and Compact Synchrotron Light Sources (Upton, NY, 1988)	88-71824	0-88318-371-4
No. 172	Advances in Laser Science–III: Proceedings of the International Laser Science Conference (Atlantic City, NJ, 1987)	88-71879	0-88318-372-2
No. 173	Cooperative Networks in Physics Education (Oaxtepec, Mexico, 1987)	88-72091	0-88318-373-0
No. 174	Radio Wave Scattering in the Interstellar Medium (San Diego, CA, 1988)	88-72092	0-88318-374-9
No. 175	Non-neutral Plasma Physics (Washington, DC, 1988)	88-72275	0-88318-375-7
No. 176	Intersections Between Particle and Nuclear Physics (Third International Conference) (Rockport, ME, 1988)	88-62535	0-88318-376-5

No. 177	Linear Accelerator and Beam Optics Codes (La Jolla, CA, 1988)	88-46074	0-88318-377-3
No. 178	Nuclear Arms Technologies in the 1990s (Washington, DC, 1988)	88-83262	0-88318-378-1
No. 179	The Michelson Era in American Science: 1870–1930 (Cleveland, OH, 1987)	88-83369	0-88318-379-X
No. 180	Frontiers in Science: International Symposium (Urbana, IL, 1987)	88-83526	0-88318-380-3
No. 181	Muon-Catalyzed Fusion (Sanibel Island, FL, 1988)	88-83636	0-88318-381-1
No. 182	High T_c Superconducting Thin Films, Devices, and Application (Atlanta, GA, 1988)	88-03947	0-88318-382-X
No. 183	Cosmic Abundances of Matter (Minneapolis, MN, 1988)	89-80147	0-88318-383-8
No. 184	Physics of Particle Accelerators (Ithaca, NY, 1988)	89-83575	0-88318-384-6
No. 185	Glueballs, Hybrids, and Exotic Hadrons (Upton, NY, 1988)	89-83513	0-88318-385-4
No. 186	High-Energy Radiation Background in Space (Sanibel Island, FL, 1987)	89-83833	0-88318-386-2
No. 187	High-Energy Spin Physics (Minneapolis, MN, 1988)	89-83948	0-88318-387-0
No. 188	International Symposium on Electron Beam Ion Sources and their Applications (Upton, NY, 1988)	89-84343	0-88318-388-9
No. 189	Relativistic, Quantum Electrodynamic, and Weak Interaction Effects in Atoms (Santa Barbara, CA, 1988)	89-84431	0-88318-389-7
No. 190	Radio-frequency Power in Plasmas (Irvine, CA, 1989)	89-45805	0-88318-397-8
No. 191	Advances in Laser Science–IV (Atlanta, GA, 1988)	89-85595	0-88318-391-9
No. 192	Vacuum Mechatronics (First International Workshop) (Santa Barbara, CA, 1989)	89-45905	0-88318-394-3
No. 193	Advanced Accelerator Concepts (Lake Arrowhead, CA, 1989)	89-45914	0-88318-393-5
No. 194	Quantum Fluids and Solids—1989 (Gainesville, FL, 1989)	89-81079	0-88318-395-1
No. 195	Dense *Z*-Pinches (Laguna Beach, CA, 1989)	89-46212	0-88318-396-X
No. 196	Heavy Quark Physics (Ithaca, NY, 1989)	89-81583	0-88318-644-6
No. 197	Drops and Bubbles (Monterey, CA, 1988)	89-46360	0-88318-392-7
No. 198	Astrophysics in Antarctica (Newark, DE, 1989)	89-46421	0-88318-398-6
No. 199	Surface Conditioning of Vacuum Systems (Los Angeles, CA, 1989)	89-82542	0-88318-756-6

No. 200	High T_c Superconducting Thin Films: Processing, Characterization, and Applications (Boston, MA, 1989)	90-80006	0-88318-759-0
No. 201	QED Stucture Functions (Ann Arbor, MI, 1989)	90-80229	0-88318-671-3
No. 202	NASA Workshop on Physics From a Lunar Base (Stanford, CA, 1989)	90-55073	0-88318-646-2
No. 203	Particle Astrophysics: The NASA Cosmic Ray Program for the 1990s and Beyond (Greenbelt, MD, 1989)	90-55077	0-88318-763-9
No. 204	Aspects of Electron–Molecule Scattering and Photoionization (New Haven, CT, 1989)	90-55175	0-88318-764-7
No. 205	The Physics of Electronic and Atomic Collisions (XVI International Conference) (New York, NY, 1989)	90-53183	0-88318-390-0
No. 206	Atomic Processes in Plasmas (Gaithersburg, MD, 1989)	90-55265	0-88318-769-8
No. 207	Astrophysics from the Moon (Annapolis, MD, 1990)	90-55582	0-88318-770-1
No. 208	Current Topics in Shock Waves (Bethlehem, PA, 1989)	90-55617	0-88318-776-0
No. 209	Computing for High Luminosity and High Intensity Facilities (Santa Fe, NM, 1990)	90-55634	0-88318-786-8
No. 210	Production and Neutralization of Negative Ions and Beams (Brookhaven, NY, 1990)	90-55316	0-88318-786-8
No. 211	High-Energy Astrophysics in the 21st Century (Taos, NM, 1989)	90-55644	0-88318-803-1
No. 212	Accelerator Instrumentation (Brookhaven, NY, 1989)	90-55838	0-88318-645-4
No. 213	Frontiers in Condensed Matter Theory (New York, NY, 1989)	90-6421	0-88318-771-X 0-88318-772-8 (pbk.)
No. 214	Beam Dynamics Issues of High-Luminosity Asymmetric Collider Rings (Berkeley, CA, 1990)	90-55857	0-88318-767-1
No. 215	X-Ray and Inner-Shell Processes (Knoxville, TN, 1990)	90-84700	0-88318-790-6
No. 216	Spectral Line Shapes, Vol. 6 (Austin, TX, 1990)	90-06278	0-88318-791-4
No. 217	Space Nuclear Power Systems (Albuquerque, NM, 1991)	90-56220	0-88318-838-4
No. 218	Positron Beams for Solids and Surfaces (London, Canada, 1990)	90-56407	0-88318-842-2
No. 219	Superconductivity and Its Applications (Buffalo, NY, 1990)	91-55020	0-88318-835-X
No. 220	High Energy Gamma-Ray Astronomy (Ann Arbor, MI, 1990)	91-70876	0-88318-812-0
No. 221	Particle Production Near Threshold (Nashville, IN, 1990)	91-55134	0-88318-829-5